THERMODYNAMICS AND THE DESTRUCTION OF RESOURCES

This book is a unique, multidisciplinary effort to apply rigorous thermodynamics fundamentals to problems of sustainability, energy, and resource uses. Applying thermodynamic thinking to problems of sustainable behavior is a significant advantage in bringing order to ill-defined questions with a great variety of proposed solutions, some of which are more destructive than the original problem. The chapters are pitched at a level accessible to advanced undergraduate and graduate students in courses on sustainability, sustainable engineering, industrial ecology, sustainable manufacturing, and green engineering. The timeliness of the topic and the urgent need for solutions make this book attractive to general readers as well as specialist researchers. Top international figures from many disciplines, including engineers, ecologists, economists, physicists, chemists, policy experts, and industrial ecologists, make up the impressive list of contributors.

Bhavik R. Bakshi holds a dual appointment as a Professor of Chemical and Biomolecular Engineering at The Ohio State University (OSU) in Columbus, Ohio; and Vice Chancellor and Professor of Energy and Environment at TERI University in New Delhi, India. He is also the Research Director of the Center for Resilience at OSU. From 2006 to 2010, he was a Visiting Professor at the Institute of Chemical Technology in Mumbai, India. He has published more than 100 articles in areas such as process systems engineering and sustainability science and engineering.

Timothy G. Gutowski is Professor of Mechanical Engineering at the Massachusetts Institute of Technology (MIT) in Cambridge, Massachusetts. He was the Director of MIT's Laboratory for Manufacturing and Productivity (1994–2004) and the Associate Department Head of Mechanical Engineering (2001–2005). From 1999 to 2001, he was Chairman of the National Science Foundation and Department of Energy's Panel on Environmentally Benign Manufacturing. He is the author of the book *Advanced Composites Manufacturing*, holds seven patents and patent applications, and has authored more than 150 technical publications.

Dušan P. Sekulić is Professor of Mechanical Engineering at the University of Kentucky in Lexington, Kentucky, and a Fellow of the American Society of Mechanical Engineers. Dr. Sekulić is a Consulting Professor at the Harbin Institute of Technology in Harbin, China. He is the author of more than 150 research publications, more than a dozen book chapters, and the book *Fundamentals of Heat Exchanger Design* (jointly with R. K. Shah), which was published in both the United States and China. He is editor of the books *Advances in Brazing: Science, Technology and Applications* and *Handbook of Heat Exchanger Design*.

Thermodynamics and the Destruction of Resources

Edited by

Bhavik R. Bakshi
The Ohio State University and TERI University

Timothy G. Gutowski
Massachusetts Institute of Technology

Dušan P. Sekulić
University of Kentucky

CAMBRIDGE UNIVERSITY PRESS
Cambridge, New York, Melbourne, Madrid, Cape Town,
Singapore, São Paulo, Delhi, Mexico City

Cambridge University Press
32 Avenue of the Americas, New York NY 10013-2473, USA

Published in the United States of America by Cambridge University Press, New York

www.cambridge.org
Information on this title: www.cambridge.org/9781107684140

First published 2011
First paperback edition 2013

A catalogue record for this publication is available from the British Library

Library of Congress Cataloguing in Publication Data
Thermodynamics and the destruction of resources / edited by
Bhavik R. Bakshi, Timothy G. Gutowski, Dušan P. Sekulić.
   p.   cm.
Includes bibliographical references and index.
ISBN 978-0-521-88455-6 (hardback)
1. Power resources.   2. Energy consumption.   3. Conservation of natural
resources.   4. Thermodynamics.   I. Bakshi, Bhavik R.   II. Gutowski,
Timothy George Peter.   III. Sekulić, Dušan P.
TJ163.2.T428   2011
620 – dc22   2010043695

ISBN 978-0-521-88455-6 Hardback
ISBN 978-1-107-68414-0 Paperback

"I suggest that we are thieves in a way. If I take anything that I do not need for my own immediate use and keep it I thieve it from somebody else. I venture to suggest that it is the fundamental law of Nature, without exception, that Nature produces enough for our wants from day to day, and if only everybody took enough for himself and nothing more, there would be no pauperism in this world, there would be no more dying of starvation in this world. But so long as we have got this inequality, so long we are thieving."

M. K. Gandhi, *All Men Are Brothers*,
Navjeevan Trust Publication, Ahmedabad, 1960

"Then I say the earth belongs to each... generation during its course, fully and in its own right. The second generation receives it clear of the debts and encumbrances, the third of the second, and so on. For if the first could charge it with a debt, then the earth would belong to the dead and not to the living generation. Then, no generation can contract debts greater than may be paid during the course of its own existence."

Thomas Jefferson, letter written on Sept. 6, 1789

"Does the educated citizen know he is only a cog in an ecological mechanism? That if he will work with that mechanism, his mental wealth and his material wealth can expand indefinitely? But if he refuses to work with it, it will ultimately grind them to dust. If education does not teach us these things, then what is education for?"

A. Leopold, *A Sand County Almanac*,
The Oxford University Press, New York, 2001

Contents

Contributor List   *page* xi

Foreword by Herman E. Daly   xv

Foreword by Jan Szargut   xvii

Preface   xxi

Introduction 1
Bhavik R. Bakshi, Timothy G. Gutowski, and Dušan P. Sekulić

PART I. FOUNDATIONS

1. **Thermodynamics: Generalized Available Energy and Availability
   or Exergy** 15
   Elias P. Gyftopoulos

2. **Energy and Exergy: Does One Need Both Concepts for a Study of
   Resources Use?** 45
   Dušan P. Sekulić

3. **Accounting for Resource Use by Thermodynamics** 87
   Bhavik R. Bakshi, Anil Baral, and Jorge L. Hau

PART II. PRODUCTS AND PROCESSES

4. **Materials Separation and Recycling** 113
   Timothy G. Gutowski

5. **An Entropy-Based Metric for a Transformational Technology
   Development** 133
   Dušan P. Sekulić

6. **Thermodynamic Analysis of Resources Used in Manufacturing
   Processes** 163
   Timothy G. Gutowski and Dušan P. Sekulić

7. **Ultrapurity and Energy Use: Case Study of Semiconductor Manufacturing** 190
   Eric Williams, Nikhil Krishnan, and Sarah Boyd

8. **Energy Resources and Use: The Present Situation, Possible Sustainable Paths to the Future, and the Thermodynamic Perspective** 212
   Noam Lior

PART III. LIFE-CYCLE ASSESSMENTS AND METRICS

9. **Using Thermodynamics and Statistics to Improve the Quality of Life-Cycle Inventory Data** 235
   Bhavik R. Bakshi, Hangjoon Kim, and Prem K. Goel

10. **Developing Sustainable Technology: Metrics From Thermodynamics** 249
    Geert Van der Vorst, Jo Dewulf, and Herman Van Langenhove

11. **Entropy Production and Resource Consumption in Life-Cycle Assessments** 265
    Stefan Gößling-Reisemann

12. **Exergy and Material Flow in Industrial and Ecological Systems** 292
    Nandan U. Ukidwe and Bhavik R. Bakshi

13. **Synthesis of Material Flow Analysis and Input–Output Analysis** 334
    Shinichiro Nakamura

PART IV. ECONOMIC SYSTEMS, SOCIAL SYSTEMS, INDUSTRIAL
   SYSTEMS, AND ECOSYSTEMS

14. **Early Development of Input–Output Analysis of Energy and Ecologic Systems** 365
    Bruce Hannon

15. **Exergoeconomics and Exergoenvironmental Analysis** 377
    George Tsatsaronis

16. **Entropy, Economics, and Policy** 402
    Matthias Ruth

17. **Integration and Segregation in a Population – a Thermodynamicist's View** 429
    Ingo Müller

18. **Exergy Use in Ecosystem Analysis: Background and Challenges** 453
    Roberto Pastres and Brian D. Fath

19. **Thoughts on the Application of Thermodynamics to the Development of Sustainability Science** 477

Timothy G. Gutowski, Dušan P. Sekulić, and Bhavik R. Bakshi

Appendix: Standard Chemical Exergy   489

Index   495

Contributor List

Bhavik R. Bakshi,
William G. Lowrie Department of Chemical and Biomolecular Engineering,
The Ohio State University,
Columbus, Ohio, and
Department of Energy and Environment,
TERI University,
New Delhi, India

Anil Baral,
The International Council on Clean Transportation,
Washington, D.C.

Sarah Boyd,
Department of Mechanical Engineering,
University of California,
Berkeley, California

Jo Dewulf,
Department of Sustainable Organic Chemistry and Technology,
Ghent University,
Ghent, Belgium

Brian D. Fath,
Department of Biological Sciences,
Towson University,
Towson, Maryland

Prem K. Goel,
Department of Statistics,
The Ohio State University,
Columbus, Ohio

Stefan Gößling-Reisemann,
Faculty of Production Engineering,
University of Bremen,
Bremen, Germany

Timothy G. Gutowski,
Department of Mechanical Engineering,
Massachusetts Institute of Technology,
Cambridge, Massachusetts

Elias P. Gyftopoulos,
Department of Nuclear Science and Engineering, and
Department of Mechanical Engineering,
Massachusetts Institute of Technology,
Cambridge, Massachusetts

Bruce Hannon,
Department of Liberal Arts and Sciences,
University of Illinois-Urbana,
Urbana-Champaign, Illinois

Jorge L. Hau,
Catalyst Lend Lease,
London, United Kingdom

Hangjoon Kim,
Department of Statistics,
The Ohio State University,
Columbus, Ohio

Nikhil Krishnan,
McKinsey & Company,
New York, New York

Noam Lior,
Department of Mechanical Engineering and Applied Mechanics,
University of Pennsylvania,
Philadelphia, Pennsylvania

Ingo Müller,
Institute for Process Engineering,
Technical University Berlin,
Berlin, Germany

Shinichiro Nakamura,
Graduate School of Economics,
Waseda University,
Tokyo, Japan

Roberto Pastres,
Department of Physical Chemistry,
University of Venice,
Venice, Italy

Matthias Ruth,
A. James Clark School of Engineering and School of Public Policy,
University of Maryland,
College Park, Maryland

Dušan P. Sekulić,
Department of Mechanical Engineering,
University of Kentucky,
Lexington, Kentucky

George Tsatsaronis,
Institute for Energy Engineering,
Technical University Berlin,
Berlin, Germany

Nandan U. Ukidwe,
Saflex Technology at Solutia Inc.,
Springfield, Massachusetts

Geert Van der Vorst,
Department of Sustainable Organic Chemistry and Technology,
Ghent University,
Ghent, Belgium

Herman Van Langenhove,
Department of Sustainable Organic Chemistry and Technology,
Ghent University,
Ghent, Belgium

Eric Williams,
School of Sustainable Engineering and the Built Environment and
School of Sustainability,
Arizona State University,
Tempe, Arizona

Foreword

Herman E. Daly

The first and second laws of thermodynamics should also be called the first and second laws of economics. Why? Because without them there would be no scarcity, and without scarcity, no economics. Consider the first law: if we could create useful energy as it got in our way, we would have superabundant sources and sinks, no depletion, no pollution, and more of everything we wanted without having to find a place for stuff we didn't want. The first law rules out this direct abolition of scarcity. But consider the second law: even without creation and destruction of matter-energy, we might indirectly abolish scarcity if only we could use the same matter-energy over and over again for the same purposes: perfect recycling. But the second law rules that out. So it is that scarcity and economics have deep roots in the physical world, as well as deep psychic roots in our wants and desires.

Economists have paid much attention to the psychic roots of value, but not so much to the physical roots. Generally they have assumed that the biophysical world is so large relative to its economic subsystem that the physical constraints (the laws of thermodynamics) are not binding. But they are always binding to some degree and become very limiting as the scale of the economy becomes large relative to the containing biophysical system. Therefore attention to thermodynamic constraints on the economy, indeed to the entropic nature of the economic process, is now critical, as first emphasized by Nicholas Georgescu-Roegen in his magisterial *The Entropy Law and the Economic Process* (1971). The present book is a welcome and worthy contribution to this important and continuing endeavor. It deserves studious attention.

University of Maryland

Foreword

Jan Szargut

Human production activity is based on natural resources. Their usefulness results from the ability to be transformed into useful products necessary for humans. That ability may be evaluated by means of the laws of thermodynamics that express not only the conservation of energy but also its tendency to be dissipated. The dissipation of energy (resulting in entropy generation) decreases its usefulness and reduces its ability to be transformed into useful products. The mentioned ability may be evaluated by means of the maximum work attainable in a reversible transition to the equilibrium with the environment embracing that part of nature that belongs to the area of human production activity. That quantity is usually called "exergy."

A considerable part of the natural resources utilized by humans belongs to the nonrenewable ones (organic and nuclear fuels, ores of metals, inorganic minerals). The usefulness of nonrenewable resources results from their chemical composition. The chemical exergy can be accepted as a general quality measure of the mentioned resources because it expresses the maximum work attainable during the transition to equilibrium with a dead environment.

However, equilibrium with the natural environment is not possible, because it does not appear in the real natural environment. As proved by Ahrendts [1], the concentration of free oxygen in an equilibrium environment would be very small because the prevailing part of oxygen would be bound with nitrogen in nitrates. Fortunately the formation of nitrates is kinetically blocked, and they appear very seldom in nature.

To find a real but possibly low reference state for chemical exergy, the concept of reference substances is introduced [2]. For every chemical element, an individual reference substance that is most common in the real environment is accepted. The reference substances are mutually independent, and therefore the problem of equilibrium between them does not exist. To facilitate the use of chemical exergy, the concept of normal chemical exergy is introduced. It is defined at normal environmental temperature and pressure and mean concentration in the environment, resulting from the geochemical data.

Gaseous components of air, solid components of the external layer of lands, and ionic or molecular components of seawater are assumed as reference substances. The general calculation of chemical exergy may be performed in three steps:

1. calculation of normal chemical exergy of reference substances,
2. calculation of normal chemical exergy of chemical elements,
3. calculation of normal chemical exergy of chemical compounds.

The exact calculation of chemical exergy of gaseous reference substances is easy, because the considered components of atmospheric air may be treated as components of an ideal solution. The calculation of normal chemical exergy of elements having reference substances dissolved in seawater can be made with sufficient accuracy in the case of monocharged and bicharged ions or molecular substances. The calculation method was elaborated by Szargut and Morris [3]. In the case of solid reference substances, an exact calculation is not possible because they appear in the form of multicomponent solid solutions. An approximate method to evaluate the chemical exergy in such cases was suggested [4]. It is based on the assumption that solid reference substances can be treated as components of an ideal solution. The concentration of solid reference substances in the environment results from geochemical data about the total content of the considered chemical element in the solid environment and about its fraction presumably appearing in the form of a reference substance.

The exhaustion of nonrenewable natural resources can be dangerous for the future of humankind. As a measure of the depletion of nonrenewable resources, the concept of thermoecological cost (TEC) was introduced [5]. It expresses the cumulative exergy consumption of the nonrenewable resources in the total chain of processes leading to the considered useful product. The prefix "thermo" indicates that this kind of cost is expressed in exergy units, not in monetary units. The values of TEC can be used for the selection of the production technology and for the determination of optimum design and operation parameters if the minimum depletion of the nonrenewable resources represents the objective function.

The calculation of TEC can be performed by means of a set of balance equations. Each of them contains the TEC of the following delivered components: the used domestic raw materials and semifinished products, the wear of the machines and installations used in the considered production process, the imported raw materials and semifinished products, the immediate consumption of nonrenewable exergy extracted from nature, and the compensation cost of losses that are due to the emission of deleterious products. On the side of products of the considered process, there appear to be TECs of the major products and of the useful by-products. The TEC of the useful by-product should be expressed by the value of the TEC of major products fabricated in another process. The substitution ratio between the by-product and the replaced major product should be taken into account. The value of the TEC of imported semifinished products can be determined by taking into account the fact that the financial means for import are gained by export. Hence the TEC value of the imported semifinished product results from its monetary cost and from the mean TEC value of the monetary value of exported goods.

The set of balance equations should be formulated and solved only for the semifinished products used in other production processes. If the considered useful product is not used in other production processes (or used in a very small portion), its TEC may be determined individually by means of a sequence method, which begins in the final step of the production chain and goes back through all the steps until the semifinished products considered in the mentioned balance equations are obtained.

The values of the TEC depend mainly on the technology of electricity production and on the transportation system applied in the considered region or country.

Jan Szargut, full member of the
Polish Academy of Sciences,
Silesian University of Technology,
Gliwice, Poland

REFERENCES

[1] J. Ahrendts, *Die Exergie chemisch reaktionsfähiger Systeme* (VDI-Forschungsheft 579, Düsseldorf, 1977).
[2] J. Szargut, "Chemical exergies of the elements" *Appl. Energy* **32**, 269–285 (1989).
[3] J. Szargut and D. R. Morris, "Calculation of the standard chemical exergy of some elements and their compounds based upon sea water as the datum level substance," *Bull. Polish Acad. Sci. Tech. Sci.* **33**(5–6), 293–305 (1985).
[4] J. Szargut, "Standard chemical exergy of some elements and their compounds, based upon the concentration in Earth's crust," *Bull. Polish Acad. Sci. Tech. Sci.* **35**(1–2), 53–60 (1987).
[5] J. Szargut, "Depletion of the unrestorable natural exergy resources," *Bull. Polish Acad. Sci. Tech. Sci.* **45**(2), 241–250 (1997).

Preface

This book is intended to bring together a wide range of theoretical and experimental results from many different disciplines, all addressing sustainability issues using thermodynamic arguments. The disciplines involved include mechanical and chemical engineering, physics, economics, geography, ecology, and industrial ecology. The contributors from Japan, Europe, and the United States are all to some extent speaking one language, albeit with different applications, construction of arguments, and degree of rigor.

We believe that it is time for a book like this to introduce the fundamentals and applications of thermodynamics to demonstrate the richness and broad applicability of thermodynamic principles and the essential role that it can play in quantifying the impact of human activities on natural resources and the environment. Our goal has been to assemble a collection of chapters written by authorities in their own fields who share thermodynamics-inspired approaches to a diverse set of topics. Although there are many obstacles to presenting all of these different areas in a unified way, we believe we have succeeded in bringing together in one book a glimpse of the breadth of applications that can be considered from a thermodynamic perspective. Different chapters keep quite visible the flavor of a particular discipline, although some cross-fertilization and trans-disciplinary approaches to the issues are demonstrated. So, our ultimate goal has evolved into a promotion of the use of a rigorous science/engineering discipline (thermodynamics) not to solve problems beyond its realm but to assist in understanding the problem at hand, to define well the system considered, to implement conservation principles, and to appropriately merge such an approach with other disciplines. The task of establishing a unified approach to what some call sustainability science must be left for the future. We will be satisfied if this contribution becomes one of the building blocks along the path to achieving that goal.

This book is intended for professionals in diverse fields interested in resolving sustainability issues. In addition, graduate and senior-level undergraduate students across disciplines, particularly those that already have an introductory course in thermodynamics, may find this book useful for helping them appreciate and understand the broad relevance of thermodynamic principles.

Putting this book together has been a pleasure for us in large part because of all of the gifted, enthusiastic people with whom we have had the opportunity to work.

We must start by thanking our 23 co-authors and contributors who provided their authoritative works to us in a timely manner and put up with our occasionally unreasonable requests and nagging. We would like to thank the professional organizations that hosted several of our special sessions on various aspects of thermodynamics and sustainability, particularly the American Chemical Society Green Chemistry and Engineering Conference (where the idea for this book was first launched in 2005) and the Institute of Electrical and Electronics Engineers annual symposia on electronics and the environment, now called the International Symposium on Sustainable Systems and Technologies.

It was a pleasure to work with Peter Gordon of Cambridge University Press and Victoria Danahy and Peter Katsirubas of Aptara on editing this complex text. We must also acknowledge the important role played by Skype in allowing us to continue developing this book even when we were in distant corners of the globe.

We would like to express our gratitude to several colleagues and good friends including Dr. Joseph Fiksel of Ohio State University for insight into the close connection between sustainability issues and business decision making, Professor Richard Gaggioli of Marquette University, Elias Gyftopoulos and Seth Lloyd of the Massachusetts Institute of Technology, and Ingo Müller of the Technical University Berlin for many fruitful discussions involving thermodynamics. We also thank Dr. Robert Gregory of the University of Kentucky for discussing everything else but thermodynamics. A number of years of joint work with our students (too numerous to mention by name) merits a word of appreciation.

Finally, the authors would like to acknowledge the roles by their wives Mamta Bakshi, Jane Gutowski, and Gorana Sekulić, and their children, Harshal Bakshi, Laura and Ellie Gutowski, and Višnja and Aleksandar Sekulić. Their assistance and understanding provided continuous support for completing the book.

Bhavik R. Bakshi
Timothy G. Gutowski
Dušan P. Sekulić

Introduction

*Bhavik R. Bakshi, Timothy G. Gutowski,
and Dušan P. Sekulić*

1 Resources and Sustainability

This book is about the application of thermodynamic thinking to those new areas of study that are concerned with the human use of resources and the development of a sustainable society. Exactly what a sustainable society is, is a highly debated topic, somewhat subject to personal value preferences. However, what is not sustainable is easier to identify. For example, in Jared Diamond's popular book, *Collapse* [1], he identified a variety of ancient and modern societies that failed. No one would dispute this claim. Although the reasons for these failures were complex, an important and common contributing factor was an inability to manage Earth's resources and thereby meet the needs of the society. For example, the inhabitants of Easter Island apparently became consumed with building giant stone statues (called *moai*), which required large timbers for their construction and for their transport from the quarry to the installation site. Apparently this building process, along with the destruction of the seeds wrought by the Polynesian rats, led to the destruction of most, if not all, of their trees. One result from this was a loss of the primary building material for their canoes. Because most fish were some distance from the shore, a lack of building materials for canoes meant fewer fish to eat and the inability to move to other islands. This desperate situation ultimately resulted in the islanders resorting to cannibalism. Not all resource-accounting problems are as dramatic or as straightforward as just outlined (including the full accounting for Easter Island). But it is well known that a lack of resources to sustain life and to allow the members of society to prosper will lead to severe difficulties for society, and even collapse. Particular resources of concern would include building materials, food resources, water, and energy sources, including biomass, fossil fuels, geothermal heat, and sunlight. In addition, ecological systems can provide many useful services needed to support life. Examples are plants removing carbon dioxide from the atmosphere by photosynthesis, vegetation preventing soil erosion and thereby maintaining clean water, limestone soils buffering acidic deposits, and many, many others [2]. Each of these resources and processes can be analyzed as thermodynamic systems. In particular, the transformations required to provide ecosystem services or to produce or convert (or both) material resources to other forms can all be analyzed by

1

thermodynamics [3–6]. Thermodynamics therefore should be an important contributor to any new science that focuses on resource use.

Thermodynamics is a well-established phenomenological discipline used by scientists and engineers to describe and generalize empirical evidence needed for predicting the behavior of physical systems exposed to material and thermal energy interactions with their surroundings. As such, and though still under development, thermodynamics has "aged" through a sequence of phases. These would include a number of steps through which all new sciences would have to advance, for example: (1) observation and data collection, (2) data classification and quantification, (3) simplification and abstraction, (4) symbolic representation, (5) symbolic manipulation, and finally, (6) prediction and verification. Each of these formal steps is needed to enhance knowledge and understanding. Furthermore, they are repeated as new insights are obtained. In this book we focus on the application of thermodynamics to much "younger" areas of study. These would include, for example, large natural or manmade systems, such as those that would be considered in the fields of ecology and industrial ecology and that play a central role in our sustainability. These newer areas of study, in particular their application to large complex systems, such as large, integrated ecological, economic, industrial systems (i.e., systems that combine human activities with ecosystems), cannot yet make the claim of being predictive sciences. The reasons for this have to do with both the age and the complexity of the areas of study, as well as the ambition of the models. Systems of interest to industrial ecologists involve interactions among ecological, economic, industrial, and societal processes, making them highly multidisciplinary in nature. Thermodynamics can be used to analyze these systems, however, provided that we (1) very clearly define the system under study and (2) focus primarily on material and energy transformations.

Although such an approach cannot capture all the multidisciplinary aspects of the problem, it can provide tremendous insight into those aspects that have to do with resource use and transformation. Many other aspects, such as those involving human valuation and societal and cultural preferences, may also benefit from the results of thermodynamic analysis. However, thermodynamics by itself cannot capture these aspects.

The early history of thermodynamics was concerned with the problem of how to obtain work from heat. This work developed the seminal ideas that led to workable versions of the first and second laws of thermodynamics [7]. Much of this work was carried out coincidentally with the development of the steam engine, which provided many experimental opportunities. Further development of thermodynamics refined these laws and expanded beyond primarily thermal interactions with application to new systems and areas of study. This led to many new and useful concepts such as Gibb's free energy, available energy, and many others. Although this work continues, the application of thermodynamics to mechanical systems, chemical systems, and simple ecological systems is now well established [5, 6, 8–13]. It is now clear that thermodynamics deals with a very broad class of phenomena involving the so-called well-defined *thermodynamic systems* and obeying a limited number of *natural (thermodynamic) laws*. This success has encouraged the widespread application of thermodynamic thinking to many different areas including economic theory [14–19] and social systems [20], as well as applications to systems as small as one particle or as large as the ecosystem services of our planet [3, 21, 22] and even the universe [23]. Please note that some of these applications however are not without controversy.

2 Thermodynamics and Resources

The point of this book in focusing on larger complex systems is to expand and extend these principles and to look carefully at their application to problems, particularly those related to ecology, industrial ecology, and issues of sustainability. Here, as suggested in our title, the greatest leverage will be concerning problems of resource use, resource transformations, and resource destruction. Clearly the resources we use to establish and maintain life, to construct our society, to improve economic well-being, and to provide for our progeny, will all play a central role in any discussion of sustainable behavior. Of particular interest will be accounting for these resources; how available they are, how efficiently they are used, and how they become depleted, and allocation of these resources between different species and generations on this planet. Thermodynamics is particularly well suited to address the first issue, the accounting for resources, and, by providing rigorous allocation and accounting metrics, it can contribute to the second issue.

The proper domain for thermodynamics is in energy interactions, and through these interactions, the accomplishment of useful effects, the depletion of energy resources, and the generation of unavoidable wastes. Because of the necessity of energy interactions to maintain life and the often close correlations between energy resources use and economic growth as well as waste generation, emissions and pollution, energy and how it is used often dominate many discussions about the future of humankind on this planet. What is of paramount importance here is that, once a system is established as a so-called well-defined thermodynamic system, it must obey the laws of thermodynamics. It is our belief, and the central tenet of this book, that many, if not all, of the systems we care about can be interpreted as thermodynamic systems. This of course is not to say that all we care about are thermodynamic interactions. The sustainability of human activities is far more complex, and thermodynamics is only one part of it.

The first chapter of this book [24] establishes what constitutes a thermodynamic system. The definition, however, is quite broad. Basically it includes any kind of material resources from manmade to natural, and this system may be acted on by any kind of energy and heat interactions, for example, including those in power plants, industrial systems, and living systems. Once established, the behavior of these systems has to obey the laws of thermodynamics, including the laws of other scientific disciplines and related constraints. The focus of this book, on the use, efficiency of transformation, and the destruction of resources naturally leads us to the second law of thermodynamics and the concepts of entropy and available energy or exergy. Unfortunately these concepts are probably among the most mysterious and misunderstood concepts in all of science, striking fear in the hearts of engineering students and, often enough, even professionals. This misunderstanding has not gone unnoticed, however, and a debate among scientists and engineers has resulted in careful expositions of the thermodynamic theory, for example, Gyftopolous and Beretta (G&B) [25].

Let us start by admitting that there has been a proliferation of intimidating sounding e-words in thermodynamics, which in themselves can be off-putting and probably undermine their use. Among these, *energy* is probably the least off-putting and most widely used term and has been incorporated into the analysis of economic, industrial, and ecological systems' [26]. Energy includes contributions from internal energy (say, molecular motion), kinetic energy (involving velocity of a mass), and potential energy (related to displacement in a force field such as a gravitation field and others), to mention a few, and the quantity of its flow is governed by the first law of thermodynamics – sometimes called the law of energy conservation. However, energy does not capture the actual ability to do work, or even the interpretation of most laypeople or the dictionary meaning of "the ability to do work" [27]. Thermodynamicists have responded by developing other concepts such as entropy and exergy to address these issues. These off-putting e-words often represent concepts of paramount importance to our theme and must eventually be mastered. In truth, these concepts are quite accessible. Let us start, as G&B did, by establishing the idea that there is something called the "available energy" of a system. This is exactly what the name implies, the amount of energy that is available to do some useful work or have some other useful effect. This is what most laypeople and dictionaries usually mean by the word energy. The name "available work" has been supplanted by the much more esoteric-sounding term "exergy," but the two mean the same thing. And because "exergy" is short, easy to write, and now well accepted, we use it too. Exergy then measures the energy potential of a system to do some useful task. When the system interacts with the surroundings to do some useful task, it will use or transfer some of this exergy. If the process is an ideal "reversible" process, the exergy change for the system will be exactly equal to the amount of exergy used to do the task. However, no real process is ideal, and so for real systems the exergy change for the system will be greater than the exergy effectively used to do the task. This difference represents the exergy destruction. This loss is real and irretrievable. It represents a destruction of available energy potential.

The *exergy lost* is always positive for real systems and is proportional to another e-concept, the *entropy generated*. So both concepts, *exergy lost* and *entropy generated*, can be used to measure the destruction of a thermodynamic resource. These concepts derive from both the first and the second laws of thermodynamics and are used in what is called second law analysis. Second-law analysis can tell us how efficient a process is by distinguishing between what is lost and what is gained. First law analysis, on the other hand, is based on the conservation laws for energy and mass. First law analysis can tell us energy-transfer requirements needed to effect certain changes, but it cannot tell us a complete story about the efficiency of these changes. These concepts are illustrated in the Fig. I.1.

In the figure we see that the difference between the exergy supplied to the system (marked by the system boundary) and the total exergy out of the system is equal to the "exergy losses." For example, if the inputs are two separate streams of hot and cold water and the output is the mixture at an intermediate temperature established by the energy balance, the internal exergy losses represent the lost potential of the water inputs caused by mixing, i.e., by the equalization of the temperatures of the individual streams. Note that an energy balance for the mixing problem without or with "losses" to the environment would show the same total energy out as in.

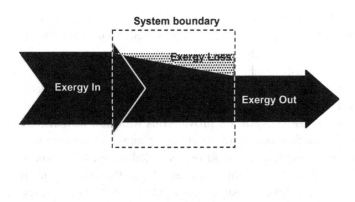

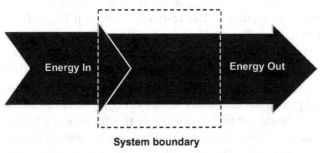

Figure I.1. Grassmann (exergy flow) and Sankey (energy flow) diagrams. In general, it is assumed that a total energy flow "supplied" to the system carries the corresponding exergy (available energy) flow. The system changes the state, and the available energy, passing through the system, is reduced because of irreversibilities inherent to the change of state. The exergy flow available for a useful effect is reduced (exergy is not conserved). However, the energy flow features the conservation principle and the total energy in must be equal to the total energy out.

Only a second law analysis shows the loss of available energy. This is one of the big advantages of exergy analysis. In the context of the title of this book, the internal and external exergy losses represent resource destruction. These are either irretrievable (such as caused by mixing) or partially retrievable (say, if some heat losses can be reduced by better system insulation). The exergy is useful because the person who frames the problem deems this application useful or the available energy at the exit from the system can be utilized for some other useful effect. For example, the useful exergy could represent the warm water (as a mixture of a hot and a cold stream) you use in the morning to shower. This would correspond to our preceding water-mixing example. However, this water will eventually go down the drain and lose all of its potential too. Hence the total "exergy input" represents a resource that is consumed. On the other hand, the warm water may be used to heat another fluid stream; hence any remaining exergy potential would be available for use in such a new application. To be more specific, the input exergy resource could be the available energy extracted from coal needed to make the hot water. After the shower, you have a clean body, but the coal and all or most of its potential is gone. The point is this: Exergy analysis will show you what will be lost (internal and external loses) to obtain some useful effect. It will also show you what useful potential remains (the exergy still available). Theoretically this remaining exergy could be used to obtain some other useful effect, but often in our society we do not use it; we throw it away. Exergy analysis will help direct our attention to losses. How humans may value the useful effect and the resources used, however, is a problem that calls for an analysis that would go beyond thermodynamics. These thermodynamics concepts are presented much more rigorously in Chaps. 1 and 2.

Recently exergy analysis has become much more prevalent in the industrial ecology literature, which has led to new results as well as new scrutiny of the concept. On the one hand, this second law concept has provided us with new insights; for example, it shows beyond a shadow of a doubt that such claims as "zero wastes" or "zero emissions" are unobtainable and should be discarded from the lexicon of industrial ecology or replaced by "zero avoidable waste." On the other hand, it is the case that first law analyses (energy balances and mass balances) can often provide some of the very same results as exergy analysis. This is the case when an analyst is not interested in the quality of the resource utilization but only in the quantity. We do not dispute these claims, but rather will attempt to show how second law analysis can be made quite accessible and that it provides new insights not obtainable from the first law. Perhaps the single most important dimension that second law analysis brings to any discussion of energy interactions is the notion of quality. For example, exergy analysis can define what the high-quality resources are by establishing a reference state, often referred to as the "dead state." That is, when the system of interest is in mutual stable equilibrium with this reference, it too is "dead" in the sense that it can do no work. The quality of a resource then is related, in energy units, to its distance from this reference state [28]. As it turns out, establishing just exactly what constitutes the reference state on Earth is not at all trivial. For one thing, the three major components of our biosphere, the crust, oceans, and atmosphere, are not themselves in stable equilibrium. For example, the high oxygen content of the atmosphere tends to oxidize exposed elements from the crust, etc. Nevertheless, and not without controversy, some reasonable reference states for these three components of our environment have been worked out. Notably the reference compositions and resulting exergy values of all the elements and many important chemical compounds have been established and several proposals for reference states exist. For example, the reference states proposed by Szargut are referred to frequently and are included here (see Appendix A). An important point then is that, armed with these data and other relevant information, such as mass balances, the work of performing a second law analysis for systems is reduced to an algebraic accounting exercise. Furthermore, the results from these exercises can provide important insights, such as whether or not a process is thermodynamically feasible, what is lost, what is gained, and what is the efficiency of the transformation.

Of a particular relevance here then is the available energy or exergy value of a resource. Intuitively, we would suspect that fuels such as coal, oil, kerosene, and natural gas should all have high values of exergy, and they do. In fact, exergy values for these fuels give numbers that are very close, if not equal, to their heating values. But, in addition, the exergy analysis gives values for all materials, valuing them for their exergy potential. Thus nonfuel resources that are typically not in focus of an energy analysis do have an exergy content and are commonly included in exergy analysis. This feature shows up particularly prominently in certain kinds of material and energy transformation and highlights the differences between first and second

Table I.1. *Overview of chapters*

Part number	Chapter number	Title	Author(s)	Description
Part I: Foundations	1	Thermodynamics: Generalized available energy and availability or exergy	Gyftopoulos	This is a classic treatment of the exergy concept by one of the foremost thermodynamicists. The chapter offers a rigorous formulation of the set of basic concepts and laws of Thermodynamics. The concept of available energy (exergy) is devised as a primitive concept, while entropy is formulated as a derivative concept. This representation is unique since it introduces thermodynamic concepts via a different sequence than what is present in most traditional expositions.
	2	Energy and exergy: Does one need both concepts for a study of resources use?	Sekulić	In this chapter, concepts of energy and exergy are reexamined in the context of balance equations and their applications to the modeling of resources use. The presentation argues a clear distinction between the concepts of energy/exergy resources and their meaning as thermodynamic properties.
	3	Accounting for resource use by thermodynamics	Bakshi, Baral, Hau	This chapter introduces and evaluates various thermodynamic methods for resource accounting. This includes methods based on energy, exergy and emergy analysis. It links these methods and demonstrates their pros and cons via an application to the life cycle of some transportation fuels.
Part II: Products and Processes	4	Materials separation and recycling	Gutowski	This chapter reviews both thermodynamic and information theory approaches to characterize separation and recycling processes. Applications areas include mining and minerals extraction as well as product recycling.
	5	Entropy-based metric for transformational technologies development	Sekulić	This chapter offers a hypothesis that metric for the evaluation of existing technologies and/or new technologies in the context of energy resources utilization can be the entropy generation, and illustrates its use for non-energy systems.

(continued)

Table I.1 *(continued)*

Part number	Chapter number	Title	Authors	Description
	6	Thermodynamic analysis of resources used in manufacturing processes	Gutowski, Sekulić	This chapter reviews both the theoretical and the actual thermodynamic performance of many different manufacturing processes from conventional processes such as machining, casting and injection molding through advanced machining processes as well as micro electronics and nanotechnology processes.
	7	Ultrapurity and energy use: Case study of in semiconductor manufacturing	Williams, Krishnan, Boyd	This chapter provides a broad overview of the energy requirements of purification processes using a number of techniques, but primarily relying on cost of ownership models and environmental input/output modeling.
	8	Energy resources and use: The present situation, possible sustainable paths to the future, and the thermodynamic perspective	Lior	This chapter is a brief summary of the state of current energy resources and use, and of their limitations and consequences (2008), and of possible paths to the future, including energy research funding trends, especially in the U.S.
Part III: Life-Cycle Assessments and Metrics	9	Using thermodynamics and statistics to improve the quality of life-cycle inventory data	Bakshi, Kim, Goel	Errors in life cycle inventory data are common and affect the quality of the results from LCA. This chapter describes an approach based on imposing the laws of thermodynamics to the available data along with statistical information to improve their quality. The resulting data are reconciled with the laws of thermodynamics and expected to be more accurate.
	10	Developing sustainable technology: Metrics from thermodynamics	Van der Vorst DeWulf, Langenhove	This chapter shows how thermodynamics can be used for defining metrics that provide insight into the sustainability of technological activities and products. With the help of various examples, it illustrates the use of concepts such as cumulative exergy consumption and abatement exergy.
	11	Entropy production and resource consumption in life-cycle assessments	Gößling-Reisemann	This is a well argued promotion of the entropy generation concept for modeling resource consumption.

Part number	Chapter number	Title	Authors	Description
	12	Exergy and material flow in industrial and ecological systems	Ukidwe, Bakshi	This chapter uses exergy as a common currency for industrial and ecological systems. It develops an integrated economic-ecological model of the U.S. economy by combining an economic input-output model with physical data about ecological inputs to various economic sectors.
	13	Synthesis of material flow analysis and input–output analysis	Nakamura	Material flow analysis has been popular to account for the use of resources in a life cycle. However, it does not consider waste streams. This chapter describes the approach of waste input-output material flow analysis to address this shortcoming. Applications of this approach to the Japanese economy are also described.
Part IV: Economic Systems, Social Systems, Industrial Systems, and Ecosystems	14	Early development of input–output analysis of energy and ecologic systems	Hannon	This is an interesting and historical look at the early development of energy resource input-output modeling by one of the pioneers.
	15	Exergoeconomics and exergoenvironmental Analysis	Tsatsaronis	This chapter elaborates two methodologies relevant for resources use and sustainability studies: (i) exergoeconomics as an exergy-aided cost reduction approach that uses the exergy costing principle, and (ii) a new approach based on the so-called exergoenvironmental costing.
	16	Entropy, economics, and policy	Ruth	This chapter gives a high level tour of physical modeling applied to economics.
	17	Integration and segregation in a Population – A thermodynamicist's view	Müller	This groundbreaking research attempts to uncover the hidden kinship between thermodynamic processes such as the phase separation in a solution and social phenomena such as segregation in a society. Both may be viewed as thermodynamic – or socio-thermodynamic – equilibria with homogeneous Gibbs free energies – or socio-chemical potentials.

(continued)

Table I.1. *(continued)*

Part number	Chapter number	Title	Authors	Description
	18	Exergy use in ecosystem analysis: Background and challenges	Pastres, Fath	Ecologists have used thermodynamics for understanding ecosystems for many decades. This chapter describes the challenges in using the exergy concept for understanding ecosystems and introduces eco-exergy as a way of quantifying the work capacity of living systems. Relationship of these concepts with similar concepts in engineering is also discussed.
	19	Thoughts on the application of thermodynamics to the development of sustainability science	Gutowski, Sekulić, Bakshi	This chapter explores the role that thermodynamics can play in sustainability by accounting for resource use and availability. It considers closed and open systems and how thermodynamics can help in using insight from ecosystems for designing technological systems.

law analysis. For example, when metal oxides are reduced to produce pure metals for manufacturing, both energy and exergy accounting will take into account the fuels used in this process. But only exergy accounting will take into account the material transformation and highlight the fact that the process has greatly improved the value of the metal oxide by converting it to pure metal. That is, most pure metals have very high exergy values. This means that they are a thermodynamically valuable energy "investment," and their destruction in subsequent manufacturing, use, and end-of-life phases should be avoided. But more than that, exergy provides the tool whereby the results from these activities can be calculated. That is, our knowledge of thermodynamics in general and exergy analysis in particular is sufficient to allow the evaluation of new or previously unexplored systems and estimate on a theoretical basis their resource needs, conversion efficiencies, and losses. For example, the exergy analysis of any chemical reaction allows for a rapid evaluation of alternative energy-production schemes, or various chemical-conversion schemes. In fact, these tools are so powerful we believe that they should be included in the standard toolkit for any engineer, industrial ecologist, and many other professionals dealing with sustainability issues.

In this book we use the tools available from thermodynamics, including first law and second law analyses, as well as the applicable science from other domains to gain insight into the functioning of resource use and transformations. There is a particular emphasis on those resources that are needed to support life and human society on this planet. Furthermore, we are also interested in the application of thermodynamic thinking to new, somewhat "remote" areas that may not necessarily be considered as being thermodynamic systems per se. Nevertheless, by demonstrating strict analogies among the thermodynamic variables, constraints, and internal forces with these new systems, it can be possible to portray these "remote" systems as having thermodynamic behavior. These "remote" areas are addressed in various chapters throughout the book.

3 Organization of the Book

This book is broken down into four parts. A brief description of the chapters in each part is provided in Table I.1.

REFERENCES

[1] J. Diamond, *Collapse – How Societies Choose to Fail or Survive* (Viking Penguin, New York, 2005).

[2] G. C. Daily, *Nature's Services: Societal Dependence on Natural Ecosystems* (Island Press, Washington, D.C., 1997).

[3] S. E. Jorgensen, *Integration of Ecosystem Theories: A Pattern* (Kluwer Academic, Dodrecht, The Netherlands, Boston, 1997).

[4] R. U. Ayres, *Information, Entropy, and Progress: A New Evolutionary Paradigm* (American Institute of Physics, College Park, MD, 1994).

[5] E. P. Odum, "Strategy of ecosystem development," *Science* **164**, 262 (1969).

[6] J. Szargut, D. R. Morris, and F. R. Steward, *Exergy Analysis of Thermal, Chemical and Metallurgical Processes* (Hemisphere, New York, 1988).

[7] S. Carnot, *Reflections of the Motive Power of Fire* (Dover, New York, 1988; originally published 1890).

[8] E. P. Gyftopolous, "Presentation of the foundations of thermodynamics in about twelve one-hour lectures," in *Thermodynamic Optimization of Complex Energy Systems*, edited by A. Bejan and E. Mamut (Kluwer, Dordrecht, The Netherlands, 1998), pp. 1–44.

[9] M. J. Moran, *Availability Analysis – A Guide to Efficient Energy Use* (ASME Press, New York, 1989).

[10] A. Bejan, *Advanced Engineering Thermodynamics* (Wiley, New York, 1988).

[11] J. de Swaan Aarons, H. van der Kooi, and K. Sankaranarayanan, *Efficiency and Sustainability in the Energy and Chemical Industries* (Taylor & Francis, New York, 2004).

[12] D. M. Gates, *Energy and Ecology* (Sinauer, Suderland, MA, 1985).

[13] J. M. Smith, H. C. Van Ness, and M. M. Abbott, *Introduction to Chemical Engineering Thermodynamics*, 6th ed. (McGraw-Hill, New York, 2001).

[14] N. Georgescu-Roegen, *The Entropy Law and Economic Progress* (Harvard University Press, Cambridge, MA, 1971).

[15] R. U. Ayres, "Eco-thermodynamics: Economics and the Second Law," *Ecol. Econ.* **26**, 189–209 (1998).

[16] A. Valero, A. A. Lozano, L. Serra, and C. Torres, "Application of the exergetic cost theory to the CGAM problem," *Energy* **19**, 365–381 (1994).

[17] G. Tsataronis, "Strengths and limitations of exergy analysis," in *Thermodynamics Optimization of Complex Energy Systems*, edited by A. Bejan and E. Mamut (Kluwer, Dordrecht, The Netherlands, 1998), pp. 93–100.

[18] C. J. Cleveland, R. Costanza, and D. I. Stern, editors, *The Nature of Economics and the Economics of Nature* (Elgar, Cheltenham, England, 2001).

[19] M. Ruth, *Integrating Economics, Ecology and Thermodynamics* (Springer, New York, 1993).

[20] I. Müller, "Socio-thermodynamics – integration and segregation in a population," *Continuum Mech. Thermodyn.* **14**, 389–404 (2002).

[21] H. T. Odum, *Environmental Accounting: Emergy and Environmental Decision Making* (Wiley, New York, 1996).

[22] H. T. Odum and E. C. Odum, *Energy Basis for Man and Nature* (McGraw-Hill, New York, 1976).

[23] R. Penrose, "Singularities and time-asymmetry," in *General Relativity: An Einstein Centenary Survey*, edited by S. W. Hawking and W. Israel (Cambridge University Press, Cambridge, 1979), pp. 581–638.

[24] E. P. Gyftopoulos, "Thermodynamics: Generalized available energy and availability or exergy," Chap. 1 of this book.

[25] E. P. Gyftopoulos and G. P. Beretta, *Thermodynamics: Foundations and Applications* (Macmillan, New York, 1991).

[26] D. T. Spreng, *Net-Energy Analysis and the Energy Requirements of Energy Systems* (Praeger, New York, 1988).

[27] *Oxford English Dictionary*, 2nd ed. (Oxford University Press, Oxford, 1999).

[28] B. R. Bakshi, A. Baral, and J. L. Hau, "Accounting for resource use by thermodynamics," Chap. 3 of this book.

FOUNDATIONS

1   Thermodynamics: Generalized Available Energy and Availability or Exergy

Elias P. Gyftopoulos

1.1 Introduction

Ever since Clausius postulated that "the energy of the universe is constant" and "the entropy of the universe strives to attain a maximum value," practically every scientist and engineer shares the beliefs that: (i) Thermodynamics is a statistical theory, restricted to phenomena in macroscopic systems in thermodynamic equilibrium states; and (ii) entropy – the concept that distinguishes thermodynamics from mechanics – is a statistical measure of ignorance, ultimate disorder, dispersion of energy, erasure of information, or other causes, and not an inherent property of matter like rest mass, energy, etc.

These beliefs stem from the conviction that the "known laws" of mechanics (classical or conventional quantum) are the ultimate laws of physics and from the fact that statistical theories of thermodynamics yield accurate and practical numerical results about thermodynamic equilibrium states.

Notwithstanding the conviction and excellent numerical successes, the almost-universal efforts to compel thermodynamics to conform to statistical and other nonphysical explanations, and to restrict it only to thermodynamic equilibrium states [1–3] are puzzling in the light of many accurate, reproducible, and nonstatistical experiences and many phenomena that cannot possibly be described in terms of thermodynamic equilibrium states.

Since the advent of thermodynamics, many academics and practitioners have questioned the clarity, unambiguity, and logical consistency of traditional expositions of the subject. Some of the questions raised are: (i) Why is thermodynamics restricted to thermodynamic equilibrium states only, given that the universally accepted and practical statements of energy conservation and entropy nondecrease are demonstrably time dependent? (ii) Why do we restrict thermodynamics to macroscopic systems, given that Gibbsian statistics [4, 5] and systems in states with negative temperatures [6] prove beyond a shadow of a doubt that thermodynamics is valid for any system? (iii) How can any of the proposed statistical expressions of entropy be accepted if none conforms to the requirements that must be satisfied by the entropy of thermodynamics [7]? and (iv) Why do so many professionals continue to believe that thermodynamic equilibrium is a state of ultimate disorder despite the fact that both experimental and theoretical evidence indicates that such a state represents ultimate order [8, 9]?

The purposes of this chapter are as follows: (i) to present a brief summary of a novel exposition of thermodynamics, (ii) to provide the rigorous definition of generalized available energy, and (iii) to define and illustrate by specific applications the concept of availability or exergy. The novel exposition was conceived by Gyftopoulos and Beretta [10].

1.2 Summary of Basic Concepts

1.2.1 Systems, Properties, and States

A well-defined *system* is a collection of constituents determined by the following specifications:

1. the type and the range of values of the *amount* of each *constituent*,
2. the type and the range of values of the *parameters* that fully characterize the *external forces* exerted on the constituents by bodies other than the constituents, such as the parameters that describe the size and geometrical shape of an airtight container, and an applied electrostatic field,
3. the *internal forces* between constituents,
4. the *internal constraints* that characterize the interconnections between separated parts, such as the condition that the overall volume of the two variable-volume parts be fixed, and that define the modeling assumptions such as the condition that some or all chemical reactions be inactive.

Everything that is not included in the system is called the *environment* or the *surroundings* of the system.

For a system consisting of r different types of constituents, we denote their amounts by the vector $\boldsymbol{n} = \{n_1, n_2, \ldots, n_r\}$. For a system with external forces described by s parameters, we denote the parameters by the vector $\boldsymbol{\beta} = \{\beta_1, \beta_2, \ldots, \beta_s\}$. One parameter may be volume V.

Two systems are *identical* if they consist of the same types of constituents, experience the same internal and external forces, and have the same ranges of values of amounts of constituents and parameters and the same constraints. If any of these identities is not valid, the two systems are *different*.

At any instant in time, the amount of each type of constituent and the parameters of each external force have specific values within the corresponding ranges of the system. By themselves, these values do not suffice to characterize completely the condition of the system at that time. We also need the values of all the properties at the same instant in time. Each *property* is an attribute that can be evaluated at any given instant of time by means of a set of measurements and operations that are performed on the system and result in a numerical value – the *value* of the property. This value is independent of the measuring devices, other systems in the environment, and other instants in time.

Two properties are *independent* if the value of one can be varied without affecting the value of the other. Otherwise, the two properties are *interdependent*. For example, position and velocity of a molecule are independent properties, whereas speed and kinetic energy of a molecule in classical mechanics are interdependent.

For a given system, the values of the amounts of all the constituents, the values of all the parameters, and the values of a complete set of independent properties encompass all that can be said about the system at an instant in time and about the results of any measurements or observations that may be performed on the system at that same instant in time. As such, the collection of all these values constitutes a complete characterization of the system at that instant in time. We call this characterization at an instant in time the *state* of the system.

1.2.2 Changes of State in Time

The state of a system may change in time spontaneously because of the internal dynamics of the system, or as a result of interactions with other systems, or both.

A system that experiences only spontaneous changes of state, that is, a system that does not affect the state of its environment, is called *isolated*. In general, a system that is not isolated interacts with other systems in a number of different ways, some of which may result in net flows of properties from one system to another.

The relation that describes the evolution of the state of a system as a function of time is the *equation of motion*. Such an equation was discovered by Beretta et al. [11, 12] but is not discussed in this chapter.

Rather than through the explicit time dependence, which requires the complete equation of motion, here a change of state is described in terms of the *end states*, that is, the initial and the final states of the system, the *modes of interaction* that are active during the change of state, and conditions on the values of properties of the end states that are consequences of the laws of thermodynamics, that is, conditions that express, not all, but most of the general and well-established features of the complete equation of motion. Each mode of interaction is characterized by means of well-specified net flows of properties across the boundaries of the interacting systems. For example, after the properties energy and entropy are defined, we will see that some modes of interaction involve the flow of energy across the boundaries of the interacting systems without any flow of entropy, whereas other modes of interaction involve the flow of both energy and entropy. Among the conditions on the values of properties of the end states that are consequences of the laws of thermodynamics – conditions that express well-established features of time-dependent behavior of systems – we will see that the energy change of a system must equal the energy transferred into the system, and that its entropy change must be greater than or at least equal to the entropy transferred into the system.

The end states and the modes of interactions associated with a change of state of a system specify a *process*. The modes of interactions may be used to classify processes into different types. For example, a process that involves no interactions and, therefore, no flows across the boundary of the system is called a *spontaneous process*. Again, a process that involves interactions that result in no external effects other than a change in elevation of a weight (or an equivalent mechanical effect) is called a *weight process*.

Another important classification of processes is in terms of the possibility of annulling all their effects. A process may be either reversible or irreversible. A process is *reversible* if it can be performed in at least one way such that both the system and its environment can be restored to their respective initial states. A process is *irreversible* if it is impossible to perform it in such a way that both the system and its environment can be restored to their respective initial states.

We will see that any irreversible process involves the irrecoverable degradation of a valuable resource, whereas no reversible process involves such a degradation. For this reason, putting aside all economic, social, and environmental considerations, we sometimes say that a reversible process is the "best possible." For example, for a given change of state of a system, an irreversible weight process results in either a smaller raise or a larger drop in the weight than do the corresponding results of a reversible process.

In general, a system A that undergoes a process from state A_1 at time t_1 to state A_2 at time t_2 is well defined at these two times but is not necessarily well defined during the lapse of time between t_1 and t_2. The reason is that the interactions that induce the change of state may involve such temporary alterations of internal and external forces that no system A can be defined during the period t_1 to t_2. Said more formally, in the course of interactions, the constituents of a system may not be separable from the environment or, if they are, the states of the system may be correlated with the states of other systems. Nevertheless, at the end of the process, the system becomes again well defined, and its state is again uncorrelated.[1]

1.2.3 Energy and Energy Balance

Energy is a concept that underlies our understanding of all physical phenomena, yet its meaning is subtle and difficult to grasp. It emerges from a fundamental principle known as the first law of thermodynamics.

The *first law* asserts that *any two states of a system may always be interconnected by means of a weight process*[2] *and, for a given weight subject to a constant gravitational acceleration, that the change in elevation during such a process is fixed uniquely by the two states of the system.*

The main consequence of this law is that every system A in any state A_1 has a property called *energy*, denoted by the symbol E_1. The energy E_1 of any state A_1 can be evaluated by means of an auxiliary weight process that interconnects state A_1 and a reference state A_o to which is assigned a fixed reference value E_o, and the expression

$$E_1 - E_o = -Mg\,(z_1 - z_o)\,, \tag{1.1}$$

---

[1] We say that a system is well defined and its constituents are *separable from the environment* if the forces exerted on the constituents by a body not included in the system do not depend explicitly on the coordinates of constituents of that body. We say that a state is *uncorrelated* from the state of the environment if none of the values of the properties of the system depends on the values of properties of systems in the environment. All statements and conclusions in this chapter refer to well-defined systems in uncorrelated states.

[2] Other processes equivalent to a weight process are discussed in Chap. 3 of [10].

where M is the mass of the weight, g is the gravitational constant, and z is the elevation of the weight. The energy E_2 of another state A_2 can be evaluated by a similar procedure so that

$$E_2 - E_o = -Mg\,(z_2 - z_o)\,. \tag{1.2}$$

Moreover, subtracting Eq. (1.1) from Eq. (1.2), we find

$$E_2 - E_1 = -Mg\,(z_2 - z_1)\,, \tag{1.3}$$

where we keep the negative sign in front of $Mg\,(z_2 - z_1)$ in order to emphasize that, the larger value of z_2, the smaller value of E_2, and vice versa.

Energy is an *additive* property, namely, the energy of a system consisting of two or more subsystems equals the sum of the energies of the subsystems, and this holds for all combinations of states of the subsystems. Moreover, energy has the same value at the final time as at the initial time whenever the system experiences a zero-net-effect weight process, or remains invariant in time whenever the process is spontaneous. In either of these two processes, $z_2 = z_1$ and $E\,(t_2) = E\,(t_1)$ for time t_2 greater than t_1, that is, energy is *conserved*.

Because of additivity, and because any process of a system can always be thought of as part of a zero-net-effect weight process of a composite system consisting of all the interacting systems, the conclusion that, as a function of time, energy is invariant is known as the *principle of energy conservation*.

Energy can be transferred between systems by means of interactions. Denoting by $E^{A\leftarrow}$ the net amount of energy transferred from the environment to system A as a result of all the interactions involved in a process that changes the state of A from A_1 to A_2, we derive an extremely important analytical tool, the *energy-balance equation* or, simply, the *energy balance*. This equation is based on the additivity of energy and on the principle of energy conservation. It requires that, as a result of a process, the change in the energy of the system from E_1 to E_2 must be equal to the net amount of energy $E^{A\leftarrow}$ transferred into the system, namely,

$$E_2 - E_1 = E^{A\leftarrow}\,. \tag{1.4}$$

For all applications of thermodynamics, relativistic affects are negligible, and the mass of a system satisfies a *mass balance* of the form

$$m_2 - m_1 = m^{A\leftarrow}\,, \tag{1.5}$$

where m_1 and m_2 are the masses of states A_1 and A_2, respectively, and $m^{A\leftarrow}$ is the mass flow into system A from other systems in the environment.

1.2.4 Types of States

Because the number of independent properties of a system is infinite even for a system consisting of a single particle with a single translational degree of freedom – a single variable that fixes the configuration of the system in space – and because most properties can vary over a range of values, the number of possible states of a system is infinite.

To facilitate the discussion of these states, we classify them into different categories according to their evolutions in time. This classification brings forth many important aspects of physics, and provides a readily understandable motivation for the introduction of the second law of thermodynamics. We consider four types of states: unsteady, steady, nonequilibrium, and equilibrium. Moreover, we further classify equilibrium states into three types: unstable, metastable, and stable.

An *unsteady state* is one that changes as a function of time because of interactions of the system with other systems. A *steady state* is one that does not change as a function of time, despite interactions of the system with other systems in the environment. A *nonequilibrium state* is one that changes spontaneously as a function of time, that is, a state that evolves in time without any effects on or interactions with any other systems. An *equilibrium state* is one that does not change as a function of time while the system is isolated. An *unstable equilibrium state* is an equilibrium state that may be caused to proceed spontaneously to a sequence of entirely different states by means of a minute and short-lived interaction that has only an infinitesimal temporary effect on the state of the environment. A *metastable equilibrium state* is an equilibrium state that may be changed to an entirely different but compatible state without leaving net effects in the environment of the system, but this can be done only by means of interactions that have a finite temporary effect on the state of the environment. A *stable equilibrium state* is an equilibrium state that can be altered to a different but compatible state only by interactions that leave net effects in the environment of the system.

Starting either from a nonequilibrium state or from an equilibrium state that is not stable, a system can be made to raise a weight without leaving any other net changes in the state of the environment. In contrast, experience shows that from some other types of states – they turn out to be stable equilibrium states – such a raise of a weight is impossible. This impossibility is one of the most striking consequences of the first and the second laws of thermodynamics.

1.2.5 Stable Equilibrium States

The existence of stable equilibrium states is not self-evident. It is the essence of the second law first proposed by Hatsopoulos and Keenan [13]. In the absence of internal mechanisms, such as chemical reactions or internal interconnections, capable of causing spontaneous changes in the values of the amounts of constituents and the parameters, the *second law* asserts that, *among all the states of a system with given values of the energy, the amounts of constituents, and the parameters, there exists one and only one stable equilibrium state*. A more general statement of the second law is this: *Among all the states of a system that have a given value E of the energy and are compatible with a given set of values **n** of the amounts of constituents and β of the parameters, there exists one and only one stable equilibrium state. Moreover, starting from any state of a system, it is always possible to reach a stable equilibrium state with arbitrarily specified values of amounts of constituents and parameters by means of a reversible weight process.*

The existence of stable equilibrium states for various conditions of matter has many theoretical and practical consequences. One consequence is that, starting from any stable equilibrium state of any system, no energy can be transferred to a weight in a weight process in which the values of amounts of constituents and parameters of the system experience no net changes. This consequence is often referred to as the *impossibility of a perpetual-motion machine of the second kind.* In some expositions of thermodynamics, it is taken as the statement of the second law. In this chapter, it is only one aspect of the first and the second laws.

Other consequences are discussed immediately below.

1.2.6 Reservoir and Generalized Available Energy

We define a *reservoir* as an idealized kind of system with a behavior that approaches the following three limiting conditions:

1. It passes through stable equilibrium states only.
2. In the course of finite changes of state, it remains in mutual stable equilibrium with a duplicate of itself that experiences no such changes.
3. At constant values of amounts of constituents and parameters of each of two reservoirs initially in mutual stable equilibrium, energy can be transferred reversibly from one reservoir to the other with no net effects on any other system.

Two systems are in *mutual stable equilibrium* if their composite system is in a stable equilibrium state.

Given a system A in state A_1 and a reservoir R with fixed values of amounts of constituents and parameters, we consider the composite of A and R and evaluate the largest amount of energy that can be transferred to a weight in a weight process for the composite of A and R. This amount is called *available energy* and is denoted by Ω_1^R. After Ω_1^R is transferred out of the composite, A and R are in mutual stable equilibrium, that is, the composite of A and R is in a stable equilibrium state.

The first scientist who raised the question about the largest amount of energy that can be transferred to a weight in a weight process for the composite of a system A and a reservoir R was Carnot [14]. He restricted his investigation, however, to A, which was also a reservoir. His results constitute the seminal ideas – the conception event – of the science of thermodynamics. The disclosure of the available energy Ω_1^R as a property is a generalization of the results of Carnot in that system A need not be a reservoir and state A_1 need not be a stable equilibrium state. Available energy can be assigned to any system in any state.

Another property, the *generalized available energy*, may also be defined as a property of a system A in any state A_1. Its definition is identical to that of available energy except that the final state A_2 of system A corresponds to arbitrarily assigned values of the amounts of constituents and parameters that differ in general from those of state A_1. Said differently, generalized available energy involves exchanges of constituents and changes in parameters in addition to other interactions. The generalized available energy of state A_1 is defined with respect to a reservoir R and the arbitrarily assigned values of the amounts of constituents and parameters. For simplicity, we denote it by the same symbol, Ω_1^R, as that of the available energy. We distinguish it from the available energy of state A_1 with respect to reservoir R by name and context.

The difference between the generalized available energies, $\Omega_1^R - \Omega_2^R$, of two states A_1 and A_2 is equal to the energy that can be exchanged with a weight in a reversible weight process of the composite AR of system A and reservoir R as system

A goes from state A_1 to state A_2. On denoting the energy exchanged with the weight by $(W_{12}^{AR\rightarrow})_{\text{rev}}$, we have

$$(W_{12}^{AR\rightarrow})_{\text{rev}} = \Omega_1^R - \Omega_2^R. \tag{1.6}$$

The value of $(W_{12}^{AR\rightarrow})_{\text{rev}}$ is positive if energy is transferred from the composite AR to the weight, and then it is the largest energy transfer to the weight that can be achieved as system A goes from state A_1 to state A_2. It is negative if energy is transferred from the weight to the composite AR, and then it is the least energy transfer that is required for achieving the change of A from state A_1 to state A_2.

Two important relations exist between the energies E_1 and E_2 and the generalized available energies Ω_1^R and Ω_2^R of any two given states A_1 and A_2 of a system A. By virtue of the first law, the two states can always be interconnected by means of a weight process for system A alone. But the first law determines neither the direction of the weight process nor its reversibility. By contrast, a comparison between the difference in energies and the difference in generalized available energies of the two states determines both the direction and the reversibility of the process. Specifically, if

$$\Omega_1^R - \Omega_2^R = E_1 - E_2, \tag{1.7}$$

then a weight process for A alone is possible both from A_1 to A_2 and from A_2 to A_1 and is reversible. However, if

$$\Omega_1^R - \Omega_2^R > E_1 - E_2, \tag{1.8}$$

then a weight process for A alone is possible only from A_1 to A_2 and is irreversible.

For spontaneous or zero-net-effect weight processes, energy conservation implies that $E_2 = E_1$ or, emphasizing the time dependence, $E(t_2) = E(t_1)$ for $t_2 > t_1$. If applied to these processes, Eq. (1.7) and relation (1.8) reveal the following results. If the process is reversible, then $\Omega_2^R = \Omega_1^R$ or, emphasizing the time dependence, $\Omega^R(t_2) = \Omega^R(t_1)$ for $t_2 > t_1$, namely, the generalized available energy is conserved. If the spontaneous or zero-net-effect weight process is irreversible, then $\Omega_2^R < \Omega_1^R$ or $\Omega^R(t_2) < \Omega^R(t_1)$ for $t_2 > t_1$, namely, the generalized available energy is not conserved. Said differently, in the course of an irreversible, zero-net-effect weight process a system loses some of its potential ability to transfer energy to a weight. Whereas energy is conserved, the amount of energy that can be transferred to a weight in a weight process – the potential of a system to perform useful tasks – is not conserved. This potential cannot be created but may be dissipated to a lesser or larger degree, depending on whether the process is a little or a lot irreversible. A quantitative measure of irreversibility can be expressed in terms of the property entropy discussed in the next section.

A noteworthy feature of E and Ω^R is that both are defined for any state of any system, regardless of whether the state is unsteady, steady, nonequilibrium, equilibrium, metastable equilibrium, or stable equilibrium, and regardless of whether the system has many degrees of freedom or one degree of freedom, or whether its size is large or small.

A disadvantage of Ω^R is that it depends both on the state of the system and on the reservoir R. As discussed in the next section, we gain independence of the reservoir, without losing additivity, by considering the difference between energy and generalized available energy.

1.2.7 Entropy and Entropy Balance

An important consequence of the two laws of thermodynamics is that every system A in any state A_1, with energy E_1 and generalized available energy Ω_1^R with respect to an auxiliary reservoir R, has a property called *entropy*, denoted by the symbol S_1. Entropy is a property in the same sense that energy is a property or momentum is a property. It can be evaluated by means of the auxiliary reservoir R, a reference state A_o, with energy E_o and generalized available energy Ω_o^R, to which is assigned a fixed reference value S_o and the expression

$$S_1 = S_o + \frac{1}{c_R}\left[(E_1 - E_o) - \left(\Omega_1^R - \Omega_o^R\right)\right], \tag{1.9}$$

where c_R is a well-defined positive constant. For the given auxiliary reservoir R, c_R is selected in such a way that the values of entropy found by means of Eq. (1.9) are independent of the reservoir.[3] In other words, despite the dependence of the value of the difference of generalized available energies, $\Omega_1^R - \Omega_o^R$, on the selection of the reservoir R, we can show that there is a constant property c_R of reservoir R that makes the right-hand side of Eq. (1.9) independent of R. Thus S is a property of system A only, in the same sense as energy E is a property of system A only. In due course, the concept of temperature is defined as a property of stable equilibrium states. Then we show that the temperature of a reservoir is constant and that c_R is equal to the constant temperature of the reservoir R.

The entropy S_2 of a state A_2 is given by an expression similar to that of A_1, namely,

$$S_2 = S_o + \frac{1}{c_R}\left[(E_2 - E_o) - \left(\Omega_2^R - \Omega_o^R\right)\right]. \tag{1.10}$$

Moreover, subtracting Eq. (1.9) from (1.10), we find

$$S_2 = S_1 + \frac{1}{c_R}\left[(E_2 - E_1) - \left(\Omega_2^R - \Omega_1^R\right)\right], \tag{1.11}$$

or, equivalently,

$$\Omega_2^R - \Omega_1^R = E_2 - E_1 - c_R\left(S_2 - S_1\right). \tag{1.12}$$

Like energy, entropy is an additive property, namely, the entropy of a system consisting of two or more subsystems equals the sum of the entropies of the subsystems and this holds for all combinations of states of the subsystems. Whereas energy remains constant in time whenever the system experiences either a spontaneous process or a zero-net-effect weight process, Eqs. (1.7) and (1.11) show that the entropy remains constant in time if the process is reversible. In the course of an irreversible either spontaneous or zero-net-effect weight process, relation (1.8) and Eq. (1.11) show that the entropy increases in time, and part of the potential ability of the system

---

[3] The precise definition of c_R and the proof that S_1 is independent of the reservoir are not summarized here for brevity. They are given in [10].

to transfer energy to a weight is destroyed. Because of additivity and because any process of a system can always be thought of as part of a spontaneous process of a composite system consisting of all the interacting systems, the conclusion that, as time proceeds, entropy can either be created, if the process is irreversible, or remain constant, if the process is reversible, but can never be destroyed is of great generality and practical importance. It is known as the *principle of entropy nondecrease*. The entropy created as time proceeds during an irreversible process is called *entropy generated by irreversibility* or *entropy production due to irreversibility*. It is positive.

Like energy, entropy can be transferred between systems by means of interactions. Denoting by $S^{A\leftarrow}$ the net amount of entropy transferred from systems in the environment to a system A as a result of all the interactions involved in a process in which the state of A changes from A_1 to A_2, we derive another extremely important analytical tool, the *entropy-balance* equation. This equation is based on the additivity of entropy and on the principle of entropy nondecrease. It requires that the change in the entropy of the system from S_1 to S_2 be equal to the net amount of entropy $S^{A\leftarrow}$ transferred into the system, plus the positive amount of entropy S_{irr} generated by irreversibility inside A in the course of the process, that is,

$$S_2 - S_1 = S^{A\leftarrow} + S_{\mathrm{irr}}. \tag{1.13}$$

The value of $S^{A\leftarrow}$ is positive if entropy is transferred into A and negative if entropy is transferred out of A.

It is worth repeating that S is defined for any state of any system because energy E and generalized available energy Ω^R are defined for any state of any system. Thus, like energy, entropy is defined for all states, that is, unsteady, steady, nonequilibrium, equilibrium, metastable equilibrium, and stable equilibrium, and for all systems, that is, systems with many degrees of freedom and systems with few degrees of freedom, including a single particle with a single translational degree of freedom or a single spin because both energy and generalized available energy are defined for all these systems, and for all these states.

The dimensions of entropy are determined by the dimensions of both energy and the property c_R of the auxiliary reservoir. We can show that the dimension c_R is independent of the dimensions of mass, length, and time, but the same as the dimension of temperature (defined later). The unit of c_R chosen in the International System of units is the Kelvin, denoted by K. Another unit is the Rankine, denoted by R, where $1\ \mathrm{R} = (5/9)\ \mathrm{K}$. Entropy values are expressed in many different units such as joules per Kelvin (J/K), kilocalorie per Kelvin (kcal/K), and British thermal unit per Rankine (Btu/R). In particular, it turns out that $1\ \mathrm{Btu/lb\ R} = 1\ \mathrm{kcal/kg\ K}$.

1.2.8 The Fundamental Relation

In the absence of internal mechanisms capable of altering the values of the amounts of constituents and the parameters, that is, in the absence of chemical reactions, nuclear reactions, and other types of internal interconnections, a system admits an indefinite number of states that have given values of the energy E, the amounts of constituents $n_1, n_2, \ldots, n_r$, and the parameters $\beta_1, \beta_2, \ldots, \beta_s$. Most of these states are nonequilibrium, metastable equilibrium, and equilibrium, and, according to the second law, only one is a stable equilibrium state. It follows that the value of any property P of the system in a stable equilibrium state is uniquely determined by the values of $E, n_1, n_2, \ldots, n_r$ and $\beta_1, \beta_2, \ldots, \beta_s$, that is, can be written as a function of the form

$$P = P(E, n_1, n_2, \ldots, n_r, \beta_1, \beta_2, \ldots, \beta_s). \tag{1.14}$$

This result, known as the *stable-equilibrium-state principle* or simply the *state principle*, expresses a fundamental physical feature of the stable equilibrium states of the system and implies the existence of interrelations among the properties at each of these states.

A system in general has a very large number of independent properties. When we focus on the special family of states that are stable equilibrium, however, the state principle asserts that the value of each of these properties is uniquely determined by the values of $E, \boldsymbol{n}, \boldsymbol{\beta}$. In contrast, for states that are not stable equilibrium, the values of $E, \boldsymbol{n}, \boldsymbol{\beta}$ are not sufficient to specify the values of all the independent properties.

When written for the entropy S of stable equilibrium states, Eq. (1.14) becomes

$$S = S(E, n_1, n_2, \ldots, n_r, \beta_1, \beta_2, \ldots, \beta_s) \tag{1.15}$$

and is known as the *fundamental stable-equilibrium-state relation for entropy* or simply the *fundamental relation*. We can show that the function $S(E, \boldsymbol{n}, \boldsymbol{\beta})$ admits partial derivatives of all orders and therefore that any difference between the entropies of two stable equilibrium states may be expressed in the form of a Taylor series in terms of the partial derivatives of $S(E, \boldsymbol{n}, \boldsymbol{\beta})$ at one stable equilibrium state, and differences in the values of the energy, amounts of constituents, and parameters of the two stable equilibrium states.

The function $S(E, \boldsymbol{n}, \boldsymbol{\beta})$ is concave in each of the variables $E, n_1, n_2, \ldots, n_r$. It is concave in each of the parameters $\beta_1, \beta_2, \ldots, \beta_s$, which are additive, like volume, and it is also concave collectively with respect to all the variables $E, n_1, n_2, \ldots, n_r$, and the parameters $\beta_1, \beta_2, \ldots, \beta_s$, which are additive. Concavity implies that $(\partial^2 S / \partial E^2)_{n, \beta} \leq 0$, $(\partial^2 S / \partial n_i^2)_{E, n, \beta} \leq 0$ for each i, $(\partial^2 S / \partial \beta_j^2)_{E, n, \beta} \leq 0$ for each additive β_j, and some other necessary conditions on all the second-order derivatives of the fundamental relation.

Using the second law, we assert that the entropy of each unique stable equilibrium state is larger than that of any other state with the same values of $E, n_1, n_2, \ldots, n_r$, and $\beta_1, \beta_2, \ldots, \beta_s$. This assertion is known as the *highest-entropy principle*. This principle is extremely useful in establishing conditions that must be satisfied by properties of systems in stable equilibrium states.

Equation (1.15) may be solved for E as a function of $S, n_1, n_2, \ldots, n_r$ and $\beta_1, \beta_2, \ldots, \beta_s$ so that

$$E = E(S, n_1, n_2, \ldots, n_r, \beta_1, \beta_2, \ldots, \beta_s). \tag{1.16}$$

The function $E(S, \boldsymbol{n}, \boldsymbol{\beta})$ admits partial derivatives of all orders, and therefore any difference between the energies of two stable equilibrium states may be expressed in the form of a Taylor series in terms of the partial derivatives of $E(S, \boldsymbol{n}, \boldsymbol{\beta})$ at one of the stable equilibrium states, and differences in the values of the entropy, amounts of constituents, and parameters of the two stable equilibrium states.

Among all the partial derivatives, each first-order partial derivative of either the function $S(E, \boldsymbol{n}, \boldsymbol{\beta})$ or the function $E(S, \boldsymbol{n}, \boldsymbol{\beta})$ represents an important and practical property of the family of stable equilibrium states of a system. It is important because each such property enters a condition for mutual stable equilibrium with other systems, and practical because it can be relatively easily related to simple measurements. It should be emphasized that each such property is defined only for the stable equilibrium states of the system.

1.2.9 Temperature

The partial derivative of $E(S, \boldsymbol{n}, \boldsymbol{\beta})$ with respect to entropy, or the inverse of the partial derivative of $S(E, \boldsymbol{n}, \boldsymbol{\beta})$ with respect to energy, that is,

$$T = \left(\frac{\partial E}{\partial S} \right)_{\boldsymbol{n}, \boldsymbol{\beta}} = \frac{1}{(\partial S / \partial E)_{\boldsymbol{n}, \boldsymbol{\beta}}}, \qquad (1.17)$$

is defined as the *absolute temperature* or, simply, the *temperature*. The first of Eqs. (1.17) defines T as a function of $E, \boldsymbol{n}, \boldsymbol{\beta}$ and the second as a function of $S, \boldsymbol{n}, \boldsymbol{\beta}$. Two units of temperature are the Kelvin and the Rankine, denoted by K and R, respectively.

If two systems A and B in states A_0 and B_0 are in mutual stable equilibrium, then the temperature T_0^A of system A must be equal to the temperature T_0^B of system B. Said differently, equality of temperatures of the two systems is a necessary condition for the two systems to be in mutual stable equilibrium.

By virtue of the definition of a reservoir, it follows that all its states have the same temperature T_R, and this temperature is equal to the constant c_R.

1.2.10 Total Potentials

The *total potential of the ith constituent*, μ_i, is defined by either of the two relations

$$\mu_i = \left(\frac{\partial E}{\partial n_i} \right)_{S, \boldsymbol{n}, \boldsymbol{\beta}} = -T \left(\frac{\partial S}{\partial n_i} \right)_{E, \boldsymbol{n}, \boldsymbol{\beta}}. \qquad (1.18)$$

The dimensions of total potential are energy per unit of amount. The first of Eqs. (1.18) defines μ_i as a function of $S, \boldsymbol{n}, \boldsymbol{\beta}$ and the second as a function of $E, \boldsymbol{n}, \boldsymbol{\beta}$. If volume is the only parameter, each total potential is called a *chemical potential*.

If two systems A and B in states A_0 and B_0 are in mutual stable equilibrium, both contain the ith type of constituent for $i = 1, 2, \ldots, r$, and the amount of that constituent may both increase and decrease in each system, then the total potential $(\mu_i)_0^A$ of the ith constituent of A must be equal to the total potential $(\mu_i)_0^B$ of the ith constituent of B. Said differently, in addition to temperature equality, equality of total potentials for every constituent is a necessary condition for two systems to be in mutual stable equilibrium.

1.2.11 Pressure

The *generalized force conjugated to the jth parameter, f_j*, is defined by either of the two relations

$$f_j = \left(\frac{\partial E}{\partial \beta_j}\right)_{S,n,\beta} = -T\left(\frac{\partial S}{\partial \beta_j}\right)_{E,n,\beta}. \tag{1.19}$$

If volume V is a parameter, the negative of the generalized force conjugated to V is called *pressure*, denoted by p, and given by either of the two relations

$$p = -\left(\frac{\partial E}{\partial V}\right)_{S,n,\beta} = T\left(\frac{\partial S}{\partial V}\right)_{E,n,\beta}, \tag{1.20}$$

where here $\beta = \{V, \beta_2, \beta_3, \ldots, \beta_s\}$. The first of Eqs. (1.20) defines p as a function of S, n, β and the second as a function of E, n, β. The dimensions of p are energy per unit volume and, as is any other generalized conjugated force, it is a property of stable equilibrium states only. Pressure can also be thought of as force per unit area. However, force per unit area is not pressure if the state is not a stable equilibrium state [10].

1.2.11.1 First-Order Taylor Series Expansions

In terms of $T_0, p_0, (\mu_i)_0$, and $(f_j)_0$ of an arbitrary stable equilibrium state A_0 of a system A, small differences in energy, $dE = E_1 - E_0$, entropy, $dS = S_1 - S_0$, volume $dV = V_1 - V_0$, other parameters, $d\beta_2 = (\beta_2)_1 - (\beta_2)_0, d\beta_3 = (\beta_3)_1 - (\beta_3)_0, \ldots, d\beta_s = (\beta_s)_1 - (\beta_s)_0$, and amounts of constituents, $dn_1 = (n_1)_1 - (n_1)_0$, $dn_2 = (n_2)_1 - (n_2)_0, \ldots, dn_r = (n_r)_1 - (n_r)_0$ between two neighboring stable equilibrium states are related by a first-order Taylor series expansion or differential energy relation

$$dE = T_0 dS - p_0 dV + \sum_{i=1}^{r} (\mu_i)_0 dn_i + \sum_{j=2}^{s} (f_j)_0 d\beta_j. \tag{1.21}$$

On solving Eq. (1.21) for dS, and writing dS as a first-order Taylor series expansion in terms of dE, dV, dn_i, and d_j, we find

$$dS = \frac{1}{T_0} dE + \frac{p_0}{T_0} dV - \sum_{i=1}^{r} \frac{(\mu_i)_0}{T_0} dn_i - \sum_{j=2}^{s} \frac{(f_j)_0}{T_0} d\beta_j \tag{1.22a}$$

$$= \left[\left(\frac{\partial S}{\partial E}\right)_{V,n,\beta}\right]_0 dE + \left[\left(\frac{\partial S}{\partial V}\right)_{E,n,\beta}\right]_0 dV + \sum_{i=1}^{r}\left[\left(\frac{\partial S}{\partial n_i}\right)_{E,V,n,\beta}\right]_0 dn_i$$

$$+ \sum_{j=2}^{s}\left[\left(\frac{\partial S}{\partial \beta_j}\right)_{E,V,n,\beta}\right]_0 d\beta_j. \tag{1.22b}$$

On comparing the coefficients of dE, dV, dn_i, and $d\beta_j$ in (1.22a) and (1.22b), we find

$$\left(\frac{\partial S}{\partial E}\right)_{V,n,\beta} = \frac{1}{T}, \tag{1.23}$$

$$\left(\frac{\partial S}{\partial V}\right)_{E,n,\beta} = \frac{p}{T}, \tag{1.24}$$

$$\left(\frac{\partial S}{\partial n_i}\right)_{E,V,n,\beta} = -\frac{\mu_i}{T} \quad \text{for } i = 1, 2, \ldots, r, \tag{1.25}$$

$$\left(\frac{\partial S}{\partial \beta_j}\right)_{E,V,n,\beta} = -\frac{f_j}{T} \quad \text{for } j = 2, 3, \ldots, s, \tag{1.26}$$

where, in writing these equalities, we simplify them by dropping the subscript 0, which specifies the particular stable equilibrium state about which we make the Taylor series expansion and at which we evaluate the partial derivatives. Each of Eqs. (1.23), (1.24), (1.25), and (1.26) proves the second of Eqs. (1.17), (1.20), (1.18), and (1.19), respectively.

1.2.12 Energy Relation of a Reservoir

We recall that a reservoir is an idealized kind of system that passes through stable equilibrium states only and remains in mutual stable equilibrium with a duplicate of itself that experiences no changes of state. If reservoir R has only volume as a parameter, the specifications just cited imply that all states of R have the same value of the temperature T_R, the same value of the pressure p_R, and the same values of the chemical potentials of the r constituents, $\mu_{1R}, \mu_{2R}, \ldots, \mu_{rR}$, so that the necessary conditions of temperature equality, pressure equality, and chemical potential equality for all constituents are satisfied. It follows that, for a reservoir, Eq. (1.21) may be stated in terms of differences – large or small – between properties of any two states R_1 and R_2 so that

$$E_2^R - E_1^R = T_R \left(S_2^R - S_1^R\right) - p_R \left(V_2^R - V_1^R\right) + \sum_{i=1}^{r} \mu_{iR} \left[(n_i)_2^R - (n_i)_1^R\right]. \tag{1.27}$$

1.2.13 Work and Heat Interactions

Interactions result in the exchange of properties across the boundaries of the interacting systems. Various combinations of exchanges are used to classify interactions into different categories. An interaction between two systems that results in a transfer of energy only between two systems is classified as a *work interaction*. The amount of energy exchanged as a result of such an interaction is called *work*. All interactions that result in the exchange of energy and at least one more property, for example entropy, between the interacting systems are called *nonwork interactions*. A process of a system experiencing only work interactions is called an *adiabatic process*. Any process that involves nonwork interactions is called a *nonadiabatic process*.

   In the course of an adiabatic process, system A changes from state A_1 to state A_2, the energy exchange $E_{12}^{A\leftarrow}$ is work, that is, $E_{12}^{A\leftarrow} = -W_{12}^{A\rightarrow}$, where $W_{12}^{A\rightarrow}$ denotes the *work done* by system A on systems in its surroundings with which it interacts.

In the course of an adiabatic process, the entropy exchange $S_{12}^{A\leftarrow} = 0$. Therefore the energy and entropy balances are

$$E_2 - E_1 = -W_{12}^{A\rightarrow}, \tag{1.28}$$

$$S_2 - S_1 = S_{\text{irr}}, \tag{1.29}$$

where S_{irr} denotes the entropy generated by irreversibility inside A during the process.

A special example of nonwork interaction entirely distinguishable from work is one between two systems initially differing infinitesimally in temperature. It results in no other effects except a transfer of energy and a transfer of entropy between the two systems such that the ratio of the amount of energy transferred and the amount of entropy transferred equals the almost common temperature of the interacting systems. It is called a *heat interaction*. The amount of energy transferred as a result of such an interaction is called *heat*.

Often, in applications, system A consists of many subsystems, one of which, A', is in a stable equilibrium state at a temperature T_Q. Similarly, system B consists of many subsystems, one of which, B', is in a stable equilibrium state at temperature almost equal to T_Q. If the two subsystems A' and B' experience a heat interaction, we say that systems A and B experience a heat interaction at temperature T_Q, even though A and B are not necessarily in stable equilibrium states.

In the course of a process that involves only a heat interaction at temperature T_Q, system A changes from state A_1 to state A_2, the energy exchange $E_{12}^{A\leftarrow}$ is heat and is denoted by $Q_{12}^{A\leftarrow}$, that is, $E_{12}^{A\leftarrow} = Q_{12}^{A\leftarrow}$, and the entropy exchange is $Q_{12}^{A\leftarrow}/T_Q$. So the two balances are

$$E_2 - E_1 = Q_{12}^{A\leftarrow}, \tag{1.30}$$

$$S_2 - S_1 = \frac{Q_{12}^{A\leftarrow}}{T_Q} + S_{\text{irr}}, \tag{1.31}$$

where S_{irr} is the entropy generated by irreversibility inside A during the process. It is noteworthy that $Q_{12}^{A\leftarrow}$ is not a function of T_Q.

If a process of a system A involves both work and heat but no other interactions, the energy exchange is $E_{12}^{A\leftarrow} = Q_{12}^{A\leftarrow} - W_{12}^{A\rightarrow}$, the entropy exchange $S_{12}^{A\leftarrow} = Q_{12}^{A\leftarrow}/T_Q$, and

$$E_2 - E_1 = Q_{12}^{A\leftarrow} - W_{12}^{A\rightarrow} \tag{1.32}$$

$$S_2 - S_1 = \frac{Q_{12}^{A\leftarrow}}{T_Q} + S_{\text{irr}} \tag{1.33}$$

where S_{irr} is the entropy generated by irreversibility inside A during the process. On dropping some self-evident subscripts and superscripts, we may rewrite Eqs. (1.32) and (1.33) in the form

$$E_2 - E_1 = Q^\leftarrow - W^\rightarrow, \tag{1.34}$$

$$S_2 - S_1 = \frac{Q^\leftarrow}{T_Q} + S_{\text{irr}}, \tag{1.35}$$

or, for differential changes,

$$dE = \delta Q^{\leftarrow} - \delta W^{\rightarrow}, \tag{1.36}$$

$$dS = \frac{\delta Q^{\leftarrow}}{T_Q} + \delta S_{\text{irr}}. \tag{1.37}$$

It is noteworthy that the prefix d denotes infinitesimal differences between the values of a property at two different states of the system, whereas the prefix δ denotes infinitesimal amounts of quantities that are not properties, such as work, heat, and entropy generation by irreversibility.

For processes in which the end states of the system are stable equilibrium states, energy and entropy changes, and therefore work, heat, and entropy generation by irreversibility may be related to changes of other properties and variables, such as temperature, pressure, and volume.

Work and heat interactions are most frequently encountered in engineering applications. Other interactions, involving transfers of energy, entropy, and amounts of constituents, are discussed later.

1.3 Availability Functions

1.3.1 General Remarks

To accomplish almost every practical task, we exploit resources in our natural environment. Some resources are used as energy sources, others as raw materials. *Energy sources* are substances not in permanent mutual stable equilibrium with the environment that can be used to power the energy-conversion systems required by various tasks. Typical sources are coal, oil, natural gas, uranium, and solar energy. Typical tasks are locomotion; motive power and process heat for manufacturing; space conditioning, such as heating, cooling, and ventilation; and electric power for communication devices, computers, industrial machines, home appliances, and lighting. *Raw materials* are substances used as feedstocks in manufacturing tasks – in materials-processing installations that produce different products. Examples of manufacturing tasks are the making of steel out of iron ore and the making of aluminum out of bauxite. Most raw materials are in mutual stable equilibrium with the environment. They are reduced to desired products at the expense of energy sources. Other raw materials are in only partial mutual stable equilibrium with the environment and remain so if prevented from chemical (or nuclear) interactions with other environmental materials. Also, these raw materials are reduced to desired products at the expense of energy sources, such as in a refinery where crude oil is processed to yield petroleum products that are subsequently used as energy sources, or in an enrichment plant where natural uranium is processed to yield fissile uranium that is subsequently used as an energy source.

Each task is accomplished by means of an arrangement of devices, materials-processing systems, and energy-conversion systems interacting with each other, with resources, and with the natural environment. The selection, evaluation, and adoption of a particular arrangement involves the resolution and reconciliation of many complex and conflicting scientific, technical, economical, environmental, social, and safety questions. A complete discussion of these issues is of decisive importance but beyond the scope of this chapter, except for the following questions that are related to thermodynamics:

1. What are the actual inlet, outlet, and end states of both the task and the energy sources used in a particular arrangement?
2. What are the optimum interactions required by the specified inlet, outlet, and end states of the task?
3. What are the optimum interactions that could be supplied by the energy sources employed if these sources were used in the best way physically possible?
4. If the answers to questions 2 and 3 differ, what aspects of the arrangement are the causes of the difference?
5. What can be done to change the difference between the answers to questions 2 and 3?
6. What is a universal measure of such a difference that characterizes how effectively the task is accomplished by a given arrangement?

It is clear that the concept of optimum that we use here is delimited only by the laws of physics, not by restrictions imposed either by economic, social, and environmental considerations or by current technology. As such, it may well be secondary to all the other concerns. Nevertheless, it does provide a limit that cannot be exceeded under any circumstances.

1.3.2 The Environment as a Reservoir

It is shown [10] that, for given inlet, outlet, and end states of a system, a process is optimum if it is reversible. Accordingly, if the inlet, outlet, and end states are not matched so as to yield zero net differences in entropy, the entropy balance must be achieved by exchange of entropy with another system; otherwise the process cannot be carried out reversibly. The only system that is readily available and can exchange large amounts of entropy at no cost is the environment. Similarly, the environment is a readily available, no-cost[4] source of certain substances, such as the air we breathe, the water we drink, and the air intake of our automobile engines. It is also an easy-access sink of substances, such as from automobile engines and energy-conversion systems and many wastes from residential, commercial, and industrial activities.

For analyses of optimum processes, we model all substances in our natural environment, except energy sources, as a system behaving as a reservoir. We call it the *environmental reservoir* and denote it by R^*. Depending on the application, to focus our attention on the phenomena that are most prevalent, we find it convenient to impose different restrictions on the values of the amounts of constituents and the volume of the environmental reservoir.

For example, in applications in which the system and the environment exchange entropy and energy but neither amounts of constituents nor volume, we model the environment as a reservoir R^* with fixed values of amounts of constituents and

---

[4] The recent decades of heavy exploitation of our natural environment show that an unregulated use of the environment may cause a variety of serious alterations that result in enormous costs to our society and impacts on the quality of our lives. Thus the cost-free use of the environment should be allowed only for purposes that are unavoidable.

volume and denote its constant temperature by T_{R^*}. Under these restrictions, the relation between entropy and energy differences [Eq. (1.27) for $E = U$] becomes

$$S_2^{R^*} - S_1^{R^*} = \frac{1}{T_{R^*}}\left(U_2^{R^*} - U_1^{R^*}\right), \qquad (1.38)$$

where R_1^* and R_2^* are any two stable equilibrium states of R^*.

Again, in applications in which the system and the environment exchange entropy, energy, and volume but no amounts of constituents, we model the environment as a reservoir R^* with variable volume and fixed values of the amounts of constituents, and we denote its constant temperature by T_{R^*} and its constant pressure by p_{R^*}. Under these restrictions, the relation among differences in values of energy, entropy, and the volume of any two stable equilibrium states R_1^* and R_2^* is

$$S_2^{R^*} - S_1^{R^*} = \frac{1}{T_{R^*}}\left(U_2^{R^*} - U_1^{R^*}\right) + \frac{p_{R^*}}{T_{R^*}}\left(V_2^{R^*} - V_1^{R^*}\right). \qquad (1.39)$$

Finally, in applications in which the system and the environment exchange entropy, energy, volume, and amounts of constituents, we model the environment as a reservoir R^* with variable volume and variable amounts of constituents, and we denote its constant temperature by T_{R^*}, its constant pressure by p_{R^*}, and the constant chemical potentials by $\mu_{1R^*}, \mu_{2R^*}, \ldots, \mu, \mu_{rR^*}$. Here the relation among differences in values of energy, entropy, volume and amounts of constituents of any two stable equilibrium states R_1^* and R_2^* becomes

$$S_2^{R^*} - S_1^{R^*} = \frac{1}{T_{R^*}}\left[U_2^{R^*} - U_1^{R^*}\right] + \frac{p_{R^*}}{T_{R^*}}\left[V_2^{R^*} - V_1^{R^*}\right] - \sum_{i=1}^{r}\frac{\mu_{iR^*}}{T_{R^*}}\left[(n_i)_2^{R^*} - (n_i)_1^{R^*}\right].$$

$$(1.40)$$

Any state of the environmental reservoir is sometimes called a *passive* or *dead state* because, starting from such a state and using no energy sources, we can accomplish no useful task. Indeed, we cannot build a perpetual-motion machine of any kind using the environmental reservoir as a system.

In addition, the state of any system A in mutual stable equilibrium with R^* is sometimes called a passive or dead state and is denoted by A_{0^*} because, once in such a state, system A is useless as well. In particular, if the environmental reservoir R^* is modeled as having variable volume and variable amounts of constituents, the dead state A_{0^*} of system A has the same values of temperature, pressure, and chemical potentials as the respective values of R^*, that is, $T_{0^*} = T_{R^*}$, $p_{0^*} = p_{R^*}$, and $\mu_{i0^*} = \mu_{iR^*}$ for $i = 1, 2, \ldots, r$. Again, if R^* is modeled as having fixed values of the volume and the amounts of constituents, the dead state A_{0^*} of A has temperature $T_{0^*} = T_{R^*}$, but values of pressure and chemical potentials not necessarily equal to the corresponding values of R^*.

Given a composite of system A and the environmental reservoir R^*, spontaneous changes of state can occur only until A reaches mutual stable equilibrium with R^*, that is, only until A is in state A_{0^*}. After state A_{0^*} is reached, no further change in the state of the composite of A and R^* is possible without expenditure of an energy source because A_{0^*} has null available energy with respect to R^*, and no reservoir other than the environmental is readily available.

In principle, it is always possible to create a reservoir at conditions of temperature, pressure, and chemical potentials different from those of the natural environment. But the creation of such a reservoir requires the expenditure of energy sources, and any benefit that could result would be at best equal to, but usually less than, the expenditure.

1.3.3 Availability or Exergy

In previous discussions, we encountered some answers to questions related to optimum interactions. For example, it was shown that, in changing the state of system A from state A_1 to state A_2 while the system is in combination with a reservoir R that has fixed amounts of constituents and parameters and the composite AR experiences a weight process, the optimum work done on the weight is

$$(W_{12}^{AR\rightarrow})_{\text{optimum}} = \Omega_1^R - \Omega_2^R, \tag{1.41}$$

where Ω_1^R and Ω_2^R are the generalized available energies of the two states of A with respect to R and to some reference values n and β of the amounts of constituents and the parameters of A. Hence, if $\Omega_1^R > \Omega_2^R$, then $(W_{12}^{AR\rightarrow})_{\text{optimum}}$ is the largest work that the composite of A and R could do in a weight process under the specified conditions, whereas if $\Omega_2^R > \Omega_1^R$, then $-(W_{12}^{AR\rightarrow})_{\text{optimum}} = (W_{12}^{AR\leftarrow})_{\text{optimum}}$ is the least work required in a weight process for the composite of A and R to change the state of A from A_1 to A_2, again under the specified conditions.

If the system is simple[5] and the reservoir environmental, Eq. (1.41) may be expressed in terms of energy and entropy in the form

$$(W_{12}^{AR*\rightarrow})_{\text{optimum}} = (U_1 - T_{R*} S_1) - (U_2 - T_{R*} S_2), \tag{1.42}$$

where U and S represent the internal energy and the entropy of system A, respectively, and in writing the equation we use the relation between energy, generalized available energy, and entropy introduced in Section 1.2. Moreover, under the specified conditions, we recall that the available energy is zero if A and R^* have the same temperature and conclude that

$$\Omega_1^{R*} = (W_{10*}^{AR*\rightarrow})_{\text{rev}} = (U_1 - T_{R*} S_1) - (U_{0*} - T_{R*} S_{0*}), \tag{1.43}$$

where U_{0*} and S_{0*} are the energy and entropy of system A in the dead state A_{0*}, respectively, with temperature $T_{0*} = T_{R*}$, and the values of the amounts of constituents and the parameters are equal to the respective reference values n and β.

The expression $U - T_{R*} S$ is called an *availability function* or *exergy function*. As Eq. (1.42) indicates, for the specified conditions the difference in the values of this function at two states yields the optimum work in a weight process for the composite of A and R^*.

The expression $(U_1 - T_{R*} S_1) - (U_{0*} - T_{R*} S_{0*})$, that is, the generalized available energy of state A_1, is also called the *availability* or *exergy* of state A_1. Under the specified conditions, it represents the optimum work that can be done as a result of the state of system A changing from A_1 to the state A_{0*} with the reference values

---

[5] The definition of a *simple system* is given in Chap. 17 of [10].

n and β and temperature T_{0*} equal to that of the environmental reservoir. It turns out that this work is not sign definite, namely, it can be either positive or negative.

Expressions analogous to Eqs. (1.42) and (1.43) can be derived for conditions other than those involved in the definition of generalized available energy. Ideally, we should define a distinct name for each set of conditions and the corresponding function and its differences. Because there are innumerable conditions that we must examine, we would then have so many names that it would be questionable whether the richness of the vocabulary would be of any help. To avoid this linguistic pileup, we proceed as follows.

First, we consider a system A and the environmental reservoir R^* with given specifications regarding whether the values of their respective amounts of constituents and volume are variable or fixed. We define as the *availability function* or *exergy function* corresponding to the given specifications that expression the differences of which yield the optimum work in a weight process for the composite of A and R^* as system A changes from a given state A_1 to another given state A_2. Moreover, we define as the *availability* or *exergy* corresponding to the given specifications and to state A_1 that expression which yields the optimum work in a weight process for the composite of A and R^* as system A changes from state A_1 to state A_{0*} in which A and R^* are in mutual stable equilibrium. For example, for the conditions discussed at the beginning of this section, we summarize the results by writing, in addition to Eq. (1.42)

$$\text{availability function} = U - T_{R^*}S, \tag{1.44}$$

$$\text{availability or exergy} = (U - U_{0*}) - T_{R^*}(S - S_{0*})$$
$$= (U - T_{R^*}S) - (U_{0*} - T_{R^*}S_{0*}). \tag{1.45}$$

Other examples are discussed in the following subsection.

Next, we consider a given type of interaction, such as work, heat, or bulk flow, and the environmental reservoir R^* with given specifications regarding whether the values of its amounts of constituents and volume are variable or fixed. We define as the *availability rate function* or *exergy rate function* corresponding to the given specifications that expression whose differences yield the optimum work rate in a process for the composite of reservoir R^* and a system A maintained in a steady state by two given interactions of the same type. Moreover, we define as the *availability rate* or *exergy rate* corresponding to the given specifications and associated with a given interaction that expression which yields the optimum work rate in a process for the composite of reservoir R^* and a system A maintained in a steady state by the given interaction and an interaction of the same type with the reservoir R^*. A discussion of availability rate functions is given in [10, Chap. 22].

1.3.4 Different Availabilities or Exergies

Here we consider a simple system A, changing from state A_1 with volume V_1 to state A_2 with a different volume V_2, and surrounded by the environmental reservoir R^*, modeled as having variable volume but fixed values of the amounts of constituents, so that the reservoir experiences an equal and opposite change in volume.

At the moving boundary between A and R^*, a volume exchange occurs according to

$$V_2^{R^*} - V_1^{R^*} = -(V_2 - V_1).\qquad(1.46)$$

The motion of the boundary against the constant reservoir pressure p_{R^*} results in a work interaction between A and R^*, and the work done by A on R^* is $p_{R^*}(V_2 - V_1)$. In a weight process for the composite of A and R^*, the work $p_{R^*}(V_2 - V_1)$ represents just an internal exchange between A and R^* and not work $W_{12}^{AR^* \rightarrow}$ done on the weight. To evaluate the optimum work $(W_{12}^{AR \rightarrow})_{\text{optimum}}$ done on the weight under the specified conditions, we begin by writing the energy and entropy balances:

$$\left(U_2 + U_2^{R^*}\right) - \left(U_1 + U_1^{R^*}\right) = -\left(W_{12}^{AR^* \rightarrow}\right),\qquad(1.47)$$

$$\left(S_2 + S_2^{R^*}\right) - \left(S_1 + S_1^{R^*}\right) = S_{\text{irr}}.\qquad(1.48)$$

On combining (1.46)–(1.48) with (1.38) and setting $S_{\text{irr}} = 0$ for optimality, we find that

$$\left(W_{12}^{AR^* \rightarrow}\right)_{\text{optimum}} = (U_1 - T_{R^*} S_1 + p_{R^*} V_1) - (U_2 - T_{R^*} S_2 + p_{R^*} V_2),\quad(1.49)$$

$$\text{availability function} = U - T_{R^*} S + p_{R^*} V,\qquad(1.50)$$

$$\text{availability or exergy} = (U - U_{0^*}) - T_{R^*}(S - S_{0^*}) + p_{R^*}(V - V_{0^*})$$

$$= (U - T_{R^*} S + p_{R^*} V) - (U_{0^*} - T_{R^*} S_{0^*} + p_{R^*} V_{0^*}),\quad(1.51)$$

where U_{0^*}, S_{0^*}, and V_{0^*} are the energy, entropy, and volume, respectively, of A in mutual stable equilibrium with the reservoir and therefore in state A_{0^*} with temperature $T_{0^*} = T_{R^*}$ and pressure $p_{0^*} = p_{R^*}$. Although the value of $U_{0^*} + p_{R^*} V_{0^*} - T_{R^*} S_{0^*}$ equals that of the Gibbs free energy of state A_{0^*} because $T_{0^*} = T_{R^*}$ and $p_{0^*} = p_{R^*}$, it is noteworthy that $U + p_{0^*} V - T_{0^*} S$ is not a Gibbs free energy because U, S, V, T_{0^*}, and p_{0^*} are not all associated with the same state of system A.

Here we consider a simple system A, changing from state A_1 with values V_1 and $(\boldsymbol{n})_1$ of the volume and the amounts of constituents to state A_2 with values V_2 and $(\boldsymbol{n})_2$ and surrounded by the environmental reservoir R^*. The reservoir is modeled as having variable values of volume and amounts of constituents, so that it experiences changes in values of volume and in each of the amounts of constituents equal and opposite to the respective changes in values of A. Thus Eq. (1.46) also holds here and, in addition, we have

$$(n_i)_2^{R^*} - (n_i)_1^{R^*} = -\left[(n_i)_2 - (n_i)_1\right]\quad\text{for } i = 1, 2, \ldots, r,\qquad(1.52)$$

where $(n_i)_1^{R^*}$ and $(n_i)_2^{R^*}$ are the values of the amount of the ith constituent of R^* at states R_1^* and R_2^*, respectively, and $(n_i)_1$ and $(n_i)_2$ are the values of the amount of the same constituent at states A_1 and A_2 of A, respectively.

The energy and entropy balances for a weight process of the composite of A and R^* under the specified conditions are still given by Eqs. (1.47) and (1.48).

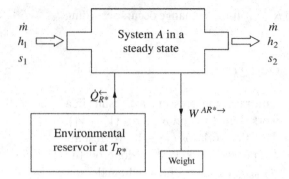

$\dot{m}$
h_1
s_1

$\dot{m}$
h_2
s_2

$\dot{Q}_{R*}^{\leftarrow}$

$W^{AR*\rightarrow}$

Figure 1.1. Schematic of system A maintained in steady state by two bulk-flow interactions, shaft work, and heat with the environmental reservoir.

Combining Eqs. (1.39), (1.46–1.48), and (1.52) and setting $S_{\text{irr}} = 0$ for optimality, we find that

$$(W_{12}^{AR*\rightarrow})_{\text{optimum}} = \left[U_1 - T_{R*}S_1 + p_{R*}V_1 - \sum_{i=1}^{r} \mu_{iR*}(n_i)_1 \right]$$

$$- \left[U_2 - T_{R*}S_2 + p_{R*}V_2 - \sum_{i=1}^{r} \mu_{iR*}(n_i)_2 \right], \quad (1.53)$$

$$\text{availability function} = U - T_{R*}S + p_{R*}V - \sum_{i=1}^{r} \mu_{iR*}n_i, \quad (1.54)$$

$$\text{availability or exergy} = (U - U_{0*}) - T_{R*}(S - S_{0*}) + p_{R*}(V - V_{0*})$$

$$- \sum_{i=1}^{r} \mu_{iR*}[n_i - (n_i)_{0*}]$$

$$= \left[U - T_{R*}S + p_{R*}V - \sum_{i=1}^{r} \mu_{iR*}n_i \right]$$

$$- \left[U_{0*} - T_{R*}S_{0*} + p_{R*}V_{0*} - \sum_{i=1}^{r} \mu_{iR*}(n_i)_{0*} \right]. \quad (1.55)$$

As a third example, we consider system A maintained at steady state by two bulk-flow interactions and the environmental reservoir R^* modeled as having variable volume and amounts of constituents. We assume no changes in mass flow rate and composition and negligible changes in kinetic and potential energies between the bulk-flow states of the inlet and outlet streams. For example, for the arrangement shown in Fig. 1.1, bulk-flow states 1 and 2 are the states of the inlet and outlet streams of a steady-state device A that, in addition to these two bulk-flow interactions, is surrounded by the environmental reservoir and connected to a weight.

For the conditions just specified for the composite of the device A and the reservoir R^*, we find that

$$(W_{12}^{AR*\rightarrow})_{\text{optimum}} = \dot{m}(h_1 - T_{R*}s_1) - \dot{m}(h_2 - T_{R*}s_2), \quad (1.56)$$

$$\text{availability rate function} = \dot{m}(h - T_{R*}s), \quad (1.57)$$

$$\text{availability rate or exergy rate} = \dot{m}(h - T_{R*}s) - \dot{m}(h_{0*} - T_{R*}s_{0*}), \quad (1.58)$$

where h_{0*} and s_{0*} are the specific enthalpy and the specific entropy, respectively, of a bulk-flow state O^* at temperature $T_{0*} = T_{R*}$ and pressure $p_{0*} = p_{R*}$. It is noteworthy that the stream in bulk-flow state O^* is not in mutual stable equilibrium with R^*. The reason is that a bulk-flow state is not stable equilibrium unless the kinetic and potential energies are zero, and even if these energies are zero, the condition of chemical potential equality cannot be met because of the specification that no changes in compositions can occur. In other words, the bulk-flow interaction at state O^* introduces into the environment substances that do not correspond to the environmental composition and therefore cause a subsequent irreversible mixing. This irreversibility is built into the system specifications we are considering. If the specifications are different, such as when chemical reactions are allowed, the availability rate has a different expression.

In processes involving many streams, the availability rate function and the availability rate are given by expressions similar to Eqs. (1.57) and (1.58) except that here each rate is a sum over many streams. Specifically,

$$(\dot{W}_{12}^{AR^* \to})_{\text{optimum}} = [\dot{H}_{\text{in}} - T_{R*}\dot{S}_{\text{in}}] - [\dot{H}_{\text{out}} - T_{R*}\dot{S}_{\text{out}}], \tag{1.59}$$

$$\text{availability rate function} = \dot{H} - T_{R*}\dot{S}, \tag{1.60}$$

$$\text{availability rate or exergy rate} = [\dot{H} - T_{R*}\dot{S}] - [\dot{H}(T_{R*}, p_{R*}) - T_{R*}\dot{S}(T_{R*}, p_{R*})] \tag{1.61}$$

where $\dot{H}$ and $\dot{S}$ represent summations of flow rates of enthalpies and entropies over many streams. For details see [10, Chap. 22].

Under the same conditions as specified in Subsection 1.3.4.3, except that the changes in kinetic and potential energies of the bulk-flow streams are not negligible, Eqs. (1.56)–(1.58) become

$$(\dot{W}_{12}^{AR^* \to})_{\text{optimum}} = \dot{m}\left[h_1 - T_{R*}s_1 + \frac{\xi_1^2}{2} + gz_1\right]$$

$$- \dot{m}\left[h_2 - T_{R*}s_2 + \frac{\xi_2^2}{2} + gz_2\right], \tag{1.62}$$

$$\text{availability rate function} = \dot{m}\left[h - T_{R*}s + \frac{\xi^2}{2} + gz\right] \tag{1.63}$$

$$\text{availability rate or exergy rate} = \dot{m}\left[(h - h_{0*}) - T_{R*}(s - s_{0*}) + \frac{\xi^2}{2} + g(z - z_{0*})\right]$$

$$= \dot{m}\left[h - T_{R*}s + \frac{\xi^2}{2} + gz\right] - \dot{m}(h_{0*} - T_{R*}s_{0*} + gz_{0*}), \tag{1.64}$$

where h_{0*}, s_{0*}, and z_{0*} refer to a bulk-flow state with temperature $T_{0*} = T_{R*}$, pressure $p_{0*} = p_{R*}$, bulk-flow speed $\xi_{0*} = 0$, and the lowest elevation z_{0*} in the environment.

It is clear that many more availability (availability rate) or exergy (exergy rate) functions can be defined, each associated with a particular set of conditions.

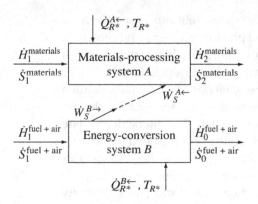

Figure 1.2. The burning of fuel–air mixture in the energy-conversion system provides the work needed to process the bulk-flow stream through the materials-processing system.

1.3.5 Availability or Exergy Analysis

An illustration of the usefulness of Eqs. (1.59) and (1.61) is provided by the bulk-flow processes in Fig. 1.2. Various substances enter a materials-processing system A in various bulk-flow streams with overall enthalpy rate $\dot{H}_1^{\text{materials}}$ and entropy rate $\dot{S}_1^{\text{materials}}$. The plant is designed to operate in steady state and to transform the entering streams into products having overall enthalpy rate $\dot{H}_2^{\text{materials}}$ and overall entropy rate $\dot{S}_2^{\text{materials}}$. The transformation requires shaft work from a power plant at a rate $\dot{W}_s^{A\leftarrow}$ and heat from the natural environment at temperature T_{R*} at a rate $\dot{Q}_{R*}^{A\leftarrow}$. The rate balance for the materials-processing system is

$$\dot{W}_s^{A\leftarrow} = \left[\dot{H}_2^{\text{materials}} - \dot{H}_1^{\text{materials}} - T_{R*} \left(\dot{S}_2^{\text{materials}} - \dot{S}_1^{\text{materials}} \right) \right] + T_{R*} \dot{S}_{\text{irr}}^A, \quad (1.65)$$

where $\dot{S}_{\text{irr}}^A$ is the rate of entropy generation by irreversibility in the materials-processing system A. It is noteworthy that $\dot{W}_s^{A\leftarrow}$ is optimum if $\dot{S}_{\text{irr}}^A = 0$ and that to each different value of $\dot{W}_s^{A\leftarrow}$ there corresponds a different value of $\dot{Q}_{R*}^{A\leftarrow}$.

The shaft work is provided by energy-conversion system B (Fig. 1.2), which converts a fuel and air stream into products of combustion. The fuel and air enter the system as bulk-flow streams with overall enthalpy rate $\dot{H}_1^{\text{fuel+air}}$ and overall entropy rate $\dot{S}_1^{\text{fuel+air}}$. The energy conversion system does shaft work at a rate $\dot{W}_s^{B\rightarrow}$ and interacts with the natural environment at T_{R*} with heat at a rate $\dot{Q}_{R*}^{B\leftarrow}$. Moreover, we assume that the products of combustion exit in bulk-flow streams at temperature $T_{0*} = T_{R*}$ and pressure $p_{0*} = p_{R*}$. For these streams, we denote the overall enthalpy rate by $\dot{H}_{0*}^{\text{fuel+air}}$ and the overall entropy rate by $\dot{S}_{0*}^{\text{fuel+air}}$. If no entropy is generated by irreversibility in the energy-conversion system, the work rate is given by the availability rate of the fuel–air mixture, Eq. (1.61). However, if entropy is generated by irreversibility, $\dot{W}_s^{B\rightarrow}$ satisfies the relation

$$\dot{W}_s^{B\rightarrow} = \left[\dot{H}_1^{\text{fuel+air}} - \dot{H}_{0*}^{\text{fuel+air}} - T_{R*} \left(\dot{S}_1^{\text{fuel+air}} - \dot{S}_{0*}^{\text{fuel+air}} \right) \right] - T_{R*} \dot{S}_{\text{irr}}^B, \quad (1.66)$$

where $\dot{S}_{\text{irr}}^B$ is the rate of entropy generation by irreversibility in the energy-conversion system.

On subtracting Eq. (1.65) from Eq. (1.66), recognizing that $\dot{W}_s^{B\rightarrow} = \dot{W}_s^{A\leftarrow}$, and rearranging terms, we find that

$$\left[\dot{H}_1^{\text{fuel+air}} - \dot{H}_{0*}^{\text{fuel+air}} - T_{R*} \left(\dot{S}_1^{\text{fuel+air}} - \dot{S}_{0*}^{\text{fuel+air}} \right) \right]$$
$$= \left[\dot{H}_2^{\text{materials}} - \dot{H}_1^{\text{materials}} - T_{R*} \left(\dot{S}_2^{\text{materials}} - \dot{S}_1^{\text{materials}} \right) \right] + T_{R*} \dot{S}_{\text{irr}}, \quad (1.67)$$

where $\dot{S}_{\mathrm{irr}} = \dot{S}_{\mathrm{irr}}^{A} + \dot{S}_{\mathrm{irr}}^{B}$, that is, the rate of entropy generation by irreversibility in both the materials-conversion and the energy-conversion systems.

The left-hand side of Eq. (1.67) is the availability rate of the fuel–air mixture, that is, the largest rate at which work could possibly be done by processing the mixture in the natural environment under the specified conditions on the types of interactions between the energy-conversion system and the environment. The bracketed term on the right-hand side is the least work rate required for achieving the change of state of the bulk-flow streams processed by the materials-processing system under the specified types of interactions with the environment. The term $T_{R*}\dot{S}_{\mathrm{irr}}$ is the work rate equivalent of the rate of entropy generation by irreversibility. It represents a loss of availability. It is a partial loss of the ability of the fuel source to perform a useful task. This loss is incurred because the processes in the materials-processing and the energy-conversion systems are not the best achievable under the specified conditions, that is, the processes are not reversible. So, in contrast to energy, availability is not conserved. It is destroyed or consumed by the generation of entropy due to irreversibility.

For the arrangement in Fig. 1.2, Eq. (1.67) provides answers to questions raised in Subsection 1.3.1. Specifically, it includes the inlet and outlet states of the task and the energy sources and therefore answers question 1: "What are the actual inlet, outlet, and end states of both the task and the energy sources used in the arrangement?" It specifies the optimum interactions required by the task and, therefore, answers question 2: "What are the optimum interactions required by the specified inlet, outlet, and end states of the task?" And it specifies the optimum interactions that could be supplied by the energy sources and thus provides an answer to question 3. It is important to emphasize here that the only optimum dictated by the laws of thermodynamics (physics) is reversible processes.

To answer question 4, "What aspects of the arrangement are the causes of the difference between the answers to questions 2 and 3?," we must look into the detailed design characteristics of the equipment used in the process. The answer to question 4 indicates the difference between the answers to questions 2 and 3 and is suggestive of steps that might be taken to answer question 5.

An analysis of a system based on considering the energy, entropy, and combined balances for each component of the system and computing the availability or exergy consumption, that is, the entropy generation by irreversibility, is called an *availability analysis* or *exergy analysis*.

Availability analyses, as well as energy and other analyses, require that the inlet, outlet, and end states of the task be specified and that the changes in availability or availability rates of feedstocks, products, and energy sources be evaluated. Because of practical considerations related to existing knowledge and technology, the specification of a desired task is very often relative to existing knowledge and technology and not absolute, and therefore availability and other analyses yield results that are relative to existing knowledge and technology.

For example, a common process encountered in industry is the heat treating of alloy steel parts to produce a locally hard surface, such as the surface of a steel ball for a bearing or the surface teeth of a gear. Although only a very small fraction of the material of each part needs to be hardened, conventional technology has required that the entire part be heated to about 900 °C. So the task is defined according to this requirement. Another way to specify the task, however, is to say that only a small fraction near the surface of the material need be hardened. The availability change required by the first task is much larger than that required by the second. Moreover, the results of the two availability analyses are not comparable to each other, just as the task of making pig iron in a blast furnace is not comparable to that of making aluminum in an electrolytic cell.

In the example of steel hardening, the second specification of the task has of course little practical significance if we do not know how to treat just the surface without affecting the bulk of the processed piece. However, the lower availability change required by this specification in the framework of the conventional technology of the task can provide useful guidance for innovative approaches to the problem of metal hardening. In fact, recent developments in high-power lasers and electron-beam accelerators have led to the development of practical processes for localized heat treating. In one carburizing application, for example, electron-beam heat treating reduced the energy needed for a particular part from 1 kWh to only 2 Wh. Thus, by redefining the task, the required availability was lowered well below the level that was previously thought to be optimum.

1.3.6 Thermodynamic Efficiency or Effectiveness

Associated with each task, such as heating a room or making a specified amount of steel out of iron ore, is the least amount of work that must be done to accomplish the task. This least amount of work is equal to the change in availability of the substances processed to achieve the task and is independent of any details of the arrangement of devices and engines used in the task.

In practice, however, each specific arrangement consumes a certain, not necessarily optimum, amount of fuel or energy source to accomplish the task. Associated with this amount of fuel or energy source is the largest amount of work that can be delivered to a weight. This largest amount of work is equal to the availability of the fuel or energy source consumed and is independent of any details of the energy-conversion systems and devices used to convert the fuel or energy source to work.

For emphasis, we denote the least work rate required by a specified task production rate as $\dot{W}_{\text{least}}^{\leftarrow}$ and the largest availability rate of the fuel or energy source consumption as $\dot{W}_{\text{largest}}^{\rightarrow}$, and we define the *thermodynamic efficiency* or *effectiveness*[6] ε of the actual arrangement as the ratio of these two rates, that is,

$$\varepsilon = \frac{\dot{W}_{\text{least}}^{\leftarrow} \text{ required by the actual task production rate}}{\dot{W}_{\text{largest}}^{\rightarrow} \text{ of the actual energy source consumption rate}}. \tag{1.68}$$

The effectiveness is a measure of the degree to which the processes involved in carrying out the task and in converting the energy source are reversible. If the processes are reversible, $\varepsilon = 1$. If the processes are irreversible, $\varepsilon < 1$.

---

[6] In some literature on this subject, the concept of effectiveness defined here is called second-law efficiency. Such terminology, however, is misleading because the concept is based on not just the second law but on the first law as well, and on many other concepts, such as work, heat, and bulk-flow interactions, and energy and entropy balances. All these concepts are certainly related to but not derivable solely from the second law.

The concept of effectiveness is applicable to any process and can always be expressed in the form

$$\varepsilon = 1 - \frac{T_{R*}\overleftarrow{\dot{S}}_{\mathrm{irr}}}{\overrightarrow{W}_{\mathrm{largest}} \text{ of the actual energy source consumption rate}} \qquad (1.69)$$

because the difference between the denominator and the numerator in Eq. (1.68) is always $T_{R*}\dot{S}_{\mathrm{irr}}$, where $\dot{S}_{\mathrm{irr}}$ is the total rate of entropy generation by irreversibility in the process. For example, the effectiveness of the materials-processing plant discussed in Subsection 1.3.5 is of the form of Eq. (1.69) as we can readily verify by using Eq. (1.67).

The effectiveness may assume even negative values. A negative value signifies that ideally the task can be accomplished while the processed streams transfer energy to a weight rather than consume energy sources, that is, the processed streams can be used as energy sources themselves. Instead, because of large irreversibilities in the actual materials-processing and energy-conversion systems, not only is the contribution from the processed streams wasted but other energy sources are consumed. The term $T_{R*}\dot{S}_{\mathrm{irr}}$ is sometimes called the *lost work rate*. It represents the work rate that could be produced in the absence or irreversibility, but is not produced because of irreversibility.

We can express the thermodynamic efficiency or effectiveness also in terms of batch quantities rather than rates. Then,

$$\varepsilon = \frac{\overleftarrow{W}_{\mathrm{least}} \text{ required by the actual task production}}{\overrightarrow{W}_{\mathrm{largest}} \text{ of the actual energy source consumption}}$$

$$= 1 - \frac{T_{R*} S_{\mathrm{irr}}}{\overrightarrow{W}_{\mathrm{largest}} \text{ of the actual energy source consumption}}. \qquad (1.70)$$

This effectiveness behaves exactly in the same way as that defined by Eq. (1.68).

Subject to the qualifications discussed in the next subsection, the concept of thermodynamic efficiency or effectiveness is the answer to question 6 posed in Subsection 1.3.1, namely, the universal measure of how effectively the task is accomplished by a given arrangement.

1.3.7 Thermal Efficiency

In practically every textbook, work-delivering engines are analyzed, and the ratio of the work output $\overrightarrow{W}$ over the heat input $\overleftarrow{Q}$ per cycle is evaluated under the assumption that in each cycle all processes are reversible. For example, the cycles that are considered are the Carnot, Rankine, Otto, Diesel, Joule–Brayton, and Stirling. The ratio just cited is called the *thermal efficiency*, is smaller than unity, and, more often than not, its difference from unity is interpreted as indicative of the margin for improvement of the processes involved in the cycle. Such interpretation is faulty because there are no processes that are better than reversible. Said differently, the correct measure of perfect use of the heat from a high-temperature reservoir is the thermodynamic efficiency or effectiveness for every one of the cycles listed earlier, and this efficiency is equal to unity. For example, for the Carnot cycle, the largest thermal efficiency of use of heat from the hot reservoir is

$$\eta_{\text{thermal}} = \frac{W^{\rightarrow}}{Q^{\leftarrow}} = \frac{T_1 - T_2}{T_1}. \tag{1.71}$$

But exergy of $Q^{\leftarrow}$ with respect to the low-temperature reservoir is $Q^{\leftarrow}(T_1 - T_2)/T_1$ so that the effectiveness or thermodynamic efficiency of the Carnot cycle under ideal (reversible) conditions is equal to unity.

Similar conclusions are obtained for all the other types of cycles if the processes of use of heat from the hot reservoir are reversible.

1.3.8 Practical Limitations

The construction of each machine, engine, and device is in itself a task that involves materials-processing and energy-conversion systems and therefore the consumption of energy sources. When it is sizable, this consumption must be accounted for. An important requirement of any installation used in primary energy processing, such as the production of electricity from various energy sources, is that the installation be capable of extracting more availability from the sources than the availability consumed for the construction of the machinery.

In many applications it may be technically impossible to take full advantage of the availability of the energy sources utilized. This may happen because some of the availability is either lost in processes outside the application or remains intact for use in subsequent applications. When this occurs, defining the effectiveness of the application in terms of the availability of the energy sources is misleading.

For example, if the only known method to carry out a process is by means of electrolysis, and electricity is generated from coal, it is hopeless to expect to improve the electrolytic process so as to take full advantage of the fuel availability. Electricity is not available in nature, and its generation entails losses. These losses should not be charged to the imperfections of the electrolytic process because it requires electricity to operate and the losses cannot be recovered no matter how perfect the electrolytic process is. To avoid this difficulty, the reasonable thing to do is to consider the availability of electricity as a source of input and evaluate the effectiveness of the electrolytic process with respect to electricity rather than with respect to coal.

Again, in each stage of a steam turbine, only some of the availability of the flowing steam is consumed. The remaining availability is ready for use in subsequent stages. Hence it is misleading to compute the effectiveness of one stage of the turbine with respect to the full availability of the steam flow.

In some applications, the properties of the materials of the equipment do not permit the full utilization of the fuel availability. For example, in oil-fired power plants, the exhaust combustion gases contain water vapor. If cooled to environmental temperature, the vapor condenses and corrodes the equipment. So exhaust gases are not cooled to such low temperatures. Correspondingly, the availability of the fuel should be evaluated with respect to a final state, not in temperature equality with the environment, but at a temperature such that vapor condensation cannot occur.

1.3.9 Comments

In contrast to other measures of efficiency, each specifically designed for a class of applications, the concept of thermodynamic efficiency or effectiveness is applicable to any task without conceptual modifications. For example, miles per gallon of gasoline is a measure of thermodynamic performance of a transportation task by an automobile, and the larger the value of this measure, the better the performance. Again, equivalent barrels of oil per ton of steel are a measure of thermodynamic performance of a steelmaking task at a steel plant, and the smaller the value of this measure, the better the performance. Clearly these two measures are not interchangeable and have different limiting values. In contrast, the concept of effectiveness can be applied to both an automobile and a steelmaking plant. The result for each of these two tasks would be a number less than unity, with an upper limit equal to unity. The upper limit of unity corresponds to perfect thermodynamic performance, namely, to all processes involved in the task that are reversible, and it is *the only limit imposed by the laws of thermodynamics* or, more generally, *the laws and theorems of nonstatistical quantum thermodynamics*.

Being directly related to irreversibility, the thermodynamic efficiency or effectiveness provides a realistic measure of the degree to which the performance of a task can be improved. Other measures of efficiency may be misleading. To illustrate the last assertion, we consider a perfectly insulated heat exchanger in which all the energy change of the primary stream is transferred to the secondary stream. On defining efficiency as the energy increase of the secondary stream divided by the energy decrease of the primary stream, we would find that this heat exchanger is 100% efficient. Such a result is correct but misleading. It implies that the heat exchanger is perfect and cannot be improved. However, if we define the effectiveness as the ratio of the availability increase of the secondary stream divided by the availability decrease of the primary stream, we find that the best exchanger is less than 100% efficient and subject to improvement by reduction of the temperature differences between the two streams. Clearly the second answer is realistic and, more important, relevant to our concerns about efficient use of resources.

Another important characteristic of thermodynamic efficiency or effectiveness is that it provides a realistic evaluation of tasks with dissimilar outputs. To see this point, we consider a cyclic device that produces work $W^{\rightarrow}$ and heat $Q^{\rightarrow}$ at temperature T_Q, while using heat $Q_1^{\leftarrow}$, from a source at temperature $T_{Q'} > T_Q$. If these are the only interactions, the energy and entropy balances are

$$Q_1^{\leftarrow} = W^{\rightarrow} + Q^{\rightarrow}, \tag{1.72}$$

$$\frac{Q_1^{\leftarrow}}{T_{Q'}} + S_{\mathrm{irr}} = \frac{Q^{\rightarrow}}{T_Q}. \tag{1.73}$$

If efficiency were defined as the energy out divided by the energy in, then this efficiency would be unity here, regardless of whether most of $Q_1^{\leftarrow}$ is provided as work or low-temperature heat $Q^{\rightarrow}$. We know, however, that heat is not equally valuable as work. For example, if T_Q were equal to the environmental temperature $T_{R^{\circ}}$, then $Q^{\rightarrow}$ would be entirely useless and yet the energy ratio would count it as equally useful as work.

These difficulties are eliminated if we compare the availability of the two outputs with the availability of the input, because then all interactions are evaluated on a comparable basis. Specifically, the effectiveness of the cyclic device is

$$\varepsilon = \frac{W^{\rightarrow} + Q^{\rightarrow}(1 - T_{R^*}/T_Q)}{Q_1^{\leftarrow}(1 - T_{R^*}/T_Q)} = 1 - \frac{T_{R^*} S_{\text{irr}}}{Q_1^{\leftarrow}(1 - T_{R^*}/T_Q)}, \qquad (1.74)$$

where, in writing the second form of Eqs. (1.74), we use Eqs. (1.72) and (73).

REFERENCES

[1] H. B. Callen, *Thermodynamics and an Introduction to Thermostatistics*, 2nd ed. (Wiley, New York, 1984).

[2] E. H. Lieb and J. Yngavson, "The physics and mathematics of the second law of thermodynamics," *Phys. Rep.* **310**, 1–96 (1999).

[3] E. H. Lieb and J. Yngavson, "A fresh look at entropy and the second law of thermodynamics," *Phys. Today*, 32–37 (April, 2000).

[4] R. C. Tolman, *The Principles of Statistical Mechanics* (Oxford University Press, New York, 1962), p. 1.

[5] E. B. Wilson, "Application of probability to Mechanics," *Ann. Math.* **10**, 129–148 (1909), and "Thermodynamic analogies for a simple dynamical system," *Ann. Math.* **10**, 149–166 (1909).

[6] N. F. Ramsey, *A Critical Review of Thermodynamics*, edited by E. B. Stuart, A. J. Brainard, and B. Gal-Or, (Mono Book Corp., Baltimore, 1970), pp. 217–233.

[7] E. P. Gyftopoulos and E. Çubukçu, "Entropy: Thermodynamic definition and quantum expression," *Phys. Rev. E* **55**, 3851–3858 (1997).

[8] D. F. Styer, "Insight into entropy," *Am. J. Phys.* **68**, 1095–1096 (2000).

[9] E. P. Gyftopoulos "Entropies of statistical mechanics and disorder versus the entropy of thermodynamics and order," *J. Energy Resources Technol.* **123**, 110–123 (2001).

[10] E. P. Gyftopoulos and G. P. Beretta, *Thermodynamics: Foundations and Applications* (Macmillan, New York, 1991; Dover, Mineola, NY, 2005).

[11] G. P. Beretta, E. P. Gyftopoulos, J. L. Park, and G. N. Hatsopoulos, "Quantum thermodynamics: A new equation of motion for a single constituent of matter," *Nuovo Cimento* **82B**, 169–191 (1984).

[12] G. P. Beretta, E. P. Gyftopoulos, and J. L. Park, "Quantum thermodynamics: A new equation of motion for a general quantum system," *Nuovo Cimento* **87B**, 77–97 (1984).

[13] G. N. Hatsopoulos and J. H. Keenan, *Principles of General Thermodynamics* (Wiley, New York, 1965).

[14] S. Carnot, *Reflections on the Motive Power of Fire* (Dover, New York, 1960).

2 Energy and Exergy: Does One Need Both Concepts for a Study of Resources Use?

Dušan P. Sekulić

2.1 Introduction

In Chapter 1, the concepts of thermodynamics were introduced, and a consistent exposition of thermodynamics' basic laws and associated system properties was offered. Out of an infinite set of properties and a limited number of defined interactions, only a few are relevant in any given study, and these include a limited number of independent system properties. This chapter is devoted to a further elaboration of two such fundamental concepts that belong to a class of system properties: (1) *energy* and (2) *exergy*. The main purpose of this exposition is threefold: (1) to offer a discussion of physical *meanings* of these concepts; (2) to present their analytical structure within the traditional thermodynamics framework; which is useful for applications; and (3) to emphasize the importance of balancing them. These topics are relevant for a variety of situations in diverse fields of interest, all identifiable in complex systems involving destruction of resources.

The notion of a *concept* is used to describe abstract theoretical constructs of classical thermodynamics theory. The concept of energy [1] is reintroduced in this chapter by means of a notion of a change of a system *property*; the concept of exergy [2] is reintroduced by means of the magnitude of an *energy interaction*. This magnitude would be expressed in energy units, but would be measured as an interaction extracted from the system changing the state between the given state and the state of thermodynamic equilibrium with the referent surroundings.[1] It is shown that knowing the *change* of energy as a property of a well-defined system that evolves between two states is what is important for an analysis of energy-resource utilization (i.e., not absolute values of energy at given states). One must realize that in classical thermodynamics only the magnitude of an energy change as a system property change has a practical meaning for the analysis of energy conversion, rather than an

---

[1] That is, the environment (defined as the so-called thermal reservoir; see Chapter 1 by Gyftopoulos for related definitions). As will become apparent later, the distinction between defining exergy as a specific energy interaction vs. defining it as a property of a system (involving the state of the system's environment) can also be interpreted as an argument for keeping a distinction between exergy functions and long ago introduced traditional properties, such as Gibbs free energy. If both were evaluated at the state of equilibrium with the referent environment, there would be no difference between the two. The collection of well-established properties may carry all relevant information, thus making an introduction of yet another esoteric thermodynamics concept seemingly superfluous.

absolute value of energy quantity at a given state. However, in the case of exergy, not only the change but the absolute value of its quantity may be of importance. Moreover, it must be noted that lay interpretations often equate the concept of energy with a notion of *energy resource* (or an equivalent energy-conversion value of a given energy-resource quantity).[2] An energy resource is measured in equivalent mass units of "energy carriers," converted into energy units assuming certain conversion efficiency. So an energy carrier has to participate in a thermodynamic process executed by a system within which it would represent a subsystem changing its state and through which its energy (as a property) would change. Through such interactions, the energy change of the subsystem can be converted into an energy change of other associated subsystems that have to accomplish certain useful tasks. It must be clear that, along the path of energy conversions involved, the energy changes of the involved subsystems would be converted into associated energy interactions and vice versa, suffering inevitable conversion imperfections at each conversion step. Consequently, any discussion of energy resources must assume system's energy interactions and energy changes in the presence of process inefficiencies.

Therefore the main objective of the material presented in this chapter is to enrich not only the *conceptual* background, but to gain an understanding of similarities and differences as well as applications of both energy and exergy concepts used in the analysis of complex systems. These complex systems span from the realm of man-made systems to the realm of natural systems, including their mutual interactions. This understanding becomes relevant for analyses of resources utilization and system efficiency in light of the synergy between different segments of complex systems, e.g., industrial processes within a technosphere, ecology, or even in the context of societal discourses (under certain well-defined conditions).[3]

2.1.1 The Scope of the Chapter

This chapter is not necessarily intended to offer a new or different, or both, point of view adopted within dominant and well-accepted treatments of the subject, as presented in a number of excellent fundamental and applied thermodynamics texts [3–11]. Although such presentations differ from one to the other, often even with regard to the level of rigor, they constitute the common core of a well-established scientific inquiry involving primarily both equilibrium and nonequilibrium thermodynamics. So our implied position would be that for a proper utilization of these concepts in an analysis of resources management we need not new and esoteric interpretations but proper implementation of available approaches.

---

[2] In other words, in colloquial language the word *energy* is used instead of energy *resource*. These two concepts, however, imply a nontrivial distinction. Quantities of energy resources, say, a barrel of oil, an equivalent ton of coal, a kilogram of nuclear fuel, are all relevant and necessary pieces of information in an energy-related study, but all as the *quantities of resources* to be utilized in an energy-conversion process, characterized by a preassumed conversion efficiency. Energy, in a rigorous, thermodynamic sense though, is a *system property* (in a given state).

[3] All of the systems traditionally considered in such studies are called "physical systems." So we would assume that thermodynamic analysis is implemented on systems beyond, say, social–economic (or nonphysical) realms. Somewhat controversial issues of whether thermodynamics can be used for study of nonphysical systems is addressed later; see also S. Lloyd, et al., "Discussion on frontiers of the second law," in *Meeting the Entropy Challenge*, edited by G. P. Beretta, A. F. Ghoniem, and G. N. Hatsopoulos (American Institute of Physics, Danvers, MA, 2008) pp. 253–264, Ref. 22.

A somewhat more visible difference between our material selection and interpretation and many other reviews can be seen within the scope of applications emphasized in this chapter. Concepts of energy, exergy, or both, can be used for a study of *any* well-defined system prone to thermodynamic analysis, i.e., they are not necessarily intended to be used only for traditional thermal–process systems, or even more restricted, for so-called energy systems. Furthermore, it will be shown that the *semantic content* of these concepts is, in addition to related mathematical formalisms, very important and should never be underestimated.

However, one should be careful not to overestimate capability of a single specialized discipline for an analysis of complex systems. It would be naive to claim that a particular concept, for example exergy, has a potency to convey definite information about a variety of (or even major) aspects of a system behavior relevant for resources-utilization studies. Such an attempt would be based on a fallacy because any approach to modeling large complex systems must depend on a consequent and rigorous system definition, consistent with the laws of the discipline used in conjunction with the analysis objectives, not only thermodynamics. Some of these laws are universal, natural laws, but some are not. This holds in particular for the analysis of nonthermal systems, complex natural and man-made systems (including a human society), or both, driven not only by thermodynamic laws (but indeed obeying them), but also governed by specific laws of living systems' sciences, ecology, economy, etc. For example, it would be inappropriate to search for an intrinsic value of the concept of exergy in economic systems' studies without converting material or energy carriers of this entity into involved economic metrics. Setting up a coherent system of analogies may not necessarily be sufficient; rather, a consistent theory of the phenomena under consideration should be developed in the same manner as is done for physical systems analyzed by thermodynamics.

2.1.2 Relevance of an Energy–Exergy Analysis

For any application of energy–exergy analysis [12–14], the system under consideration must be *well defined*. This is a known fact, but, surprisingly, it has not always been acted on. A complete set of conditions for defining a system should always be explicitly stated. Any system, regardless of whether it represents, say, a traditional thermal system engineering setting [14] or a large hybrid complex system in the living world, is suitable for a thermodynamic analysis – if the conditions imposed by the requirements for defining the system well[4] are satisfied.

When a system is defined, entities crossing its boundary can be balanced by use of appropriate property changes and interactions balancing. These involve both (1) conserved (e.g., energy) and (2) nonconserved (e.g., entropy) flows. It is often assumed that a notion of conservation of energy is somehow self-evident and easy to grasp. However, it is important to understand that the conservation principle per se and the property change and energy–entropy interaction balancing are not necessarily synonyms for thermodynamic laws. In operational terms (i.e., for practical applications) this distinction may sometimes be considered as being only a semantic issue, but in the light of the axiomatization offered in Chapter 1, the relevant

---

[4] The notion of the "well-defined system" has a precise meaning; see Subsection 2.1.4. A rigorous definition is given in Chapter 1 (Gyftopoulos) and in Subsection 2.1.4 of this chapter.

laws (i.e., the first and the second laws of thermodynamics) actually *define energy and entropy* as system properties that may feature conservativeness (in the case of energy) or not (in the case of entropy). The concept of exergy is associated with both the notions of energy and entropy, which in turn are related to both the first and the second laws of thermodynamics (through the existence of the stable equilibrium state) – therefore it is not conserved [14, 15].

The consensus among scientists regarding the importance and applicability of energy balancing for modeling of energy-conversion phenomena or resources optimization or both has been undisputed [16–22]. In the case of the concept of exergy and its balancing, authors sometimes do not agree on either the domain or on the importance of its use [23, 24].[5] Moreover, the concept of exergy had not been used extensively until the late 1970s (and when used, it was promoted primarily by engineers). This was most likely because of a need to involve the second law of thermodynamics and the concept of entropy in energy-conversion studies, traditionally believed to be much more elusive than energy. Regardless, a case for the use of exergy was made quite some time ago [25, 26].

2.1.2.1 Traditional Applications of Energy–Exergy Analysis

Concepts of energy and exergy have traditionally been used in the so-called energy or exergy analyses, mostly in various subfields of *thermal–process engineering* dealing with energy resources. An example of a broadly accepted system optimization methodology based on the concepts of exergy is exergoeconomics, discussed later in this book. However, the usefulness of such or similar analysis techniques may be claimed to be much broader.

Notions of available-energy-related functions, formally introduced way back in the last years of the 19th century and advanced in the first half of the 20th century (although at that time not under the name of exergy) have never been *theoretically* controversial. Reservations about their usefulness though have been in the domain of interpretations of the exergy concept. For example, an argument may be advanced that exergy-balancing accounting is less potent than a full-scale irreversibility (entropy-generation) analysis because, say, it most often lumps the individual irreversibility contributions through exergy losses. Moreover, the exergy concept inevitably promotes a "hybrid" of a quasi system property that depends on the parameters and state properties of surroundings, hence making it impossible to consider this quantity rigorously as a true system property – unless the external reservoir-based properties are declared to be fixed (but still based on a selection by convention). Finally, exergy analysis based on accounting of chemical exergy uses by definition only the Gibbs free-energy function, identical to the exergy function if evaluated at the state of the environment (as in ecological studies). Consequently the concept of exergy has been somewhat marginalized beyond engineering applications – until the energy-resources crisis of the 1970s promoted development of thermoeconomics. This concept's potential to reflect a quality of an energy interaction and a

---

[5] See also the other discussions in *Meeting the Entropy Challenge*, edited by G. P. Beretta, A. F. Ghoniem, and G. N. Hatsopoulos (American Institute of Physics, Danvers, MA, 2008), p. 314, Ref. 22.

system's ability to utilize what is called *available energy* was immediately recognized, but its utility did not attract a wider attention until the mid-1980s.[6]

The energy crisis of early 1970s dramatically increased the attention of the engineering community's exergy analysis and directed these efforts toward evaluation of energy-resources utilization. Since then, a number of new thermodynamics tools and metrics have been uncovered – some of which were either entropy-generation or exergy based [27, 28]. Still, the concept of exergy has been used much more frequently in energy-systems research and optimization than in any other engineering or nonengineering field, in particular for the formulation of a discipline of thermoeconomics, currently known under the name of exergoeconomics.[7]

2.1.2.2 Nonenergy-Generating Systems Analysis

Attempts to formulate thermodynamics-inspired concepts for applications encompassing multiple disciplines, i.e., not necessarily strictly within classical engineering thermodynamics context, rarely obtain endorsements of mainstream thermodynamics [29, 30]. The problem may sometimes be an approach to the definition of a system, such as that even for physical systems [31–34]. On other occasions, an application of thermodynamics analogies may be questionable, or, even more often, the concepts may be introduced as metaphors in fields of study different from physical sciences [35, 36]. Still, the use of thermodynamics concepts may be justified in "remote fields" of inquiry.[8]

On the other side, some important fields of *engineering applications*, say materials processing, manufacturing, and a number of related nonenergy-generating-systems disciplines (such as recycling and recovery of materials in industrial ecology), have not been advanced in a way to promote methods of thermodynamics until relatively recently (notably the approaches involving exergy-balancing and entropy-generation studies).[9] This has been the case even though thermodynamics has been used extensively for process analysis and modeling at the involved phenomena level. Most recently, increased concerns involving an impact of engineering on environmental and societal domains, say involving ecology (in particular industrial ecology) and resources management have been promoted in the domain of environmental studies [37]. Many of these efforts have been popularized first through a notion of green engineering and, most recently, a somewhat broader field of "sustainability engineering." These new fields of inquiry have been a bit friendlier toward the use of thermodynamics concepts (though not without controversies) [38–43]. For example, an increased awareness of interrelations between human actions and environmental balance, within what has become the field of sustainability (although still vaguely delineated from, to a large extent, heuristic and mostly speculative predictions, such

---

[6] The first international meeting in the series of efficiency, costs, optimization and symulation of energy systems (ECOS) conferences sponsored by the advanced energy systems division of the American Society of Mechanical Engineers was held in 1987, May 25–29, in Rome (*4th International Symposium on Second Law Analysis of Thermal Systems*); *Second Law Analysis of Thermal Systems*, edited by M. J. Moran and E. Sciubba (American Society of Mechanical Engineers, New York, 1987).

[7] See Chap. 15 by Tsatsaronis in this book.

[8] See the Introduction to this book.

[9] See Chap. 6 by Gutowski and Sekulić in this book. The term "non-energy-generating-system" denotes a system which function is not to deliver energy as a useful outcome of its operation (i.e., a manufacturing system vs. a power plant system).

as within the studies of population trends of living organisms), has promoted the use of thermodynamics concepts for nonenergy-generating-systems analysis. Still, the wide use of these concepts has not been followed with a unified approach that is sufficiently rigorous for related efforts to be regarded as (predictive) science.

2.1.3 Quality Versus Quantity of Energy

One of most abstract and at the same time most widely used concepts of physics is the concept of energy. As famously stated by Feynman [44] and often cited, "there is a certain quantity, which we call energy, that does not change in the manifold changes which nature undergoes... it is just a strange fact that we calculate some number, and when we finish watching nature go through her tricks and calculate the number, it is the same...." This "strange fact," though, underpins a natural law on the existence of the property called energy. This law, the first law of thermodynamics, is associated with the conserved property of energy (although the first law definition and the property of conservation are not necessarily equivalent).[10]

The notion of conservation of such an abstract entity as energy has been of profound practical importance for engineers and scientists. Engineers have been using the concept of energy conservation for more than the past 150 years without contemplating the degree of abstraction of the related natural law. On the other hand, any practical mind would most likely fully agree with the proposition of Nagel [45] that

> the label "law of nature"... is not a technical term clearly defined in any empirical science.... There is therefore more than an appearance of futility in the recurring attempts to define with great logical precision what a law of nature is – attempts often based on the tacit premise that a statement is a law in virtue of its possessing and inherent "essence" which the definition must articulate.

Still, the existence of a conserved entity called energy is, except possibly for conservation of mass (in cases for which it can be applied), the most widely used tool of any engineering or science discipline. So investing time in understanding this concept is of a paramount importance.

An attempt to define with undisputed logical precision and no circularity the semantic content of the basic concepts of thermodynamics is, indeed, not an easy task. But, in this chapter, that task is not the principal objective – the objective is to discuss operational aspects of the *use* of energy-related concepts, starting with the conservation property of energy that deals with its *quantity* and completing the discussion with the notion of energy *quality* through the concept of exergy. While fulfilling this objective, we follow a rigorous path of formal definitions and precise formulations. Occasionally, an intricate narrative – which carries a rigorous approach – needs proper use of advanced concepts borrowed from one discipline and implemented to another. In the case of transdisciplinary topics of our interest (issues of resources destruction and sustainable development), this exact language is even more important because the contemporary literature in the field tends toward a somewhat lax use of thermodynamic concepts.

---

[10] See Chap. 1 by Gyftopoulos in this book.

Let us consider an example. One of the phenomena most often present in materials processing involves *heat exchange* between a material and a heating source or sink. The objective of materials processing would not necessarily be to do an energy conversion to generate mechanical work (or its equivalent) as a useful outcome. The objective would be to expose certain material to a change of its state – to, say, modify its properties (or to keep them from being modified to an undesired level) or to promote change of some other physical features, including the shape. Heat exchange is occurring in an engineering setting (while the object of study is considered as being either an open or closed system). Simultaneous change of energy–enthalpy levels of any of the multiple subsystems in the thermal contact we identify as (thermal) *energy exchange (transfer)*. This exchange may be manifested, say, through heat interactions. It may take place between a product to be processed and a working fluid or a distant participating medium. This can be illustrated by a host of different processes: (1) heat treatment to facilitate solid phase change of a material, (2) forming by deformation along with thermal energy dissipation, (3) casting assisted by a liquid-to-solid phase change, (4) machining to execute material separation and associated cooling, etc. In a word, any manufacturing technology deals with this process to a certain extent. Thus the task *is not* necessarily to transform an energy flow into heat (or vice versa) per se, but to initiate a change in or of the system through a heat interaction.

The very nature of materials processing (as is also true for all natural and man-made processes) implies inherently nonideal execution. That is, in thermodynamics terms, any such process would be irreversible – it would require more resources to be invested into running it in reverse than would be invested in executing it forward. Hence this irreversibility would imply that some energy resources would be "lost." Because energy is a conserved property (across all the participants of that exchange), that loss may not be *energy* quantity (rather, it would have an impact on the *energy-resource* quantity and quality).

So transformation of an energy resource may require an additional consideration beyond determination of a mere energy quantity. For example, a question may be whether a task should be performed by involving a higher or a lower heat source–sink temperature level but involving exactly the same energy change of participating systems (or the same energy-interaction level). It can be assumed that in both of these alternative cases sufficient energy resources may be available for the product treatment. However, the *quality* of such transformation may not be the same! It is very important to understand that this quality depends not only on the manner of execution of the transformation, but on the temperature level of the thermal source used (which must, indeed, differ from the product temperature level – i.e., it must be higher). The notion of quality of energy interaction vs. temperature level is not at all apparent from the point of view of the energy-conservation principle and energy balancing at the system interface. Moreover, as long as the proper amount of an energy resource can be used, and as long as that is the only concern, it seems to be irrelevant to consider implications of the impact of temperature levels on energy change or delivery (of course, an impact on other system or process aspects may still be profound). Hence, in such a case, the related "quality" of energy (as a resource) used in the process may not be of interest. In other words, in such an analysis the assessment of the "quantity" of the transfer of energy interactions and energy changes of systems and energy resources are going to be in focus. In contrast to such a view, an assessment of the quality of the resource (and its conversion) would be needed and related to consequences of both the first law *and* the second law of thermodynamics. This is because the latter involves both the quality of the change of state (i.e., the level of irreversibility) and the quality of the utilized energy-resource delivery (the available energy with respect to an a priori determined referent state; see Chap. 1). As we will see soon, the entropy balance, or exergy balance, or both, (in addition to the energy balance), should be considered. In more colloquial terms, the utilization of resources must include the consequences of *all* laws of thermodynamics. The idea of quality may be introduced for a heat transfer interaction through a notion that a delivery of thermal energy from a thermal source at a higher temperature level (instead from a source at a lower temperature level but still above the level of its use) would imply a failure to use the available potential for a delivery of a heat at any intermediate temperature level (because it is already used for delivery at the lower level).

An additional word of caution would be in place here. The concept of quality in the preceding discussion is related to the level of thermal energy delivery needed for a change of state of a system, i.e., in terms of thermodynamic quantities. The notion of quality may, however, transcend the thermodynamics realm. We do consider these aspects.

In conclusion, one must say that not only the implications of the change in *quantity* but also of the change in *quality* of energy resources (in a process of their utilization) must be considered.

2.1.4 System Definition

A necessary starting point in any consideration of both energy and exergy concepts is a rigorous definition of a system under study, as postulated in Chap. 1. However, how does one go about completing this task properly? It is important to note that an inexperienced analyst may not necessarily be fully aware of all the assumptions to be adopted or justified for a particular analysis. This almost always leads to a poor (or, most often, not rigorous enough) definition of the system or inadequate understanding of the objectives of the analysis in light of the problem formulation. So we start our study by revisiting unambiguous definitions of three interrelated entities: (1) a system, (2) its surroundings, and (3) the system–surroundings boundary. The three entities are interrelated; consequently a definition of any of the three will identify (but not necessarily define) the remaining two. For example, the definition of the system boundary that separates the system from its environment is the necessary first step in identifying what would belong to the system, but would not necessarily define it well.

In Chap. 1, the system is defined as a collection of constituents. The notion of constituents must be supported with a predefined set of additional descriptions. How does one formulate this collection of attributes? A system is defined by a *mental abstraction* applied to a given natural or engineering setting. This abstraction is done effectively by extracting a certain domain from an observed real setting (note, it may not be restricted to an *engineering* setting). The extracted domain, i.e., the collection of entities in terms of constituents, is what will be considered as the system. An often held but ill-conceived perception is that delineating the boundary of the system from the rest of the universe is both necessary and sufficient to define the system. Note also that the system boundary does not represent a real physical entity (it is also an abstract construct of the theory). Hence this boundary identification defines what is within the system and what is not, but also makes an identification of interactions that cross the system boundary possible.

Therefore, for a system to be well defined, additional information about constituents is needed. Moreover, the system extraction from the surroundings must be accompanied with a specification of certain features involving system's structure. A system is a physical entity (a collection of constituents) of a more or less complex structure. So a system is defined by (1) the features of its complexity, (2) its constituents, and (3) the specification of the nature and magnitudes of external and internal forces acting on its constituents or its constituents and the surroundings. So the system is well defined if its description presents fully its complexity and its constituents and offers identification of external and internal forces (including all related assumptions).

With both the system and the system boundary defined, the surroundings represent the rest of the universe. In engineering and natural settings this is often reduced to only the immediate surroundings. The immediate surroundings (environment) sense the interactions. Among all possible surroundings, one category deserves particular attention. In the case in which the surroundings remain in thermodynamic equilibrium with themselves and with their duplicate throughout any change of state, they would be defined as a thermal reservoir (see Chap. 1). That means, for example, that, in spite of interactions with a considered system, the state variables of the thermal reservoir stay unchanged. This is certainly an idealization. In the first approximation such idealization is often interpreted as the environment being sufficiently large; hence the state variable changes inflicted by interactions would be negligible. Note that the term environment is taken as a thermodynamic concept, not necessarily as the environment in its ecological sense.

Let us consider an example of a system description suited for analysis of materials processing in manufacturing, i.e., in a *nonenergy* generation application.

In Fig. 2.1, a continuous furnace for a state-of-the-art process called controlled atmosphere brazing (CAB) is presented. The description of its basic functions is given in the figure caption.

Obviously multiple material flows through the furnace can be identified at inlet and outlet ports. These ports are identified by the locations where the flows cross the system boundary. These include two primary material-flow types: (1) a discrete flow of product units and (2) a set of continuous flows of various material streams involved with processing. In addition, a set of nonmass flow interactions (discussed later in this chapter) has to be identified. Material flows include gas, liquid, and solid mass flow rates, i.e., pure nitrogen for facilitating the controlled atmosphere conditions, effluent fumes leaving the process, water and air for cooling the products, and mass transfers involving material of products and material of the conveyor. All these primary and auxiliary material flows are essential for executing the required materials processing. These flows pass through some or all of the segments (zones) of the furnace. In addition, within various zones of the furnace (in a direction of the principal product material flow), a succession of electrical heaters delivers thermal

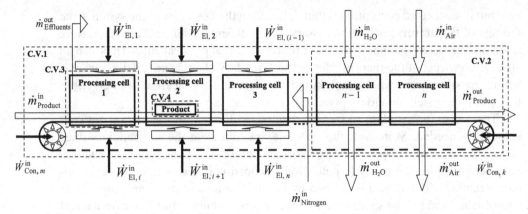

Figure 2.1. A complex materials-processing system: engineering setting and system definition. The schematic represents a continuous CAB furnace for mass production of automotive compact heat exchangers (manufactured products). Note that this is not an "energy-generating" system (such as a power plant or an engine). The products undergo a sequence of a fine-tuned heating and cooling steps in a succession while traveling through the furnace on a conveyor (i.e., passing through a succession of heating and cooling zones facilitated by heating and cooling thermal sources and sinks positioned from both sides of the product flow). These facilitate heat radiation delivery (cells 1, 2, 3, ...), while cooling is executed in two steps, by a water-jacketed cell first (cell $n-1$) and subsequently a direct air-blast cooling (cell n). Thus the materials-processing sequence involves a product's solid-phase heating (and later cooling), phase-change phenomena (melting and solidification of the clad and filler within mating surfaces brazed joint zones), solid and liquid state diffusion during brazing, reactive flow that is due to capillary surface phenomena, etc. Each of these phenomena in the sequence is triggered and terminated by imposed temperature history. C.V., control volume.

radiation heating (generated through use of electrical, i.e., work, interactions, i.e., nonmass interactions); Fig. 2.1.

One should note that the considered system is associated with just one manufacturing processing step (called brazing) along the life cycle of the primary product material, the step starting with assembled material parts and ending with the desired product. Multiple other steps may exist and are prone to similar analysis and identification of their interactions with resources flows (i.e., extraction of raw material, material processing prior to manufacturing, etc.; see Fig. 2.5 in Section 2.4).

Identifying the system boundary is the first step in an analysis. This action (including well defining it) fully depends on an intention of the analyst to study a certain aspect of the system design, operation, or both, and it is closely related to an identification of interactions to be crossed by the selected system boundary. So the second step in analysis may be an identification of the interactions between the *well*-defined collection of constituents and the rest of the universe. From the preceding example, it must be clear that these (resources related) interactions may be, in general, identified as (1) mass flow interactions (for all but closed systems) and (2) other interactions (associated with either mass flows or nonmass interactions as heat or work interactions). Simultaneously, one must define attributes of the system in terms of the constituents' interactions (internal, external forces acting on the constituents) and the complexity of the system structure.

If a system possesses an unchanged number or amount of constituents (say, a fixed-mass system), and if these constituents do not cross the system boundary, it is called a *closed* system. Note that a closed system is not necessary an *isolated* system. Otherwise (when the constituents of the system are crossing the system boundary) the system is defined as *open*. In the latter case, the concept of the control-volume (CV) boundary is synonymous with the system's boundary and sometimes (incorrectly) referred to as the system itself. As opposed to many typical examples of closed systems, say, so-called "energy systems," the systems involving manufacturing and materials processing (as well as many other engineering, but also nonengineering, natural settings, as in ecological studies) are open systems, exposed also to transients, often with variable amounts of constituents and the presence of chemical or other reactions between the constituents spatially distributed in a three-dimensional (3D) domain. One should understand that the classification of systems into closed and open is not a fundamental and necessary distinction needed for building the theory from a methodological point of view (but it is indeed of cardinal importance with respect to a system's behavior).

An important point to be emphasized in any system analysis of an engineering or natural setting is that an analyst may identify not a single, but a number of different systems from the same setting – depending on which segment(s) of the setting is(are) to be extracted. This decision also defines, in each of these cases, the interactions that would cross the identified systems' boundaries. Figure 2.1 illustrates four (out of virtually infinite) possible cases that may be generated in an analysis of a given manufacturing setting. These are marked as "control volumes," i.e., C.V.1–C.V.4.

To illustrate the variety of decisions an analyst can make, let us discuss a few associated examples. For the sake of obviousness, we identify only mass flow and nonmass flow interactions uncovered by a system boundary selection for the engineering setting taken from the example of Fig. 2.1. Each system can be extracted from the engineering setting by positioning the adequate CV boundaries (the C.V.s), hence by extracting the elements of the engineering setting from the rest of the universe *and* identifying the interactions crossed by these boundaries. At the same time, as already suggested, an analyst *must* define well what the system actually is (including *all* the previously listed features, i.e., defining not only its boundary). One should note that the set of interactions would, as a rule, differ from one system definition to another system definition, regardless of the fact that all systems (i.e., subsystems) to be considered would be extracted from the same engineering or natural setting. Extractions of various systems in these particular cases have to be motivated, as previously indicated, by different major analysis tasks, One may be a study of resource utilization and the effluents delivery manifested through interactions between the *engineering setting as a whole* and the environment (C.V.1; see Fig. 2.1). Such a study would require the system boundary to encircle the whole engineering setting, as is the case with C.V.1 in Fig. 2.2. Note that the system in this case has a very complex structure and multiple constituents. Each of these must be defined with corresponding constitutive relations, and the structure of the system must be defined. The other may be a study of *a part* of the overall engineering setting, say, a study of a set of *a few processing cells* for the cooling sections. The related interactions must be formulated to describe this system's behavior (C.V.2, Fig. 2.1). The next may be a study of a *single processing cell*, say, a single heating cell that would require certain mass flow and involve some other interactions for materials processing (say, processing cell 1,

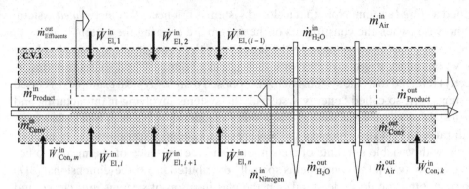

Figure 2.2. C.V.1 (an open system).

C.V.3, Fig. 2.1). Finally, but not limited to, yet another may be a study of a single product's state change to evaluate, say, product quality (C.V.4, Figs. 2.1 and 2.3).

The first control volume, C.V.1, identifies as the system the whole engineering setting represented by Fig. 2.1, i.e., a complete manufacturing system including all its auxiliary functions; see Fig. 2.2 for the main identified interactions. Such system's interactions include the set of mass flow rates as follows: (1) the set of discrete flows at inlet–outlet ports (the product flows, and the conveyer mass flow), $\dot{m}_{Product}^{in}$ and $\dot{m}_{Product}^{out}$ (which include any chemicals – brazing flux – added) and $\dot{m}_{Conv}^{in}$ and $\dot{m}_{Conv}^{out}$; (2) the bulk mass flow rates of the controlled atmosphere gas (nitrogen), $\dot{m}_{Nitrogen}^{in}$, and effluents $\dot{m}_{Effluents}^{out}$ (which include $\dot{m}_{Nitrogen}^{out}$); and (3) coolant flows (water and air), $\dot{m}_{H_2O}^{in}$, $\dot{m}_{H_2O}^{out}$, $\dot{m}_{Air}^{in}$, $\dot{m}_{Air}^{out}$. Depending on the objective of an analysis, the change of state of individual bulk flows requires an identification of a subsystem's mass flow rates, representing each of the bulk flows. Note that the system as a whole has its structure and internal interactions between its parts, which would be necessary to define. The interactions between these parts though would not constitute an overall system's interactions and would be reviled only on subsystem's identification as objectives of a study, thus redefining the so-defined subsystem constituents, complexity, interactions, and impact of acting forces.

In many applications (for either a closed or an open system), mass is considered as being conserved. In such cases, the system mass M must satisfy the *mass-balance* equation:

$$\sum_{i=1}^{m} \dot{m}_i^{in} - \sum_{i=1}^{n} \dot{m}_i^{out} = \frac{\partial M}{\partial t} \begin{cases} \neq 0, & \text{transient} \\ = 0, & \text{steady} \end{cases}. \tag{2.1}$$

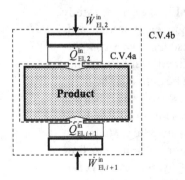

Figure 2.3. C.V.4 of Fig. 2.1; note two options for identifying the control volume: (1) C.V.4a includes the product as a system exposed to thermal interactions in form of thermal radiation (this is a heat interaction), whereas (2) C.V.4b uncovers interactions through electrical work, i.e., not heat (this is a work interaction). If one intends to study the product only, the proper system boundary is C.V.4a.

In the case of the system identified by C.V.1, Eq. (2.1), the CV defines an open system operating in a steady-state mode. Therefore the balance can be rewritten as follows (auxiliary mass flow rates such as those related to the mass of conveyer are not included):

$$\sum_{i=1}^{4} \dot{m}_i^{in} - \sum_{j=1}^{n} \dot{m}_j^{out} = \left(\dot{m}_{Product}^{in} + \dot{m}_{Nitrogen}^{in} + \dot{m}_{H_2O}^{in} + \dot{m}_{Air}^{in} \right)$$

$$- \left(\dot{m}_{Product}^{out} + \dot{m}_{Nitrogen}^{out} + \dot{m}_{H_2O}^{out} + \dot{m}_{Air}^{out} \pm |\Delta \dot{m}| \right) = 0. \quad (2.2)$$

In Eq. (2.2), the lumped term $|\Delta \dot{m}|$ closes the balance if creation or annihilation of constituents must be included (say, because of a chemical reaction involving the effluents and flux-interaction gaseous products). If such effluents or reaction products are of negligible importance or do not exist, this term is equal to zero. Otherwise, a set of additional mass flow rates (such as harmful effluents created during the process) must be included [this is signified by an unspecified upper bound designator of the second summation in Eq. (2.2) on the left-hand side of the first equality]. It should be noted that Eq. (2.2) represents a single analytical relationship, so it can serve as an analytical tool for verifying the balance validity or for determination of a single mass flow entity at any of the ports if the others are known. In such form it is in most cases of limited use. Most often, such balances must be written for individual material flows (bulk flows), hence generating a set of mass conservation equations.

In addition to the listed bulk mass flow rates, a series of nonmass flow interactions can be identified (work and heat, but also the enthalpy rates' flows associated with mass flow rates).

A similar identification of interactions would be performed for any other CV separating other identified systems from corresponding surroundings. As an illustration, let us consider the control volume positioned around the product only, C.V.4a, Fig. 2.3.

Note that C.V.4a is a limiting case for which the CV encloses a single product exposed to heating (a heat interaction) in the processing cell, say, cell 2 of Fig. 2.1, but with *no mass flow interactions* present (at this stage of processing, say, chemical reactions are not considered as being significant), as well as no other nonmass interactions (such as work). So only the constituents of the solid mass of the product represent the system. The system is closed and exposed to a thermal treatment – e.g., heating of a solid mass. The heating is characterized by system changes of state manifested in certain property changes. There is no mass flow rate crossing the boundary of the system because it includes only the constant mass of the product. So the interaction that leads to an increase of system temperature (a representative property) is a nonmass flow interaction called the heat transfer rate, and it is delivered from electrical heaters at the given instant of time. It is important to note that, although the heaters themselves are supplied by a nonmass flow rate interaction called electrical work (electrical current flowing across the imposed voltage difference overcomes the conductor's electric resistance), the *net* thermal interaction delivered across the boundary (C.V.4a) to the system considered here is in the form of thermal radiation (i.e., emitted from the heater to the product). If an analyst would consider as the system a CV featuring ports of entry of the nonmass interactions beyond the heaters (C.V.4b), the conversion originating in Joule heating would be inside the system boundary and would contribute to a system's internal irreversibility; hence heat transfer would *not* be manifested as the interaction (unless heat losses to the environment are identified). However, the relevant interactions of the heater itself would be an electrical work on one side and heating on the other.

A careful reader will have noted that, in the description of system state changes and interactions, the word *energy* has not been mentioned explicitly (although it has been intuitively present through notions of "nonmass flow interactions" and state changes manifested through certain property changes). The important point is that the electrical current flow across the electrical resistance involves a conversion (performed in a heater, beyond the system under consideration in C.V.4a) of one nonmass interaction entity into the other (an electrical current flow across the voltage difference through the heater resulting in "Joule heating"). This phenomenon is manifested as a *work interaction* whereas the thermal radiation delivered to the product is interpreted as a *heat interaction*. Another conversion (a bit more subtle) involving a change of system properties (an energy change of the system, for example) is present as well. Such thermal conversions are inherently irreversible phenomena. Therefore the appearance of heat (thermal) interactions would always raise a question of the quality of a conversion that such a phenomenon features. So there is an important, but not necessarily apparent, quality aspect. The heat flow crosses a temperature "gap" between the high temperature of the heater and the low temperature levels of the product. The presence of the temperature gap and the fact that heat transfer is inherently associated with finite temperature differences bring into consideration an inherent irreversibility of this process. Hence the associated energy-resource quality deterioration appears because of the exchange of the given thermal energy (in the form of heat interaction, generated by an energy-resource conversion system beyond the considered system boundary) across that temperature gap.

A very important point to be made is that the system boundary features crossings of *energy interactions* (heat and work) – but not of *energy* as a *system property*. In this statement, the bulk flows are excluded because such interactions are "carriers" of energy as a property of a material stream in terms of corresponding enthalpy rates! As argued here neither heat nor work is energy *per se*. They both may manifest themselves by influencing (increasing, decreasing, or both) the energy of a system as its property (i.e., within the system boundary) and an energy resource (outside the system boundary) – but in both cases as *energy in transit*. Both interactions do not possess inherent features of system properties:

$$Q = \int_{\text{State–1}}^{\text{State–2}} \delta Q \neq \Delta Q \overset{\Delta}{=} Q_2 - Q_1(!), \quad W = \int_{\text{State–1}}^{\text{State–2}} \delta W \neq \Delta W \overset{\Delta}{=} W_2 - W_1(!). \quad (2.3)$$

Equations (2.3) state just that neither the heat transfer (Q) nor the work (W) are the changes of the state properties, i.e., they *cannot be identified as property changes*, but only as *energy-interactions'*. Stated in more formal, mathematical terms, these entities do not have total differentials. One can infer that the following equation also holds for a *nonadiabatic* system interacting with the environment only through heat and work while undergoing a *cyclic* change of state:

$$\oint \delta Q = \oint \delta W. \tag{2.4}$$

That is, any property change in a cyclic change of state is equal to zero; hence the change of a system's energy is equal to zero and the energy interactions must be balanced.

Finally, let us consider a vastly different, complex natural system, called System Ω.[11] Let us identify within its structure a *community* of two *populations* of constituents, conveniently here called components ω_i, $i = 1$ or 2. Let us assume that these components represent two different species from the world of social beings engaged in a competition for resources. One may postulate that their interactions *within* the system are controlled by a specified set of behaviors called social strategies. These strategies force a particular behavior of the species to accommodate a desired level of comfort, regardless of what that behavior may represent (say, a tendency to move within the habitat to gain a resource). These behaviors are specified as consequences of corresponding constitutive relations for the system. The satisfaction of the comfort level may be measured by a rate of getting to the resource versus the maximum possible rate of the same if there were no competitions (one may call that trade-off a value of the net benefit). The rate of the net benefit to be extracted by a species would certainly depend on the size of the habitat inhabited by the community of two populations that one may call a *society*. With its increase under the action of population pressure, a work would be performed to accomplish the task. Also, an influx of a resource across the system boundary would influence the net rate of the benefit. Such a description of the system should lead to a postulation of a set of states that the system of the considered set of populations can visit. If the set of states is established and its changes identified (existence of the species, presence of strategies, system properties such as resource value or population pressure, net benefits, and influx of resources, etc.),[12] a hypothesis may be formulated that the system considered may well be analyzed by using the tools identical to the tools of thermodynamics, implemented to study a physical system characterized by its constituents, phases, state variables, and interactions. If the system is well defined, if the set of properties is postulated, and if these system properties in conjunction with interactions obey the imposed laws, the analysis performed may describe or at least approximate (to the extent of how sizable the consequences of assumptions would be) the system behavior.[13] It is interesting to note that some far-reaching predictions (such as integration, segregation, and even evolutionary trends) may be identified for such a system (within the realm of transposed variables). However, one still should be careful not to declare the system of such a set of populations that forms a

[11] A rigorous study of an example of such a system is given in a chapter devoted to "sociothermodynamics"; see Chap. 17 by Müller in this book.

[12] The terminology used in describing the system follows the one adopted in I. Müller, "Sociothermodynamics – Integration and segregation in a population," *Continuum Mech. Thermodyn.* **14**, 389–404 (2002).

[13] See Chap. 17 by Müller in this book.

community or a society as a thermodynamic system.[14] What can be said is that a set of necessary (but not sufficient) conditions for implementing an analysis analogous to a thermodynamic analysis has been identified. That analysis is not necessarily a thermodynamic analysis. In other words, one may argue that physical, "nonphysical," or hybrid (social, in this case) systems at best may feature behaviors that can possibly be described by a theory that is more general than the ones developed separately, exclusively for physical or (in this case) social systems. One must be careful to state this as a hypothesis (that must be proven in any application by a rigorous study of the outcomes of the analysis).

In conclusion, we see that a system definition implies identification of various interactions, which in the case of an open system involve mass flow interactions (or, more generally, the flows of constituents – whatever these may be for a given well-defined system) and nonmass interactions (in the form of, say, mechanical, electrical,..., work, or other nonmass interactions, such as heat). In either case, these interactions involve resources (material or energy resources) and a depletion of them in accord with their availability and conversion–processing systems efficiency.

The mass and energy conservation principles do not necessarily support a consideration involving a *quality* of an energy resource, but offer important constraints on the conservation of energy *quantity*. However, an identification of nonmass interactions, such as the entropy rate accompanying heat and associated property changes, brings the issue of the quality of the use of resources, or the quality of the associated process, to the forefront of an analyst's interest. For that aspect to be studied, one would need to include in the analysis, in addition to a conservation principle, a consideration of the nonconserved entities. This will bring (within the realm of physical systems) both the first (promotes the concept of energy) and the second (promotes the concept of entropy) laws of thermodynamics.

2.2 Concept of Energy as a State Property

From the discussion offered so far, any analyst involved with a study of the behavior of a system (let us restrict our scope to physical systems) would realize that to describe the system at any instant of time (i.e., to define the *state* of the system) one

---

[14] The author of this chapter argues that there is a fuzzy line between (a) defining a physical system as a thermodynamic system and (b) defining a "nonphysical," hybrid system *by analogy* as a thermodynamic system. The latter, if properly defined, may be analyzed with the tools of thermodynamics through the set of analogies, but it may be argued that it would be more appropriate to define a theory without using analogies, i.e., defining the system and its variables in accord with the developed theory. This may be justified by the utility of the results. See the reference in footnote 12, as well as J. Kalisch and I. Müller, "Strategic and evolutionary equilibria in a population of hawks and doves," *Rend. Circolo Mat. in Palermo*, Serie II, Supplemento, **78** 163–172 (2006); I. Müller, Personal communication, Vienna, 2006, Berlin, 2008. There is a deeper argument yet for a further study of such applications of thermodynamics within the realm of a remote area of inquiry; see, I. Müller, *A History of Thermodynamics* (Springer, Berlin, 2006), pp. 159–164. An eventual validity of the results of the analysis based on well-defined nonthermodynamic systems may be a consequence of either of the two following reasons: (1) There is an underlying common theoretical background of a higher hierarchy that encompasses the realms of both considered disciplines but not yet discovered, or (b) the well-defined system using a system of analogies *is* a system that can be declared as being a thermodynamic system within the realm of a new discipline. The third option (achieving the outcomes by coincidence) is not considered worth discussing.

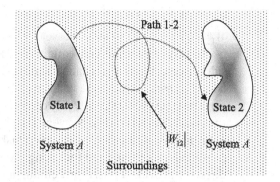

Figure 2.4. Adiabatic System A experiences a work interaction while changing from State 1 to State 2.

needs to know, in addition to the information about the constituents and system parameters (like the realm of the system, i.e., volume), the *properties* of the system, i.e., the attributes that can be presented quantitatively for any instant of time.[15] Among these properties, one deserves a central position, and it is known as *energy*. We use capital and lowercase italic e as symbols for its extensive or intensive values (E or e).

Energy is an additive *system property* defined uniquely for any given state of the system in a given system of reference. This concept is introduced in Chap. 1 by the formulation of the first law of thermodynamics. The energy of a system in a given state is equal to the sum of energies of all the subsystems that may be identified within the given system. An operational definition of energy as a system property can be introduced by a statement that this property's values at the end states of an adiabatic process exercised by the system are fully determined by the magnitude of the work interaction inflicted on or by the system. For the sake of this illustration, let us consider a System A that changes the state between State 1 and State 2, undergoing only a work interaction (i.e., no heat interaction, that is, the system undergoes an adiabatic change of state); see Fig. 2.4. The state of the system is signified symbolically by its geometric shape, and the change of the state is marked by path 1–2 and change of the shape.

The quantity of work used during the change of the system state $\left|W_{1,2}\right|$ will be equal to the change of energy of the system featured by the given two terminal states:

$$\left|\Delta E\right| = \left|E_2 - E_1\right| = \left|W_{1,2}\right|. \tag{2.5}$$

The left-hand side of this relationship signifies a finite change of the property called energy, whereas the right-hand side signifies the energy interaction recognized as work. This equation may also be written in terms of energy and work rates. The work in Eq. (2.5) may not necessarily be considered as the single mechanical work interaction; instead, it may represent the sum of all effects manifested during the interaction of the system with its surroundings that *can be expressed in terms* of an equivalent mechanical work. This definition is a consequence of the first law of thermodynamics in the spirit of the axiomatics advocated in Chap. 1.

The relationship of Eq. (2.5) may conveniently be presented as

$$(E_2 - E_1)_{\text{System } A} = W_{1,2}^{A\leftarrow} = \Delta E_{1,2}^{A\uparrow}, \tag{2.6}$$

---

[15] See Chap. 1 by Gyftopoulos for more details.

where $W_{1,2}^{A\leftarrow}$ and $\Delta E_{1,2}^{A\uparrow}$ are marked with $A\leftarrow$ and $A\uparrow$ to signify that a value of work $W_{1,2}^{A\leftarrow}$ corresponds to an increase of energy of the system, i.e., $\Delta E_{1,2}^{A\uparrow}$ is positive.[16]

Several important consequences of Eq. (2.6) should be noted. First, in the considered case, the energy change is fully defined only by the quantity of work exercised by the *system* based on the validity of the first law of thermodynamics. So, by definition, the energy-change *quantity* cannot provide any information about the *quality* of the system's state change between the given states (as well as the quality of the available energy resource vs. an intended task of the change of state). This statement has profound importance and for a long time has been overlooked. Second, this relation does not provide a basis for defining a specific value of energy as a state property in a given state – only the energy change is strictly defined (i.e., energy is defined up to the arbitrary constant value at the referent state). In other words, an absolute value of the property called energy in classical phenomenological thermodynamics has no practical meaning; rather, the meaning is associated with the energy-level departure with respect to a selected (by convention) state of reference. Third, as a state variable, the change of energy does not depend on the state of the environment. These three important insights were selected to emphasize the difference between energy change as a change of a property and a change of exergy (see the formulation of the definition later).

So the most general definition of energy in terms of Eq. (2.6) is as follows:[17]

$$E = E_0 + W^{A\leftarrow} = E_0 - Mg\,(z - z_0), \tag{2.7}$$

where $W^{A\leftarrow} = -Mg(z - z_0)$ represents the simplest mechanical interpretation of an equivalent work interaction, with the negative sign introduced by convention [1]. In Eq. (2.7) the work interaction is expressed in "equivalent mechanical work terms," say, by shifting a weight M in the gravity field $g = |\mathbf{g}|$ between the height z_0 and the height z–a quite obvious notion of work. The reference state (and the value of energy in it, E_0) can be selected arbitrarily! That is, energy is defined only up to an additive constant that ultimately depends on the choice of the analyst (i.e., a convention). Certainly, in actual applications, energy change involves different modes, not only mechanical.

An abstract definition of energy as a state variable in terms of the formulation of the first law of thermodynamics and a definition of the single work interaction have

---

[16] Note that $W_{1,2}^{A\leftarrow} = -W_{1,2}^{A\rightarrow}$ and $\Delta E^{A\leftarrow} \equiv \Delta E^{A\uparrow}$. This symbolism, in the former case, agrees with the one offered in Chap. 1, but not in the latter case, i.e., with respect to the symbol in the exponent of the energy-change term. The arrow direction indicates here an increase of the energy within the system boundary for this adiabatic process exposed to work only (hence an arrow sign oriented upward). In this chapter, we do not necessarily interpret a change of the system property that is due to energy interaction as a *flow* of a property *over the system boundary*, as in Chap. 1 (in Chap. 1 the symbol $\Delta E_{1,2}^{A\leftarrow}$ signifies the flow of energy into system A). Instead, the energy interaction $W_{1,2}^{A\leftarrow}$ does act on the system (therefore the arrow sign directed toward the letter A, signifying the system being exposed to a work interaction), and therefore causes an increase of its energy $\Delta E_{1,2}^{A\uparrow}$ (note, no other interaction acts on the system). The difference between the two symbolisms appears to be only semantic, but this distinction is important for correctly understanding the meaning of the concept of energy as a system property. After gaining sufficient experience, an analyst may interpret this energy change more loosely as a "flow of energy" into or out of the system as widely used in literature (and omit the arrow sign and the system's name).

[17] We follow the approach provided in Chap. 1 for the sake of simplicity of the energy concept exposition.

great axiomatic importance but are of a limited operational value for an actual study of a complex system. An arbitrary system would be, as a rule, exposed to transient behavior, and naturally would extend within a 3D spatial domain under the influence of multiple interactions.

The equation expressing energy change may be formulated in a differential form, quantifying the elemental energy change through the elemental interconversion transfers. Let us focus on a general form of the *energy equation* (an equation that would include various modes of energy change). Such an energy equation, as we will subsequently see, would have a very complex structure – if one includes explicitly all (or at least most often considered) (1) modes of energy transfer and (2) interconversions from one mode to the other [46]. In virtually all applications, an analyst does not need to include all the modes and all the interconversions. This is simply because only the most dominant influences are considered first (or exclusively). For the sake of as general but still as compact a formulation of the complex explicit relation between the time rate of energy change and possible contributing energy-interaction modes, in Table 2.1 a compilation of the terms of such a relationship is presented.[18] The first column of the table offers in each row the additive terms to be summed to formulate an expression for the change of the given energy mode. Different modes do not have the same number of transfer and interconversion terms, so for the sake of compactness of the table, the background shade of a table cell, rather than the row in its entirety, indicates attribution of the terms to the modes.

The underlined energy equation in its rather general form is complex (but still not necessarily complete, i.e., one may consider additional terms, such as the contributions of electromagnetic fields). The important interconversions between different modes of energy may be included, as presented in Table 2.1. The time rate of energy change can be written as follows (potential energy not included explicitly):

$$
\left\{ \begin{array}{c} \text{Time rate of} \\ \text{energy change} \end{array} \right\} = \left\{ \begin{array}{c} \text{Change of} \\ \text{kinetic energy} \end{array} \right\} + \left\{ \begin{array}{c} \text{Change of} \\ \text{thermal energy} \end{array} \right\}
$$
$$
+ \left\{ \begin{array}{c} \text{Change of} \\ \text{strain energy} \end{array} \right\} + \left\{ \begin{array}{c} \text{Change of} \\ \text{chemical energy} \end{array} \right\}. \tag{2.8}
$$

The system considered is assumed to have a large enough mass per unit volume so that each differential element of the system represents a simple thermodynamic system, as defined in Chap. 1 [1].

Table 2.1 offers a collection of mathematically explicit time-rate expressions for each energy-mode change through corresponding energy interactions. Energy is presented as an intensive property, i.e., the symbol "e" signifies energy per unit

---

[18] Note on the physical units in Table 2.1. An energy mode denoted as e and a subscript (k, t, s, and ch denoting kinetic, thermal, strain, and chemical energy modes, respectively) is given in units of energy (J) per unit mass (kg) so that the rate of any mode is in W/kg. The physical entities in energy interaction expressions are as follows: ρ (kg/m^3) mass density; t (s), time; τ_s (N/m^2), shear stress; $\mathbf{v}$ (m/s^2) velocity; $\pi_{\mathbf{n}}$ (N/m^2) pressure tensor; $\mathbf{n}$ (kg/m^2 s), mass diffusion flux with respect to stationary coordinates; $\mathbf{g}$ (N/kg), body force per unit mass; $\mathbf{j}$ (kg/m^2 s), mass diffusion flux with respect to the mass average velocity; T (K), temperature; s (J/kg K), specific entropy; σ (J/K m^{2}s), entropy flux; μ (J/kg), chemical potential; v (m^3/kg), specific volume; p (N/m^2), pressure; λ (J/kg), chemical affinity; R (kg/m^3 s), reaction rate; χ (dimensionless), mass fraction. Table 2.1 offers the same information as given in Ref. 46 and R. Gaggioli, personal communication, 2009, but in a different format.

Table 2.1. *Energy modes and interactions*

Energy modes change	Change of the given energy mode because of specific energy interactions				
$\dfrac{De_k}{Dt}$	Net transport of kinetic energy via shear stresses $-\dfrac{1}{\rho}(\nabla \cdot [\tau_s \cdot \mathbf{v}])$	Irreversible conversion of kinetic energy to thermal via shear stresses $\dfrac{1}{\rho}(\tau_s : \nabla\mathbf{v})$	Net reversible interconversion between kinetic and strain energy $-\dfrac{1}{\rho}(\mathbf{v} \cdot [\nabla \cdot \pi_n])$	Net transport of kinetic energy via work done by body forces $\dfrac{1}{\rho}\sum_i (\mathbf{n}_i \cdot \mathbf{g}_i)$	Irreversible conversion of energy associated with body forces to thermal energy via diffusion $-\dfrac{1}{\rho}\sum_i (\mathbf{j}_i \cdot \mathbf{g}_i)$
Thermal $\dfrac{De_t}{Dt} = \dfrac{D(Ts)}{Dt}$	Net transport via diffusion and heat transfer $-\dfrac{1}{\rho}(\nabla \cdot T\sigma)$	Reversible interconversion between thermal, strain and/or chemical energy $s\dfrac{DT}{Dt}$	Irreversible conversion of kinetic energy to thermal via shear stresses $-\dfrac{1}{\rho}(\tau_s : \nabla\mathbf{v})$	Irreversible conversion of strain energy to thermal via normal stresses $-\dfrac{1}{\rho}(\tau_n : \nabla\mathbf{v})$	Irreversible conversion of chemical energy to thermal energy via diffusion $-\dfrac{1}{\rho}\sum_i [\mathbf{j}_i \cdot \nabla(\mu_i - \mu_n)]$
Strain $\dfrac{De_s}{Dt} = \dfrac{D(-pv)}{Dt}$	Net transport of strain energy via normal stresses $-\dfrac{1}{\rho}(\nabla \cdot [\pi_n \cdot \mathbf{v}])$	Reversible interconversion of strain energy with thermal and/or chemical energy $-v\dfrac{Dp}{Dt}$	Net reversible interconversion between strain and kinetic energies $\dfrac{1}{\rho}(\mathbf{v} \cdot [\nabla \cdot \pi_n])$	Irreversible conversion of strain to thermal energies via normal stresses $\dfrac{1}{\rho}(\tau_n : \nabla\mathbf{v})$	Irreversible conversion of chemical to thermal via chemical reaction $\dfrac{1}{\rho}\sum_i \lambda_i R_i$
Chemical $\dfrac{De_{ch}}{Dt}$	Net transport of chemical energy via diffusion $-\dfrac{1}{\rho}\sum_i (\nabla \cdot \mu_i \mathbf{j}_i)$	Reversible conversion between chemical and/or thermal and/or strain energy $\sum_i \chi_i \dfrac{D\mu_i}{Dt}$	Irreversible conversion of chemical to thermal via diffusion $\dfrac{1}{\rho}\sum_i [\mathbf{j}_i \cdot \nabla(\mu_i - \mu_n)]$	Irreversible conversion of chemical to thermal via chemical reaction $-\dfrac{1}{\rho}\sum_i \lambda_i R_i$	Irreversible conversion of energy associated with body forces to thermal energy via diffusion $\dfrac{1}{\rho}\sum_i (\mathbf{j}_i \cdot \mathbf{g}_i)$

Notes: In this interpretation, the change of energy does not feature explicitly the potential energy changes. The effects of body forces are accounted for by the conversion of energy associated with body forces.

mass. This table represents a compilation of terms of component equations of energy defined for a continuum system [46].

One should read the structure of Table 2.1 as follows. Each of the energy modes (kinetic, thermal, strain, and chemical) of a system changes because of either net transport or both reversible and irreversible interconversions between a given mode and the other modes. For example, the kinetic energy change in time is a result of "net transports" of kinetic energy by means of (1) shear stresses and (2) work done by body forces, as well as a series of conversions from or to other modes to kinetic energy involving (3) thermal energy, (4) strain energy, and (5) diffusion-involved conversions. Hence the kinetic energy-mode change (the first column and the second row intersection of Table 2.1), for example, stands for the time rate of the kinetic energy. Correspondingly, the terms to the right of the identified kinetic energy time rate (within Table 2.1 the fields featuring the same shade of background) represent the five involved energy-interaction terms (two transport terms, two irreversible conversions, and one reversible conversion). Analogously, the change of thermal energy (the third row in the first column) involves the net transport by heat transfer diffusion and a series of six conversion–interconversion terms between thermal energy and other modes of energy interactions collected in the subsequent six cells (the four ones with the same background shade in the third row and the remaining two down the last column of Table 2.1). Note that terms signifying *interconversions* appear as corresponding pairs of the terms in their respective modes of energy change (but with different signs – to cancel when lumped together). For further details on both physical meanings and mathematical readings of the terms of the energy equation presented in Table 2.1, one should consult the relevant literature [47, 48].

The practical significance of the representation as given by Table 2.1 for modeling purposes can best be shown by illustrating a couple of examples. Let us assume that one has to evaluate what would be the energy rate in the form of heat transfer to be delivered to a constant-properties material in a thermal treatment manufacturing process. Heating of the material system is in general a 3D transient process, and one can determine the *actual* energy rate and time for the process to be accomplished by using the appropriate energy governing equation for this transient process. In addition, an argument can be made that a *theoretical* minimum of energy rate in form of heat would be associated with a heat transfer from a given thermal source to a material in such a way that temperature of the material throughout the process would be, in a limit, spatially uniform. In either case, one must write the corresponding models' energy equations that would include the energy rate change and respective conducting–convective terms. Consider the transport term $-(1/\rho)(\nabla \cdot T\sigma)$, the first term of the portion of energy change responsible for thermal energy transports and interconversions (Table 2.1). This term indicates net energy transport associated with heat diffusion. The entropy flux associated with the heat flux rate q is $\sigma \sim q/T$ ([47], note 18, p. 702). If the heat transfer rate is of a simple Fourier type, $\mathbf{q} \sim -k\nabla T$, one can clearly see that this transport term becomes $(1/\rho)k\nabla^2 T$. In the frequent case of a transient heat transfer by conduction, in which kinetic energy change can be neglected, as well as when energy changes that are due to strain and chemical causes are not significant, Eq. (2.8) for the rate of energy change reduces to

$\rho c \, (dT/dt) = k\nabla^2 T$, a very well-known spatially distributed transient conduction. So, in this case, a complex energy rate balance equation (including its intricate structure, as presented by Table 2.1) reduces to a relevant governing equation that can be used for further modeling of the process (after being complemented by boundary and initial conditions, constitutive relations, and the remaining set of governing equations). In a case of idealized, spatially uniform transient conduction and convection-driven (thermal diffusion) heating, the right-hand side of the balance equation would have a term that would be proportional to the temperature difference rather than to the temperature gradient (the well-known "lumped heat capacitance model" of transient heating). Certainly, in both cases, an appropriate set of boundary and initial conditions would accompany these governing equations.

Consider now the term $(1/\rho)\,(\tau_s : \nabla \mathbf{v})$, an irreversible conversion of kinetic energy to thermal energy that signifies an interconversion between kinetic energy and thermal energy by means of shear stresses. This term appears two times (see Table 2.1) but with opposite signs, once within the kinetic energy mode and once within the thermal energy mode (so that it cancels itself within the overall energy balance as it must because of the requirement that the overall energy balance be satisfied). This term reduces to a viscous dissipation term in an energy equation (for a general form written explicitly one may refer to [47] and [48]). With further simplifications, this term reduces ultimately (for a one-dimensional situation) to a product of viscosity and a square of the velocity gradient.

Finally, let us correlate the *energy change* with the system *energy interactions* through an energy balance. The overall energy balance, if written for a simple open system interacting with its surroundings *only* through *i heat interactions* ($Q_i^{in} > 0$) *a lumped work* ($\dot{W}_i^{in} < 0$) *interaction, and a series of bulk mass flow interactions* (featuring *j* inlet and *k* outlet ports), in a gravity field would lead to an expression given by

$$\frac{dE}{dt} = \dot{E} = \sum_{i=1}^{n} \dot{Q}_i - \dot{W} + \sum_{j=1}^{m} \dot{m}_j^{in} \left(h + \frac{1}{2}v^2 + gz \right)_j - \sum_{k=1}^{l} \dot{m}_k^{out} \left(h + \frac{1}{2}v^2 + gz \right)_k.$$

(2.9)

Note that Eq. (2.9) offers a macroenergy balance for a system in which the value of energy change, (dE/dt), may be, in general, recast in the form given by Eq. (2.8), depending on the system considered and the physical conditions to which it is exposed. Equation (2.9) expresses the following statement: The energy change (time rate) of the system is caused by the existence of at least three types of interactions: (1) interactions *i* that include heat transfer (net), (2) interactions that include net work exchanged, and (3) interactions *j* and *k* that include mass (bulk) flow of constituents to and from the system carrying thermal (expressed by the enthalpy rate difference), kinetic (expressed through the kinetic energy rate difference), and potential (expressed through the potential energy difference that is due to gravity) energy contributions of the constituents. Of course, this structure is imposed (for the sake of simplicity) by the emphasized set of assumptions that eliminates all the other possible interaction causes.

2.3 Concept of Exergy as an Interaction, a System Property, or Both

A rigorous definition of exergy within the context of promoted thermodynamics axiomatic is given in Chap. 1. As inferred there, it can be devised from the concept of the so-called *availability*.[19] In the present discussion, a descriptive characterization of this concept is offered in terms of a system's *energy interaction*. We use a script capital and a lowercase script letter e as symbols for its extensive or intensive values, respectively ($\mathscr{E}$ or $\mathscr{e}$).[20]

The concept of exergy was introduced by engineering thermodynamicists to assist an assessment of a maximum quantity of energy change available for a given task manifested in form of an *energy interaction*.[21] This interaction may be entertained by a system residing, in the most general case, in a given arbitrary state. Such a system may deliver a *maximum* possible *energy interaction in form of work* only if its change of state is ideal (reversible). But, even then, the energy potential available for change from a given state to the end state can be exhausted only up to the point of reaching stable equilibrium state between the system and its surroundings, regardless of an actual "remaining" energy level of the system at the point of reaching that equilibrium. In terms of real applications, the surroundings are often described in terms of the notion of "environment," but in more rigorous thermodynamic terms one has to interpret the surroundings for defining exergy in terms of the so-called *thermal reservoir* characterized with a very specific set of requirements (see Chap. 1).

So the energy of a system that would actually be *available* for performing a given task in the form of work (an energy interaction) while in a contact with the environment would be equal only to engaging a portion of system's energy. That is, not all the energy a system possesses would actually be available for performing a task. Energy interaction would be available only until the system reaches equilibrium with the surroundings. Not all energy interactions possess the same "capacity" to change system's energy (e.g., heat vs. work) as imposed by the second law of thermodynamics. In other words, different interactions differently convert system's energy from one mode to the other. The difference may be interpreted through a presence (for a heat interaction) or an absence (for a work interaction) of an associated (simultaneous) entropy change within interacting systems (i.e., the associated

---

[19] For those interested in the history of the development of this concept, the concept of exergy in terms of availability has been around for long time (R. W. Haywood, "A critical review of the theorems of thermodynamic availability, with concise formulations," *J. Mech. Eng. Sci.* **16**, Part 1, No. 3, 160–173 and Part 2, No. 4, 258–267 (1974; Ref. 26). Haywood attributes the first notion of available work to Maxwell (1871), and the first mathematical formulation to Gibbs [J. W. Gibbs, "A method of geometrical representation of the thermodynamic properties of substances by means of surfaces," *Trans. Comm. Acad. Arts. Sci.* **11**, 382 (1973)], and Keenan [J. H. Keenan, *Thermodynamics* (Wiley, New York, 1941)] is credited as presenting Gibbs results in more practical terms [J. H. Keenan, "Availability and irreversibility in thermodynamics," *Br. J. Appl. Phys.* **2**, 183–192, Ref. 25 (1951)]. A notion of exergy (available energy) was clearly defined by F. Bošnjaković earlier than 1935 {F. Bošnjaković, *Technische Thermodynamik* (Part 1 and Part 2) [T. Steinkopf, Dresden-Leipzig, 1935 (Part 1), 1937 (Part 2)]; an early edition of Ref. 21}.

[20] Elsewhere in this book, multiple symbols for exergy are used, such as Ex.

[21] Sometimes this maximum value of energy interaction is characterized as an "optimum."

"entropy flow" between the system and the source of interaction would be nonzero for a heat interaction or zero for a work interaction).

The consequences just mentioned are associated with the presence of irreversibilities during any real change of state of a system, hence a decrease in *energy-conversion quality* along the process execution. An inability to exploit all the energy of a given state (even for a reversible process) would be manifested because the energy change is constrained by the energy level associated with the stable equilibrium state of the environment. In a word, there is, in addition to quantity, a *quality* aspect of energy associated with its conversion. The notion of quality is revealed during energy conversion, that is, during the change of a system's state in the presence of acting interactions. So the energy interaction quality must be related to (1) the given state of the system, (2) the referent state, and (3) system's state change between the terminal sates.

A system's ability to change its state, and in the process to deliver a maximum useful effect $\dot{W}_{max}$, is constrained by the given starting departure of the initial state from the ultimate stable equilibrium state of the associated reservoir with which the former would ultimately be in equilibrium (e.g., *thermodynamic surroundings in a virtual interaction with the system*) upon delivering the useful effect. In many cases (but not all) this thermodynamics surroundings may be interpreted as the environment (as otherwise considered in ecology).

The change of the given state into a new state of a real process is inherently irreversible; hence certain energy rate difference that a system features versus thermodynamic surroundings (not energy rate per se) would not be available for an interaction effect, i.e., will always be "lost" $\dot{W}_{lost}$ because of irreversibilities. Therefore,

$$\dot{W}_{actual} = \dot{W}_{max} - \dot{W}_{lost}. \tag{2.10}$$

Equation (2.10) is valid quite generally.

In our further discussion, we assume that the system considered is an open system. The "lost work" rate, $\dot{W}_{lost}$, as thermodynamics teaches us, must be proportional to the entropy-generation rate, converted into energy units per unit time through a coefficient of proportionality T_0 (that is, the temperature of the surroundings – thermal reservoir). This relationship is called the Gouy–Stodola theorem:[22]

$$\dot{W}_{lost} = T_0 \dot{S}_{gen}. \tag{2.11}$$

In Eq. (2.11), $\dot{S}_{gen}$ represents the magnitude of the entropy-generation rate produced during the change of state of the system. Note that the entropy generation is *not* a

---

[22] Attributed (in particular in earlier European and current U.S. sources) to Gouy and Stodola, M. Gouy, "Sur les transformations et l'equlibre en thermodynamique," *Comptes rendus*, Paris **108**, 509 (1889); M. Gouy, "Sur l'energie utilizable," *J. de Phys.* 2^e Serie, **8**, 501–518 (1889); A. Stodola, "Die Kreisprocesse der Gasmachine," *Z. Ver. Dtsch. Ing.* **42**, 1088 (1898). Note that, through the validity of Eq. (2.11), the two alternative methodologies for system-state-change quality evaluation may be devised. One may be based on an evaluation and analysis of entropy generation, and the other on considerations of available energy destruction (exergy). In that sense, the school of thought inclined to consider the exergy approach as a somewhat superfluous technique may have an argument. However, a strong counterargument is in the fact that the energy units of exergy offer a clear physical meaning (vs. a somewhat esoteric energy per unit temperature entropy metric). For an alternative attribution of Eq. (2.11), see Haywood, op. cit., footnote 19.

system property change! Keeping that in mind, one should not consider it necessary to signify the quantity of entropy generation by using the Δ symbol.

From Eqs. (2.10) and (2.11) it becomes clear that the maximum work (read, "the maximum available energy change to be extracted from the system") can be expressed as a sum of the actual work rate performed, and the work rate lost (the one that cannot be used, i.e., the portion not available because of associated irreversibilities) during the change of state, i.e.,

$$\dot{W}_{\max} = \dot{W}_{\text{actual}} + T_0 \dot{S}_{\text{gen}}. \tag{2.12}$$

The notion of the *associated* exergy[23] rate is identical to the maximum value of this work rate, i.e.,

$$\dot{\mathscr{E}} = \dot{W}_{\max}. \tag{2.13}$$

2.3.1 Concept of Material Flow Exergy (Physical and Chemical)

In Eq. (2.13), the quantity $\dot{\mathscr{E}}$ signifies the exergy rate (in watts), which is, clearly devised as an entity that must be related to an energy interaction between the system and the referent environment. It is just a matter of applied thermodynamics rigor to express explicitly the left-hand side of Eq. (2.13) in terms of the state properties (taking into account both the system and its environment), regardless of whether the system would have been open or closed. In this text, we do not invest time in executing such a derivation (one may consult any comprehensive applied thermodynamics source [1–21]). Rather, a set of analytical formulations is offered that may be of practical use in exergy balancing for an analysis of resources utilization and intrinsic exergy analysis at the process level.

One important point must be emphasized up front. A careful reader has most likely already noticed that $\dot{W}_{\max}$ must depend on both the state of the system *and*

---

[23] A note regarding the name and the symbolism for the concept of exergy is warranted. The term exergy was coined by Rant [Z. Rant, "Exergie, ein neues Wort für 'technische Arbeitsfähigkeit,'" *VDI-Forschungsh.* **22**, 36–37 (1956)], who attributes the term "technical work capability" to Bosnjakovic (F. Bošnjaković, *Technical Thermodynamics*, T. Steinkopf, Dresden/Leipzig, 1935), Vol. 1. This quantity either as an entity associated with a state of a fluid stream $[h-h_0 -T_0(s-s_0)]$ or as a heat interaction $[Q(T - T_0)/T]$ has been in the English language literature associated with the word "availability." Rant offered a very detailed argument for coining the term, but did not suggest a symbol. Subsequently, a number of symbols have been used, such as Λ, E, E_x, Ξ, E, B, X, Ψ, Φ, with a similar string of lowercase symbols for specific exergies. A consensus regarding the symbolism has not been reached in spite of multiple efforts to resolve this issue [see J. Kestin, "Availability: The concept and associated terminology," *Energy* **5**, 679–692 (1980); T. J. Kotas, Y. R. Mayhew, and R.C. Raichura, "Nomenclature for exergy analysis," *Proc. Inst. Mech. Eng.* **209**, 209–280 (1995), G. Tsatsaronis, "Definitions and nomenclature in exergy analysis and exergoeconomics," *Energy* **32**, 249–253 (2007)]. We accept a notion that exergy, regardless of whether it is defined for a closed or an open system or whether it indicates the exergy equivalent of a quantity of heat, work, or mass interaction, may be denoted by the same symbol. A more precise further specification (say, for designating chemical or physical exergy) is facilitated with superscripts, subscripts, or both. In this chapter a script uppercase or lowercase $\mathscr{E}$ (or ε) is used to make a distinction from the E (or e) usually reserved for energy, a solution in agreement with the suggestion offered by the Working Party for Nomenclature set up in Cambridge, England, in September 1984 and summarized in the previously cited paper of Kotas et al. Note that, elsewhere in this book, different symbolism may be used for the same entity.

on the state of the environment. That dependence makes the physical meaning of the concept of *exergy* rate rather impossible to grasp exclusively in terms of system properties (such as energy).

Two major types of material flow exergy may be identified: (1) physical and (2) chemical exergies. It can be shown that the *physical exergy* rate, scaled by the mass flow rate of a given system bulk flow (taken as an example in this consideration), can be represented as a difference of flow availability function values in the end states of a system's change of state (between the given state and the state of the stable equilibrium state of the environment; see Chap. 1 for details):

$$e^{\text{ph}} = \frac{\dot{\mathscr{E}}}{\dot{m}} = b - b_0, \tag{2.14}$$

where b, and b_0 represent the so-called (1) flow availability function of the system in the given state and (2) the flow availability in the so-called restricted dead state, respectively. The latter corresponds to the stable equilibrium state of the environment; see Eqs. (2.15) and (2.16). Note that a system considered here represents an open system; hence flow availability would be a property of that system, say, a stream of a fluid, but defined in relation to the so-called "restricted" dead state (in that state the system will be at the state of stable equilibrium with the thermodynamic surroundings with respect to both temperature and pressure, i.e., at T_0 and p_0). The availability functions are defined (see Chapter 1) as[24]

$$b = h - T_0 s, \tag{2.15}$$

$$b_0 = h_0 - T_0 s_0. \tag{2.16}$$

So the specific physical material flow exergy of a system in the given state would be

$$e^{\text{ph}} = h - h_0 - T_0(s - s_0) \tag{2.17}$$

or, in molal representation,

$$\overline{e}^{\text{ph}} = \overline{h} - \overline{h}_0 - T_0(\overline{s} - \overline{s}_0). \tag{2.18}$$

In Eqs. (2.15)–(2.18) the enthalpy included is the bulk enthalpy, i.e., it is assumed that no kinetic energy and potential energy contributions to the enthalpy should be taken into account (that is, the problems considered would not involve, for example, gas dynamics and compressible fluid mechanics applications).[25]

Before we proceed, an important aspect of physical meaning of the new concept should be stressed. Exergy may be interpreted as a property of *both the system and the environment compounded together*, and if considered so, one may interpret it as a thermodynamic property!

The material flow exergy, as defined in Eqs. (2.17 and 2.18), represents the so-called "physical exergy." The justification for this name is as follows. Both the

---

[24] Note that the specific Gibbs free energy is defined as $g = h - Ts$, whereas the availability functions involve T_0. Hence the flow availability becomes the same as the Gibbs free energy in the state of stable equilibrium with the surroundings. Note also that molal Gibbs free energy is equivalent to the chemical potential μ, i.e., for the state in equilibrium with the environment $\mu_{0,i} = \overline{h}_{0,i} - T_0 \overline{s}_{0,i} = \overline{g}_{0,i}$.

[25] If one has to take into account the stagnation enthalpies rather than the bulk flow enthalpies, instead of h the flow availability would involve $h^\circ = h + (1/2) v^2 + gz$, sometimes called methalpy. See J. Kestin, *A Course in Thermodynamics* (Hemisphere, Washington, D.C., 1979) Ref. 18.

temperature and the pressure differences (between the given state and the state of the same system in the restricted dead-state equilibrium with the environment) are responsible for its magnitude level. Hence "physical" potentials (rather than chemical; see subsequent discussion) are relevant.

However, as is known from thermodynamics fundamentals, an additional and very important driving potential available for achieving a given task is still not taken into consideration. This is the reason why the equilibrium state so far mentioned is called the restricted dead state. This remaining potential for extracting exergy from a system is related to a chemical potential difference. One can show [14] that this chemical flow exergy can be written as

$$\overline{e}^{ch} = \sum_{i=1}^{n} (\overline{\mu}_i^* - \overline{\mu}_{0,i}) x_i, \tag{2.19}$$

where $x_i = \dot{N}_i / \dot{N}$ represents the molal fraction of the ith chemical component in the system consisting of n components, each having $\dot{N}_i$ number of moles per unit time: $\dot{n} = \sum_{i-1}^{n} \dot{N}_i$. The chemical potential $\overline{\mu}^*$ is evaluated at the "restricted dead state," i.e., it represents the chemical potential of a component evaluated at (T_0, p_0), and $\overline{\mu}_{0,i}$ represents the chemical potential of the ith component at the state of equilibrium with the environment ("ultimate dead state"). Combining Eqs. (2.18) and (2.19), one can state that the total molal flow exergy of the given flow system consisting of n chemical constituents at the given state versus the environment is defined as follows:

$$\overline{e} = \overline{e}^{ph} + \overline{e}^{ch} = \left[\overline{h} - \overline{h}^* - T_0 \left(\overline{s} - \overline{s}^* \right) \right] + \sum_{i=1}^{n} (\overline{\mu}_i^* - \overline{\mu}_{0,i}) x_i. \tag{2.20}$$

Equation (2.20) includes material flow *available energy (exergy)* of the given system versus the environment in terms of both physical exergy [maximum useful work delivered from the system (i.e., material flow) to a user (i.e., to another system beyond the system boundary considered) as the considered system reaches the so-called "restricted dead state" (T_0, p_0)] and chemical material flow exergy [maximum work delivered to a user associated with the system's transition from the restricted dead state to the ultimate dead state, that is $(T_0, p_0, \mu_{0,i}, i = 1, 2, \ldots, n)$].

2.3.2 Concepts of Heat and Work Exergies

So far, we have discussed the flow exergy associated with the given state of the system (e.g., a fluid stream) that corresponds to a material flow interaction. This exergy is due to the state properties difference between the given state and the state of the system if it would have been in stable equilibrium with the environment [either in the restricted (physical) or the ultimate (chemical and physical) dead state]. This available energy at any state is carried by a bulk material flow. However, a system may interact with its surroundings through the two energy interactions, i.e., heat and work. By the very definition of exergy as the maximum available energy in terms of work [Eq. (2.13)] that a system may exercise, it must be clear that the magnitude of any work interaction corresponds exactly to the same magnitude of corresponding exergy that this energy interaction carries. In other words, any work is available to be converted into a useful effect entirely, i.e.,

$$\dot{\mathscr{e}}_W = \dot{W}. \tag{2.21}$$

This "equality" outcome, however, is not to be expected in case of a heat transfer interaction! The heat transfer rate has a lower exergy rate than magnitude in terms of the heat rate! To extract that statement from the discussion previously offered may not be apparent, but to prove it is quite straightforward. So we are stating it as a hypothesis, and we prove it as follows. Imagine a system interacting with the environment at a temperature T_0 through a work interaction and n heat transfer rates delivered (or received) at n temperature levels $T_i, i = 0, 1, \ldots, n$. For the sake of obviousness, let us assume that no bulk flow interactions exist. If we formulate both energy and entropy balances for the system in a steady state,[26] we can write the flowing relations as

$$0 = \sum_{i=0}^{n} \dot{Q}_i - \dot{W} \quad \text{and} \quad 0 = \sum_{i=0}^{n} \frac{\dot{Q}_i}{T_i} + \dot{S}_{\text{gen}}. \tag{2.22}$$

Combining the two balances by eliminating $\dot{Q}_0$, we get

$$\dot{W} = \sum_{i=0}^{n} \left(1 - \frac{T_0}{T_i}\right) \dot{Q}_i - T_0 \dot{S}_{\text{gen}}. \tag{2.23}$$

In the case of a reversible process, the lost work $T_0 \dot{S}_{\text{gen}}$ is equal to zero. The maximum $\dot{W}$ becomes $\sum \dot{\mathscr{E}}_{Qi}$, and therefore [see Eq. (2.13)]

$$\sum_{i=0}^{n} \dot{\mathscr{E}}_{Qi} = \sum_{i=0}^{n} \left(1 - \frac{T_0}{T_i}\right) \dot{Q}_i, \tag{2.24}$$

or, for any heat transfer rate $\dot{Q}_i$, $\dot{\mathscr{E}}_{Qi} = (1 - T_0/T_i)\dot{Q}_i$. The result given by Eq. (2.24) can be interpreted as follows: Any heat transfer rate carries the exergy rate that is in its magnitude (in energy terms) smaller by a factor of $(1 - T_0/T)$ than is the heat transfer rate magnitude.

The system's CV form of the exergy rate balance is as follows:

$$\underbrace{\frac{\partial \mathscr{E}_{\text{CV}}}{\partial t} = \dot{\mathscr{E}}_{\text{CV}}}_{\substack{\text{Time rate of} \\ \text{systems (CV)} \\ \text{exergy change}}} = \underbrace{\sum_{i=1}^{n} \left(1 - \frac{T_0}{T_i}\right)\dot{Q}}_{\substack{\text{Time rate of exergy} \\ \text{transfer by heat}}} \quad \underbrace{-\dot{W}}_{\substack{\text{Time rate} \\ \text{of exergy} \\ \text{transfer} \\ \text{by work}}}$$

$$+ \underbrace{\sum_{j=1}^{j=m} \dot{m}_j^{\text{in}} e_j - \sum_{k=1}^{k=l} \dot{m}_k^{out} e_k}_{\substack{\text{Time rate of exergy transfer} \\ \text{by bulk flow exergy inputs} \\ \text{or outputs}}} - \underbrace{T_0 \dot{S}_{\text{gen}}}_{\substack{\text{rate of exergy} \\ \text{destruction}}}. \tag{2.25}$$

This balance, however, does not provide detailed insight into the physical mechanisms responsible for the exergy rate [the left-hand side of Eq. (2.25)). This insight

---

[26] An assumption of steady-state conditions does not restrict generality of the consideration of exergy associated with heat.

can be obtained from detailed analysis of transfers and conversions of exergy-mode changes, similarly to the energy analysis summarized in Table 2.1 and Eq. (2.8). The result of that analysis is summarized in Table 2.2. Individual entries in this table follow the same sequence of contributions as those offered for energy presented in Table 2.1, i.e,

$$\left\{ \begin{array}{l} \text{Time rate of} \\ \text{exergy change} \end{array} \right\} = \left\{ \begin{array}{l} \text{Change of} \\ \text{kinetic exergy} \end{array} \right\} + \left\{ \begin{array}{l} \text{Change of} \\ \text{potential exergy} \end{array} \right\}$$

$$+ \left\{ \begin{array}{l} \text{Change of} \\ \text{thermal exergy} \end{array} \right\} + \left\{ \begin{array}{l} \text{Change of} \\ \text{strain exergy} \end{array} \right\} + \left\{ \begin{array}{l} \text{Change of} \\ \text{chemical exergy} \end{array} \right\} .$$

$$(2.26)$$

These entries are assembled following the representation of the time rate of exergy that includes the potential exergy [46]. As argued by Lior et al. [49], "most of the exergy analysis is nowdays conducted on the system level development, by evaluating the exergy values and changes of component input and output streams and energy interactions." However, as will be demonstrated in Chap. 5, advances in resources utilization through the development of new technologies for given tasks does require identifying exergy destructions at the process level, i.e., evaluating entropy generations within the system for all process phenomena. These phenomena involve interconversions and net transfer of exergy-change components, as marked in Eq. (2.26). This is the reason why the explicit analytical expressions of all the components of the modes of exergy changes are summarized in Table 2.2.

2.3.4 The Restricted and the Ultimate Dead States

The concept of available energy in terms of exergy carries its full meaning of a system property only in conjunction with the partial, restricted ("physical") and ultimate ("physical and chemical") dead states, i.e., states of stable equilibrium of a considered system with the selected referent surroundings, the environment. So it must be clear that the available energy potential for achieving a useful task would depend on state variables of the system in equilibrium with the environment. Because various systems exposed to various processes may be considered as naturally evolving toward equilibrium with various environments, the legitimate question is this: What would be the most convenient environment to refer to when calculating exergy? As far as the physical exergy is concerned, a consensus has been made to select T_0 and p_0 as $T_0 = 298.15$ K and $p_0 = 1.0 \times 10^5$ Pa (sometimes, 1 atm instead of 1 bar is also used). Note that, in principle, the referent state may be defined with respect to another set of equilibrium state's properties – if the referent environment is selected differently for a given application.

The ultimate dead state requires a full description of the environment's composition that would identify the chemical exergy potentials on the path from the restricted to the ultimate dead state. Two more widely referenced approaches to that definition should be mentioned. The first [50] approach[27] assumes thermodynamic equilibrium (with a few exceptions) for all chemical components of the atmosphere, hydrosphere, and a portion of the lithosphere. The second approach assumes an existence of a

---

[27] In this approach $p_0 = 1.019 \times 10^5$ Pa.

Table 2.2. *Exergy modes and interactions*[a]

Exergy-modes changes	Change of given exergy mode because of specific energy interactions					
Kinetic $\dfrac{De_k}{Dt}$	Net transport of kinetic exergy via shear stresses $-\dfrac{1}{\rho}(\nabla \cdot [\tau_s \cdot \mathbf{v}])$	Net transport via work by nonconservative body forces $\dfrac{1}{\rho}\sum_i (\mathbf{u}_i \cdot \mathbf{f}_i)$	Irreversible conversion of kinetic exergy to thermal via shear stresses[b] $\dfrac{1}{\rho}(\tau_s : \nabla \mathbf{v})$	Reversible interconversion between kinetic and strain exergy $-\dfrac{1}{\rho}(\mathbf{v} \cdot [\nabla \cdot \boldsymbol{\pi}_n])$	Reversible interconversion between kinetic and potential exergy $-\dfrac{1}{\rho}\sum_i (\mathbf{u}_i \cdot \nabla \phi_i)$	Irreversible conversion to thermal exergy associated with nonconservative body forces $-\dfrac{1}{\rho}\sum_i (\mathbf{j}_i \cdot \mathbf{f}_i)$
Potential $\dfrac{De_p}{Dt}$	Net transport of potential exergy via diffusion $-\dfrac{1}{\rho}\sum_i (\nabla \cdot \phi_i \mathbf{j}_i)$	Interconversion between kinetic and potential exergy (reversible) $\dfrac{1}{\rho}\sum_i (\mathbf{u}_i \cdot \nabla \phi_i)$	Potential exergy change as a consequence of reactions $\dfrac{1}{\rho}\sum_i \phi_i r_i$	Reversible interconversion of exergy associated with body forces to thermal $\dfrac{1}{\rho}\left(1 - \dfrac{T_0}{T}\right)\sum_i (\mathbf{j}_i \cdot \mathbf{f}_i)$		Irreversible conversion of potential-to-kinetic to thermal exergy $\dfrac{1}{\rho}\sum_i (\mathbf{j}_i \cdot \nabla \phi_i)$
Thermal $\dfrac{D}{Dt}\dfrac{e_t}{}$ $= \dfrac{D}{Dt}(T - T_0)s$	Net transport of thermal energy via diffusion, conduction, and radiation $-\dfrac{1}{\rho}(\nabla \cdot (T - T_0)\sigma)$	Reversible interconversion thermal with strain and/or chemical exergy $\dfrac{DT}{Dt}s$	Destruction of exergy due to entropy flow toward lower temperatures $\dfrac{T_0}{\rho T}(\sigma \cdot \nabla T)$	Irreversible conversion of strain to thermal exergy via shear stresses $-\dfrac{1}{\rho}\left(1 - \dfrac{T_0}{T}\right)(\tau_n : \nabla \mathbf{v})$	Irreversible conversion to thermal of a portion of potential to kinetic conversion $-\dfrac{1}{\rho}\left(1 - \dfrac{T_0}{T}\right)\sum_i (\mathbf{j}_i \cdot \nabla \phi_i)$	Irreversible conversion of chemical exergy to thermal via chemical reaction $\dfrac{1}{\rho}\left(1 - \dfrac{T_0}{T}\right)\sum_i \lambda_i R_i$
Strain $\dfrac{De_s}{Dt}$ $= \dfrac{D}{Dt}[-(p - p_0)v]$	Net transport of strain energy via normal stresses $-\dfrac{1}{\rho}(\nabla \cdot [\boldsymbol{\pi}_n - \mathbf{p}_0\delta] \cdot \mathbf{v})$	Reversible interconversion with thermal and/or chemical exergy $-v\dfrac{Dp}{Dt}$	Net reversible interconversion between strain and kinetic $\dfrac{1}{\rho}(\mathbf{v} \cdot [\nabla \cdot \boldsymbol{\pi}_n])$	Irreversible conversion of strain to thermal via normal stresses $\dfrac{1}{\rho}(\tau_n : \nabla \mathbf{v})$	Change due to change in environmental pressure $v\dfrac{Dp_0}{Dt}$	Change of exergy due to a change of the intensive property of the environment (temperature) $-s\dfrac{DT_0}{Dt}$
Chemical $\dfrac{De_{ch}}{Dt}$ $= \dfrac{D}{Dt}\sum_i (\mu_i - \mu_{i0})\chi_i$	Net transport of chemical exergy via diffusion $-\dfrac{1}{\rho}\sum_i [\nabla \cdot (\mu_i - \mu_{i0})\mathbf{j}_i]$	Reversible conversion between chemical and thermal and/or strain energy $\sum_i \chi_i \dfrac{D\mu_i}{Dt}$	Irreversible conversion of chemical to thermal via diffusion[c] $-\dfrac{1}{\rho}\sum_i [\mathbf{j}_i \cdot \nabla(\mu_i - \mu_{in})]$	Irreversible conversion of chemical to thermal via chemical reaction $-\dfrac{1}{\rho}\sum_i \lambda_i R_i$	Change in chemical exergy due to changes in reference chemical potentials $-\sum_i \chi_i \dfrac{D\mu_{i0}}{Dt}$	Irreversible conversion of chemical exergy to thermal energy via diffusion $\dfrac{-1}{\rho \tau}\sum_i [\mathbf{j}_i \cdot \nabla(\mu_i - \mu_n)]$ $\tau = \left(1 - \dfrac{T_0}{T}\right)$

[a] In addition to the nomenclature and symbolism presented in footnote 18, p. 63, the following physical entities are included here: δ unit tensor, dimensionless; r_i (kg/s m^3), chemical species reaction rate; ϕ_i (J/kg), potential of species i per unit mass that is due to conservative force field (note: $\nabla \phi_i = f_i - g_i$); e (w/kg) exergy rate per unit mass; f_i (N/kg) nonconservative body force per unit mass.

[b] Fraction $(1 - T_o/T)$ of this quantity becomes thermal exergy and a fraction T_o/T is destroed in conversion.

[c] The fraction $(1 - T_o/T)$ becomes thermal exergy and the T_o/T fraction is destroyed.

Table 2.3. *Molal chemical exergies*[a]

Substance	$\bar{e}^{ch}$ (kJ/kmol), Ahrendts[b]	$\bar{e}^{ch}$ (kJ/kmol), Szargut[c]	$\bar{e}^{ch}$ (kJ/kmol),[d] Baehr and Schmidt[d]	$\bar{e}^{ch}$ (kJ/kmol), Gaggioli and Petit[e]
Oxygen, O_2	3947	3973	3953	3947
Nitrogen, N_2	639	718	692	691
Water, H_2O (v)	8636	11,758	8595	8668
Water, H_2O (l)	45	3168	0	0
Carbon, C (Graphite)	405,000	411,000	411,000	411,000
CO_2	14,174	20,189	20,108	20,108
SO_2	240,633	304,300	–	287,600
Ammonia, NH_3 (g)	337,000	340,000	336,000	337,000
Heptane C_7H_{16}	4,716,000	4,783,000	4,757,000	4,757,000
CH_3OH (l)	711,000	723,000	717,000	717,000
Methane, CH_4 (g)	824,000	837,000	830,000	830,000

[a] The numerical values presented in this table differ slightly from the values in some other sources, e.g. [14]. The values listed here are taken from J. Ahrendts, "Reference states," *Energy* **5**, 667–677 (1980).
[b] J. Ahrendts, "Die Exergie chemish reaktionsfähiger Systeme. Erklärung und Bestimmung," dissertation (Rurh Universität, Bochum, 1974).
[c] J. Szargut, *Freiberger Forschungsh.* **B68**, 81 (1962).
[d] H. D. Baehr and E. F. Schmidt, *Brenstoff-Wärme-Kraft* **15**, 375 (1963).
[e] R. A. Gaggioli and P. J. Petit, "Second law analysis for pinpointing the true inefficiencies in fuel conversion systems in Vol. 21 of the *ACS Symposium Series* (American Chemical Society, Washington, D.C., 1976), pp. 56–76.

reference substance for each chemical element from among substances abundantly present in the environment although not necessarily in mutual equilibrium among themselves [15].[28] Based on either of the conventions, standard chemical exergies of chemical substances may be determined (at the restricted dead state) [15]. Table 2.3 offers molal chemical exergies for a few substances calculated by different authors that show, in some cases, significant differences, hence indicating a need to establish a standardized approach to calculating these values. The early calculations assumed that an environment consisting of water and air was sufficient. Subsequently the reference environment has been enriched, as suggested in [15] and [50]. For a detailed list of more recently updated chemical exergy data based on the second approach previously listed see Appendix.

2.4 An Illustration of Energy-, Entropy-, and Exergy-Balance Analyses at the System Level

As a demonstration of energy or exergy balancing, let us consider a *generalized* setting that includes materials extraction, processing, manufacturing, and recycling subsystems along a life cycle of a generalized manufacturing process path, starting with a raw resource and ending with a product. Let us assume that each of these subsystems has a separate energy supply (Fig. 2.5), and, for the sake of obviousness, that no auxiliary material streams are added along the processing except for energy-supply subsystems. The former does not restrict analysis in any way, and it is assumed only to somewhat simplify an already very intricate set of material and energy flows; the latter can easily be generalized by adding all the auxiliary flows.

[28] In this approach $p_0 = 1.0 \times 10^5$ Pa.

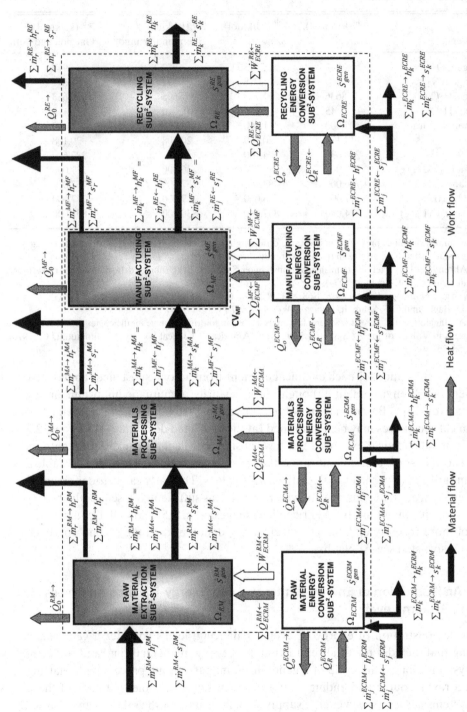

Figure 2.5. Combined materials extraction, processing, manufacturing, and recycling systems resources flow. See Section 2.6 for definitions of acronyms used.

Raw material enters the system through the materials extraction subsystem Ω_{RM} and is subsequently processed in Ω_{MA}. A portion of the material, after processing, goes to the manufacturing subsystem, Ω_{MF}, while a portion exits materials processing and ends up in the environment as a waste. Similarly, during the prior extraction, some residue may be dumped into the environment. After the manufacturing process is completed, the related subsystem delivers a stream of products and a stream of a residue waste that, again, ends up with either a user of the residue or in the environment. Close to the end of the product's life, recycling may take place within the subsystem Ω_{RE}. *Note that the product-use segment was not included for simplicity.* All subsystems are fed by required energy resources, such as work and heat interactions, through corresponding energy-conversion subsystems, Ω_{ECRM}, Ω_{ECMA}, Ω_{ECMF}, and Ω_{ECRE}. Each of these subsystems delivers residue flows into the surroundings. These subsystems are supplied by the respective input resources and may interact with the environment through additional energy communications such as so-called waste heat losses.

The objectives of an analysis may be multiple. In this illustration, we consider defining the total exergy loss of the system as a whole. The resources involved would be the materials or energy resources, but all expressed in terms of exergy. The system considered represents a modified and significantly expanded version of a single materials-processing function, manufacturing system functions, or both [1].[29]

The analysis must start with writing mass, energy, and entropy balances. Each of these balances can be written for each of the subsystems (one for executing the primary objective and the other serving as an energy-supplying unit) marked as SUB^2-SYSTEMS, for each subsystem if energy conversion is included, or for the system as a whole. In addition to delineating control volume of each of these systems or subsystems, each must be well defined. For the manufacturing sub-system (see the control volume CV_{MF} around the subsystem Ω_{MF}), mass, energy, and entropy balances are given by Eqs. (2.27)–(2.29), see also Eqs. (2.1) and (2.9).

Mass balance, CV_{MF}:

$$\left\{ \begin{array}{l} \text{The rate of } \Omega_{MF} \\ \text{subsystem's} \\ \text{mass change} \end{array} \right\} = 0 = \left\{ \begin{array}{l} \text{The rate of} \\ \text{mass flow} \\ \text{rates into } \Omega_{MF} \end{array} \right\} - \left\{ \begin{array}{l} \text{The rate of} \\ \text{mass flow} \\ \text{rates out of } \Omega_{MF} \end{array} \right\}. \quad (2.27)$$

Energy balance, CV_{MF}:

$$\left\{ \begin{array}{l} \text{The rate of } \Omega_{MF} \\ \text{subsystem's} \\ \text{energy change} \end{array} \right\} = 0 = \left\{ \begin{array}{l} \text{The work} \\ \text{rate} \\ \text{into } \Omega_{MF} \end{array} \right\} - \left\{ \begin{array}{l} \text{The heat} \\ \text{rate out of} \\ \Omega_{MF} \text{ at } T_0 \end{array} \right\}.$$

$$+ \left\{ \begin{array}{l} \text{The heat} \\ \text{rates into} \\ \Omega_{MF} \text{ at } T_i, \\ i = 1, \ldots, n \end{array} \right\} + \left\{ \begin{array}{l} \text{The total} \\ \text{enthalpy} \\ \text{rate into} \\ \Omega_{MF} \end{array} \right\} - \left\{ \begin{array}{l} \text{The total} \\ \text{enthalpy} \\ \text{rate} \\ \text{out of } \Omega_{MF} \end{array} \right\}.$$

$$(2.28)$$

[29] See Chap. 6 by Gutowski and Sekulić in this book; M. Branham, T. Gutowski, A. Jones, and D.P. Sekulić, *A Thermodynamic Framework for Analyzing and Improving Manufacturing Processes, Electronics and the Environment* (IEEE San Francisco, 2008), CD Edition.

Entropy balance, CV_{MF}:

$$
\left\{ \begin{array}{l} \text{The rate of } \Omega_{MF} \\ \text{subsystem's} \\ \text{entropy change} \end{array} \right\} = 0 = \left\{ \begin{array}{l} \text{The entropy} \\ \text{flow into} \\ \Omega_{MF} \text{ through} \\ Q \text{ at } T_0 \end{array} \right\} + \left\{ \begin{array}{l} \text{The entropy} \\ \text{flow into} \\ \Omega_{MF} \text{ through} \\ Q_i \text{ at } T_i \end{array} \right\}
$$

$$
+ \left\{ \begin{array}{l} \text{The entropy} \\ \text{rates into} \\ \Omega_{MF} \text{ through} \\ \text{raw mass} \end{array} \right\} - \left\{ \begin{array}{l} \text{The entropy} \\ \text{rates out of} \\ \Omega_{MF} \text{ through} \\ \text{all flows} \end{array} \right\} - \left\{ \begin{array}{l} \text{The total} \\ \text{entropy} \\ \text{generation} \\ \text{within } \Omega_{MF} \end{array} \right\}.
$$

$$(2.29)$$

Equations (2.27)–(2.29), in explicit form, read as follows.

Mass balance:

$$
\frac{dM^{MF}}{dt} = 0 = \sum_{j} \dot{m}_{j}^{MF\leftarrow} - \sum_{k,r} \dot{m}_{k,r}^{MF\rightarrow}. \tag{2.30}
$$

Energy balance:

$$
\frac{dE^{MF}}{dt} = 0 = -\sum_{n} \dot{W}_{ECMF}^{MF\leftarrow} - \dot{Q}_{0}^{MF\rightarrow} + \sum_{i} \dot{Q}_{ECMF}^{MF\leftarrow} + \sum_{j} \dot{m}_{j}^{MF\leftarrow} h_{j}^{MF}
$$
$$
- \sum_{k} \dot{m}_{k}^{MF\rightarrow} h_{k}^{MF} - \sum_{r} \dot{m}_{r}^{MF\rightarrow} h_{r}^{MF}. \tag{2.31}
$$

Entropy balance:

$$
\frac{dS^{MF}}{dt} = \frac{\dot{Q}_{0}^{MF\leftarrow}}{T_0} + \sum_{i} \frac{\dot{Q}_{ECMF}^{MF\leftarrow}}{T_i} + \sum_{j} \dot{m}_{j}^{MF\leftarrow} s_{j}^{MF} - \sum_{k} \dot{m}_{k}^{MF\rightarrow} s_{k}^{MF}
$$
$$
- \sum_{r} \dot{m}_{r}^{MF\rightarrow} s_{l}^{MF} + \dot{S}_{gen}^{MF}. \tag{2.32}
$$

Note that, in general, waste-material streams may not be in equilibrium with the environment when leaving the subsystem. However, if the state of the bulk stream leaving subsystems is at the temperature and pressure of the environment (i.e., in the restricted dead state), these values should be marked with subscript "0," as presented in Fig. 2.5. In this example, chemical exergy contributions are not considered for the sake of obviousness of the general approach to mass–energy–exergy balancing.

The set of equations Eqs. (2.30)–(2.32) is written specifically for the manufacturing subsystem (i.e., for the control volume CV_{MF} around the subsystem Ω_{MF}). Analogously, one can write respective balances for the energy conversion for the manufacturing subsystem (i.e., for the control volume CV_{ECMF} around the subsystem Ω_{ECMF}), as well as for the subsystems involving material extraction, materials processing, and product recycling (Ω_{RM}, Ω_{MA}, and Ω_{RE}) and their energy-conversion counterparts (Ω_{ECRM}, Ω_{ECMA}, and Ω_{ECRE}).

For a steady state, i.e., $d/dt(.) = 0$, the energy and entropy balances expressed by Eqs. (2.31) and (2.32) may be combined by multiplying the entropy-balance equation,

Eq. (2.32), by T_0 and eliminating (from the resulting equation and the energy equation) the heat transfer rate dumped into the environment $\dot{Q}_0^{MF\rightarrow}$. After a straightforward algebraic manipulation, one gets

$$\sum_i \dot{W}_{ECMF}^{MF\leftarrow} = \sum_i \left(1 - \frac{T_0}{T_i}\right) \dot{Q}_{ECMF}^{MF\leftarrow} + \sum_j \dot{m}_j^{MF\leftarrow}\left(h_j^{MF} - T_0 s_j^{MF}\right)$$
$$- \sum_k \dot{m}_k^{MF\rightarrow}\left(h_k^{MF} - T_0 s_k^{MF}\right) - \sum_r \dot{m}_r^{MF\rightarrow}\left(h_r^{MF} - T_0 s_r^{MF}\right) - T_0 \dot{S}_{gen}^{MF}.$$

$$(2.33)$$

An inspection of Eq. (2.33) indicates that each but the last term on the right-hand side of the expression may further be reinterpreted or recast to represent the exergy rate of (1) heat transfer energy interactions, and (2) inlet and (3) outlet bulk mass flows' physical exergies, respectively. Note that the flow exergy rates on the right-hand side of Eq. (2.33) can be formally introduced in explicit form by algebraically manipulating the terms of availability functions [see Eq. (2.15) and Eq. (2.16) involving the streams $\dot{m}_j^{MF\leftarrow}$, $\dot{m}_k^{MF\rightarrow}$, and $\dot{m}_r^{MF\rightarrow}$], evaluated at the state of environment (note $\dot{m}_r^{MF\rightarrow} = \dot{m}_j^{MF\leftarrow} - \dot{m}_k^{MF\rightarrow}$). The last term of Eq. (2.33) is the lost exergy, i.e., the lost availability of the involved resources. In the case in which all the processes within the manufacturing subsystem are executed ideally, the last term vanishes. A real process, however, is always characterized by the loss of exergy within the subsystem Ω_{MF} (equal to $T_0 \dot{S}_{gen}^{MF} \geq 0$). So Eq. (2.33) can be written in the exergy rate form as follows (see Subsections 2.3.1 and 2.3.2):

$$\dot{\mathscr{e}}_W^{MF\leftarrow} = \dot{\mathscr{e}}_Q^{MF\leftarrow} + \dot{\mathscr{e}}^{MF\leftarrow} - \dot{\mathscr{e}}^{MF\rightarrow} - T_0 \dot{S}_{gen}^{MF},$$

$$(2.34)$$

where exergy rates signify the sums of all mass and energy-interaction contributions.

One can implement the same energy and entropy balancing for the remaining set of subsystems or sub^2-systems and determine the total loss of energy resources. For example, if one considers both materials-processing and manufacturing subsystems (including both processing and energy sub^2-systems) the loss becomes equal to

$$T_0 \left(\dot{S}_{gen}^{ECMA} + \dot{S}_{gen}^{MA} + \dot{S}_{gen}^{MF} + \dot{S}_{gen}^{ECMF}\right) = \Delta\dot{\mathscr{e}}_{total},$$

$$(2.35)$$

i.e., it is directly proportional to the total entropy generation within the overall set of subsystems!

Equation (2.34) offers a powerful tool for the analysis.[30] Namely, the exergy rates of Eq. (2.34) can be used to determine the exergy loss as the residue of the exergy balance. This can be straightforwardly calculated by measuring system properties and performing mass, energy, and exergy balances after identifying the system's mass or energy-interaction ports.

The actual exergy rate of the left-hand side of the equality of Eq. (2.34) is always smaller than the ideal one whenever the last term is assumed *not* to be zero. Hence a comparison of these quantities may offer a metric for evaluating exergetic efficiency. The exergy flow analysis represents a rigorous framework for capturing physical resources conversion and their use in considered systems.

---

[30] See Chaps. 5 (Sekulić) and 6 (Gutowski and Sekulić) in this book.

2.5 Conclusion

The notion of *exergy* is quite an old concept of classical thermodynamics. It was defined through the concept of *available energy*. Its semantic content differs from the concept of energy – it can be measured by establishing the maximum (equivalent) work effect that can be extracted from a given system while reaching reversibly the equilibrium with respect to a given reference state. Exergy *is not* a conserved property of a system. Rather, it is destroyed in all real, irreversible processes. This destruction of an energy-resource availability is expressed in energy units, and it is directly proportional to entropy generation manifested within the considered system.

Energy is a system property. Energy-resource quantity is a "carrier" of available energy, and it can be utilized through an energy-conversion process. One should clearly distinguish the concept of an *energy resource* from the concept of energy as a system property. The path from the former into the latter involves energy conversion characterized by a certain *process efficiency*.

Exergy can also be formally defined as a property of a system *and* environment through a notion of availability functions, but in our interpretation its quantity is expressed through a concept of energy interaction that would be manifested during a system's change of state from a given state to the state of thermodynamic equilibrium with the selected reference system (thermodynamic surroundings).

Energy does not carry a notion of quality, as opposed to exergy. In a process, energy "flow" may be arbitrarily complex, but at the ports of any imaginable subsystem, it obeys the conservation principle. Note that the colloquial metaphor "energy flow" signifies, at any identified boundary, either a simultaneous change of the property called energy on the both sides of the boundary (with different signs so that it would be interpreted as flow because of the conservation principle) or alternatively the energy interaction.

Figure 2.6 illustrates this point for a hypothetical system. Such a Sankey (energy) diagram does not offer any insight into the quality level of individual energy flows. 1 kW of the low-temperature "waste heat" dumped into the environment appears to be equal to 1 kW of electrical power in such representation. This difference can be identified only in terms of exergy. The inset in Fig. 2.6 extracts a domain of energy-resource conversion and illustrates this difference schematically. The energy flow (Sankey diagram segment extracted) is characterized by the conservativeness of energy. The exergy flow (Grassman diagram segment) though reveals quite different picture that allows for exergy destruction and clearly illustrates apparent available energy differences versus the corresponding energy flows.

2.6 Nomenclature

Note: The symbols for the variables included in Tables 2.1 and 2.2 are also provided in corresponding footnotes.

b	Availability function, J/kg
c	Specific heat, J/kg K
E	Energy, J
e	Specific energy, J/kg

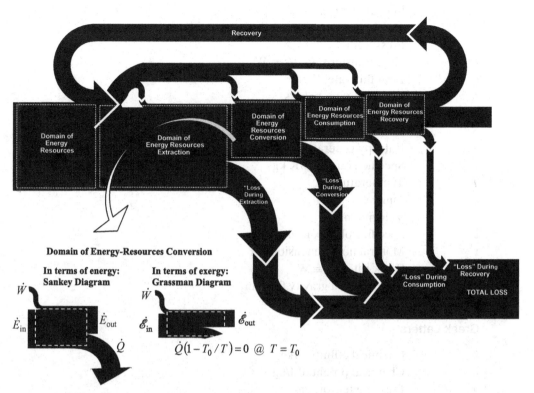

Figure 2.6. A generic (not to scale) Sankey diagram of an energy flow, starting with the energy-resources extraction and ending with the energy recovery (a fictitious system). The energy flow through extraction, conversion, and consumption to the recovery, including dissipation in form of various losses, features the principle of conservation (total IN equals total OUT – at any level of complexity). The exergy flow though can be reduced literally to zero because of the available energy (exergy) destruction, see the inset: Exergy flows of corresponding energy flows may feature radically different magnitudes depending on energy quality, including a complete destruction. Input equals output for energy flows, but not for exergy flows. The exergy equivalent of a high-quality energy interaction (such as work rate or electrical power) is identical to the corresponding energy rate, but the exergy equivalent of an energy interaction in the form of a low-quality heat transfer rate dumped at the temperature of the environment to surroundings is equal to zero; see Eqs. (2.13) and (2.24). The exergy rates of bulk mass flow rates differ from the corresponding energy rates; see Eq. (2.20).

$\mathscr{E}$	Exergy, J
e	Specific exergy, J/kg
$\mathbf{f}$	Nonconservative body force per unit mass, N/kg
g	Gibbs free energy, J/kg
g	Acceleration of gravity, 9.81 m/s^2
$\mathbf{g}$	Body force per unit mass, N/kg
h	Specific enthalpy, J/kg
i, n, m	index
$\mathbf{j}$	Mass diffusion flux (with respect to mass average velocity), kg/m^{2}s
k	Thermal conductivity, W/mK
$\mathbf{n}$	Mass diffusion flux (with respect to stationary coordinates), kg/m^{2}s
M	Mass of the system, kg

$\dot{m}$	Mass flow rate, kg/s
$\dot{N}$	Number of moles per unit of time, mol/s
p	Pressure, Pa
$\dot{Q}$	Heat transfer rate, J/s $\equiv$ W
$\mathbf{q}$	Heat flux rate, W/m^2
R	Reaction rate, kg/m^{3}s
$\boldsymbol{r}$	Chemical species production rate, mol/s
$\dot{S}$	Entropy rate, J/Ks
$\dot{S}_{gen}$	Entropy generation, J/Ks
s	Specific entropy, J/K kg
T	Temperature, K
t	Time, s
$\mathbf{v}$	Velocity, m/s
v	Specific volume, m^3/kg
x	Mol fraction, dimensionless
$\dot{W}$	Work rate, J/s $\equiv$ W
z	Location in a gravity field, m

Greek Letters

λ	Chemical affinity, J/kg
μ	Chemical potential J/kg
$\boldsymbol{\pi}_n$	Pressure tensor, Pa
ρ	Density, kg/m^3
$\boldsymbol{\sigma}$	Entropy flux, J/K m^{2}s
$\boldsymbol{\tau}_{s,n}$	Shear/normal stress tensor, Pa
χ	Mass fraction, dimensionless
ϕ	Potential exergy per unit mass, J/kg
Ω	Denotes a system, dimensionless
ω	Denotes the component constituents, dimensionless

Superscripts

A, B	Denotes systems A, B
ch	Chemical
in	Inlet port
out	Outlet port
ph	Physical
*	Restricted dead state
$-$	Molal quantity
$\leftarrow$	Signifies a direction of an interaction
$\uparrow$	Signifies the change of a property

Subscripts

1, 2	Denoting State 1 or State 2 or a path from 1 to 2
0	Denoting environmental or reference state

actual   Nonideal, real
adiabat   Adiabatic
air   Denoting air
Con   Conveyer
ch   Chemical
Effluents   Associated with effluents
El   Electrical
gen   Generation
H_2O Denoting water
i, j, k, n, m, r Denoting an index
k Kinetic energy
lost   Not utilized
max   Maximum, reversible
Nitrogen   Denoting N_2
n Normal (stress)
Product   Associated with a product
Q Heat
System   Denotes a system (A, B, ...)
s Strain
t Thermal
W Work

Acronyms

CAB   Controlled atmosphere brazing
C.V., CV   Control volume (a system boundary)
ECMF   Energy conversion for manufacturing
ECMA   Energy conversion for materials processing
ECRE   Energy conversion for recycling
ECRM   Energy conversion for raw material extraction
MF   Manufacturing
MA   Materials processing
RE   Recycling
RM   Raw material

REFERENCES

[1] E. P. Gyftopoulos and G. P. Beretta, *Thermodynamics, Foundations and Applications* (Dover, Mineola, NY, 2001).
[2] M. J. Moran, *Availability Analysis: A guide to efficient energy use* (ASME Press, New York, 1989).
[3] L. Tisza, *Generalized Thermodynamics* (MIT, Cambridge, MA, 1960).
[4] G. N. Hatsopoulos and J. E. Keenan, *Principles of General Thermodynamics* (Wiley, New York, 1965).
[5] I. Müller and W. Weiss, *Entropy and Energy* (Springer, Berlin, 2005).
[6] M. J. Moran and H. N. Shapiro, *Fundamentals of Engineering Thermodynamics*, 3rd ed. (Wiley, New York, 1995).
[7] Y. A. Cengel and M. A. Boles, *Thermodynamics. An Engineering Approach*, 6th ed. (McGraw-Hill, Boston, 2008).

[8] G. Falk and H. Jung, Axiomatik der Thermodynamik [Axiomatic of Thermo-dynamics, in German], in *Handbook der Physik*, edited by S. Flüge (Springer, Berlin, 1959), Vol. 3/2.

[9] M. P. Vukalovitch and I. I. Novikov, *Termodinamika* [Thermodynamics, in Russian] (Mashinostroyenie, Moscow, 1972).

[10] H. B. Callen, *Thermodynamics* (Wiley, New York, 1985).

[11] E. F. Obert and R. A. Gaggioli, *Thermodynamics* (McGraw-Hill, New York, 1963).

[12] T. J. Kotas, *The Exergy Method of Thermal Plant Analysis* (Krieger, Melbourne, Fl, 1995).

[13] A. Bejan, *Advanced Engineering Thermodynamics* (Wiley, Hoboken, NJ), 2006.

[14] A. Bejan, G. Tsatsaronis, and M. Moran, *Thermal Design & Optimization* (Wiley, New York, 1996).

[15] J. Szargut, D. R. Moris, and F. R. Steward, *Exergy analysis of Thermal, Chemical and Metallurgical Processes* (Hemisphere, New York, 1988).

[16] G. N. Hatsopoulos and J. H. Keenan, *Principles of General Thermodynamics* (Wiley, New York, 1965).

[17] E. F. Obert, *Concepts of Thermodynamics* (McGraw-Hill, New York, 1960).

[18] J. Kestin, *A Course in Thermodynamics* (Hemisphere, Washington, D.C., 1979).

[19] H. B. Callen, *Thermodynamics and an Introduction Thermostatics* (Wiley, New York, 1985).

[20] I. Müller and W. H. Müller, *Fundamentals of Thermodynamics and Applications* (Springer, Berlin, Heidelberg, 2009).

[21] F. Bosnjakovic, *Technical Thermodynamics* [in German] (Verlag Theodor Steinkopff, Dresden, 1972).

[22] S. Lloyd, A. Bejan, C. Bennett, G. P. Beretta, H. Butler, L. Gordon, M. Grmela, E. P. Gyftopoulos, G. N. Hatsopoulos, D. Jou, S. Kjelstrup, N. Lior, S. Miller, M. Rubi, E. D. Scheider, D. P. Sekulić and Z. Zhang, "Discussion on 'Frontiers of the Second Law,'" in *Meeting the Entropy Challenge*, edited by G. P. Beretta, A. F. Ghoniem, and G. N. Hatsopoulos (American Institute of Physics, Melville, NY, 2008), AIP 1033, pp. 253–261.

[23] G. Alefeld, "Probleme mit der Exergie," *BWK* **40**, 72–80 (1988).

[24] H. D. Baehr, "Probleme mit der Exergie?," *BWK* **40**, 450–457 (1988); G. Alefeld, "Die Exergy und der II. Hauptsatz der Thermodynamik," *BWK* **40**, 458–464 (1988).

[25] J. H. Keenan, "Availability and irreversibility in thermodynamics," *Br. J. Appl. Phys.* **2**, 183–192 (1951).

[26] R. W. Haywood, "A critical review of the theorems of thermodynamic avail-ability, with concise formulations," Part 1. Availablity, Part 2. Irreversibility, *J. Mech. Eng. Sci.* **16**, 160–173 and 258–267 (1974).

[27] M. J. Moran, *Availability Analysis: A Guide to Efficient Energy Use* (ASME Press, New York, 1989).

[28] A. Bejan, *Entropy Generation Through Heat and Fluid Flow* (Wiley, New York, 1982).

[29] E. Sciubba and S. Ulgiati, "Energy and exergy analyses: Complementary meth-ods or irreducible ideological options," *Energy* **30**, 1953–1988 (2005).

[30] J. L. Hau and B. R. Bakshi, "Promise and problems of emergy analysis," *Ecol. Model.* **178**, special issue in honor of H.T. Odum. 215–225 (2004).

[31] E. P. Gyftopoulos, "Fundamentals of analyses of processes," *Energy Convers. Mgmt.* **38**, 1525–1533 (1997).

[32] D. P. Sekulić, "A fallacious argument in the finite time thermodynamics concept of endoreversibility," *J. Appl. Phys.* **83**, 4561–4565 (1998).

[33] E. P. Gyftopoulos, "Infinite time (reversible) versus finite time (irreversible) thermodynamics: A misconceived distinction," *Energy* **24**, 1035–1039 (1999).

[34] L. Chen, C. Wu, and F. Sun, "Finite time thermodynamics optimization or entropy generation minimization of energy systems," *J. Non-Equilib. Thermodyn.* **24**, 327–359 (1999).

[35] G. Guillaume and E. Guillaume, *Sur le fundements de l'economique rationelle* (Gautier-Villars, Paris, 1932).

[36] J. L. McCaulei, "Thermodynamic analogies in economics and finance: Instability of markets," *Physica A* **329**, 199–212 (2003).

[37] T. Gutowski, M. Branham, J. Dhamus, A. Jones, A Thiries, and D. Sekulić, "Thermodynamic analysis of resources used in manufacturing processes," *Environ. Sci. Technol.* **43**, 1584–1590 (2009).

[38] R. U. Ayres, L. W. Ayres, and K. Martinas, "Exergy, waste accounting, and life cycle analysis," *Energy* **23**, 355–363 (1998).

[39] R. U. Ayres and B. Warr, "Accounting for growth: The role of physical work," *Structural Change Econ. Dyn.* **16**, 181–209 (2005).

[40] E. Sciuba, "Beyond thermoeconomics? The concept of Extended Exergy Accounting and its application to the analysis and design of thermal systems," *Exergy, Intl. J.* **1**(2), 68–84 (2001).

[41] J. L. Hau and B. R. Bakshi, "Expanding exergy analysis to account for ecosystem products and services," *Environ. Sci. Technol.* **38**, 3768–3777 (2004).

[42] S. E. Jorgensen and H. F. Majer, "Ecological buffer capacity," *Ecol. Model.* **3**, 39–45 (1977).

[43] L. Susani, F. M. Pulselli, S. E. Jorpensen, and S. Bastianoni, "Comparison between technological and ecological exergy," *Ecol. Model.* **193**, 447–456 (2006).

[44] R. P. Feynman, R. B. Leighton, and M. Sands, *The Feynman Lectures of Physics* (Addison-Wesley, Reading, MA, 1989), Vol 1, 4–1.

[45] E. Nagel, *The Structure of Science, Problems in the Logic of Scientific Explanation* (Harcourt, Brace & World, New York, 1961), p. 49.

[46] W. R. Dunbar, N. Lior, and R. A. Gaggioli, "The component equations of energy and exergy," *J. Energy Resources Technol.* **114**, 75–83 (1992).

[47] J. O. Hirschfelder, C. F. Curtiss, and R. B. Bird, *Molecular Theory of Gases and Liquids* (Wiley, New York, 1954).

[48] R. B. Bird, W. E. Stewart, and E. N. Lightfoot, *Transport Phenomena* (Wiley, New York, 2002).

[49] N. Lior, W. Sarmiento-Darkin, and H. S. Al-Sharqawi, "The exergy fields in transport processes: Their calculation and use," *Energy* **31**, 553–578 (2006).

[50] J. Ahrendts, *Die Exergie chemisch reactionfähiger Systeme* [*Exergy of Chemically Reacting Systems*], VDI-Forschungsheft No. 579, (VDI Verlag, 1977, pp. 26–33.

ADDITIONAL READINGS

Note: References listed here illustrate the breadth of the literature on the crossroads between thermodynamics and other fields relevant for resources analysis involving engineering, social sciences, ecology, and economy.

R. U. Ayres, L. W. Ayres, and K. Martinas, "Exergy, waste accounting, and life cycle analysis," *Energy* **23**, 355–363 (1998).

A. Bejan and E. Mamut, *Thermodynamic Optimization of Complex Energy Systems* (Kluwer, Dordrecht, The Netherlands 1998).

R. B. Evans, "Thermoeconomic isolation and essergy analysis," *Energy* **5**, 805–821 (1980).

N. Georgescu-Roegen, *The Entropy Law and the Economic Process* (Harvard University Press, Cambridge, MA, 1971).

P. Glandsdorff and I. Prigogine, *Thermodynamic Theory of Structure, Stability and Fluctuations* (Wiley Interscience, London, 1971).

B. Linnhoff et al., *A User Guide on Process Integration for the Efficient Use of Energy* (The Institution of Chemical Engineers, Rugby, Warwickshire, England, 1982); B. Linnhoff, "Pinch analysis and exergy – a comparison," in *Energy Systems and Ecology*, edited by J. Szargut (ASME, New York, 1993), Vol. 1, pp. 43–52.

A. J. Lotka, "Contribution to the energetics of evolution," *Proc. Natl. Acad. Sci. USA* **8**, 147–151 (1922).

J. Mimkes, "Binary alloys as a model for the multicultural society, *J. Thermal Anal.* **43**, 521 – 537 (1995).

I. Müller and W. Weiss, *Entropy and Energy, A Universal Competition* (Springer, Berlin, Heidelberg, New York, 2005).

H.T. Odum, "Self-organization, transformity, and information," *Science* **242**, 1132–1139 (1988).

E. Schrödinger, *What is Life? The Physical Aspects of the Living Cell* (Cambridge University Press, Cambridge, 1944).

R. K. Shah and D. P. Sekulić, *Fundamentals of Heat Exchanger Design* (Wiley, New York, 2003), Chap. 11.

A. Tsuchida and T. Murota, "Fundamentals in the entropy theory of ecocycle and human economy," in *Environmental economics: The Analysis of a major interface*, edited by. G. Pillet and T. Murota (R. Leimgruber, Geneva 1987), pp. 11–35.

3 Accounting for Resource Use by Thermodynamics

Bhavik R. Bakshi, Anil Baral, and Jorge L. Hau

3.1 Motivation

Given the crucial role played by natural resources in any economic activity, many efforts have focused on accounting for their contribution to individual industrial processes and broader life cycles. These include traditional engineering approaches such as pinch [1, 2], exergy (see Chap. 2) and exergoeconomic analysis (see Chap. 15 by Tsatsaronis), that are used extensively for equipment and process design. Because environmentally conscious decisions require consideration of a broader boundary that accounts for the whole life cycle, many of these approaches have also been extended from the equipment to the life-cycle scale.

Most methods that analyze the life cycle of industrial products focus on emissions and their impact and the use of nonrenewable resources such as fossil fuels and minerals. More recent efforts also account for the role of land, soil, etc. [3]. Although sophisticated methods have been developed and standardized for assessing the life-cycle impact of emissions [4, 5], methods for resource accounting have been lagging behind. Even with many *resource aggregation* methods developed over the past several decades, consensus about the most appropriate method is still nonexistent. In addition, most resource accounting methods fail to consider some of the most critical resources, namely, *ecosystem goods and services*, that sustain all life on Earth. This "*natural capital*" includes provisioning services such as minerals, fossil fuels, water, biomass; supporting services such as biogeochemical cycles, carbon sequestration, pollination; regulating services such as regulation of pests, climate, and soil fertility; and cultural services. Studies such as the recent *Millennium Ecosystem Assessment* [6] have found that human beings have altered the ecosystem more rapidly and extensively in the past 50 years than in any comparable period in human history. Among the 24 ecosystem services the study examined, it found that 15 are being degraded or used unsustainably. Unfortunately, most existing methods for resource accounting consider only some provisioning services. Ignoring other ecosystem services could encourage perverse decisions that may cause depletion of the ignored services.

Some of the challenges facing *resource accounting* include the following:

- The diverse array of ecological resources that support human activities is represented in a variety of units, making it difficult to compare and combine their roles.

- The quality of resources can be vastly different. For example, sunlight is a very dilute source of energy, whereas fossil fuels are significantly more concentrated. This is reflected in the versatility of fossil fuels and the efficiency of converting them into products such as electricity. Furthermore, the quality of material and energy resources can also be very different and difficult to compare.
- Vast differences in *resource quality* make it difficult to devise approaches for their aggregation or comparison, because these activities implicitly assume the aggregated or compared resources to be substitutable.
- Accounting for the role of ecosystem goods and services requires a boundary that includes *ecosystems*. Inadequate knowledge and understanding about these systems and their resource networks make this task very challenging.

This chapter provides an overview of and insight into various approaches developed to quantify the consumption of resources over a life cycle, with emphasis on thermodynamic methods. It introduces approaches such as net energy, cumulative exergy, and emergy analysis, along with their strengths and weaknesses. Because some of these methods were developed in different disciplines, this chapter also explains the relationship among these methods. The ability of each approach to address the challenges listed in the previous paragraph is described and illustrated by means of examples. The rest of this chapter is organized as follows. Section 3.2 contains a brief introduction to existing approaches for resource accounting. Section 3.3 discusses controversial aspects of the thermodynamic methods. This includes the implicit assumptions about energy quality and *substitutability* of resources in the calculation of aggregate metrics and controversy and misunderstanding surrounding the concept of emergy. Section 3.4 introduces the concepts of industrial cumulative exergy consumption (ICEC) and ecological cumulative exergy consumption (ECEC) and uses them to shed insight into the relationship among existing approaches described in Section 3.2. Section 3.5 illustrates the methods by means of application to the life cycle of corn ethanol. Finally, Section 3.6 presents a summary and discusses areas where further research could be directed.

3.2 Background

This section summarizes existing resource accounting methods with emphasis on those based on thermodynamic concepts. Material flow analysis (MFA) has also been popular for quantifying resource use in mass units. Such studies have been completed for specific products [7] and entire nations [8]. Resource use is quantified by aggregate indicators such as total material requirement, which adds the mass of all resources except air, water, and agricultural tillage. The material intensity per unit service is defined as the cumulative material consumption per unit of product or service [9]. Representing resources in terms of their mass provides an idea of the use of some resources, but it does not address the challenges listed in Section 3.1. In particular, many resources such as land and sunlight cannot be represented in units of mass, and even for other resources their equal masses need not imply substitutability. In addition, mass need not capture its useful properties or quality because a ton of sand and a ton of natural gas have very different availabilities, uses, and impacts. Nevertheless, the data collected for MFA can be useful for other methods.

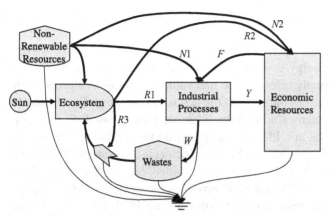

Figure 3.1. Network of energy flows beween ecosystems, nonrenewable resource storages, industrial processes, and the economy.

Resource accounting approaches used with *life-cycle impact assessment* methods include surplus energy [10] and abiotic depletion potential [11]. The former quantifies resource availability by means of the energy required for extracting the resource for future generations, whereas the latter is related to the ratio of present use to its reserves. These approaches suffer from practical challenges such as the need to know future energy requirements for extracting resources and the quantity of reserves. Furthermore, both approaches are best for nonrenewable resources and do not consider ecosystem goods and services.

Monetary valuation of ecosystem services is increasingly popular [12–14] and usually relies on valuation methods from environmental and ecological economics. Despite the appeal of monetary valuation for communication with general audiences, this approach is fraught with challenges such as subjectiveness. Furthermore, such efforts usually have a narrow focus on only the direct use of resources and lack a life-cycle view. Also, proper valuation requires knowledge about resources that can only be provided by biophysical methods.

The rest of this section describes methods based on energy and exergy under the premise that they are better suited for resource accounting because of their rigor and ability to account for a larger variety of resources than other methods. In addition, exergy is the ultimate limiting resource, so using this concept for quantifying resources can be more meaningful. Figure 3.1 shows a general diagram that is used for understanding the metrics used in all the approaches discussed in this chapter. The block "industrial processes" represent the processes or life cycle being analyzed. Direct inputs of nonrenewable and renewable resources are indicated as $N1$ and $R1$, respectively. These resources are also used by other processes in the economy and are shown as $N2$ and $R2$. Resources used from the economy to support the selected processes are shown as F. The main products from the selected processes are sold to the economy, shown as Y, and the wastes as W. Additional inputs from nature and the economy may be needed to treat the wastes or absorb their impact, as shown. All, or a part of, the resources, $N1$ and $R1$, are usually converted into the product and form the "feedstock," represented by the subscript f. The rest of the resources are used for converting the feedstock into the product and are referred to as "processing" resources, indicated by the subscript p.

3.2.1 Net Energy Analysis

Net energy analysis or *full fuel cycle analysis* accounts for the *cumulative energy consumption* (CEnC), or energy consumed directly and indirectly for making a product. Here, "*energy*" usually refers to the fuel value of various resources, with emphasis usually only on nonrenewable fossil fuels. It is among the earliest methods with a life-cycle view, and many studies have been completed based on selecting the most important processes in a life cycle or based on more comprehensive but aggregate models representing interaction between sectors in an economy. In current parlance, the former approach is akin to a *process-model-based life-cycle analysis* (LCA), whereas the latter is analogous to an *economic-model-based LCA*. Input–output models for energy analysis were developed in the 1970s and used widely for evaluating many technologies and studying energy flows in economic systems [15–18]. A historical overview of these efforts is provided in Chap. 14 by Hannon. Most net energy analyses focus on the consumption of nonrenewable fuels including coal, oil, and natural gas. Recently there have been several process-model-based energy analyses, particularly of transportation fuels from biomass [19, 20]. Some studies have also accounted for renewable energy inputs. For example, Costanza [21] and Costanza and Herendeen [22] included the contribution of sunlight to various economic sectors and explored correlations between embodied energy intensities of economic goods and services and their economic prices. Similar studies based on a hybrid model that combines process and economic models have also been completed [23] and are discussed in Section 3.6.

A variety of metrics has been used for analyzing the results of energy analysis. *Net energy* is defined as the difference between the fuel value of the desired product and the processing energy needed to convert the feedstock to the product. In terms of Fig. 3.1, it may be represented by the following equation (only nonrenewable energy resources are included):

$$E_{\text{net}} = \text{fuel value of product} - \text{cumulative processing energy}$$

$$= E_Y - C_{Y,p}^E. \tag{3.1}$$

Here $C_{Y,p}^E = C_{N1,p}^E + C_{N2,p}^E$, and it is assumed that the processing energy consists of nonrenewable resources that are used to make the economic resources, F. The cumulative energy consumed for processing, C_{N2}^E, may be written as

$$C_{N2}^E = \sum_i E_{N2,i}, \tag{3.2}$$

where i indicates the ith nonrenewable resource. The *energy return on investment* (EROI) is defined as

$$EROI = \frac{\text{fuel value of products}}{\text{cumulative processing energy}}$$

$$= \frac{E_Y}{C_{Y,p}^E}, \tag{3.3}$$

and the *energy efficiency* is

$$\eta = \frac{\text{fuel value of products}}{\text{cumulative energy}}$$

$$= \frac{E_Y}{C_Y^E}. \tag{3.4}$$

The denominator in Eq. (3.4) represents the cumulative fuel value of fossil resources consumed from all inputs, including the feedstock, and may also be written as $C_Y^E = C_{N1}^E + C_{N2}^E$. EROI and net energy are often used as indicators to determine the feasibility of a product. Net energy is analogous to an energy profit, whereas EROI is analogous to a monetary ROI. Thus the net energy should be positive and EROI should be larger than one for an energetically feasible product.

Despite the popularity of this approach, it is quite limited in its ability to meet the challenges mentioned in Section 3.1. Net energy analysis is best for quantifying the role of fossil fuels and cannot quantify the role of nonfuel materials. Although renewable resources may also be included in this approach, the vast difference in their qualities can skew the results and be misleading, as illustrated in Section 3.6. Even the quality of fossil fuels considered in energy analysis is different, and aggregating them by means of the indicators described in this paragraph can be misleading, as discussed in more detail in Section 3.5. Finally, this approach does not account for the role of nature in sustaining human activities.

3.2.2 Cumulative Exergy Consumption Analysis

Cumulative exergy consumption (CExC) analysis is similar to energy analysis, but uses exergy instead of energy. The primary benefits of using exergy are that both material and energy resources can be quantified, and some differences in quality may be captured because exergy represents only the maximum energy available for doing work. Consequently all the equations in the previous section may also be used for exergy analysis by replacing energy E with exergy $\mathscr{E}$ and C^E with $C^{\mathscr{E}}$. The CExC of a product is the sum of the exergy of all the natural resources consumed directly and indirectly in the life cycle [24]:

$$C_Y^{\mathscr{E}} = \sum_i \sum_j (\mathscr{E}_{Ri,j} + \mathscr{E}_{Ni,j}). \tag{3.5}$$

Here $\mathscr{E}_{Ri,j}$ and $\mathscr{E}_{Ni,j}$ refer to the direct and indirect exergy content, respectively, of renewable and nonrenewable resources. Metrics analogous to those defined for net energy analysis in Subsection 3.2.1 may also be defined and calculated for CExC analysis. Efficiency in CExC analysis is often called the *cumulative degree of perfection* (CDP) [24], which is defined as

$$\eta_{\mathscr{E},Y} = \frac{\mathscr{E}_Y}{C_Y^{\mathscr{E}}}. \tag{3.6}$$

CExC also enables calculation of additional metrics that indicate the renewability of products. Examples of such metrics include the *exergy breeding factor*, defined as the ratio of the exergy of the product to the nonrenewable CExC [25] and the *renewability indicator* [26].

CExC is able to account for a wider diversity of resources than CEnC because it can consider material and energy resources. However, when renewable and non-renewable resources are considered together, the former tend to dominate and the calculated metrics can be misleading, as illustrated in Section 3.6 and [23, 27]. This is due to the inability of CExC to account for quality differences between resources, as discussed in Section 3.5. The cumulative exergy extracted from the natural environment approach [27] aims to address this issue by accounting for only the used exergy, which for sunlight to biomass is about 2%. However, like energy analysis, this approach also does not account for the role of nature. Other extensions of CExC include extended exergy analysis [28] for including the role of human labor and activities.

3.2.3 Emergy Analysis

This approach has been developed by systems ecologists based on their understanding of the role and transformation of energy in ecological systems. In recent years, it has also been applied to many industrial products and processes with the intention of encouraging ecologically conscious decision making. *Emergy*, specifically solar emergy, is "the available solar energy used up directly and indirectly to make a service or product" [29]. Emergy analysis considers all systems, industrial and ecological, to be networks of exergy flow and determines the emergy value of the streams and systems involved. It characterizes all products and services in equivalents of solar energy, that is, how much energy would be needed to do a particular task if solar radiation were the only input. It considers the Earth to be a closed system with solar energy, deep Earth heat, and tidal energy as major constant energy inputs and that all living systems sustain one another by participating in a network of energy flow by converting lower-quality energy into both higher-quality energy and degraded heat energy. Because solar energy is the main energy input to the Earth, all other energies are scaled to solar equivalents to give common units. Other kinds of energy existing on the Earth can be derived from these three main sources through energy transformations. Consequently, emergy is measured in solar embodied or equivalent joules, abbreviated sej.

Emergy analysis characterizes products and systems by means of their solar transformity or transformation ratio, which is defined as "the solar emergy required to make one Joule of a service or product" [29]. *Solar transformity* is measured in solar embodied joules (sej/J). The solar transformity of a product is its solar emergy divided by its available energy, that is,

$$\mathcal{M} = \mathcal{T}_{\mathscr{E}}, \tag{3.7}$$

where $\mathcal{M}$ is emergy, $\mathcal{T}$ is transformity, and $\mathscr{E}$ is available energy. Because solar energy is the baseline of all emergy calculations, transformity of solar energy is unity. Odum [29] argues that the "energy flows of the universe are organized in an energy transformation hierarchy" and that "the position in the energy hierarchy is measured with transformities." Therefore transformity is regarded as an indicator of energy quality. From a practical point of view, transformity is useful as a convenient way

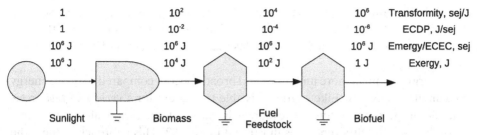

Figure 3.2. Exergy flow and transformation in a food or industrial chain.

of determining the emergy of commonly used resources and commodities. These concepts from emergy are illustrated in Fig. 3.2, which shows a simple industrial chain for transforming solar energy into biomass, fuel feedstock, and fuel. It shows the transformation of a million joules of sunlight into 1 J of fuel. This chain implies that 1 J of fuel is equivalent to 10^2 J of feedstock, 10^4 J of biomass, or 10^6 J of sunlight, resulting in the transformities as shown. Such an approach can account for differences in the quality of multiple resources, as discussed in more detail in Section 3.5.

Emergy analysis has metrics similar to those used in net energy and cumulative exergy analysis, except that all the flows are represented in terms of emergy. *Net emergy*, $\mathcal{M}_{\text{net}}$, is the emergy gained by the economy in exchange for providing its services. It is analogous to net energy, defined in Eq. (3.1), except that the product is represented in terms of its emergy, not energy:

$$\mathcal{M}_{\text{net}} = \mathcal{M}_Y - \mathcal{M}_F. \tag{3.8}$$

The *emergy yield ratio* (EYR) of products is the ratio of the emergy of the product to the emergy used from the economy:

$$\text{EYR} = \frac{\mathcal{M}_Y}{\mathcal{M}_F}. \tag{3.9}$$

The EYR is analogous to the ROI, but because the numerator is emergy instead of energy or exergy, EYR is always greater than unity because $\mathcal{M}_Y > \mathcal{M}_F$.

Emergy metrics have also been defined to evaluate the performance of industrial activity [30]. The *environmental loading ratio* (ELR) is defined as the sum of the feedback emergy from the economy and emergy from nonrenewable resources divided by the emergy from renewable resources. This index is an indicator of the stress on the local environment and is defined as

$$\text{ELR} = \frac{(\mathcal{M}_F + \mathcal{M}_{N1})}{\mathcal{M}_{R1}}. \tag{3.10}$$

Brown and Ulgiati [30] modified this metric to consider the ecosystem services needed to dissipate the emissions. However, it ignores the actual impact that is due to emissions such as dead trees that are due to acid rain [31]. The *sustainability index* (SI) is defined as the ratio of the EYR to the ELR. This index evaluates the ecological–economic-integrated performance of the activity:

$$\text{SI} = \frac{\text{EYR}}{\text{ELR}}. \tag{3.11}$$

Detailed calculations about transformities for various ecosystem goods and services are available in Odum [29, 32]. These include Earth's main processes such as the total surface wind, rainwater in streams, Earth's sedimentary cycle, and waves absorbed on shore.

Emergy is a much more ambitious and broad concept compared with net energy and cumulative exergy. Like exergy, it is able to account for a variety of resources. In addition, by representing all resources in terms of solar equivalents, it is able to consider quality differences, as illustrated in Fig. 3.2. This approach reduces the contribution of dilute resources such as sunlight compared with fossil fuels, because the latter are found to be equivalent to about 40,000 sej [29, 33]. Consequently, as illustrated in Section 3.6, the results from emergy analysis can be closer to existing intuition when renewable and nonrenewable resources are considered together. The fact that emergy analysis accounts for the exergy flow in ecosystems means that this approach is able to account, at least partially, for the role of ecosystem goods and services [34]. Despite these attractive features, this approach does face a set of unique challenges.

Representing all resources in solar equivalents is a challenging task, particularly because the network of industrial and ecological systems as well as their products is often only partially known [35, 36]. Emergy analysis relies on the best available knowledge about exergy flows in ecosystems [37–39], but cannot yet account for some ecosystem resources such as pollination, biodiversity, and many regulating services, and much is still not known about ecological networks. Consequently the emergy and transformity numbers are likely to be quite uncertain, and information about this uncertainty is not yet available. The lack of complete information about ecological networks means that emergy analysis needs to use unique *allocation methods*, which have been misunderstood and are controversial, as discussed in more detail in Section 3.3.

Emergy analysis has also been used to quantify the contribution from human resources and economic inputs. Contribution from human resources is calculated by means of information about the emergy of consumption in society and income [29, 40]. Similarly, the emergy of economic inputs is estimated by multiplying the monetary value of the inputs by the ratio of the nation's [29] or economic sector's [40] emergy to its economic throughput. Similar ideas have also been proposed in an extension of CExC called *extended exergy analysis* [28].

In addition, Odum and co-workers made claims about the relevance of emergy analysis to various nonecological or technological systems.

- Emergy is claimed to be the true measure of wealth because it reflects the contribution of goods and services that are essential for human well-being. Usually, money ignores the value of natural capital by treating the environment as being outside the market. Emergy is more comprehensive than money and is claimed to be a more complete measure of wealth. It is suggested as an ecocentric view of human activities, and even a substitute for money [41]. Such claims have a long history [18, 36, 42, 43], but have been rejected by mainstream economics because they ignore human preferences and focus on only the supply side.
- The *maximum empower principle* claims that all self-organizing systems tend to maximize their rate of emergy use or empower [29, 44]. This is based on

the work of Boltzmann [45] and Lotka [46], and some self-organizing systems have been shown to follow this principle [44]. However, there has not been enough evidence to support the generality of this claim, making this principle controversial [35], and its one-dimensional nature is considered by some to be highly simplistic [47].

- The relationship between emergy and other thermodynamic properties has not been clear until recently, causing confusion and rejection. It has led to impressions of emergy being a "very different approach" compared with exergy analysis [48], and questions about the need for emergy when conventional thermodynamic variables such as exergy are available [35]. Sciubba claims that emergy ignores the second law [49]. Recently, the relationship between exergy and emergy has been proved [50, 51], as discussed in more detail in Section 3.4.

Further discussion about the promise and problems of emergy analysis is in [52].

Despite these controversial aspects of emergy analysis, it is the only biophysical approach that overcomes the inability of many existing approaches to consider the contribution of ecological processes to human activities. The importance of accounting for nature's services is gaining wide acceptance [6, 53–55]. Claims about the ability of emergy to account for the quality of resources are appealing but need to be validated by means of empirical studies. Methods based on economics have been suggested for quantifying the quality of energy, and such approaches have been claimed to be most appropriate for capturing human preferences as well as energy quality. Recent studies of life-cycle resource accounting compare various methods and point toward the pros and cons of the methods discussed in this section and other approaches, including accounting based on mass and monetary weighting factors [23]. These results indicate the complementarity of various methods and the benefits of using multiple approaches. If the goal is to obtain a single indicator for situations in which a large variety of resources are consumed, emergy or ECEC seems to provide the most meaningful indicator. More research is needed to gain greater insight into different resource accounting methods. The relationship between exergy and emergy is discussed in more detail in Section 3.4.

3.3 Allocation

Allocation issues arise in most accounting methods when a process or production chain generates two or more products. For instance, if a process separates air to produce oxygen and nitrogen, an allocation method is needed to determine what fraction of the energy or money invested in the process should be attributed to oxygen and what fraction to nitrogen production. This task is far more complicated than it may seem at first, particularly if each product cannot be produced exclusively and independently of the others with only a fraction of the total energy input. Allocation is an artificial accounting tool to measure the value or cost of things and tends to be a controversial subject in many accounting problems. In LCA, various approaches have been suggested such as allocation in proportion to monetary value or physical units such as mass, energy, or exergy. Alternatively, allocation may be avoided completely by expanding the analysis boundary and considering substitution by the by-products [56]. Similar methods have also been used in energy, exergy, and

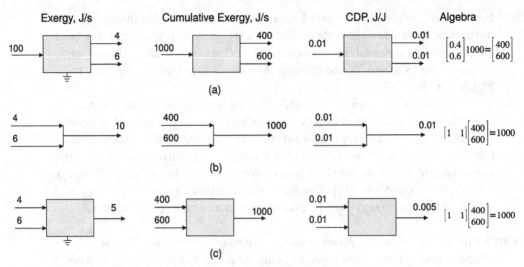

Figure 3.3. Allocation in cumulative energy and cumulative exergy analyses.

emergy analysis, as described in this section. Many studies have shown that, often, the conclusion of a study comparing alternative products is not sensitive to the allocation approach.

The *network algebra* for net energy and cumulative exergy analysis is based on partitioning the CEnC or CExC in an input stream between multiple-output streams in proportion to the energy or exergy of the output streams. Figure 3.3(a) illustrates allocation in proportion to the energy content. In this case, the CEnC of the ith product may be calculated by distributing the C_N^E of the input resources as

$$C_{Y,i}^E = \frac{E_{Y,i}}{\sum_i E_{Y,i}} C_N^E. \tag{3.12}$$

Figure 3.3b illustrates the algebra used for combining the CEnC of multiple streams. As shown, combining the streams simply involves adding the cumulative consumption of the relevant streams.

In general, the relationship between CEnC of the products, represented by the vector $\mathbf{C}_Y^E$, and CEnC of the inputs, $\mathbf{C}_N^E$, may be written as,

$$\mathbf{C}_Y^E = \mathbf{\Gamma}_i \, \mathbf{C}_N^E, \tag{3.13}$$

where $\mathbf{\Gamma}_i$ is an allocation matrix [50]. This matrix represents the flow network and the selected allocation method. Typical examples are shown in Fig. 3.3. Figure 3.3a shows allocation in proportion to the exergy of the product streams. Here, $C_N^E = 1000$, $\mathbf{\Gamma}_i = [0.4 \ 0.6]^T$, and $\mathbf{C}_Y^E = [400 \ 600]^T$. Note that metrics such as the CDP do not change for each stream, and when the products are combined by direct mixing, as shown in Fig. 3.3(b), or by some other process, as shown in Fig. 3.3(c), the inputs' cumulative exergies may simply be added to get the product cumulative exergy. For Figs. 3.3(b) and 3.3(c), $\mathbf{C}_N^E = [400 \ 600]^T$, $\mathbf{\Gamma}_i = [1 \ 1]$, and $\mathbf{C}_Y^E = 1000$.

The emergy allocation scheme differs from that of energy or exergy analysis, because emergy distinguishes between "co-products" and "splits." Co-products are products that cannot be produced independently, such as the yolk and yellow of an egg, whereas splits are products that can be produced independently, such as the components of a mixture. Figure 3.4 illustrates how emergy flows through a

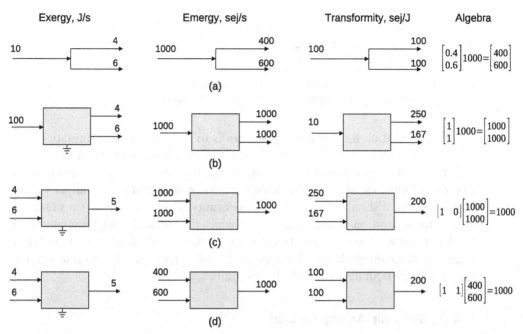

Figure 3.4. Allocation in emergy analysis.

network [29]. When a stream is divided without any energy transformation, emergy is split according to the scheme illustrated in Fig. 3.4(a), that is, in proportion to the available energy of the outputs. In this case, transformity does not change its value. This is identical to the exergy allocation scheme of Fig. 3.3. However, if there is an energy transformation, emergy is not distributed, but assigned entirely to each stream. This is justified based on the fact that they are co-products, meaning that none of the streams can be produced independently from the other. This situation is illustrated in Fig. 3.4(b). In this case transformity does change its value and emergy is not conserved through the network. This complicates the algebra because extra care is necessary to avoid double counting when emergy is combined. Figures 3.4(c) and 3.4(d) show how emergy is determined for stream intersections. Emergy is added when available energy of the same kind is joined. Emergy is also added if available energies of different kinds and sources are joined. This includes situations in which the emergy content of streams represents vastly different time periods, such as the emergy of biomass from seasonal crops with that of woody biomass from mature trees. However, if available energies of different kinds but same sources are joined, then the higher value of emergy is assigned. Figure 3.5 illustrates how emergy is determined in stream intersections with feedback loops. As shown, the emergy of loops is not considered in the calculations to avoid double counting [57]. This algebra has been the source of much confusion about emergy, but may be justified based on the fact that such decisions about allocation will remain subjective.

Figure 3.5. Allocation in emergy flow in a loop.

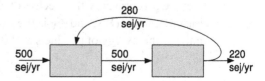

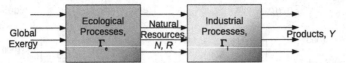

Figure 3.6. Flow of exergy in industrial and ecological systems. ICEC considers only industrial processes, whereas ECEC considers industrial and ecological processes.

The motivation for this emergy algebra is to maintain the interpretation that transformity reflects energy quality. Thus, for co-products, as shown in Fig. 3.4(b), among items being produced deliberately, the item produced in smaller quantities is likely to be more capable of influencing the environment and to do a larger variety of tasks; hence it should have a larger transformity. This makes sense for situations such as the separation of metal from ore, in which the former has higher quality than the latter and will have a higher transformity. If the allocation scheme of energy or exergy analysis shown in Fig. 3.3 is used for such separation, the overburden and mineral would both have the same transformities.

3.4 Relationship Among Methods

Understanding the relationship among the thermodynamic methods discussed in this chapter can be useful for utilizing the best features of each approach and for using thermodynamic methods to guide the development of technologies and manufacturing systems toward greater sustainability. This goal of sustainability necessitates consideration of the role of ecosystem goods and services for supporting industrial activity, because these goods and services are the foundation of all other activities. Such understanding can also shed light on the controversial aspects discussed in Subsection 3.2.3.

Among thermodynamic concepts, exergy is most widely accepted for its rigor because it accounts for both laws of thermodynamics and can be used for assessing the life-cycle aspects of systems and products. *Cumulative exergy consumption*, described in Subsection 3.2.2, expands *exergy analysis* to include the role of other industrial processes in the life cycle. Because this approach considers only industrial processes, it is called *industrial cumulative exergy consumption* (ICEC) [50]. This approach may be expanded to include the exergy consumed in ecological processes, resulting in *ecological cumulative exergy consumption* (ECEC). Figure 3.6 shows the relationship between the systems considered for ICEC and ECEC analyses. ICEC considers only the industrial processes, whereas ECEC also considers the supporting ecological processes. Methods from network analysis may be used to clearly specify the relevant algebra, allocation method, and system boundary.

Based on Eq. (3.6) in CExC analysis,

$$\mathbf{C}_Y^{\mathscr{E}} = \eta_{\mathscr{E},Y}^{-1}\, \mathscr{E}_Y, \qquad (3.14)$$

where $\mathbf{C}_Y^{\mathscr{E}}$ represents the vector of CExC of the products, $\eta_{\mathscr{E},Y}$ is the matrix of efficiencies or CDPs, and $\mathscr{E}_Y$ is the exergy vector of the products. Based on emergy analysis, the exergy of products and their emergy may be written as

$$\mathcal{M}_Y = \mathcal{T}_Y \mathscr{E}_Y, \qquad (3.15)$$

where $\mathcal{T}_Y$ is the matrix of transformities of the products. For $\mathbf{C}_Y^{\mathscr{E}}$ and $\mathcal{M}_Y$ to be equivalent, Eqs. (3.14) and (3.15) show that

$$\mathcal{T}_Y = \eta_{\mathscr{E},Y}^{-1}, \tag{3.16}$$

that is, transformity is equal to the reciprocal of the *CDP*. Additional insight into the relationship between CExC and emergy may be obtained by means of the following equations for the systems shown in Fig. 3.6. For emergy analysis, the relationship between the exergy of products $\mathscr{E}_Y$ and exergy of global inputs $\mathscr{E}_G$ may be derived as

$$\mathbf{\Gamma}' \mathcal{T}_G \mathscr{E}_G = \mathcal{T}_Y \mathscr{E}_Y \tag{3.17}$$

where $\mathbf{\Gamma}'$ is the matrix representing allocation of global exergy inputs to the products and $\mathcal{T}_G$ represents the transformities of global exergy inputs to adjust for differences in their quality. Similarly based on CExC analysis, the following equation may be derived:

$$\mathbf{\Gamma} \eta_G^{-1} \mathscr{E}_G = \eta_{\mathscr{E},Y}^{-1} \mathscr{E}_Y, \tag{3.18}$$

where $\mathbf{\Gamma}$ represents the allocation approach used by CExC analysis and η_G represents the relationship between the exergy of global inputs and their cumulative exergy. Equations (3.16–3.18) lead to

$$\mathbf{\Gamma} \eta_G^{-1} = \mathbf{\Gamma}' \mathcal{T}_G. \tag{3.19}$$

For a fair comparison between methods, the allocation methods for emergy and CExC should be identical, implying that $\mathbf{\Gamma} = \mathbf{\Gamma}'$, which leads to,

$$\eta_G^{-1} = \mathcal{T}_G. \tag{3.20}$$

Additional details about the preceding equations are in [50]. As stated in that work, these equations imply the following:

Cumulative exergy consumption and emergy are equivalent if the following are identical,
1. Analysis boundary,
2. Allocation method,
3. Approach for combining global energy inputs.

This condition shows that ECEC and emergy are very closely related. Moreover, it justifies the use of the reciprocal of transformity to estimate the CDP of natural resources, which can be used to expand ICEC analysis to account for the role of ecological systems. It is also important to point out that most of the controversial aspects of emergy analysis, summarized in Subsection 3.2.3, are not relevant for using the transformities of ecosystem goods and services to indicate the contribution from ecosystems.

The allocation approaches used in emergy and CExC analysis, as described in Section 3.3, are quite different. Allocation according to the exergy of output streams relies on detailed knowledge of the network and outputs for allocation. Its benefits are that cumulative exergy follows laws of conservation, making the algebra quite straightforward, intuitive, and consistent with widely used network algebra. The alternative allocation approach used in emergy analysis allocates the entire cumulative exergy of the inputs to all the outputs without any partitioning.

Instead of thinking of this allocation approach as a subjective decision, it may be interpreted as an approach for avoiding allocation in cases for which knowledge about the network structure and its outputs is not available. This is usually the case with ecosystems because complete knowledge about the ecological network and its goods and services is not available. Emergy analysis may be considered to avoid allocation for such partially defined systems by considering the exergy consumption of the process to be essential for making each product.

In general in *emergy analysis*, the network for most renewable resources is not known and allocation is avoided, making them nonadditive, whereas the algebra used for nonrenewables makes them additive resources. Hau and Bakshi [50] suggest the following general equation for calculating ECEC:

$$\text{ECEC}^{\text{total}}(k) = \max_{i=1,\ldots,r_1} \text{ECEC}^{R_{1,i}}(k) + \sum_{j=1}^{r_2} \text{ECEC}^{N_{2,j}}(k), \quad \forall k = 1,\ldots,n. \quad (3.21)$$

This equation is also relevant to emergy analysis if the three conditions for the equivalence of emergy and ECEC are satisfied.

There remain conceptual differences between emergy and ECEC analyses. ECEC analysis does not imply any relationship with economic value. In fact, ECEC analysis can complement economic analysis by providing the physical information that can form the basis of economic valuation. Legitimacy of Odum's *maximum empower principle* is irrelevant for the applicability of ECEC analysis. Also, as shown in this section, there are clear links between ECEC and other thermodynamic quantities. Representing global exergy inputs in equivalents of solar energy is not necessary, albeit convenient. ECEC faces similar quantification challenges as emergy, but these challenges are no different from those faced by any holistic approach, including life-cycle assessment and energy analysis. Addressing these challenges by developing practically useful and theoretically sound methods is an area of active research.

3.5 Issues in Aggregation

One of the benefits of thermodynamic methods is that they allow *aggregation* of multiple resources in a scientifically rigorous manner. This feature makes metrics such as those introduced in Section 3.2 extremely appealing for providing a quick snapshot of the life-cycle aspects of alternative products. It is important to realize that such aggregate metrics imply *substitutability* among resources that are being combined. That is, if energy from different resources, say wind, geothermal, oil, and coal are combined to obtain an aggregate metric, details about each resource are lost and the implication is that the task can be accomplished by any of these sources as long as they have the required energy content. This assumption is clearly wrong because a joule of wind energy cannot be a substitute for a joule of oil energy. Thus energy does not capture such differences in quality, and metrics that compare or combine the energy content of widely different resources such as fossil fuels, biomass, and sunlight can be severely misleading because biomass and sunlight have a very high energy content but are of lower quality than fossil fuels [58, 59]. Nonetheless, energy analysis should be appropriate only for processes that rely mainly on fuel

Table 3.1. *Efficiency and relative quality of selected resources based on exergy, CExC, and emergy*

Energy source	Exergetic efficiency η	Relative quality based on η	ICDP	Relative quality based on $\eta_{\varepsilon,Y}$	Transformity T	Relative quality based on T
Geothermal	13.8	5.0	10.1	3.8	6,055	4.0
Wind	2.8	1	2.7	1	1,496	1
Hydropotential	51.5	18.5	35.5	13.4	27,764	18.6
Natural Gas	21.0	7.6	5.6	2.1	48,000	32.1
Oil	36.1	13.0	3.0	1.1	53,000	35.4
Coal	22.2	8.0	6.2	2.3	17,242	11.5

resources of similar quality. It has even been argued that, because of the quality challenge, metrics based on *net energy analysis* are meaningless and irrelevant to decision making [60].

As discussed in this book, one of the attractions of *exergy* is that it represents the useful part of energy. This property has been very useful for identifying opportunities for improving the thermodynamic performance of equipment in a process or of processes in a life cycle. This has been a popular use of *exergy analysis* for improving process efficiency [61], and extensions to identifying opportunities of improvement in a life cycle are being explored [62]. Exergy is also able to account for fuel and nonfuel resources such as materials and account for the first and second laws of thermodynamics. Consequently, metrics based on exergy and CExC analyses are often calculated with the expectation that they would be better than energy at capturing the *quality of energy* and material resources, resulting in more meaningful metrics for decision making [25, 26]. Unfortunately, although exergy is better than energy at capturing *resource quality*, aggregation of diverse resources does assume that a joule of natural gas exergy is equivalent to and can be replaced with a joule of solar exergy. Consequently, like energy metrics, exergetic metrics may also provide misleading indicators, except if the resources being aggregated are of similar quality or are directly substitutable.

To gain greater insight into the energy and exergy quality issue, consider the data shown in Table 3.1 for various paths for generating electricity [63]. As shown, a joule of *wind* exergy produces about 0.03 J of electricity whereas a joule of *coal* exergy produces about 8 times as much *electricity*. Thus, for the purpose of producing electricity, coal is a higher-quality energy resource than wind. Furthermore, coal may be used for many other purposes, such as producing other gaseous or liquid fuels whereas wind cannot. Comparing the resource exergy required per joule of electricity in Table 3.1 indicates hydropotential exergy to be of the highest quality because the corresponding process is most efficient, whereas wind exergy is of the lowest quality because of its low process efficiency. Based on the efficiency of these processes for producing the same product, their quality may be represented in terms of wind equivalents by dividing each efficiency by that of wind electricity. The results are shown in column 3. They indicate that hydropotential is about 19 times superior to wind in terms of its ability to generate electricity. Thus 19 times more wind exergy is needed to produce 1 kWh of electricity than hydropotential exergy. A problem with this approach for quantifying resource quality is that it ignores the exergy needed for converting the feedstock into electricity. That is, if a resource needs much more process exergy for converting it into electricity, it may indicate a lower quality than that of a resource that can be converted more easily. We may address this problem by performing the quality calculations based on the *CDP* values shown in column 4 of Table 3.1. The resulting quality indicators are shown in column 5. These results indicate that converting fossil fuels to electricity relies on relatively large amounts of processing exergy. Hydropotential still seems to have the highest quality and wind the lowest, but now geothermal looks better than fossil fuels.

Unfortunately, both the approaches just proposed are too simplistic: the first because it ignores the life cycle, and the second because it calculates the CDP by combining the exergy of various resources used in the life cycle without considering their quality. It may be possible to overcome these shortcomings by formulating resource quality estimation as a regression problem, but more research is needed in this direction. Cleveland et al. [36] suggested that market prices may be used as quality multipliers. However, their goal is for the quality indicator to include a wide variety of factors such as human preferences, impact of resource use, cleanliness of resource, scarcity, and ability to do work. It makes sense that monetary value may be an appropriate quality indicator because physical quantities cannot capture subjective considerations such as human preferences. However, for applications such as LCA, the focus is usually on physical aspects and impact of emissions, as opposed to human preferences. Also, many life-cycle studies focus on emerging technologies for which monetary values are usually not available or are difficult to find. Finally, traditional economics is unable to capture the role of ecosystem goods and services that are crucial for sustainability but are outside the market.

Emergy analysis claims to provide a way of quantifying the quality of resources by means of the value of their transformity. As shown in the transformation chain in Fig. 3.2, the efficiencies along this supply chain inversely correspond to the energy quality of resources. Ecologists have observed that, in food networks, species that are more exergy intensive, that is, have a lower life-cycle efficiency or higher transformity, are more capable of influencing their surroundings and doing a larger variety of work [29]. This insight also seems to be valid for the industrial supply chain. Thus, accounting for the cumulative exergy consumed in making a product can be an effective way of considering the quality of resources. If the boundary for such analysis is restricted to industrial processes only, as is done in CExC analysis, then differences in the quality of natural resources such as coal, wood, biomass, and sunlight will not be captured because the cumulative exergy of natural resources is considered to be equal to their exergy. It seems that expanding the boundary to include ecosystems would be an effective way of capturing the quality of various natural resources. As shown in Subsection 3.2.3, this is what emergy analysis does. Emergy analysis attempts to address the energy quality issue by converting all resources to a common currency: solar equivalent joules. This idea makes conceptual sense but faces many challenges because of difficulties in getting the necessary information for ecological and industrial networks. *Transformities* of the different energy sources and their relative qualities are shown in the last two columns of Table 3.1. Fossil fuels are considered to be of higher quality than *renewable resources*, which may make sense because these resources can have a larger variety of uses than renewable resources. However, further research is required for evaluating the validity of various approaches for quantifying the quality of resources.

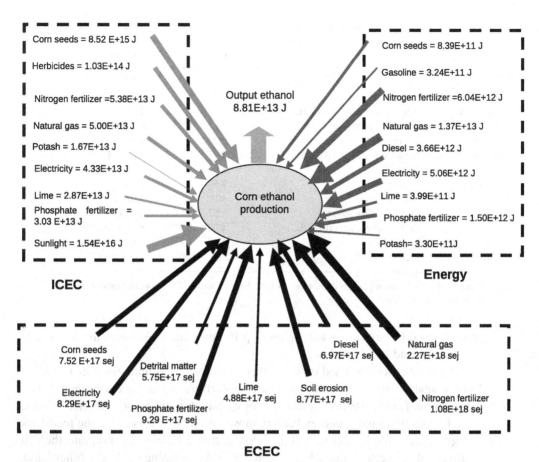

Figure 3.7. Resource flows from the life cycle of corn ethanol in terms of cumulative energy, industrial cumulative exergy, and ecological cumulative exergy. Computer notation has been used; this translates to the logarithmic notation of × 10, e.g., 8.81E + 13J = 8.81 × 10^{13} J.

3.6 Examples

This section presents the results of *energy*, *ICEC*, and *ECEC/emergy* analyses to highlight the pros and cons of these methods. The specific fuels considered are *corn ethanol* and *gasoline*, with more details in [23]. Figure 3.7 shows the 10 largest inputs based on energy, ICEC, and ECEC analyses. The differences among the results of these three methods confirm the issues discussed in this chapter. Energy analysis identifies natural gas, nitrogen fertilizer, and electricity as the three inputs with the largest life-cycle consumption of nonrenewable energy. In contrast, ICEC identifies sunlight, corn seeds, and herbicides as the most dominant. This insight from ICEC is because of the inclusion of renewable and nonrenewable energies in this approach and because the dilute nature of the former resource type causes them to dominate. This is potentially misleading because the large contribution of sunlight does not correspond to its abundance and low quality. If renewable resources are ignored in ICEC analysis, the results are closer to those of energy analysis. More details are provided in [23]. ECEC identifies natural gas, electricity, nitrogen fertilizer, and corn seeds as the largest inputs. This seems to be more sensible than the resources identified by ICEC and more comprehensive than those identified by energy analysis. It is important to note that ECEC also identifies soil erosion and detrital matter

Table 3.2. *Metrics for comparing the life cycles of gasoline and corn ethanol*

Metrics		Gasoline	Ethanol
Energy	Energy return on investment (rE)	7.75	2.22
	Efficiency	0.69	0.89
ICEC	Exergy return on investment (rEx)	7.77	1.41
	ICDP (exergy/total ICEC)	30.2×10^{-2}	0.362×10^{-2}
	Exergy/nonrenewable ICEC	0.92	2.29
	Nonrenewable ICEC/km	2.93×10^{-6}	1.48×10^{-6}
	ICEC/km	8.94×10^{-6}	560×10^{-6}
ECEC	Yield ratio	7.69	1.26
	Loading ratio	803	3.01
	Sustainability index	9.57×10^{-3}	418×10^{-3}
	ECDP (exergy/total ECEC)	4.00×10^{-6}	8.06×10^{-6}
	ECEC/km	6.75×10^{11}	3.88×10^{11}
	ECEC/ICEC ratio	755×10^2	4.5×10^2

Notes: For corn ethanol per km data, refer to E85. Energy and ICEC data are based on mass-based allocation.

(used for soil formation) as among the top 10 flows. Such resources are ignored in both energy and ICEC analyses.

Aggregate metrics based on these three methods are summarized in Table 3.2. Energy analysis shows that efficiency and EROI are lower for corn ethanol than for gasoline. It suggests that gasoline is more lucrative than corn ethanol because we get seven times more energy back than we invest. This is because the feedstock energy for gasoline, i.e., crude oil, is a concentrated form of energy, and there is relatively little energy required for extracting and processing it. On the other hand, corn ethanol is derived directly from sunlight, which is a very dilute form of energy. It requires corn farming to concentrate sunlight to more concentrated energy in the form of corn grains and further processing to convert it to ethanol, all of which adds up to more energy consumption. As process energy increases, life-cycle efficiency decreases. Because process energy comes from the economy, higher process energy may translate into higher costs of production. In the case of gasoline, as crude oil becomes scarcer, energy spent in oil exploration and extraction becomes higher such as from the need to drill the holes deeper to tap the low-lying crude oil [64]. As a result, *EROI* and efficiency for gasoline may decrease considerably.

The exergetic ROI of gasoline accounts for energy and material use and is also higher than that of ethanol. Material contribution to CExC of gasoline is minuscule, but material exergy for corn ethanol is about 23% of cumulative exergy, a large part of which is contributed by soil and water. Although water is undervalued and there is no market value for soil erosion, conserving water and minimizing soil erosion can improve the overall sustainability of corn ethanol. This insight is not available from energy analysis. Comparing efficiencies based on ICEC analysis shows that corn ethanol has considerably lower industrial CDP (ICDP) compared with gasoline. This is due to sunlight being inefficiently captured by corn plants, a characteristic of biological ecosystems. Energy analysis ignores energy from sunlight because of its focus on only nonrenewable energy. As discussed in Section 3.5, aggregate metrics based on CExC may be misleading because of their inability to account for the quality

of exergy sources. Nevertheless, accounting for sunlight energy implies that, if this
energy is utilized more efficiently, it would result in higher biomass productivity.
This implies a reduction in land area, which in turn reduces fertilizer inputs, farm
equipment, and fossil energy use. Metrics based on nonrenewable exergy are also
interesting because they are indicative of which products consume relatively more
nonrenewable exergy that we are trying to conserve. Table 3.2 shows that gasoline
consumes more nonrenewable ICEC per exergy output generated or per kilometer
traveled. This is expected because gasoline is derived from nonrenewable crude
oil. For biofuels, the exergy to nonrenewable ICEC ratio also assumes another
significance because it tells how much "renewable" biofuel exergy can be generated
by investing 1 J of nonrenewable exergy. This metric has also been referred to as an
exergy breeding factor by Dewulf et al. [25].

ECEC is a macroanalysis that deals with broader concepts such as resource
qualities, renewability, and scarcity. Unlike energy analysis, it provides an indication
of sustainable resource utilization in production processes. Some metrics based on
ECEC analysis are presented in Table 3.2. The yield ratio indicates how much work
or investment is needed by the economy to process natural resources into products.
The lower the work done by the economy, the less ecologically costly the product
may turn out to be. In essence, the yield ratio is similar to energy or EROI but is
more comprehensive. The yield ratio of gasoline is higher than that of corn ethanol,
implying that work done by the economy in sej is relatively less than the work done
by the ecosystem to process the fuel. Because the ecosystem has already invested
more work to produce high-quality crude oil, little effort is required for extracting
and processing it into gasoline. Gasoline has a *loading ratio* that is larger by a
factor of 157 than corn ethanol. This is primarily because the gasoline is derived
from nonrenewable crude oil. It is possible to expand the loading ratio by including
impact ECEC resulting from emissions and combining with nonrenewable ECEC.
In doing so, a loading ratio measures not only resource scarcity (renewability) but
also the impact on human health. When we divide the yield ratio by the loading ratio,
we get a SI. This index may be improved by increasing a yield ratio or decreasing
a loading ratio or both. Gasoline has a lower *SI* than gasoline, even though it has
a better yield ratio. This is because the contribution of the yield ratio is offset by
gasoline's higher loading ratio.

Analogous to the ICDP, the ecological CDP (ECDP) measures ecological efficiency. Unlike ICDP, ECDP assigns more weight to nonrenewable and quality
resources. Therefore a product that consumes quality resources or nonrenewable
resources has a lower ecological efficiency (ECDP). It is instructive to see that the
ICDP and the ECDP of corn ethanol follow opposite trends with respect to gasoline.
As evident from Table 3.1, gasoline has a lower ECDP than corn ethanol owing to its
reliance on crude oil, which is associated with significantly higher previously invested
ecological work. The lower ECDP also translates into higher ECEC/km – a measure
of the well-to-wheel ecological efficiency – as evident from higher ECEC/km for
gasoline. That the ECEC analysis consciously accounts for some quality aspects of
resource consumption is demonstrated by the ECEC/ICEC ratio. Alternatively, the
ratio indicates the extent to which ICEC analysis ignores the quality of resources consumed. The ECEC/ICEC ratio is larger for gasoline than for corn ethanol. It means
more quality resources are used by gasoline that are not accounted for in ICEC

analysis. ECEC analysis does quality corrections in terms of previous work done in sej to produce 1 J of exergy in a product – also known as transformity. Because quality is measured in terms of previous work done, and there are virtually thousands of qualities to choose from, ECEC is far from being a universal way of accounting for qualities, Nonetheless, ECEC analysis represents an attempt to account for some aspects of quality where energy and ICEC analyses have failed to do so.

3.7 Conclusions

This chapter provides an overview of thermodynamic methods for quantifying the cumulative consumption of resources. Methods discussed in this chapter include net energy analysis, cumulative exergy consumption analysis, and emergy analysis. Similarities, differences, and controversies surrounding these methods are discussed, and concepts of industrial and ecological cumulative exergy consumption are introduced to demonstrate the relationship between cumulative exergy consumption, derived in engineering, and emergy, derived in systems ecology. Shortcomings in using net energy or cumulative exergy for aggregating different types of energy are discussed and illustrated by means of examples. Emergy analysis is a bold approach that seems to be capable of capturing quality differences among various resources, but it faces challenges because of the inadequate understanding of ecosystems. Some areas for further research include verification of claims about approaches (exergy, money, emergy) for representing the quality of resources, addressing allocation issues, and further case studies to understand the pros and cons of thermodynamic and other methods for environmentally conscious decision making.

REFERENCES

[1] B. Linhoff, "Pinch analysis – a state-of-the-art overview," *Chem. Eng. Res. Design* **71a**, 503–522 (1993).

[2] M. M. El-Halwagi, *Process Integration* (Academic, New York, 2006).

[3] M. Stewart and B. P. Weidema, "A consistent framework for assessing the impacts from resource use – a focus on resource functionality," *Intl. J. Life Cycle Assess.* **10**, 240–247 (2005).

[4] J. C. Bare and T. P. Gloria, "Critical analysis of the mathematical relationships and comprehensiveness of life cycle impact assessment approaches," *Environ. Sci. Technol.* **40**, 1104–1113 (2006).

[5] International Organization for Standardization, *Environmental Management – Life Cycle Assessment* (International Organization for Standardization, Geneva, Switzerland, 1998).

[6] *2005 Millennium Ecosystem Assessment (MEA), Ecosystems and Human Well-Being: Synthesis* (Island Press, Washington, D.C., 2005). Available at www.maweb.org, accessed Jan. 12 2008.

[7] E. D. Williams, R. U. Ayres, and M. Heller, "The 1.7 kilogram microchip: Energy and material use in the production of semiconductor devices," *Environ. Sci. Technol.* **36**, 5504–5510 (2002).

[8] E. Matthews, C. Amanu, and S. Bringezu, *The Weight of Nations: Material Outflows from Industrial Economies* (World Resources Institute, Washington D.C., 2000).

[9] F. Schmidt-Bleek, "MIPS. A universal ecological measure?" *Fresenius Environ. Bull.* **2**, 306–311 (1993).

[10] R. Müller-Wenk, "Depletion of abiotic resources weighted on base of "virtual" impacts of lower grade deposits used in future," Tech. Rep. (Institut für Wirtschaft und Ökologie, Universität St. Gallen, Switzerland, 1998).

[11] J. B. Guinée, editor, *Handbook on Life Cycle Assessment* (Kluwer Academic, Dordrecht, The Netherlands, 2002).

[12] R. Costanza, R. d'Arge R, R. de Groot, S. Farber, M. Grasso, B. Hannon, K. Limburg, S. Naeem, R. V. O'Neill, J. Paruelo, R. G. Raskin, P. Sutton, and M. van den Belt, "The value of the world's ecosystem services and natural capital," *Nature (London)* **387**, 253–260 (1997).

[13] G. C. Daily and K. Ellison, *The New Economy of Nature* (Island Press, Washington, D.C., 2002).

[14] A. Balmford, A. Bruner, P. Cooper, R. Costanza, S. Farber, R. E. Green, M. Jenkins Feriss, V. Jessamy, J. Madden, K. Munro, N. Myers, S. Naeem, J. Paavola, M. Rayment S. Rosendo, J. Roughgarden, K. Trumper, and R. K. Turner, "Economic reasons for conserving wild nature," *Science* **297**, 950–953 (2002).

[15] S. Casler and B. Hannon, "Readjustment potentials in industrial energy efficiency and structure," *J. Environ. Econ. Manage.* **17**, 93–108 (1989).

[16] C. W. Bullard and R. A. Herendeen, "The energy cost of goods and services," *Energy Policy* **3**, 268–278 (1975).

[17] B. Hannon, "Analysis of the energy cost of economic activities: 1963–2000," *Energy Syst. Policy J.* **6**, 249–178 (1982).

[18] D. T. Spreng, *Net-Energy Analysis* (Praeger, New York, 1988).

[19] J. Hill, E. Nelson, D. Tilman, S. Polasky, and D. Tiffany, "Environmental, economic, and energetic costs and benefits of biodiesel and ethanol biofuels," *Proc. Natl. Acad. Sci. USA* **103**, 11206–11210 (2006).

[20] A. E. Farrell, R. J. Plevin, B. T. Turner, A. D. Jones, M. O'Hare, and D. M. Kammen, "Ethanol can contribute to energy and environmental goals," *Science* **311**, 506–508 (2006).

[21] R. Costanza, "Embodied energy and economic valuation," *Science* **210**, 1219–1224 (1980).

[22] R. Costanza and R. Herendeen, "Embodied energy and economic value in the United States economy – 1963, 1967 and 1972," *Resources Energy* **6**, 129–163 (1984).

[23] A. Baral and B. R. Bakshi, "Thermodynamic metrics for aggregation of natural resources in life cycle analysis: Insight via application to some transportation fuels," *Env. Sci. Technol.* **44**, 2, 800–807 (2010).

[24] J. Szargut, D. R. Morris, and F. R. Steward, *Exergy Analysis of Thermal, Chemical and Metallurgical Processes* (Hemisphere, Washington, D.C., 1988).

[25] J. Dewulf, H. Van Langenhove, and B. Van De Velde, "Exergy-based renewability assessment of biofuel production," *Environ. Sci. Technol.* **39**, 3878–3882 (2005).

[26] R. Berthiaume, C. Bouchard, and M. A. Rosen, "Exergetic evaluation of the renewability of a biofuel," *Exergy Intl. J.* **1**, 256–268 (2001).

[27] J. Dewulf, M. Bosch, B. De Meester, G. Van der Vorst, H. Van Langenhove, S. Hellweg, and M. Huijbregts, "Cumulative exergy extraction from the natural environment (CEENE): A comprehensive life cycle impact assessment method for resource accounting," *Environ. Sci. Technol.* **41**, 8477–8483 (2007).

[28] M. Belli and E. Sciubba, "Extended exergy accounting as a general method for assessing the primary resource consumption of social and industrial systems," *Intl. J. Exergy* **4**, 421–440 (2007).

[29] H. T. Odum, *Environmental Accounting: EMERGY and Environmental Decision Making* (Wiley, New York, 1996).

[30] M. T. Brown and S. Ulgiati, "Emergy-based indices and ratios to evaluate sustainability: Monitoring economies and technology toward environmentally sound innovation," *Ecol. Eng.* **9**(1–2), 51–69 (1997).

[31] B. R. Bakshi, "A thermodynamic framework for ecologically conscious process systems engineering," *Comput. Chem. Eng.* **26**, 269–282 (2002).

[32] H. T. Odum, "Folio # 2, emergy of global processes," Tech. Rep., (University of Florida, Gainesville, Florida, 2000). Available at www.emergysystems .org.

[33] S. Bastianoni, D. E. Campbell, R. Ridolfi, and F. M. Pulselli, "The solar transformity of petroleum fuels," *Ecol. Model.* **220**, 40–50 (2009).

[34] Y. Zhang, A. Baral, and B. R. Bakshi, "Accounting for ecosystem services in life cycle assessment, part ii: Toward an ecologically-based LCA," Tech. Rep. (Ohio State University, Columbus, OH, 2009).

[35] R. U. Ayres, "On the life cycle metaphor: Where ecology and economics diverge," *Ecol. Econ.* **48**, 425–438 (2004).

[36] C. J. Cleveland, R. K. Kaufman, and D. I. Stern, "Aggregation and the role of energy in the economy," *Ecol. Econ.* **2**, 301–317 (2000).

[37] W. A. Hermann, "Quantifying global exergy resources," *Energy* **31**, 1685–1702 (2006).

[38] G. Q., Chen, "Exergy consumption of the earth," *Ecol. Model.* **184**, 363–380 (2005).

[39] J. T. Szargut, "Anthropogenic and natural exergy losses (exergy balance of the Earth's surface and atmosphere)," *Energy* **28**, 1047–1054 (2003).

[40] N. U. Ukidwe and B. R. Bakshi, "Thermodynamic accounting of ecosystem contribution to economic sectors with application to 1992 US economy," *Environ. Sci. Technol.* **38**, 4810–4827 (2004).

[41] H. T. Odum and E. P. Odum, "The energetic basis for valuation of ecosystem services," *Ecosystems* **3**, 21–23 (2000).

[42] F. Soddy, *Virtual Wealth and Debt – the Solution of the Economic Paradox* (Allen and Unwin, London, 1926).

[43] K. E. Boulding, "The economics of the coming spaceship earth," in *Environmental Quality in a Growing Economy*, edited by H. Jarrett (Resources for the Future, Washington, D.C., 1966).

[44] H. T. Odum, "Self-organization and maximum empower," in *Maximum Power: The Ideas and Applications of H.T. Odum,* edited by C.A.S. Hall (University of Colorado Press, Niwot, CO, 1995).

[45] L. Boltzmann. *Theoretical Physics and Philosophical Problems: Selected Writings of Ludwig Boltzmann* (Reidel, Dordrecht, The Netherlands, 1974).

[46] A. J. Lotka, *"Elements of Physical Biology"* (Williams & Wilkins, Baltimore 1925).

[47] B. A. Mansson and J. M. McGlade, "Ecology, thermodynamics and H. T. Odum's conjectures," *Oecologia* **93**, 582–596 (1993).

[48] J. Emblemsvag and B. Bras, *Activity-Based Cost and Environmental Management: A Different Approach to ISO 14000 Compliance* (Kluwer, Dordrecht, The Netherlands, 2001).

[49] E. Sciubba and S. Ulgiati, "Emergy and exergy analyses: Complementary methods or irreducible ideological options?" *Energy* **30**, 1953–1988 (2005).

[50] J. L. Hau and B. R. Bakshi, "Expanding exergy analysis to account for ecosystem products and services," *Environ. Sci. Technol.* **38**, 3768–3777 (2004).

[51] S. Bastianoni, A. Facchini, L. Susani, and E. Tiezzi, "Emergy as a function of exergy," *Energy* **32**, 1158–1162 (2007).

[52] J. L. Hau and B. R. Bakshi, "Promise and problems of emergy analysis," *Ecol. Model.* **178**, 215–225 (2004).

[53] G. C. Daily, editor, *Nature's Services* (Island Press, Washington, D.C., 1997).

[54] K. Arrow, B. Bolin, R. Costanza, P. Dasgupta, C. Folke, C. S. Holling, B. O. Jansson, S. Lavne, K. G. Mäler, C. Perrings, and D. Pimentel, "Economic growth, carrying capacity, and the environment," *Science* **268**, 520–521 (1995).

[55] S. R. Carpenter, H. A. Mooney, J. Agard, D. Capistrano, R. DeFries, S. Diaz, T. Dietz, A. Duriappah, A. Oteng-Yeboah, H. M. Pereira, C. Perrings, W. V. Reid, J. Sarukhan, R. J. Scholes, and A. Whyte, "Science for managing ecosystem services: Beyond the millennium ecosystem assessment," *Proc. Natl. Acad. Sci.* **106**, 1305–1312 (2009).

[56] S. Suh, M. Lenzen, and G. J. Treloar, H. Honda, A. Horvath, G. Huppes, O. Jolliet, U. Klann, W. Krewitt, Y. Moriguchi, J. Munksgaard, G. Norris, "System boundary selection in life-cycle inventories using hybrid approaches," *Environ. Sci. Technol.* **38**, 657–664 (2004).

[57] M. T. Brown and R. A. Herendeen, "Embodied energy analysis and emergy analysis: A comparative view," *Ecol. Econ.* **19**, 219–235 (1996).

[58] H. Haberl, H. Weisz, C. Amann, A. Bondeau, N. Eisenmenger, K. H. Erb, M. Fischer-Kowalski, and F. Krausmann, "The energetic metabolism of the European Union and the United States: Decadal energy input time-series with an emphasis on biomass," *J. Ind. Ecol.* **10**(4), 151–171 (2006).

[59] M. Giampietro, "Comments on 'The Energetic Metabolism of the European Union and the United States by Haberl and Colleagues: Theoretical and Practical Considerations on the Meaning and Usefulness of Traditional Energy Analysis,'" *J. Ind. Ecol.* **10**, 173–185 (2006).

[60] W. Schulz, "The costs of biofuels," *Chem. Eng. News* **85**(51), 12–16 (2007).

[61] A. Bejan, G. Tsatsaronis, and M. Moran, *Thermal Design and Optimization* (Wiley, New York, 1996).

[62] G. F. Grubb and B. R. Bakshi, "Improving the improvement assessment step in LCA," Tech. Rep. (Department of Chemical and Biomolecular Engineering, Ohio State University, Columbus, OH, 2009).

[63] N. U. Ukidwe, "Thermodynamic input output analysis of economic and ecological systems for sustainable engineering," Ph.D. dissertation (Ohio State University, Columbus, OH, 2005).

[64] C. J. Cleveland, "Net energy from the extraction of oil and gas in the United States," *Energy* **30**, 769–782 (2005).

PRODUCTS AND PROCESSES

Materials Separation and Recycling

Timothy G. Gutowski

4.1 Introduction

In this chapter, we develop several models for the materials-recycling process. The focus is on the separation of materials from a mixture. This problem can be modeled by using the principles of thermodynamics, particularly the concept of mixing entropy, as well as by using some of the results from information theory. In doing this calculation we will find, from a thermodynamic point of view, that the theoretical minimum work required for separating a mixture is identical to the work lost on spontaneous mixing of the chemical components. In other words, the development in this chapter in conjunction with the results from previous chapters will allow us to track both the degradation in materials values as they are used and dispersed in society as well as the improvement and gain as materials are restored to their original values. Of course, this restoration does not come for free, and so we also look at the losses and inefficiencies involved in materials recycling. This approach allows us to look at the complete materials cycle as it moves through society and to evaluate the gains and losses at each step. The chapter starts with the development of the needed thermodynamics concepts and then moves on to the application of these ideas. This chapter also introduces an alternative way of looking at the recycling problem by using information theory.

4.2 The Thermodyamics of Materials Separation

The basic separation problem can be illustrated by consideration of the separation of a molecular mixture into its pure components. This result is then developed for the special case of an ideal mixture. Ideal mixtures include ideal gas mixtures and ideal solutions, but not necessarily many of the material-separation situations that occur in recycling, material extraction, and material purification. These cases may deviate from ideal mixtures because of specific interactions between dissimilar molecules, such as volume effects and heat effects, or because the mixtures are not actually molecular mixtures. Nevertheless, the ideal mixture result can provide guidance, for example, by suggesting concentration scaling effects that could apply to many situations, including nonideal processes. It will be shown that these results have useful applications in the fields of resource accounting and industrial ecology.

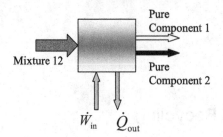

Figure 4.1. An ideal separation process.

An introduction to the thermodynamics of mixing and separation can be found in the text by Çengel and Boles [1]. The development here follows the material found in Chaps. 1 and 2 in this book and the work of Gyftopoulos and Beretta [2].

Consider the open system shown in Fig. 4.1. A mixture denoted by 12 at temperature T_0 and pressure p_0 enters on the left and the pure components 1 and 2, also at T_0 and p_0, exit on the right. Each stream has enthalpy H (measured in joules, J) and entropy S (measured in J/K), which are denoted by their subscripts. The system has a work input W and can exchange heat Q with the surroundings at temperature T_0. We can then write the rate balance equations (shown by the dot over each variable that changes with time) for constituents, energy, and entropy as given by

$$\frac{dN_{i,\text{sys}}}{dt} = \dot{N}_{i,\text{in}} - \dot{N}_{i,\text{out}}, \quad i = 1, 2, \tag{4.1}$$

$$\frac{dE}{dt} = -\dot{Q}_{\text{out}} + \dot{W}_{\text{in}} + \dot{H}_{12} - \dot{H}_1 - \dot{H}_2, \tag{4.2}$$

$$\frac{dS}{dt} = -\frac{\dot{Q}_{\text{out}}}{T_0} + \dot{S}_{12} - \dot{S}_1 - \dot{S}_2 + \dot{S}_{\text{irr}}, \tag{4.3}$$

where N_i are in moles and S_{irr} is the entropy production associated with irreversibilities in the system. This term allows us to write (4.3) as a balance even though entropy is not conserved. Assuming steady state and eliminating the heat rate $\dot{Q}_{\text{out}}$ between (4.2) and (4.3) yields an expression for the work rate for separation:

$$\dot{W}_{\text{in}} = [(\dot{H}_1 + \dot{H}_2) - \dot{H}_{12}] - T_0[(\dot{S}_1 + \dot{S}_2) - \dot{S}_{12}] + T_0\dot{S}_{\text{irr}}. \tag{4.4}$$

The mass balance is implied because the terms in (4.4) are all extensive. We can also write this result by using the molar-intensive forms of the thermodynamic properties (denoted by lowercase font) as

$$\dot{W}_{\text{in}} = -\dot{N}_{12}(\Delta h_{\text{mix}} - T_0\Delta s_{\text{mix}}) + T_0\dot{S}_{\text{irr}} \tag{4.5}$$

or

$$\dot{W}_{\text{in}} = -\dot{N}_{12}\Delta g^*_{\text{mix}} + T_0\dot{S}_{\text{irr}}, \tag{4.6}$$

where $\Delta h_{\text{mix}} = (h_{12} - x_1h_1 - x_2h_2)$ and $\Delta s_{\text{mix}} = (s_{12} - x_1s_1 - x_2s_2)$, and x_1 and x_2 are mole fractions, (N_1/N_{12} and N_2/N_{12} respectively), with $N_{12} = N_1 + N_2$. And we recognize the term within the parentheses in (4.5) as the intensive form of the Gibbs free energy of mixing at the so-called restricted dead state (T_0, p_0), i.e., $\Delta g^*_{\text{mix}} = \Delta h_{\text{mix}} - T_0\,\Delta s_{\text{mix}}$.

We can now obtain an expression for the minimum work of separation per mole of mixture by letting $\dot{S}_{irr} = 0$. This gives the minimum rate of work as

$$\dot{W}_{min} = -\dot{N}_{12}\Delta g^*_{mix} \tag{4.7}$$

and the minimum work as

$$w_{min} = \frac{\dot{W}_{min}}{\dot{N}_{12}} = -\Delta g^*_{mix}. \tag{4.8}$$

That is, the minimum work of separation is the negative Gibbs free energy of mixing. When two substances spontaneously mix, the Gibbs free energy of mixing is negative. So the minimum work required for separating these is the positive value of Δg^*_{mix}. Losses in the system, i.e., $\dot{S}_{irr} > 0$, will make the work required even larger. If we consider the reverse problem, the one of mixing two pure streams, this could be accomplished without any work input provided $\Delta g^*_{mix} < 0$ (a common enough occurrence for many systems). Now the irreversible loss on mixing is

$$T_0\dot{S}_{irr} = -\dot{N}_{12}\Delta g^*_{mix}. \tag{4.9}$$

That is, the irreversible loss on spontaneous mixing is the same as the minimum work for separation; compare with (4.7). (Note the sign change when you write the material flows in the opposite direction.)

For an ideal mixture the enthalpy of mixing is zero, i.e., $\Delta h_{mix} = 0$. Hence the minimum work for separation becomes

$$w_{min} = T_0\Delta s_{mix}. \tag{4.10}$$

The mixing entropy Δs_{mix} for noninteracting particles can be calculated from the case of mixing ideal gases or from a statistical interpretation of entropy and Boltzmann's entropy equation, with the same result (see Appendix A for this chapter):

$$w_{min} = -T_0 R\sum_{i=1}^{n} x_i \ln x_i. \tag{4.11}$$

This is the general result for a mole of mixture with n constituents, x_i is the mole fraction of the ith constituent, and R is the universal gas constant, 8.314 J/(mol K).

When there are only two components in the mixture, as is the case in Fig. 4.1, the result for the separation work required per mole of mixture then is

$$w_{min} = -T_0 R[x \ln x + (1 - x)\ln(1 - x)]. \tag{4.12}$$

Here the mole fraction of component 1 is x and for component 2 it is $(1 - x)$. This equation is symmetric, giving the largest work when $x = \frac{1}{2}$ (at $T_0 = 298.2$ K, this is 1.7 kJ/mol of mixture) and, at the end points ($x = 0, 1$), the work is zero. Equations (4.11) and (4.12) give the minimum work per mole of mixture to completely separate an ideal n component mixture or an ideal 2 component mixture, respectively. We will see that these equations describe situations that can occur in materials recycling.

Finally, we should address a situation that often occurs in materials extraction and purification problems, and might occur in recycling. This is when a mole of a valuable material, say "1" is extracted from a mixture with a nonvaluable material, "2". To differentiate this case from the previous separation problem, which separates all components, we call this "extraction." We start by writing the extensive form of

Eq. (4.12) and then subtracting from it the work of separation for a mixture at the same mole fractions but with one less mole of material 1. One physical interpretation of this situation is that the mixture is very large and extracting a mole of material 1 does not significantly change the molar concentrations of the original mixture:

$$W_{\min}^{(N_i)} = -T_0 R[N_1 \ln x + N_2 \ln (1 - x)], \qquad (4.13)$$

$$W_{\min}^{(N_i-1)} = -T_0 R[(N_1 - 1) \ln x + N_2 \ln (1 - x)], \qquad (4.14)$$

$$w_{\min,1} = T_0 R \left(\ln \frac{1}{x} \right). \qquad (4.15)$$

This is the minimum work to extract one mole of material 1 and concentrate it from mole fraction x to the pure form at $x = 1$. Note that Eq. (4.15) is monotonically increasing as one tries to extract 1 at more and more dilute concentrations. In fact, in the limit as $x \to 0$, the term $\ln(1/x)$ goes to infinity. This means the work to extract one unit of a valuable material from a dilute mixture increases without bounds as the solution becomes more and more dilute. This general result is often observed in material-extraction problems and is discussed subsequently in this chapter. Note that this result could also be interpreted as indicating that the work to extract a unit of impurity from an ultrapure material also increases; without bounds as the purity requirement increases, however this interpretation might be misleading. For the problem of purification, what is of interest is the work to extract the impurity *per unit of valuable material*, and this goes to zero as the impurity concentration becomes more dilute. In this case, the thermodynamics of separation for an ideal mixture does not agree with the common observation, that the work to extract an impurity per unit of valuable material actually increases as the concentration of impurity decreases. This issue is discussed in Chap. 7 by Williams et al. in this book.

4.3 Exergy Framework for Materials Evaluation

By using the separation and mixing results from the previous section and the results from material-transformation processes as given in Chaps. 2, 3 and 6, we may now illustrate how the exergy values of materials change as they go through the various life-cycle phases when they are used by society. In this context, the exergy difference represents the minimum work required for transforming a material to a new improved state.

Figure 4.2 shows schematically the improvements in material exergy as a material is extracted from the crust and purified and the reductions in material exergy as the material is mixed and spontaneously reacts with other materials including components of the environment (e.g., corrosion and oxidation). Also shown, as a dashed line, is the reestablishment of the material exergy value through recycling. A scheme similar to this was proposed by Connelly and Koshland [3] to measure material consumption.

To illustrate, consider how iron is transformed as it is used by society. According to Szargut et al. [4] iron is found in the Earth's crust in the form of hematite, Fe_2O_3, at an average molar concentration of 1.3×10^{-3}. In other words, when hematite is at this concentration at T_0, p_0, it is in the "dead state" with exergy equal to zero. By a combination of geological processes followed by the anthropogenic processes of

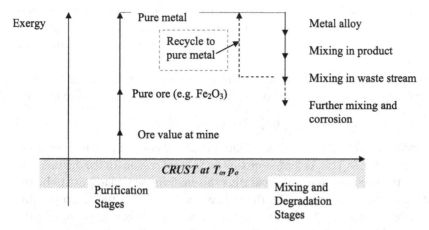

Figure 4.2. Theoretical exergy values for a metal extracted from the Earth's crust shown at various stages of a product life cycle (not to scale).

exploration, mining, and separation, the iron ore in the crust can be purified to pure Fe_2O_3. The theoretical exergy value of pure hematite is calculated with Eq. (4.15) for a process that concentrates hematite from a molar concentration of 1.3×10^{-3} to 1. The result is

$$e^*_{x\ Fe_2O_3} = T_0 R \ln \frac{1}{1.3 \times 10^{-3}} = 16.5 \, kJ/mole. \tag{4.16}$$

In the next stage of purification, the iron is reduced from a pure oxide to a pure metal. The minimum work to create pure iron from Fe_2O_3 is equal to the exergy lost when pure iron is oxidized to Fe_2O_3. The oxidation reaction for this is

$$2Fe + \frac{3}{2}O_2 \rightarrow Fe_2O_3. \tag{4.17}$$

This leads to Eq. (4.18), which shows the exergy balance for this reaction:

$$2e^*_{x\ Fe} + \frac{3}{2}e^*_{x\ O_2} - e^*_{x\ Fe_2O_3} = -\Delta g^*_{f\ (Fe_2O_3)}. \tag{4.18}$$

That is, the exergy lost on oxidation is equal to negative of the Gibbs free energy of formation of Fe_2O_3 at standard conditions (T_0, p_0); See [5, 6]. From this equation it is possible to calculate the exergy for pure iron,[1] as $e^*_{x\ Fe} = 376.4 \, kJ/mol$.

This value corresponds to the top line in Fig. 4.2. It is the minimum work to produce pure iron from hematite in the crust. Keep in mind that real processes will use much more work because of irreversibilities in the system. This additional work does not improve the value of the iron, but rather is lost. For example, Baumgarter and de Swaan Arons [8] and de Swaan Arons et al. [5] discuss the reduction of hematite with carbon to produce iron and calculate much higher exergy requirements. Now, after the material is purified, many of the other steps, as indicated in Fig. 4.2, lead to a reduction in the exergy value of the iron. For example, the iron may be alloyed and then mixed in a product and further mixed in a waste stream. If discarded in

---

[1] For example, we have just calculated the exergy $e^*_{x\ Fe_2O_3}$ in Eq. (4.16); by a similar procedure we may calculate $e^*_{x\ O_2} = 3.97 \, kJ/mol$ by concentrating it from the atmosphere. Finally, Smith et al. [7] give $-\Delta g_{f(Fe_2O_3)}$ as 742 kJ/mol.

a landfill, and the material corrodes and further mixes, the exergy of the iron may eventually approach very low values. However, this will take time to happen. To illustrate the loss in exergy that is due to mixing, consider an ideal mixture made up of many components.

The exergy of an ideal material mixture is the sum of the individual component molar exergies, minus the exergy loss on mixing, as subsequently given (see [5, 6]):

$$e_x^* {}_M = \sum x_i e_x^* {}_i + RT_0 \sum x_i \ln x_i. \qquad (4.19)$$

The difference then between the unmixed and the mixed components is identical to the result given in Eq. (4.11). As long as mixing is the main reason for the degradation of a material, then recapturing and separating this material would appear to be the main route for fully recycling it. The theoretical minimum cost for accomplishing this task would be identical to the exergy loss on mixing. This is indicated by the dashed arrow in Fig. 4.2, which returns the material to its full value.

Although the paths just described in Fig. 4.2 provide a conceptual framework for thinking about how materials are used in product life cycles, the actual evaluation of these transformations requires more detail than is provided here. One problem is that many mixtures are not ideal and therefore a more accurate version of Eq. (4.11) is needed, usually by use of empirically obtained activity coefficients. Connelly and Koshland [3] give an example of this for methanol–water mixtures, and Amini et al. [9] and Castro et al. [10] discuss this for metal solutions. Furthermore, in many cases, feasible separation methods for metal mixtures do not exist, and so new alloys are made by diluting mixtures with pure material. Although this scenario can be modeled with thermodynamics, it will deviate from the minimum work calculated here (see Amini et al. [9]). Furthermore, many mixtures of interest in recycling and separation problems are not only not ideal, they are not actually molecular mixtures at all. That is, applications of thermodynamics to problems of resource accounting sometimes ignore this distinction and consider solid macroscopic mixtures such as the components of the Earth's crust, and mixtures of shredded waste materials, as mixtures. These are not molecular mixtures, and therefore the mixing and separation results developed earlier in this chapter do not rigorously apply. Nevertheless, they may apply by analogy and provide some useful insights. These are discussed in the next section.

4.4 Recycling and Materials-Separation Analogies

The theoretical results for separation and extraction from the previous sections suggest significant limits on our ability to economically manage resources in society. For example, Eq. (4.15) suggests that as we disperse and mix materials, as in our emissions and wastes, the effort to recapture them could become enormous, possibly making them for all practical purposes lost to society. One of the earliest contributors to this line of thought was Nicholas Georgescu-Roegen [11], who applied the limits established by the second law of thermodynamics to resource accounting and economics. At about the same time, the new field of ecological economics was developing, and several other researchers emerged to make significant contributions in this area including Ayres [12, 13], Berry [14], Cleveland [15], and Ruth [16]. Also see the three chapters by E. Williams et al. (Chap. 7), S. Gößling-Reisemann (Chap. 11),

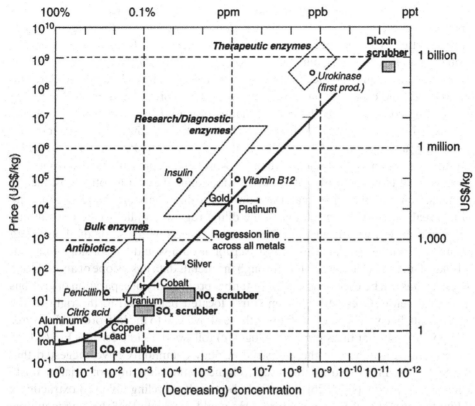

Figure 4.3. Sherwood Plot showing the relationship between the concentration of a target material in a feed stream and the market value of (or cost to remove) the target material [24]. ppm, parts in 10^6; ppb, parts in 10^9; ppt, parts in 10^{12}.

and M. Ruth (Chap. 16) in this book. In a similar fashion, Eqs. (4.11) and (4.13) suggest that there are significant differences between material mixtures and that these could affect our ability to recapture these mixed materials from our wastes. Interestingly, there is a considerable amount of physical data, which agree quantitatively with the concentration scaling given by these equations.

4.4.1 Extraction

The idea that the cost to extract a material from a dilute mixture increases with the degree of dilution (1/concentration) has been explored empirically in a variety of fields. In chemical engineering, Thomas Sherwood [17] is given credit for making this observation as early as 1959. Subsequently Sherwood and others have added additional data to a log–log plot of price versus dilution, which has come to be known in chemical engineering and industrial ecology as the "Sherwood Plot" [18]. Apparently independently, metallurgists also established a similar relationship for metal ores and minerals, starting in the 1970s [19–21]. More recently, Holland and Peterson [22], and then Johnson et al. [23] updated the price versus dilution data for metals, and in his book, *Technology and Global Change*, Grübler [24] combined much of these data in a single plot, which is reproduced in Fig. 4.3.

Note that, in the dilute region, where $c < 10^{-2}$, the data indicate that price rises linearly with $1/c$, where c is the mass fraction concentration of the target material in the mixture. The central line in the figure is the curve fit for metals provided by Holland and Petersen [22], but extended well beyond the data for metals. Note that the curve flattens in the concentrated region where $c > 10^{-1}$. Holland and Petersen showed that the cost of metals prepared from dilute ores is dominated by mining and milling costs, whereas the cost of metals prepared from concentrated ores is dominated by smelting and refining costs. Their work shows that the crossover occurs between 1% and 10%. Hence, in the dilute region ($c < 0.05$), the Sherwood Plot captures the essential cost scaling. In other words, $1/c$ represents the transport and processing of large quantities of materials that dominate over all other cost factors for the extraction of a target material from a very dilute mixture. Note also that the plot reveals three different groups in the dilute region. Parallel to the central line, one could draw a line through the biological materials with a slope of about $1/kg of mixture. Similarly a line through the scrubber data would have a slope of about 0.1¢/kg of mixture. The central line through the metals data has a slope of about 1¢/kg. These trends can be used to assess the potential profitability of separation operations in a gross sense. Ores and dilute mixtures located below and to the right of the appropriate line would be unprofitable; those above and to the left, profitable. Ayres [25] pointed out that the log(price) $\sim$ log($1/c$) follows the same form as Eq. (4.15).

Several researchers in the industrial ecology community have suggested that the Sherwood Plot could also be used to indicate the recycling potential of waste streams and products [23, 26]. This could be true if recycling involved extracting a dilute target material from a mixture. One could speculate that this line would lie above the line for obtaining metals from ores. In most cases, however, recycling is more closely related to separation than to extraction. This is because waste streams of discarded products almost always contain multiple target materials and often at high concentrations. Furthermore, materials that are not targeted will bear a disposal cost that acts to encourage targeting even marginal materials. In the next subsection we explore this analogy further.

4.4.2 Separation

This section outlines the use of an expression like Eq. (4.11) to represent the cost for separating the materials from a waste product. However, instead of using a thermodynamic analogy, an analogy is introduced from information theory. There are several advantages to the information theory approach that will become evident as the method is introduced. More details can be found in earlier publications [27–30].

The product recycling problem can be broken down into two subproblems. The first is to represent the materials-separation system. If you were to visit various recycling systems such as for automobiles or computers you would find that they have a similar pattern. After some useful or potentially dangerous components are removed, the products are shredded into smaller pieces and then separated by various means, for example magnetics, eddy currents, electrostatic charges, and density. There are multiple separation steps, and the whole system could be presented as a tree diagram, as shown in Fig. 4.4.

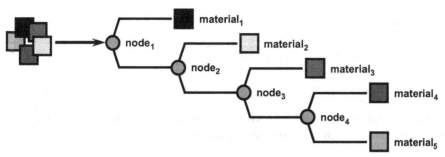

Figure 4.4. Examples of a tree diagram showing a material-separation system for five materials using four binary separation steps labeled as $node_1$ through $node_4$.

In this diagram, the product or mixed waste stream (or both) enters at the trunk and the separated materials exit at the branches. Separations take place at the nodes, and in the simplest scheme with binary separation steps and M incoming materials, there are $M - 1$ nodes. More complicated systems can and do exist, but most can be represented as tree diagrams. In fact, the tree diagrams themselves can be used as a representation of how large and complex the recycling system is. The simplest metric to measure the size of the systems could be just to count the nodes. A slightly more sophisticated measure would be to weight each node by the amount of material that passes through the node. As it turns out, tree diagrams are used in information theory to represent how large a code is for representing a message. In this case each branch would represent a word. At the same time (and this is the second part of this problem), information theory quantifies just exactly how much information is in a message. This means that any given code could be compared with the actual information the code is trying to represent. This comparison would give an estimate of just how efficient the code is. A famous result from Claude Shannon's information theory shows that the information content of the message is a lower bound on all possible code lengths. This result is called the "Noiseless Coding Theorem" [31–33]. It is introduced here because the recycling problem can be represented in a directly analogous manner. The parts of the problem are (1) the representation of the recycling system as a tree diagram, and (2) the representation of the information content or "complexity" of the mixture. The first part has already been mentioned. We measure the cost for this system as the weighted-average node count (number of separation steps) $\bar{n}$. That is, if there are n_i number of separation steps to isolate material i with initial mass concentration c_i, then

$$\bar{n} = \sum_{i=1}^{M} c_i n_i,\qquad(4.20)$$

where M is the number of materials in the input stream. In information theory n_i would represent the number of characters in a word and c_i would be the probability of the occurrence of the word. $\bar{n}$ is then the average code-word length.

The measure of the information content of the mixture (or degree of "material mixing") can be formulated in a manner quite similar to how Shannon formulated the information content of a message. The assumptions and results are the same. To paraphrase Shannon, there are three requirements for this measure of mixing

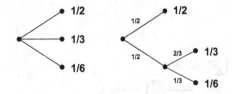

Figure 4.5. Illustration of the weighted-sum property for mixtures of mixtures.

(which we will represent here as $\mathcal{H}$ to avoid confusion with enthalpy H), but still use a similar symbol as used in communications theory:

1. $\mathcal{H}$ should be continuous in the c_i.
2. If all the c_i are equal, $c_i = 1/M$, then $\mathcal{H}$ should be a monotonic increasing function of M (with equal concentrations, there is more mixing when there are more materials).
3. $\mathcal{H}$ should be additive. Thus, if a mixture can be broken down as a mixture of mixtures, then the final $\mathcal{H}$ should be the weighted sum of the individual component values of $\mathcal{H}$. The meaning of this is illustrated in the equality and in Fig. 4.5. On the right-hand side of Fig. 4.5 we see a mixture of mixtures. On the left we see the final three-component mixture. For this special case, making the two representations of this problem equal requires that

$$\mathcal{H}\left(\frac{1}{2}, \frac{1}{3}, \frac{1}{6}\right) = \mathcal{H}\left(\frac{1}{2}, \frac{1}{2}\right) + \frac{1}{2}\mathcal{H}\left(\frac{2}{3}, \frac{1}{3}\right).$$

Shannon showed that the only $\mathcal{H}$ satisfying these three assumptions is of the form

$$\mathcal{H} = -K \sum_{i=1}^{M} c_i \log c_i, \qquad (4.21)$$

where K is a constant, M is the number of materials, and c_i is the concentration of material i. By convention, we set $K = 1$ and take logarithms to base 2, yielding $\mathcal{H}$ in bits.

For our purposes, we use $\mathcal{H}$ as a measure of material mixing. $\mathcal{H}$ can be interpreted as the average number of binary separation steps needed to obtain any material from the mixture. Of course, this function is also quite similar to the thermodynamic work of separation for an ideal solution [Eq. (4.11)], in which case we would use mole fractions instead of mass fractions and the constant would be different. The final step is to show that $\mathcal{H}$ is a lower bound for $\bar{n}$. As already mentioned, Shannon shows this in his Noiseless Coding Theorem. The assumptions in information theory can be applied directly to recycling [27]. One important assumption is that each branch end result is a unique material. In fact, the sequence of separation steps defines the material. The result is that

$$\sum_{i=1}^{M} c_i n_i \geq -\sum_{i=1}^{M} c_i \log_2 c_i \qquad (4.22)$$

or

$$\bar{n} \geq \mathcal{H}. \qquad (4.23)$$

Table 4.1. *Product data used in Figs 4.6 and 4.7*

Product	0m K^i ($)	$\mathcal{H}$ (bits)	Recycling rate (%)
Automobile battery	10.95	1.30	96
Automobile	358.61	2.22	95
Catalytic converter	107.54	0.399	95
Refrigerator	34.69	1.67	90
Newspaper	0.028	0.095	70
Automobile tire	1.85	0.575	66
Steel can	0.004	0.060	63
Aluminum can	0.019	0.001	45
HDPE bottle (#2)	0.012	0.163	27
PET bottle (#1)	0.008	0.476	23
Glass bottle	0.002	0.003	20
Desktop computer	17.69	2.36	11
Television	7.05	2.09	11
Laptop computer	2.79	2.89	11
Aseptic container	0.005	1.10	6
Cell phone	0.908	2.91	1
Work chair	12.19	2.27	0
Fax machine	6.43	2.09	0
Coffee maker	0.535	1.93	0
Cordless screwdriver	0.130	1.80	0

Notes: Recycled-material values from [34, 35]. References for the product bills of material and recycling rates can be found in [29, 30].

That is, our material mixing metric is a lower bound on separation cost. We use $\mathcal{H}$ as our estimate of the cost of separation. From a practical point of view, this result greatly simplifies the cost calculation, because $\mathcal{H}$ requires only knowledge about aspects of the material-counting scheme for the product and no detailed knowledge about the nature of the recycling system.

With these results, our profitability requirement for product recycling is then

$$\sum_{i=1}^{M} m_i k_i > -k_b \sum_{i=1}^{M} c_i \log_2 c_i, \tag{4.24}$$

where m_i is the mass of material i (kg), k_i is the value of material i ($/kg), and k_b is the processing cost per bit ($/bit).

To test this assumption, we analyzed 20 products with widely different material compositions and recycling rates in the United States [29, 30]. The data are plotted in Fig. 4.6, and Table 4.1 gives the details. The ordinate in Fig. 4.6 is the single-product recycled-material values ($\sum m_i k_i$) and the abscissa is the material mixing parameter $\mathcal{H}$. The area of the circles around each data point represents the degree of recycling for the product (i.e., fraction of retired products that enter the recycling system). Note from the table that automobiles are recycled at about 95%.

Figure 4.6 was constructed in a manner somewhat similar to the Sherwood Plot, in that products that would appear to be profitable to recycle should appear in the top left-hand corner of the figure, and the unprofitable ones should be in the bottom right. It appears that the data do line up that way, with a very rapid decay in recycling

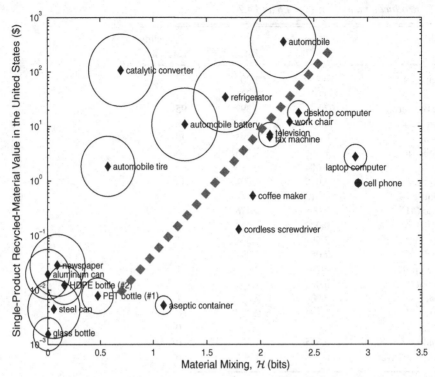

Figure 4.6. Single-product recycled-material values ($\sum m_i k_i$) and material mixing ($\mathcal{H}$) with recycling rates (indicated by the area of the circles) for 20 products in the United States. The "apparent recycling boundary" is shown as a dashed line; see [29].

rates (indicated by the size of the circles around each data point) as one moves from upper left to lower right. To highlight this abrupt transition a diagonal line is added to represent the apparent recycling boundary. Also, the data are replotted in a 3D fashion in Fig. 4.7 to lend another perspective on this dramatic change in recycling rates as one moves from high-value, low-mixing products to low-value, high-mixing products. The definitive organization of the data in this figure suggests that an entropy function like Eq. (4.11) or (4.21) is a good surrogate for the cost to separate materials. Additional information on the effects of alternative counting schemes on the calculation of $\mathcal{H}$ can be found in [29, 30].

4.4.3 Historical Design Trends

Given the insights provided by Figs. 4.6 and 4.7, it is now possible to plot the historical trends for product designs and make quantitative statements about how they have changed in terms of recycleability. This is done for three products: automobiles, refrigerators, and computers; the data are presented in Fig. 4.8.

The results illustrate a rather significant design trend. In general, all products have become materially more complex, which is shown as a large displacement along the $\mathcal{H}$ axis. The ironic exception is the sports utility vehicle (SUV), which has both increased in material value and decreased in material mixing, both for the same reason: the addition of many kilograms of aluminum and steel. In addition to changes

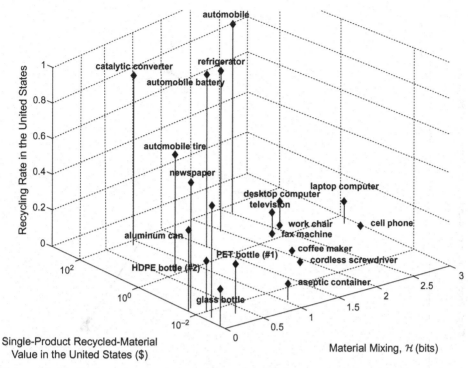

Figure 4.7. Single-product recycled-material values ($\sum m_i k_i$) and material mixing ($\mathcal{H}$) versus recycling rate for 20 products in the United States (i.e., 3D version of Fig. 4.6).

in material mixing, we also see changes in material value. These are due mostly to changes in product size and, to a lesser extent, material composition. In general, refrigerators and SUVs have gotten bigger, whereas computers and 1950s- to 1980s-era cars have gotten smaller. Overall, the trends show an apparent remarkable reduction in the recyclability of products that is due primarily to greater material mixing. Given the rather significant resources devoted to developing complex material mixtures for products compared with the rather modest resources focused on how to recapture these materials, it appears that there is reason for concern. As a consequence, recent policy actions such as take-back laws and "extended producer responsibility" appear to be clearly warranted in order to reclaim the materials in these products.

4.4.4 How to Differentiate Between Extraction and Separation Cases

In the previous sections of this chapter, cost-scaling schemes were presented for material extraction and material separation. Here it is shown how to differentiate between material mixtures that are dilute, and therefore can be treated on the Sherwood Plot, and the concentrated mixed-material systems typical of product recycling. To explore the differences between these two situations, we start by writing an expression for the largest value of $\mathcal{H}$ obtainable for a very general mixture made up of M materials. Of the M materials in this mixture, $M - 1$ materials are considered of value, and together have a mass concentration c_v, and the one remaining material is waste, with a mass concentration $1 - c_v$. The largest value of $\mathcal{H}$ obtainable for any

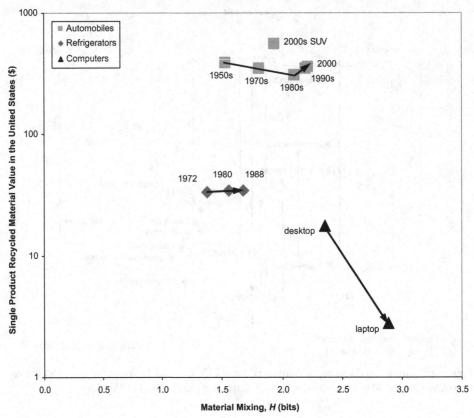

Figure 4.8. Design trends in refrigerators, automobiles, and computers. Note that material value refers to 2007 dollars; see [29, 30].

given mixture of this type would be the one with the $M - 1$ valuable materials evenly distributed within their mass fraction c_v. Using the additive property of $\mathcal{H}$, this can be written as

$$\mathcal{H} \leq (1 - c_v)\log_2 \frac{1}{1 - c_v} + c_v \, \log_2 \frac{(M - 1)}{c_v}. \tag{4.25}$$

This equation says that as a solution or ore gets increasingly dilute (the kind that the Sherwood Plot can treat), $\mathcal{H}$ becomes smaller and smaller. In fact, in the limit as $c_v \to 0$, $\mathcal{H} \to 0$ for any value of M. As a practical example, consider a relatively nondilute mixture near the lower bound of what can be treated as a "Sherwood material" with $c_v = 10^{-3}$ but with 9 co-mined valuable materials, i.e., $M = 10$. Equation 4.8 above gives the upper bound on $\mathcal{H}$, as 0.015 bits. This is a very small value of $\mathcal{H}$, about two orders of magnitude smaller than the typical complex products shown in Figs. 4.6 and 4.7 and well below the range where we draw the "apparent recycling boundary," i.e., $\mathcal{H} > 0.5$. Nevertheless, this value does overlap with some of the simple products in the lower left-hand corner of Fig. 4.6.

We can gain further insight into this problem if we create a plot for the two different material types. This is shown in Fig. 4.9. Here we have plotted a large number of mixed-material products (diamonds) and co-mined ores (squares). The x axis is as plotted in Fig. 4.3 and corresponds to the dilution parameter $1/c_v$. This

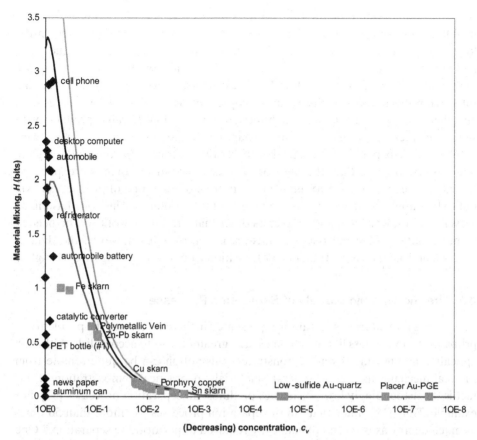

Figure 4.9. Plot showing various products and metal ore deposits in terms of $\mathcal{H}$ and $1/c_v$ along with Eq. (4.25) for $M = 4$ (nearest to origin), 10, and 50 (farthest from origin). Select data points are labeled with the names of products, as in Fig. 4.6, or names of mineral deposit models per Cox and Singer [36].

number implies a value statement concerning what part of a mixture is of value and worth capturing and what part is not. The y axis is the measure of material mixing $\mathcal{H}$, measured in bits. The material-counting scheme for calculating $\mathcal{H}$ includes all of those materials that are separated, including the waste stream.

The results show that the two material systems we are analyzing essentially lie along different axes. That is, in general, dilute mixtures are confined to the x axis with only very small values of $\mathcal{H}$, whereas the products are confined to the y axis and are quite concentrated. Near the origin, however, there is a third region of concentrated, relatively simple material systems. This region includes concentrated ores such as iron and aluminum, and simple products such as bottles and cans. We can also plot Eq. (4.25) with different values of M to show the upper bounds on $\mathcal{H}$. This is done for $M = 4, 10$, and 50. As can be seen, the three lines converge in the dilute region, eliminating the possibility of a dilute solution with a large value of $\mathcal{H}$.

It is in the third region, however, where we eventually run into the limits of both models. As already mentioned, the lower bound for the Sherwood Plot for metals is in the range of concentrations between 1% and 10%, whereas for the products it is probably somewhere below $\mathcal{H} = 0.5$. It is in this region ($\mathcal{H} < 0.5$) where we are unable to resolve the details of the recycling rates for the simple products in the lower left-hand corner of Fig. 4.6. This is likely to be related to the low values of $\mathcal{H}$ and the growing importance of other factors. Nevertheless, the fact that there are many very low-monetary-value, low-$\mathcal{H}$-value products with essentially zero recycling rates below this group, e.g., Styrofoam^{TM} cups, paper cups, plastic bags, staples, etc., and there are several relatively high-monetary-value, low-$\mathcal{H}$-value products with significant recycling rates within this group, e.g., aluminum cans 45%, steel cans 63%, and newspapers 70%, means that there still is a transition zone in the region, but it is less definite than the one for $\mathcal{H} > 0.5$. Tentatively then, we expect the transition zone for $\mathcal{H} < 0.5$ to run near the bottom of the group of products shown, but to be rather broad, encompassing many of those products. This zone has mixed recycling rates, which depend on factors other than $\mathcal{H}$. Future work will be needed to more fully resolve the issues in this area. For $\mathcal{H} > 0.5$ however, the data in Figs. 4.6 and 4.7 indicate a rather clear transition from low recycling rates to high.

4.5 Thermodynamic Models of Separation Processes

The models for materials separation as presented in this chapter cast the problem as a probabilistic event; less likely outcomes take greater effort to accomplish. Although appealing and useful, as was demonstrated, this notion can be quite remote from the real mechanisms used to separate materials. A practical demonstration of this fact is the astoundingly low efficiencies of almost all separation processes [37]. For example, King [18] states that "most separation processes use more than 50 times as much energy as is thermodynamically required to perform the separation." One could interpret this statement with optimism, emphasizing the great gains that are possible. However, a more sober interpretation would have to take into account how inadequate the ideal mixing results are for describing real separation processes. This situation can be remedied, however, by the construction of more appropriate but rather detailed thermodynamic models. Although this approach can be quite intensive, it can also be very rewarding. The problem is outlined here. For example, when various materials-separation mechanisms are reviewed, they generally have a similar theme: find a major material property difference between the targeted materials and then exploit it. Often the exergy difference on separation is positive; the materials go from a high exergy state to a low one. That is, the separation itself is free; the exergy that is expended goes into materials preparation and recovery. As a specific example, consider the separation of plastics by gravity separation in a water-based sink–float tank. The water density can be adjusted, for example, by adding salts, such that some plastics float and some plastics sink. The exergy expenditures for this process are primarily related to initially grinding up the plastics and, after separation, drying the plastics. Therefore the appropriate detailed thermodynamic models for identifying the minimum work needed for a sink–float separation process would include at least a model of the grinding (or shredding) process and a model of the drying process. Other steps may need to be added as well, for example, pumping and conveyance. Still other separation processes would require different thermodynamic models appropriate to the mechanisms and equipment employed. These would be more realistic models, but quite specific to the application.

4.6 Life-Cycle Assessment of End-of-Life Options Using Thermodynamic Variables

In this chapter we focused primarily on modeling and minimum work (or cost) for materials extraction and separation and then applied these result to the economic circumstances that shape real-world extraction and recycling processes. There are indeed still other applications of thermodynamics, information theory, and economics to materials separation and recycling. Of particular interest is the analysis of end-of-life options for retired products. This is an area that can inspire considerable controversy and where economics and thermodynamics may not be as quite compatible as was previously implied.

The environmental analysis of end-of-life options for retired products usually takes the form of a life-cycle assessment that may look at, among other variables, energy resources used (usually the sum of the lower heating values for all of the fuels used plus other energy resources; see for example [38]), or may look at cumulative exergy consumed (see for example [9, 10, 39]). These schemes almost always apply some credit for the avoided losses associated with presumably displaced primary materials. The results frequently show the clear advantages of recycling, particularly of metals. The allocation of these benefits either to the product that uses recycled materials or the product that provides the materials for recycling (or both) is not without controversy [40, 41].

On the other side of the coin, however, are the economic studies, usually from the community's perspective, that may in fact show recycling to be a marginally economic activity. Often these results hinge on transportation and collection costs [42, 43].

Hence resource accounting (using thermodynamic variables) and economic accounting do not always agree. Although this may occur for a variety of reasons, one recurring theme is that external and social costs are rarely included in economic calculation, whereas to some extent they can more readily be included in resource accounting.

Appendix A

Derivation of the Statistical Entropy of a Mixture

We may characterize the entropy of a mixture by using Boltzmann's entropy equation $S_M = -k \ln \Omega$ and counting the number of arrangements Ω of a species with r members on a lattice with n sites. We obtain the number of ways r species can be arranged on n lattice locations without double counting as

$$C(n, r) = \frac{n!}{r!(n-r)!}. \tag{4A.1}$$

Hence there are $\Omega = C(n, r)$ combinations. The other species with $(n - r)$ members would occupy the remaining sites. Now, using Stirling's approximation, $n! \cong \sqrt{(2\pi n)}$ $n^n e^{-n}$ so that $\ln n! \cong n \ln n - n$, we get

$$\ln C(n, r) = n \ln n - n - r \ln r + r - (n-r) \ln (n-r) + (n-r)$$
$$= r \ln n + (n-r) \ln n - r \ln r + r - (n-r) \ln (n-r) + (n-r)$$

$$= [r \ln n/r + (n - r) \ln (n/n - r)] \cdot [n/n]$$
$$= n[c \ln 1/c + (1 - c) \ln 1/1 - c],$$

where c (the concentration of r) is r/n.

This gives

$$S_M = -kn[c \ln c + (1 - c) \ln (1 - c)]. \tag{4A.2}$$

Note that entropy S_M is an extensive property (scales with n). Compare this with Eq. (4.12) in the text. This result can be generalized beyond two species. For example, with j species, $n_1, n_2, \ldots, n_j$, then

$$\Omega = \frac{n!}{n_1! \, n_2! \ldots n_j!}.$$

By the same procedure this gives

$$S_M = -nk \sum_1^j c_j \ln c_j.$$

This can also be written as

$$S_M = -k \sum_1^j n_j \ln c_j,$$
$$S_M = -k[n_1 \ln c_1 + n_2 \ln c_2 + \cdots + n_j \ln c_j].$$

REFERENCES

[1] Y. A. Çengel and M. A. Boles, *Thermodynamics An Engineering Approach*, 5th ed. (McGraw-Hill, New York, 2006).
[2] E. Gyftopoulos and G. Beretta, *Thermodynamics – Foundations and Applications* (Dover, New York, 2005).
[3] L. Connelly and C.P. Koshland, "Two aspects of consumption: Using an exergy-based measure of degradation to advance the theory and implementation of industrial ecology," *Resources, Conservation and Recycl.* **19**, 199–217 (1997).
[4] J. Szargut, D. R. Morris, and F. R. Steward, *Exergy Analysis of Thermal Chemical and Metallurgical Processes* (Hemisphere and Springer-Verlag, New York, 1988).
[5] J. de Swaan Arons, H. Van Der Kooi and K. Sankaranarayanan, *Efficiency and Sustainability in the Energy and Chemical Industries* (Marcel Dekker, New York, 2004).
[6] N. Sato, *Chemical Energy and Exergy – An Introduction to Chemical Thermodynamics for Engineers* (Elsevier, New York, 2004).
[7] J.M. Smith, H.C. Van Ness, and M.M. Abbott, *Introduction to Chemical Engineering Thermodynamics*, 6th ed. (McGrall-Hill, New York, 2001).
[8] S. Baumgartner and J. de Swaan Arons, "Necessity and inefficiency in the generation of waste – A thermodynamic analysis," *J. Ind. Ecol.* **7**, 113–123 (2003).
[9] S.H. Amini, J.A.M. Remmerswaal, M.B. Castro, and M.A. Reuter, "Quantifying the quality loss and resource efficiency of recycling by means of exergy analysis," *J. Cleaner Prod.* **15**, 907–913 (2007).

[10] M.B.G. Castro, J.A.M. Remmerswaal, J.C. Brezet, and M.A. Reuter, "Exergy losses during recycling and the resource efficiency of product systems," *Resources, Conversation Recycl.* **52**, 219–233 (2007).

[11] N. Georgescu-Roegen, *The Entropy Law and Economic Process* (Harvard University Press, Cambridge, MA, 1971).

[12] R.U. Ayres, "Eco-thermodynamics: Economics and the second law," *Ecol. Econ.* **26**, 189–209 (1998).

[13] R.U. Ayres, "The second law, the fourth law, recycling and limits to growth," *Ecol. Econ.* **29**, 473–483 (1999).

[14] S. R. Berry, "Summary of recycling, thermodynamics, and environmental thrift," *Bull. Atom. Sci.* **28**, 8–15 (May 1972).

[15] Cutler J. Cleveland, "Biophysical economics: From physiocracy to ecological economics and industrial ecology," In *Bioeconomics and Sustainability: Essays in Honor of Nicholas Gerogescu-Roegen*, edited by J. Gowdy and K. Mayumi (Elgar, Cheltenham, England, 1999), pp. 125–154.

[16] M. Ruth, *Integrating Economics, Ecology and Thermodynamics* (Kluwer Academic, Dordrecht, The Netherlands, 1993).

[17] Y.K. Sherwood, *Mass Transfer Between Phases* (Phi Lambda Upsilon, Pennsylvania State University, University Park, PA, 1959).

[18] C.J. King, *Separation and Purification: Critical Needs and Opportunities* (National Research Council, National Academy Press, Washington D.C., 1987).

[19] R.E. Cech, "The price of metals," *J. Metals* 21–22 (December 1970).

[20] W.G.B. Phillips and D.P. Edwards, "Metal prices as a function of ore grade," *Resour. Policy* **2**, 167–178 (1976).

[21] P.F. Chapman and F. Roberts, *Metal Resources and Energy* (Butterworth London, 1983).

[22] H. D. Holland and U. Petersen, *Living Dangerously: The Earth, Its Resources, and the Environment* (Princeton University Press, Princeton, NJ, 1995).

[23] J. Johnson, E.M. Harper, R. Lifset, and T. E. Graedel, "Dining at the periodic table: Metals concentrations as they relate to recycling," *Environ. Sci. Technol.* **41**, 1759–1765 (2007).

[24] A. Grübler, *Technology and Global Change* (Cambridge University Press, Cambridge, 1998).

[25] R.U. Ayres, *Information, Entropy and Progress* (American Institute of Physics, Melville, NY, 1994).

[26] D.T. Allen and N. Behmanesh, "Wastes as raw materials," in *The Greening of Industrial Ecosystems* (National Academy Press, Washington, D.C., 1994).

[27] T. Gutowski and J. Dahmus, "Mixing entropy and product recycling," presented at the IEEE International Symposium on Electronics and the Environment, New Orleans, LA, May 16–19, 2005.

[28] J. Dahmus and T. Gutowski, "Materials recycling at product end-of-life," presented at the IEEE International Symposium on Electronics and the Environment, San Francisco, CA, May 8–11, 2006.

[29] J.B. Dahmus and T. G. Gutowski, "What gets recycled: An information theory based model for product recycling," *Environ. Sci. Technol.* **41**, 7543–7550 (2007).

[30] J. B. Dahmus, "Applications of industrial ecology: Manufacturing, recycling, and efficiency," Ph.D. dissertation (Department of Mechanical Engineering, Massachusetts Institute of Technology, Cambridge, MA, 2007).

[31] C. E. Shannon, "A mathematical theory of communication," *Bell Syst. Tech. J.* **27**, 379–423, 623–656 (July and October, 1948).

[32] C. E. Shannon and W. Weaver, *The Mathematical Theory of Communication* (University of Illinois Press, Urbana, IL, 1964).

[33] R.B. Ash, *Information Theory* (Dover, New York, 1965).

[34] New York Spot Price, Kitco, Champlain, NY. Available at http://www
.kitco.com/market/ (accessed March 19, 2007).

[35] *Recycler's World*. Recycle Net Corporation, Richfield Springs, NY. Available
at http://www.recycle.net (accessed March 19, 2007).

[36] D. P. Cox and D.A. Singer, *Mineral Deposite Models* (U.S. Department of the
Interior, U.S. Geological Survey, Washington, D.C., 1986).

[37] T. Gutowski, "Thermodynamics and recycling, a review" presented at the IEEE
International Symposium on Electronics and the Environment, San Francisco,
CA, May 19–20, 2008.

[38] A.L. Craighill and J. C. Powell, "Lifecycle assessment and economic valuation
of recycling: A case study," *Resources, Conservation Recycl.* **17**, 75–96 (1996).

[39] J. P. DeWulf and H. R. Van Langenhove, "Quantitative assessment of solid
waste treatment systems in the industrial ecology perspective by exergy analysis," *Environ. Sci. Technol.* **36**, 1230–1135 (2002).

[40] A. Nicholson, "Methods for managing uncertainty in material selection decisions: Robustness of early stage life cycle assessment," M.S. thesis (Department
of Mechanical Engineering, MIT, Cambridge, MA, 2009).

[41] R. Frischknecht, "LCI modelling approaches applied on recycling of materials
in view of environmental sustainability, risk perception and eco-efficiency,"
in *Recovery of Materials and Energy for Resource Efficiency*, R'07 World
Congress, Davos, Switzerland.

[42] F. Ackerman, *Why Do We Recycle?* (Island Press, Washington, D.C., 1997).

[43] R. Porter, *The Economics of Waste* (RFF Press, 2002).

5 An Entropy-Based Metric for a Transformational Technology Development

Dušan P. Sekulić

5.1 Introduction

Discussion about sustainable development should start with a rigorous definition of what sustainability is in the context of utilization of natural, human, and manufactured resources. A necessary next step in implementing any such definition should be a selection of metrics for measuring a departure of a given system from sustainability thus defined. However, two main interrelated weaknesses of most efforts devoted to formulating definitions and establishing sustainability metrics have been quite visible. First, sustainability is often defined in only a somewhat qualitative manner across several disciplines, and, second, as a consequence, too many and too diverse sets of proposed metrics have been introduced – without any consensus among the proposers. Thus proclaiming a sustainable development analysis as a well-structured discipline would be premature. In general, it has been assessed that "sustainable development has broad appeal and little specificity" [1] Hence my position would be that this field of inquiry in general cannot yet be interpreted as a coherent "science of sustainability," although a need for establishing such a science (science of sustainability) has already been promoted [2].

Reasons for that state of affairs are multiple (and inevitably involve a significant level of misconceptions among participants in this notoriously transdisciplinary domain). Let us illustrate this statement by a prominent example of the Bruntdland definition of sustainable development.[1] This definition has inspired most of the early development in this field and still is the one most cited. However, it can hardly be described as precise and rigorous. On the contrary, its qualitative nature and unspecified underlined system of values (expressed through vague terms of the "needs of the present" and "[needs] of future generations"), although intuitively understandable, offer very little in terms of precise, quantitative measures for establishing science-based criteria.[2]

---

[1] Sustainable development is defined as "development that meets the needs of the present without compromising the ability of future generations to meet their own needs. "Our common future," in *Report of the World Commission on Environment and Development* (World Commission on Environment and Development, General Assembly Resolution 42/187, 11 December 1987, United Nations, New York, 1987).

[2] The effort to establish a coherent set of sustainability metrics worldwide started by the adoption of "Agenda 21" at the UN Conference on Environment and Development in Rio de

Furthermore, one may convincingly argue that a single criterion cannot define
the state of sustainability inherently mirrored through multiple sets of state variables
in diverse realms. Consequently either multiple sets of less-aggregate metrics or a
single set of highly aggregate metrics need to be established. Efforts in both directions
have been attempted, and a large number of metrics do exist in various domains of
sustainability inquiry.

To make this general inquiry palatable, our focus is restricted here. That is, we
are interested in identifying a metric that would be able to evaluate *a performance
level* of a technology but in terms of resources utilization. In other words, a better
state of sustainability would be related a priori to a selection of a transformational
technology with lower resources utilization. Hence we are interested in confronting
old (possibly obsolete) technologies with transformational (new) technologies. More
important, we are interested in uncovering maximum margins of improvements in
resource utilization for a given task – irrespective of the technologies utilized to
accomplish it.

The resolution of such an ambitious objective requires not only gradual, evolutionary improvements of the existing technologies, but radical changes achieved
through implementation of novel transformational technologies characterized with
significantly smaller impacts related to resources use. In this context, the transformational technology is "a technology that is a game-changing, as opposed to merely
incremental" [3].

The question to address in this chapter may be formulated as follows: How can
thermodynamics assist us in identifying and evaluating a need for promoting a new,
transformational technology?

Therefore this chapter has two main objectives: (1) to offer a brief, critical
review of existing metrics used to evaluate sustainable development in general, and
(2) to explore whether thermodynamics may offer a relevant, rigorous metric for
a technology evaluation in the context of promoting new, transformational technologies for sustainabile development.[3] We focus on the identification of both (1)
system properties and (2) process metrics[4] that are uniquely connected to the state
of a technology performance for sustainability.[5] In anticipating what is considered a
heuristic justification, we will, following in the footsteps of a long line of researchers

Janeiro, June 3–14, 1992, UN Department of Economic and Social Affairs, Division of Sustainable Development. An integral text of the document is available at http://www.un.org/esa/
sustdev/documents/agenda21/english/agenda21toc.htm (accessed March, 2009).

[3] This chapter does not intend to provide a comprehensive discussion on thermodynamics metrics
and, more specifically, energy resources metrics. Rather, it offers a discussion of entropy generation
as a metric, which can be related to energy metrics (like exergy) through rigorous implementation
of thermodynamics. For a discussion of the methods promoting such accounting for resource use,
refer to Chap. 3 in this book.

[4] In thermodynamics, for example, entropy is defined as a system property. Entropy production,
however, is not a system property but a process metric. In that context, we search for its use in a
sustainability study.

[5] The task of developing technology for sustainable development through the use of exergy metric is
addressed in Chap. 10 of this volume (by Van der Vorst et al.) from the point of view of balancing
exergy equivalents of resources use within both the technosphere and at the intersection of the
technosphere and the ecosphere.

in primarily engineering thermodynamics,[6] argue that such an effort should involve the concept of entropy generation.[7]

We proceed as follows. First, the concept of the state of sustainability is introduced. A formulation of a problem to be considered is offered next (i.e., a promotion of the need for establishing metrics in the given context). Subsequently, a brief review of a set of the most prominent indicators for general assessment of sustainability is offered. Next, a single system of indicators is briefly discussed in a bit more detail to establish the domain of inquiry (resources-utilization minimization). Furthermore, a single indicator (one of many) is selected from that set (e.g., an energy-resources use metric) and considered in the context of its thermodynamics definition. Finally, the entropy-generation-based metric is formulated to illustrate the rigor to be implemented for the selection of new versus obsolete technologies.

5.2 State of Sustainability

We define a *sustainability change* within the context of sustainability analysis as the evolution of the *state of sustainability* of a well-defined system. The state of sustainability is defined with a complete characterization of the system in terms of *sustainability properties* (to be defined). The state of sustainability is the state that, when altered by a resource interaction, does not leave a permanent change in the environment while at the same time allows for the preservation of all system functions. The sustainability properties are the attributes that can be evaluated at any instant of time by measurements or mathematical calculations (or both). The measured or calculated value must not depend on the tool of measurements, surroundings, and other instants of time. We postulate that the state of sustainability is characterized by the system's *and* the environment's capacity to restore their original respective states when modified by an interaction (passing through a sustainable change of state by analogy to reversibility within a thermodynamic framework). In thermodynamics, a state can be altered, leaving or not leaving the net effects in the environment. The associated process is reversible if it can be performed in at least one way such that both the system and its environment are restored to their respective initial states [4]. The process's irreversibility (i.e., the level of nonideality of the considered system's change of state vs. initial states of the system and its environment) is not necessarily the sustainability metric, but we postulate it as an adequate, *related* metric. Namely, irreversibility is related to the degree of perfection of the system evolution while being affected by the resources used.[8] Of course, what the degree of perfection is in the sustainability realm may not be the same as it has been in the thermodynamics realm. However, in the thermodynamics realm, we do know that determining either

---

[6] See Chap. 2 (by Sekulić).

[7] Although the metrics used for sustainable development assessments are discussed, the primary objective is not to define a sustainability metric upfront. It is argued that such metrics will be possible to define in a rational manner only after a rigorous definition of the state of sustainability and its properties are established. Our objective is to define a metric for transformational technologies development that would replace the state-of-the-art technologies still utilizing excessive energy resources and featuring large footprints.

[8] We will consider open systems; hence system interactions involve bulk mass flow interactions.

entropy generation or a derivative entity (such as exergy destruction) through mass, energy, and entropy balancing would lead to a metric that can be related to efficiency of the resource use (although it would not necessarily offer an insight into the overall impact on the surroundings). This is well researched, and the irreversibility analysis has been used for a long time for analyzing individual processes or complex technical systems changes of state. The formulation of the efficiency of resource use can subsequently be incorporated into a sustainability study enriched with additional metrics focused on transdisciplinary objectives.[9]

An important point has to be made as a distinction between our focus and the existing expositions of the topic. Most of existing well-established (older) studies of resources use in the context of sustainability (if related to thermodynamics-based metrics) consider energy- and material-resources flows at inlet–outlet ports of a given system only. Moreover, if for example expressed by using exergy, these are related to the assumed virgin state of the resource at the inlet ports of the system, expressed usually by chemical exergy. Such an approach cannot offer causal, cause–consequence guidance for required changes of the state of sustainability versus the resource use (although it can identify its change). The structure of the system and its relation to the resources conversion within the scope of considered technologies must be observed and *efficiencies* of related processes considered. This is the reason why it is argued that the state of sustainability must be identified.

The scope of the discussion in this chapter illustrates an inadequacy of adopting any single metric to fully identify the sustainability level (the state of sustainability). Furthermore, it promotes entropy generation as *one* of multiple important indicators of the change of sustainability – not as *the* metric.

5.3 Problem Formulation

As indicated, a complex phenomenon spanning different disciplines of inquiry cannot possibly be evaluated with a single metric. This implies that an evaluation of such a phenomenon will always be an intricate task. That is, it is to be expected that multiple metrics (referring to multiple aspects of the problem in multiple domains of its manifestation) would be needed. However, a large number of metrics per se are not the fundamental problem an analyst would face. The fundamental problem would typically be a selection of a limited set of independent properties of the system, capable of describing the system fully.

On the other hand, one should remember that a physical system in a given state, when considered in terms of thermodynamics basic laws, can be defined by use of an infinite number of properties. This is simply because any combination of properties is a property as well.[10] However, identification of *independent* properties that define the state of the system is a must in any such analysis. Therefore we may assume that the same holds for a sustainable system considered as a thermodynamic system.

---

[9] One should be careful in characterizing the proposed formalization. Here only a framework for a path to establishing what a rigorous definition of sustainability should be is proposed. An ultimate answer is not offered because it seems that, at the present state of affairs in the related field of inquiry, much more understanding and better correlation between the related disciplines are needed.

[10] Of course, a state of the system to be described needs only a limited number of independent properties, in accord with the *state principle*; see Chap. 1.

Such a system would be exposed to multiple interactions and would evolve following its state changes in response to existing interactions during a given process. This evolution would be identifiable in terms of a set of independent properties. So each of the ultimately identified sustainability metrics has to be related deterministically to these properties or process variables. In this context, let us reformulate the problem but at the higher hierarchical level, i.e., not at the process but at the technology level.

If a considered task involves an outdated technology, traditional approaches to increasing the resources-use performance of such a system would be to search for imperfections within the given technological realm.[11] For example, a resource use for materials processing (say heating), such as that related to a manufacturing process (say thermal treatment), may be improved by increasing the heat delivery system performance, reducing the temperature differences across the heat transfer paths within the system, improving insulation, and other similar efficiency-related steps – often while keeping the given basic technology for accomplishing the task (making a given product) unchanged. Any improvement under such a scenario would be a "within-the-box" solution and would most likely be effective. But if we separate the task from the technology for accomplishing it by identifying the theoretical minimum resource use for the task – not for the given technology – the so-determined margin for improvement may represent a quantum leap – not incremental improvement.[12] The problem we face is to identify a metric that would be related to the task.

5.4 Properties and Indicators

Within the framework of this discussion, the distinction among three often interchangeably used terms should be clarified: (1) a sustainable system *property*, (2) an *indicator* of sustainability, and (3) a sustainability *metric*. To provide a consistent approach to distinctions among the three, an analogy with thermodynamics is used. A system in a given state is described in terms of its properties. As elaborated in Chaps. 1 and 2, a property is a system's attribute in a given state. If we would like to describe a complex system in terms of sustainability, we must define that system's properties so that system's state of sustainability would be fully described by the set of independent properties.[13] This requirement implies not only an existence of the (state) correlation among the variables featured by the system, but also assists in defining the causality relationship in describing the change of system's state exposed to interactions. This is a very important feature of a system property, especially in

---

[11] See Section 10.2 in Chap. 10 or Subsection 15.8.3 in Chap. 15 on extensions of traditional exergy analysis into the broader domain at the intersection of technosphere and environment.

[12] See Chap. 6 for a series of applications of this approach within the realm of materials processing for manufacturing.

[13] In my opinion, a key weakness of a number of efforts promoting the use of advanced concepts of thermodynamics in the context of sustainability studies has been a lack of realization that these complex systems (taken out of their domains such as the technosphere, a society, an economy, or an industrial sector) must still be defined rigorously in the thermodynamic sense – if a thermodynamic property is used to characterize them. Even when the related concepts (such as exergy) are adequately defined at the conceptual level in the context of the involved physics, an implementation of the analysis may sometimes be offered without clearly defining *what the system is, what its properties are, how the state is defined*, and *how the interactions crossing the so-defined system's boundaries change the system's state.*

the context of sustainability studies in which, traditionally, causality relations among properties or interactions (or both) may not be clearly established, regardless of a possibly very high correlation.

An indicator of sustainability is not a property of a system in a given state. In *economy*, an indicator may be a statistics-based entity or a figure of merit for economic performance [such as the gross domestic product (GDP) in macroeconomics]. An *ecological* indicator offers information about the impact of human activity on ecosystems (such as a biological species indicator – apparently also not a property of the ecosystem). An efficiency in *engineering* is an indicator defined as a figure of merit that usually obeys three main characteristics: (1) it reflects the main feature of the process considered in the selected context, (2) it is dimensionless, and (3) it ranges, as a rule, between 0 and 1 [such as a thermal efficiency figure of merit, say indicating a ratio of an actual and an ideal (defined in the given context) power rate – apparently not a property of the working fluid of a power plant system, but a well-defined indicator of its function vs. the "design objective"].

Metrics are standardized indicators of measure, used within the selected system of measured indicators in an effort of assessing a system or a process (or both). In the context of a metric's semantic content, the use of either term, metric or indicator, may be a matter of personal preference. In our considerations we will distinguish the meaning of the *indicator* from the *metric* in the context of selecting a metric's subset out of the set of indicators to serve the analysis as a key, narrower subset; otherwise these terms will denote the same concept.

We will clearly distinguish a system property in a given state from a numerical value of a metric (or, for that matter, an indicator in general). In thermodynamics, for example, it must be clear that entropy *is* a system property, but entropy generation within the system *is not* a system property; it should be considered as an indicator of process quality and may be selected as a metric that indicates the level of imperfection.

5.5 Nonthermodynamics Sustainability Indicators

The driving force behind a number of efforts in defining sustainability indicators has been a need to monitor economic, social, and environmental impacts of human activities. Therefore, many indicators (there is no a single generally accepted system of indicators yet) have been devised from, as a rule, uncoupled studies of natural, social, or economic processes (or all three). This state of affairs is very unfortunate, but possibly a natural consequence of rapid development of related disciplines. It must also be clear that a number of indicator systems have been defined within different sustainability frameworks (say, ecological or social) with a multifaceted structure, hence leading to an impressive set of a total number of indicators (in the hundreds). Unfortunately, most of them are not necessarily suited for aggregation across the disciplines. The prominent issue involved with these assessments is devising a methodology for building metric(s) that can be used to make a comprehensive sustainability relevant decision at a given instant of time.[14]

---

[14] Reemphasized from J. McGlade, "Finding the right indicators for policymaking," in *Sustainability Indicators. A Scientific Assessment*, edited by T. Hak, B. Moldan, and A.L. Dahl (Scope 67, Island Press, Washington, D.C., Covelo, London, 2007), pp. 17–21.

Still, some general principles should be easy to formulate. For example, sustainability metrics should offer (1) reliable, (2) low uncertainty level, and (3) quantitative measures of processes of human–environmental systems that (4) stay valid across time [5].

As previously indicated, a variety of sustainable metrics systems have been proposed [1], of which only a few of the most prominent are listed here. These include the United Nations Commission on Sustainable Development (UNCSD) development indicators [6], the human development index (HDI) [7], well-being indicator [8], environmental performance index (EPI) [9], ecological footprint (EF) [10], and the genuine progress indicator [11] (an economic indicator inspired by a need to overcome shortcomings of more traditional ones [12]), to mention only a few. These measures are diverse in level of aggregation, scope, and domains of applicability. Some possess a high level of aggregation, merging multiple indices into a framework of assessment, but some are fully disaggregated; see Table 5.1. Except for the HDI and EF, all offer sets of multiple indicators with low aggregation and are primarily statistical in nature.

Let us select one of the most comprehensive sets and consider it in more detail, as in Table 5.2.[15] Table 5.2 offers a summary of one such set, the UNCSD metrics. These include metrics split across four aspects of sustainable development: (1) social, (2) environmental, (3) economic, and (4) institutional. Methodology for determining quantitative values of these metrics is well defined,[16] which is not necessarily the case with the other metrics sets.

Careful study of the subsets of the metrics in Table 5.2 indicates that the given UN set has not been structured within a coherent framework. Namely, some metrics are expressed in terms of absolute measures (e.g., abundance of selected key species), whereas the others are included as relative (e.g., forest area percentage). Some are dimensionless (e.g., protected areas), and some are not (e.g., energy intensity). Most are quantitative (e.g., GDP per capita), but some are also qualitative (e.g., sustainable development strategy). Because high aggregation cannot be consistently implemented, any assessment task requires careful weighing of individual impacts. Further study of these metrics reveals that many of them (metrics related to population, economics, and institutional indicators) cannot easily be related to processes governed by conservation principles of physics, such as has firmly been established in a thermodynamic analysis for materials and energy flows.

Several general, easy-to-observe problems involving these broad metrics definitions, are as follows:
1. different physical scales or dimensions of measured entities,
2. different time rates of observed phenomena,

---

[15] We are not interested in a detailed study of these indicator systems, a topic beyond the limited scope of this chapter. The consideration offered serves only a purpose of providing an insight into their scope and emphasizes a need to approach to the metrics definition in a system-based rigorous manner.

[16] "Indicators can provide crucial guidance for decision-making in a variety of ways. They can translate physical and social science knowledge into manageable units of information that can facilitate the decision-making process. They can help to measure and calibrate progress towards sustainable development goals. They can provide an early warning, sounding the alarm in time to prevent economic, social and environmental damage." From *Indicators of Sustainable Development: Framework and Methodologies*, UN Commission on Sustainable Development, 9th Session, Background Paper No. 3 (DESA/DSD/2001/3, 2001), p. 2.

Table 5.1. *A selection of systems of sustainability indicators*

System of indicators	Description	Comments	WWW source[17]
UNCSD [14]	Total of 58 indicators organized into four sets of metrics (social, environmental, economic, and institutional)	No aggregation between the sets aggregated at the indicator level, data depend on the source	http://www.un.org/esa/sustdev/csd/csd9_indi_bp3.pdf
European Environmental Agency (EEA [15])	Total of 345 indicators (climate change, air pollution, water, waste, energy, transport, agriculture, biodiversity, terrestrial, fisheries)	Criteria for selecting: relevance, policy target, data sets, spatial coverage, temporal coverage, national scale. Understandable, well founded methodology, priority topics in strategy	http://themes.eea.europa.eu/indicators
EPI [13]	Total of 20 core indicators (2–8 variables, total of 62 variables). Overall progress toward sustainability.	Cross-national evaluation. High aggregation. Databases vulnerability. Selection criteria: relevance, performance orientation, transparency, data quality	http://sedac.ciesin.columbia.edu/es/esi/
Eurostat Sustainable Development Indicators (SDI [16])	Total of 10 themes (socioeconomic, consumption and production, social, demographic, health, climate change and energy, transport, resources, global partnership, governance)	Three levels of indicators (overall – robust, operational objectives, actions). Inspired by UNCSD but expands on European Union level. Number of indicators: 11 in level 1, 33 in level 2, and 80 in the level 3	http://www.insee.fr/fr/publications-et-services/dossiers_web/dev_durable/eurostat_report2007.pdf
Indicator for Sustainable Energy Development (ISED [17])	Total of 41 indicators organized to economic, social, and environmental dimensions	Includes 15 indirect driving force indicators, 14 direct driving force indicators and 12 state indicators	http://www.iea.org/dbtw-wpd/textbase/papers/2001/csd-9.pdf
HDI [18]	A highly aggregate indicator that includes (1) life expectancy, (2) educational level through adult literacy and enrollment, and (3) standard of living	The upper limit of the index is predefined by assumed individual achievement limits. It measures three basic aspects of human development. High data quality. No environmental impact	hdro@undp.org
EF [19]	The EF uses the yields per unit area of primary product flows to calculate the area necessary to support a given activity.	Highly aggregate indicator (carrying capacity, overconsumption, biocapacity) but includes only environmental aspects.	http://www.footprintnetwork.org/en/index.php/GFN/page/methodology/

[17] All www sites in Table 5.1, last column, were accessed in Spring, 2009.

Table 5.2. *UN Sustainability metrics (general), UNCSD*[18]

Social metrics	Environmental metrics	Economic metrics	Institutional metrics
% of population below poverty line; Unit: %; Ravallion (1994)	Emission of greenhouse gases; Unit: Gg; IPCC 2nd Assessment Report (1995); UNFCCC review	GDP per capita; Unit: $US; SNA (1993)	National sustainable development strategy; Unit: qualitative; Carew-Reid (1994)
Gini index of income inequality; Unit: [-] ∈ (0,1); Chen et al. (1992)	Consumption of ozone-depleting substances; Unit: tonnes of ODS by ODP; Ozone Secretariat UNEP (2000)	Investment share in GDP; Unit: %; System of National Accounts (1993)	Implementation of ratified global agreements: Unit: ratio of legislated and ratified agreements; www.basel.int
Unemployment rate; Unit: %; Yearbook of Labor Statistics, ILO, Geneva	Ambient concentration of air pollutants (urban); Unit: ppm or ppb; WHO (2000)	Balance of trade in goods and services; Unit: $US; System of National Accounts (1993)	Number of Internet subscribers per 1000 inhabitants: Unit: number per 1000 of the population http://www.itu.int/ti
Average female wage to male wage ratio; Unit: %; Current Int. Recomm. On Labor Statistics, Geneva 1988	Arable and permanent crop land area; Unit: 1000 ha; FAO Stat. Dev. Ser., (2000)	Debt to GNP ratio; Unit: %; The World Bank, Global Development Finance (2000)	Expenditure on research and development as a % of GDP; Unit: %; OECD (1995)
Nutrition statistics of children; Unit: %; WHO/NUT/ 97.4, Geneva 1997	Use of fertilizers; Unit: kg/ha; FAO (1998)	Total ODA given or received as a % of GDP; Unit: %; The World Bank, Global Development Finance (2000)	Economic and human loss due to natural disasters; Unit: number of fatalities per $US; CRED (1994)
Mortality rate of children under 5 years old; Unit: per 1000 live births; Pressat (1972)	Use of agricultural pesticides; Unit: metric tons of active ingredients per 10 km^2; http://www.fao .org/	Intensity of material use; Unit: kg, or m^3 per $1000 of GDP; Hammond (1995)	Main telephone lines per 1000 inhabitants; Unit: number per 1000; http://www.itu.int
Life expectancy at birth; Unit: years of life; DESA, ESA/P/WP.156 (1999)	Forest area as a % of land area; Unit: %; FAO (2000)	Annual energy consumption per capita; Unit: GJ/capita; UN Energy Stat (1991)	

(continued)

[18] Details and reference citations listed are provided in *Indicators of Sustainable Development: Framework and Methodologies*, UN Commission on Sustainable Development, 9th Session, Background Paper No. 3 (DESA/DSD/2001/3, 2001).

Table 5.2 *(continued)*

Social metrics	Environmental metrics	Economic metrics	Institutional metrics
% of population with adequate sewage disposal; Unit: %; WHO, WSSSMR (1990)	Wood harvesting intensity; Unit: %; FAO State of World Forests (1999)	Share of consumption of renewable energy resources; Unit: %; World Energy Council, Energy Statistics Yearbook	
Population with access to safe drinking water; Unit: %; WHO, WSSSMR (1990)	Land affected by desertification; Unit: km^2 and % of land affected; S.W Bie (1990)	Intensity of energy use (energy use per unit of GDP); Unit: MJ/$; http://www.iea.org; Intensity: energy use – service; Intensity: energy use – manufacturing; Intensity: energy use – residental; Intensity: energy use – transport	
Population with access to primary health care; Unit: %; El-Bindary and Smith (1992) Immunization Against Infections Childhood Diseases; Unit: %; WHO/EPI/GEN99 .012 (1999)	Area of urban formal or informal settlements; Unit: km^2; World Bank Housing (1993)	Generation of industrial and municipal solid waste; Unit: tons per capita per annum; UNEP (1993), OECD (1995)	
Contraceptive prevalence rate, Unit: %; World Population. Monitoring, ESA/P/WP.131 (1996)	Algea concentration in coastal waters; Units: mg of chlorophyll per m^3 or g of carbon per m^2 per year; UNEP 55 (1995)	Generation of hazardous waste; Unit: metric tons or tons per GDP; Bakes (1994)	
Secondary or primary school completion rate; Unit %; World Education Report UNESCO (1998)	% of total population in coastal areas; Unit %; UNDP, UNEP, WB, WRI (2000)	Generation of radioactive waste; Unit: m^3 per annum; IAEA Safety Series 115 (1996)	
Adult literacy rate; Unit: %; UNESCO Stat. Yearbook (1995)	Annual catch by major species; Unit: metric tons; FAO (1999)	Waste recycling and reuse; Unit: %; Rees and Wakernagel (1994)	
Floor area per person; Unit: m^2; World Bank, Housing (1993)	Annual withdrawal of water as % of total available water; Unit: %; Shiklomanov (1990)	Distance traveled per capita by mode of transport: Unit: km/yr; EEA (2000)	

Social metrics	Environmental metrics	Economic metrics	Institutional metrics
Number of recorded crimes per 100,000; Unit: police reported per year; ICVS, UNICR (1998)	Biochemical oxygen demand (BOD) in water bodies; Unit: mg/l of O_2 for oxidation; ISO 6060, (1989)		
Population growth rate; Unit: % DESA (1998)	Concentration of fecal coliform in freshwater; Unit: %; WHO (1993)		
Population of urban settlements; Unit: number of inhabitants; World Bank (1993)	Area of selected key ecosystems; Unit: km^2 or ha; Loh et al. (1999)		
	Protected area as a % of total area; Unit: %; McNeely (1993)		
	Abundance of selected key species; Unit: no, if mature individuals; Caro (1998)		

Notes: ppm, parts in 10^6; ppb, parts in 10^9; Acronyms explained in Nomenclature, pp. 159–160.

3. different physical domains of the observed phenomena,
4. different measurability or difficulty in quantifying the metrics,
5. difficulty in gathering and validation of data, and
6. difficulties in establishing causality.

An implied domain size imposes up front a corresponding scale for the change of rate of a particular metric, leading to an assumption of time scales that may not be appropriate within all domains (i.e., quite different time rates may exist). Natural, ecological-change phenomena versus economics perturbations or technology development level versus resources consumption clearly refer to different domains of observed phenomena and different time scales. The serious problems with these weaknesses are not only the difficulties in forming accurate databases for prediction of future trends (because the time scale of historical data is short), but the trustworthiness of aggregate data because of the differing methodologies used in gathering them (involving measurability, data uncertainty, validation, and causality).

It should be clear that, given these problems, diverse measures based on statistics of the values of multiple indicators and with correlations featuring not necessarily rigorously proven causality (often with many outliers) would not support reliable models. However, some metrics related to material flows and energy resources may be acceptable for rigorous science (thermodynamics) modeling. Let us consider a couple of prominent measures from Table 5.2 from that point of view.

5.6 Energy-Resources-Based Metrics

Arguably among the most important metrics (classified as "economic" in Table 5.2, but clearly with a much broader relevance) are the energy resources and material-use intensities. These metrics are defined as the *material use* in the units of mass (kilograms) or volume (cubic meters) per monetary unit (say 10^3 US$) of a GDP, and *energy use per capita* (in units of energy, joules, per population size) or *energy use per GDP* (aggregate). These hybrid metrics (in noncoherent physical and eco-nomical units)[19] are widely used. These metrics are arguably (by the proponents) correlated very well to a system's sustainability state. In Fig. 5.1,[20] a correlation between HDI and energy consumption per unit of GDP per capita for most of the world countries (between 1980 and 2006) is presented. The inset includes the data set for a single year, 2006. The HDI (Table 5.1), may be assumed to be well correlated by GDP per capita. Despite the presence of outliers, the correlation is quite strong ($r^2 = +0.9$).[21] However, the issue is whether the deterministic causality between the energy-resources consumption has a true cause–consequence relation with the aggregate HDI metric. Moreover, a simple metric, such as total energy consumption per capita as previously defined, does not reflect the quality of energy resource used (say, a type of converted energy) or the quality of energy transformations along the path of energy-resource use.

The quality of both the energy-resource type and the energy-conversion effi-ciency one, can be captured by thermodynamic irreversibility featured by an energy conversion system. The argument is as follows.

A level of sustainability with regard to a use of a particular resource can be measured by the margin of the selected resource's use in terms of its available energy versus its total availability, say ΔR. This margin depends on the current level of depletion of resources, and hence depends on the total available resource reserves on one side and on the rate of resource use on the other. The latter depends on the system or process resource-conversion efficiency, imposed by the level of technological development. So this margin rate of change (a time-rate derivative, $\partial R/\partial t$) must be nonnegative, i.e., $\partial R/\partial t \geq 0$. The magnitude of this derivative directly depends on system dynamics (a need for resources) on one side and on the efficiency of the resource utilization imposed by the technology used on the other. So we would benefit from assessing the quality of performance of a technology for a given task. This performance may be assessed in terms of entropy production along the path of accomplishing a *task* because entropy production signifies a departure of a real process from the ideal process between given two states of the system.

Ultimately the margin ΔR would depend on physical limits of the resource use efficiency. The ultimate physical limit of the resource use for a given task is inde-pendent of the current technology level or, for that matter, any future technology.

---

[19] Therefore such metrics would not be declared as being "thermodynamics metrics," regardless of the dominant role of the thermodynamics concepts used.

[20] Data used to construct this plot were handled by Z. Pu of the University of Kentucky (2009).

[21] The two independent verifications of the correlation were conducted by S. Yang and Z. Pu (2009) of the University of Kentucky to correlate HDI and the logarithm of the energy consumption per capita. Both calculations (the Parson product-momentum and Spearman's correlation coefficient) are virtually the same.

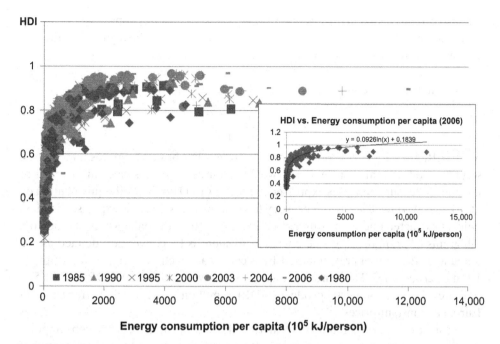

Figure 5.1. HDI vs. energy consumption per capita for various countries.

It is constrained by natural laws involving the task and not by the state-of-the-art technology. For that limit to be revealed, an analysis of ideal resource use for a given task must be performed.[22] Such an analysis can be done within the thermodynamics framework. So we postulate that the known thermodynamic irreversibility level, manifested during the process under consideration, may offer an insight. This analysis assumes the entropy-balance evaluation and a quantification of the entropy generation.[23]

How large the material-use margin and energy-resources consumption would be in a given case cannot be assessed by determining the available material- or energy-resource use only. An in-depth study of the process efficiency of technologies used for the considered tasks would be mandatory. For example, material use may be defined as primary or secondary and include data from available databases on, say, minerals and metals consumption. However, these quantities may not necessarily include information on process efficiencies in material-resources extraction and transformation. Furthermore, the use of the units of volume, as an alternative, may lead to an even less-rigorous metric formulation (vs. the use of mass). More important, the "use of energy," as suggested, is not necessarily related to the efficiency of its conversion from primary resources (which would, however, be related to a given technological level of the used technology).

Energy-resources use within this set of metrics is defined as the total energy consumption (the so-called "apparent consumption") of energy resources. This is not a thermodynamically rigorous concept. That is, "energy consumption" is not an

---

[22] See Chap. 6 by Gutowski and Sekulić on "Thermodynamic analysis of resources used in manufacturing processes" in this book.

[23] It should be clear that the mass and energy balances would be performed as well.

energy (property) of a given energy resource (at a given state) considered as a thermodynamic system, but a quantity of a resource aggregated through a hypothetical conversion into energy units (through a specified heating value of the resource).

Let us review in even more detail the total primary energy consumption (TPEC) as listed in the EEA core set (Table 5.1). It is defined as follows:

$$\text{TPEC} = \frac{\sum_{i=1}^{5} \text{ER}_i}{\text{GDP}}, \tag{5.1}$$

where ER_i, $i = 1, 2, \ldots, 5$, represent consumption of five basic energy-resources types: solid fuel, oil, gas, nuclear, and renewable resources.[24] In this calculation, each of the energy-resource terms is expressed in terms of 1000 tons of "oil-equivalent" (i.e., kton) energy units. Clearly the primary form of these resources is expressed in mass units of extracted energy-resources material flows (tons of coal, barrels of crude oil, kilograms of uranium ...). The energy equivalent is by convention defined as the amount of thermal energy released by a chemical reaction during combustion of 1 ton of crude oil.[25] This amount is approximately 42 GJ.[26] The GDP in Eq. (5.1) indicates the gross domestic product (GDP) in million monetary units (for EEA, in Euros) at constant prices.

From Eq. (5.1) it should be clear that the total primary energy-consumption metric (and a number of similar ones) offers very limited insight into a potential of the resource use (i.e., a quality in terms of higher conversion efficiency). Because of its rather heuristic nature and lack of connection to technology of conversion and in particular vastly variable availability (measured by exergy level, for example), it has a limited utility for the study of system sustainability.

More sophistication can be incorporated into the thermodynamics-based metrics in two main directions. First, the aggregation may be performed within the context of life-cycle assessment (involving natural resources), and second, instead of the energy-resources quantity accounting, the actual available (exergy) accounting may be promoted. The aggregation can be performed in several ways [20, 21], for example, an accounting for total exergy use along the life cycle:

$$\mathscr{E}_{\text{LC}} = \sum_i e_i \chi_i, \tag{5.2}$$

where e_i represents a specific energy-resource-utilization metric, i.e., the exergy use of ith resource per unit of resource quantity χ_i (expressed in mass units, for example). There are several schemes of structuring such metrics[27] based on the selection of e_i and χ_i quantities. These metrics suffer from imperfect representation of a distinction between renewable and nonrenewable energy resources, but, more important, such metrics are not well suited for uncovering locations of improvements of processes

---

[24] Note that the careful use of the wording "energy resource" instead of "energy" (a usage so prevalent in related literature) is dictated by a need to clearly distinguish the resources from energy as a state property; see Chap. 2 for more insight.

[25] This energy release is equal to the positive value of enthalpy of combustion (a lower heating value) under stoichiometric conditions (complete, theoretical combustion).

[26] 1 toe $= 41.868$ GJ, toe $=$ ton of oil equivalent; 77th ed. D.R. Lide, editor-in-chief, *Handbook of Chemistry and Physics*, (CRC Press, Boca Raton, FL, 1996); available at http://www.aps.org/policy/reports/popa-reports/energy/units.cfm.

[27] B. Bakshi, personal communication (2009). See more details elsewhere in this book.

along the life cycle because no correlation between the actual resource use and internal physical mechanisms of these processes would be revealed as long as the aggregation hides both the process efficiencies and their locations.

So, although such metrics do offer a clear idea about energy-resource use, they are not uncovering the margin of improvement of this or any other resource use for accomplishing a certain task.

One possible venue for identifying such a margin, as well as for identifying the imperfections within a process, can be an evaluation of entropy generation along the energy-conversion path. How that can be done is discussed in the next section.

5.7 An Entropy-Based Metric

In our consideration of metrics the discussion will be restricted to physical systems that are part of a technosphere and that interact with the ecosphere through fluxes of resources and products (mass and energy rates), as well as effluents and residues (mass and energy) interactions. In all such cases, defining the system well in a thermo-dynamic sense would be straightforward (but not necessarily trivial; see Chaps. 2 and 6). The system though should not be considered as an energy system; on the contrary, most systems of interest are not energy systems (i.e., would not be like a closed system of a power plant, a prime mover engine, etc.), but broadly are open manufacturing or materials-processing systems (see Fig. 2.5, Chap. 2). Any subsystem in such a system performs a certain task in accord with an underlined technology. Any so-identified subsystem (a system per se for the reminder of our analysis) changes state in both thermodynamic *and* sustainability realms and performs in the context of resources use according to the energy-conversion efficiency, dictated by the implemented tech-nology level. This performance would be in agreement with the conservation laws of mass, energy, and entropy (the latter case includes entropy changes, entropy flows across the system boundary, and entropy generation). In other words, such systems change their thermodynamic state (i.e., change at least one of the properties). This change may also be taking place simultaneously with the change of state of sus-tainability. Note that the associated change of entropy may not be telling an analyst anything explicitly about sustainability. For that matter, we should keep in mind that the entropy change per se may not be telling an analyst about process reversibility in the thermodynamic sense either (i.e., a change of entropy may be associated with a reversible change of state). However, a nonstate variable, entropy generation, may offer such an information. So we declare the *entropy generation* (i.e., not entropy as the property, but the "residue" needed to close the entropy balance) to be a metric. As we learned in Chaps. 1 and 2, this quantity is not a property of the system; it is an indicator of a "system's performance" manifested during a system's change of state. Let us consider the relevance of this metric through an example.[28]

---

[28] In this context, this example is considered recently in T. Gutowski, D.P. Sekulić, and B. Bakshi, "Preliminary thoughts on the application of thermodynamics to the development of sustainability criteria, ISSST 2009 – The International Symposium on Sustainable Systems and Technology, IEEE, Phoenix, AZ, May 18–20, 2009, CD edition. The same engineering task, a separation of chemical species from a mixture in general, and desalination in particular, has frequently been a subject of a large number of studies over a long period of time. For some relevant details, see Y.M. El-Sayed and A.J. Aplenc, "Application of the thermoeconomic approach to the analysis and optimization of

Let us assume that we are interested in determining the physical limit – the smallest possible energy-resource use – for a specific task. Chapter 6 explores this idea in more detail for various manufacturing processes. Let us consider an extraction of a certain chemical species from a mixture. We must define the sustainability realm of the considered process first. Let us assume that the objective would be to explore how much energy resources we would need to invest to separate two constituents of a mixture. For the sake of obviousness, we consider a depletion of water resources. Let us hypothesize that freshwater depletion would lead to a need to extract it from seawater (a mixture of water and salts). The associated process in an engineering context is called desalination. So let us assess the minimum energy use for separation in a desalination process in order to substitute for all the freshwater withdrawals. Furthermore, let us assume that one of the perceived scenarios requires that a total water supply be secured by desalinating the water. This task can apparently be accomplished by a number of processes known within related engineering art. In a *sustainability* study, however, we are not necessarily interested in an implementation or a gradual improvement of existing technologies, but in a projection of minimum resources use for a given task *regardless* of technology used (so that an assessment of cardinal trends of resources use would be determined). Therefore the technologies to be used must be assessed in light of the *physical limits* of resources use for a given task (that is, regardless of current, or for that matter any future technologies). Hence, for a sustainability study, we need a metric to measure a margin of resource use between the competing current technologies and the ultimate physical minimum of a resource (say, energy-resource) use for a given task. Such a physical limit for performing a given task (a separation process in which H_2O is separated from $H_2O + NaCl$) would be the minimum energy interaction (available work) needed to accomplish the separation. So we may reformulate the task as follows: We separate water from salts present in seawater, but use the minimum needed amount of energy resources irrespective of an engineering technology. We may add an additional constraint by specifying the energy resource to be channeled from renewable resources. The minimum work, as thermodynamics teaches us (see Chap. 2), is the exergy expenditure for the process – that is, an exergy use irrespective of present or any other future technology to be implemented for accomplishing this task. We can determine this minimum exergy expenditure, however, by implementing the Gouy–Stodola theorem, Eq. (2.10), Chap. 2, i.e., $\dot{W}_{used} = T_0\dot{S}_{gen}$, where $\dot{S}_{gen}$ represents the entropy generation rate associated with the process of separation. Therefore the entropy generation inherent in the process of separation must be determined first.

To make this determination, we can use the exergy-balance equation for an open system that executes separation of pure water from salts in the saline water and utilizes solar energy with an ideal conversion into the exergy rate. This process takes place between a hydrosphere and a technosphere and involves an open subsystem that accepts saline and delivers freshwater.

a vapor-compression desalinating system, *J. Eng. Power* **92**, 17–26 (1970); R.B. Evans, G.L. Crellin, and M. Tribus, "Thermoeconomic considerations of sea water demineralization," in *Principles of Desalination* (Academic, New York, 1980), pp. 1–54; W.T. Andrews, W.F. Pergande, and G.S. McTaggart, "Energy performance enhancements of a 950 m³/d seawater reverse osmosis unit in Grand Cayman," *Desalination* **135**, 195–204 (2001); T. LeTourneau, L. Liang, A. Jha, and R. Knauf, *Desalination: Current Technologies and Future Solutions* (Siemens, 2009, personal communication).

Under a hypothetical scenario, we assume that the withdrawal rate would be defined based on the withdrawal rate in the United States for the year 2000, but projected to 2030 for the whole world. Under this scenario,[29] it would be interesting to determine whether the available seawater would suffice and what the amount of renewable energy resources would be for the task [22].

Assuming that no other heat and work interactions exist, in a steady state, entropy balance reduces to the equality of $\dot{S}_{gen}$ and the difference between inlet and outlet entropy rates of the open system for the desalination task. As shown by Gyftopoulos and Beretta [23], separation of salts from saline water, considered at any instant of time as a removal of a small amount of mineral constituents from a large quantity of seawater, leads to the entropy change in the process:

$$\Delta S = -[(\mathcal{R}/M_{H_2O})\ln(y_{H_2O})] = -[(\mathcal{R}/M_{H_2O})\ln(1 - y_{NaCl})] \sim [(\mathcal{R}/M_{H_2O})y_{NaCl}], \tag{5.3}$$

where y's represent mol fractions of the constituents (it is assumed that saline water has 35,000 ppm of NaCl in it). The entropy rate change, under the described conditions, is equal to the entropy-generation rate $\dot{S}_{gen} = \Delta \dot{S}$. By multiplying the entropy-generation rate by the surroundings temperature (assumed for the sake of calculation to be 25 °C), we can determine the exergy rate to be supplied to the separation process. Let us assume that in that separation process no output brine stream exists (i.e., the inlet saline water is entirely separated into the freshwater and solid salts). The following data are used. The withdrawal of freshwater [24, 25] in the United States is assumed to be 1.3×10^{12} kg/day (data from 2000). If we take into account that the U.S. population in 2000 was 285 million [26], and under the condition of the same withdrawal per capita for the world population in 2030 (9 billion) [26], the required withdrawal per year would be 1.5×10^{16} kg/yr. The amount of saline water is far larger than this amount [24], so the availability of this resource (from this source) would not be in question. However, what energy resource would be needed to get the required amount from the saline water is a clear sustainability issue. The exergy expenditure (a product of entropy generation and surroundings temperature) would be 1.41 kJ/kg.[30] So the exergy rate use for this task in the year 2030 would be ~1 TW. This is indeed a substantial amount of energy resources, with possibly a significant impact on sustainable development. More specifically, the projected consumption in the year 2030 of each of the nonrenewable energy-resource types (oil, gas, coal, nuclear, and other) is of the same order of magnitude, i.e., between 1 and 8 TW [27]. For the sake of understanding the physical scale of these numbers, note that about 1 TW of the chemical exergy rate from the plants of the biosphere is consumed by humans [20].

---

[29] The selection of the scenario is not within the scope of this consideration. We may consider multiple as well as more relevant scenarios in this context. This scenario is selected for the sake of obviousness.

[30] Note that, by following the same thermodynamic calculation and taking into consideration data related to the global water cycle, we can determine that the total input of exergy associated with rain and snow in a form of an essentially freshwater resource amounts to the order of magnitude of 6 TW [J. Szargut, "Anthropogenic and natural exergy losses (exergy balance of the Earth's surface and atmosphere)," *Energy* **28**, 1947–1054 (2003)]. But the use of this resource may have serious logistics issues (availability of the resource vs. location of use).

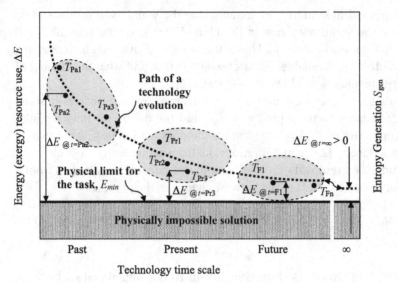

Figure 5.2. Physical limit of energy use for a given task vs. technology evolution.

The calculated physical limit of energy resources used constitutes not the *current* technological limit for a given task at a given instant of time and a given level of technological development, but the absolute minimum, *theoretically possible* amount, regardless of the technology level. From the sustainable development point of view, it is of interest to know that the most advanced desalination process currently under development requires at least four times larger exergy use[31] than the assessed minimum. A standard technology of today would require approximately eight times more than the most efficient state-of-the-art system. Finally, we may note that the technology of the mid-20th century would require at least an order-of-magnitude larger exergy expenditure [28, 29].

5.8 Energy Resource Use versus Technology Evolution Time

Let us now consider in more detail the uncovered margins of energy resources use for a state-of-the-art technology versus any future technology, but in light of a physical limit of energy-resource use for the given task. The energy resources use may be measured by any of the established metrics for accounting for resources use.[32] In Fig. 5.2 these margins are depicted, in principle, for a generalized task (to be accomplished by a technology at the given instant of time).

The task (say, a separation of two constituents of a system as in the previously discussed case, or material removal, as in a machining process, etc.) can be accomplished by using a state-of-the-art technology (say, T_{Pr3}) selected from a set of currently available technologies (T_{Pr1}, T_{Pr2}, and T_{Pr3}). Each of these technologies requires certain energy (exergy) resource utilization, which is smaller than for the technologies used in the past (T_{Pa1}, T_{Pa2}, or T_{Pa3}), but most likely larger, or significantly larger when compared with a hypothetical technology of the future

---

[31] Siemens technology, as reported in Siemens Research and Development, Ref. No. RN 2008.11.01e.
[32] See Chap. 3 by Bakshi et al.

(say, T_{Fn}). The key advantage of determining entropy generation as a metric in such a case would be a rigorous determination of the physical limit for energy (exergy) use – regardless of a possible development of any future technology. In a traditional characterization of sustainable development, the lack of precision related to the value system relevant for the future has now been mitigated by the determination of a physical limit *below which any future technology would never be able to reduce the resource use.* In this way, a large energy- (exergy-) resource use margin for an obsolete technology (say T_{Pa2}) measured as $\Delta E_{@\ t=Pr2}$, which is improved by the state-of-the art technology T_{Pr3}, i.e., $\Delta E_{@\ t=Pr3}$, would be replaced with a smaller margin in the future, $\Delta E_{@\ t=F1}$. Knowing that margin, we may plan energy-resource use and characterize a physical aspect sustainability state versus the given physical process. However, one should know that the given task would *never* be able to be performed with the margin equal to zero, i.e., $\Delta E_{@\ t=\infty} > 0$, a direct consequence of a limit determined by the second law of thermodynamics because entropy generation of a real process is never zero.

Finally, let us consider the selection of technologies for the analyzed task (e.g., desalination) in the context of the selection of a technology for a sustainable development but from the point of view of energy-resources use only.[33] The objective is straightforward. A number of technologies for the same task are available. On a technology time line, some are obsolete, some may be considered as the state of the art, and some may not yet exist, but most likely will be considered in the future. Through the assessment of entropy generation we may uncover the absolute margin of improvement.

It would be instructive to see in practical terms how the related technologies are engineered. A number of possible technologies may be used for the considered task (e.g., desalination). These include (1) multistage flash, (2) multieffect distillation, (3) reverse osmosis, (4) electrodialysis reversal, (5) electrodialysis and continuous electrodeionization, and (6) various hybrid processes. Each of these technologies requires different system engineering and inevitably different resource utilization. Each of these systems must be considered within its respective science and engineering field. In Fig. 5.3 a traditional vapor-compression desalination process is presented schematically along with the most recent novel desalination process. It should be clear that the two processes and underlined technologies are dramatically different – the first one combines thermocompression, phase change, and regeneration to facilitate separation whereas the second one does the same through electrochemical processes. The first system consists of a separating zone (evaporator–condenser) and regenerative zone (two- and three-fluid heat exchangers). The process is driven by the work energy resource delivered to the compressors and pumps. The second process uses electrodialysis and electrodeionization, including ion-exchange softening, and it is driven by the applied voltage difference.

Table 5.3 offers calculated and assessed or measured energy-resources use needed for several outdated, state-of the art, and novel technologies (including the two presented in Fig. 5.3) compared with the calculated physical limit energy

---

[33] It must be clear that multiple criteria would exist in any real application so that an optimization of the constructed objective function under a set of imposed constraints would be necessary.

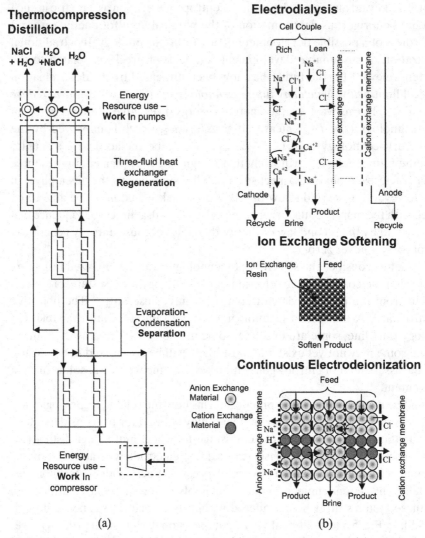

Figure 5.3. Desalination problem solved by (a) thermocompression; (b) electrodyalysis, ion exchange and Electrodeionization.

resources needed for the process of desalination. The analysis of data fully confirms the trends implied by Fig. 5.2.

Thermodynamic limit calculation is the first physical analysis step required for defining the resource needs and, more important, as previously illustrated, for assessing the physical limits for a task regardless of the present or future technology options. This aspect of the relevance of the entropy-generation metric or its exergy equivalent in a thermodynamic assessment is the most important one. No other metric assessment can offer this insight. Expressing it through energy units further offers exergy accounting. Finally, pondering such exergy quantities with monetary units (economic weighing) or environmental impact factors may offer (again through balancing conserved thermodynamic entities in the background) the tools for building objective functions to be used in exergoeconomics or environomic studies in the domain of sustainability at any hierarchy level.

Table 5.3. *Energy-resources use vs. technology and the physical limit*[34]

Technology	Energy-resource use, kJ/kg	Technology description	Comments
Thermal compression	30[a]	Separation and regeneration	Mid-20th century (old technology)
Reverse osmosis	16[b]	Typical	Electricity use only
	11[c]	SWRO, 105 membranes; 15 membrane vessels; hydraulic turbocharger	High-pressure pump–turbocharger pump
	8[c]	Upgrade SWRO; 196 membranes; 28 membrane vessels; DWEER exchanger	The same SWRO facility as above with upgrades
Multistage flash	14	Successive condensation and boiling	Approximately 85% plants in the world
Electrochemical	5[b]	Electrodialysis and Electrodeionization	The state-of-the-art proposed technology
Physical limit	1.4	Irrespective of technology	Zero recovery

[a] [28]; [b] LeTourneau et al., 2009, Loc. cit.; [c] Andrews et al., 2001, Loc. cit see footnote 28.

5.9 Technology Selection

We now find out how we can go about analyzing the given technological solution and improving it within the confines of the given technology through evaluation of entropy generation. In Fig. 5.4 a schematic of an idealized distillation system is presented, featuring internally fully reversible operation, and hence characterized with no sources of thermally caused entropy generation in any of the associated internal processes. Saline water enters the system at the state of equilibrium with the environment [in a restricted dead state, i.e., at (T_0, p_0) but including the presence of NaCl salinity] through a three-fluid heat exchanger.[35] This stream gets reversibly heated by two outgoing streams: (1) the brine (in case of a recovery ratio of less than 100%), and (2) pure water (both brine and pure water are at differing states vs. the environmental ultimate dead state because of different NaCl concentrations). The

---

[34] The physical limit value of the energy-resource use was calculated for the case of $T_0 = 25\ °C$ and saline water having 35,000 ppm NaCl, assuming that the recovery is 0%. Calculation is based on the entropy generation calculation as discussed in Gyftopoulos and Beretta, 1991, Loc. cit. Several other sources provide alternative calculations of the physical limit: R. W. Stoughton and M. H. Lietzke, "Calculation of some thermodynamic properties of sea salt solutions at elevated temperatures from data on NaCl solutions," *J. Chem. Eng. Data* **10**, 254–260 (1965); R. B. Evans and M. Tribus, "Thermo-economics of saline water conversion," *I&EC Proc. Design Develop.* **4**, 195–206 (1965); Y. M. El-Sayed and A. J. Aplenc, loc. cit.; *Desalination. A National Perspective* (National Academies Press, Washington D.C., 2008). It is relevant to notice that all these values are calculated to be around 2.5 kJ/kg. Note that some reported empirical data on energy use in desalination plants involve only the electrical energy used.

[35] For details of three-fluid heat exchanger modeling see D.P. Sekulić and R.K. Shah, "Thermal design theory of three-fluid heat exchangers," in Vol. 26 of *Advances in Heat Transfer* (Academic, San Diego, 1995), pp. 219–328.

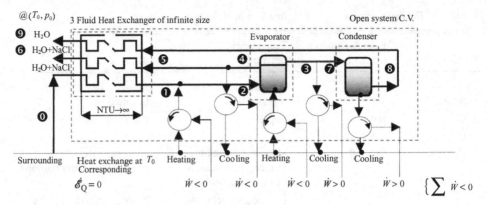

Figure 5.4. Idealized ($\Delta \dot{S}_{gen} = 0$) separation of H_2O from $H_2O + NaCl$ by a distillation system.

incoming water is subsequently heated, and in the evaporator it is separated into pure water vapor and saline water with an increased salinity (which is subsequently delivered out of the system as brine), after recovering exergy from that stream bringing it up to the initial thermal state (except for an increased salinity). The vapor of the pure water is condensed in the condenser and, after the recovery of its exergy up to the thermal state of equilibrium with the surroundings (restricted dead state, note the elimination of salinity), is leaving the system as the product (H_2O). Detailed descriptions of individual processes of similar systems are given in the literature [30, 31].

To make this engineering solution reversible, all heating and cooling processes must be declared as being reversible! This could be managed by hypothetical, idealized operations: (1) by utilizing an infinitely large, adiabatic three-fluid counterflow heat exchanger that operates at zero temperature differences; and (2) all heating and cooling tasks must be performed by using Carnot's cycles (either heat engines or heat pumps), operating between the environmental and targeted temperature. Note that by positioning the system boundary so that each of the Carnot engines stays within the system, the input heat transfer rates would be acquired at T_0, and hence the corresponding exergy equivalent of heat transfer rates would be zero and the only exergy interactions present would be the work rates (into or out of the system). The work rates' algebraic sum would offer (for an otherwise reversible system) the minimum exergy of energy resources needed for this operation. If the recovery rate is total (no brine removal), this minimum exergy becomes $\sim$1.4 KJ/kg, as calculated earlier. Figure 5.5 illustrates the flow of exergy of the system of Fig. 5.4 for a case of $N\%$, $0 < N < 100$, recovery [30], inlet temperature of $H_2O + NaCl$ of $T = T_0$, input mass flow rate of $\dot{m}$ kg/s, and standard salinity of $y\%$.

Note that the system of Fig. 5.4 for which the exergy flow of Fig. 5.5 is constructed operates reversibly except for the separation of the naturally appearing saltwater into the constituents. This leads to an overall work rate input for heating and evaporation that is larger than the total work output so that the per unit of mass difference becomes equal to $w_{min} \geq 1.4$ kJ/kg.

The exergy flow through the system of Fig. 5.4 presented in Fig. 5.5 is constructed assuming that all energy inputs into or out of the system are reversible, as are the heat transfer interactions between fluids in the heat exchanger (and no pressure drop

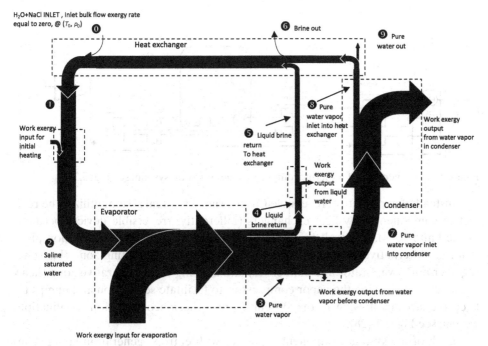

Figure 5.5. Exergy flow rate for the idealized operation of the system of Fig. 5.4.

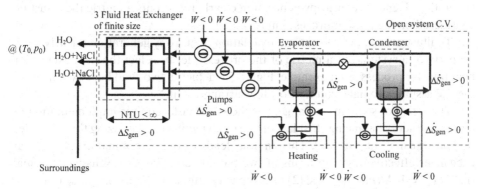

Figure 5.6. Distillation system (irreversible, entropy generation, $\Delta \dot{S}_{gen} > 0$, associated with any process).

losses included). Also, the system does not feature any fluid mixing. Hence no internal irreversibilities take place during any of energy interactions (work interactions). The process of separation though may be interpreted as executed with an entropy generation that would be in the exergy terms given by the Gouy–Stodola theorem [23].[36] Namely, the process of separation cannot be executed spontaneously and the use of exergy is needed.

A real system distillation technology would require energy inputs across finite temperature differences. For example, in principle, the system of Fig. 5.4 may be designed in the real world (irreversible processes present) as presented in Fig. 5.6.

---

[36] See Eq. (5.3) in this chapter.

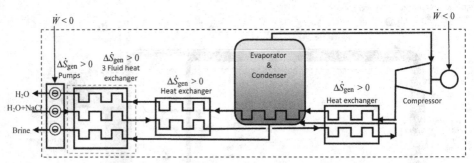

Figure 5.7. A schematic of a thermocompression desalination system; see Fig. 5.3(a).

Instead of hypothetical reversible Carnot engines, a real system utilizes the real system components, say heat exchangers (inherently irreversible devices because of the finite sizes),[37] as well as nonideal pumps and the presence of friction phenomena. Alternatively, the system design may utilize a separating zone (bringing together both evaporator and condenser) and keeping the regenerative zone (heat exchangers), including the vapor compressor, to facilitate separation and pumps to keep the streams flowing in the system, as in a thermocompression desalination system; see Fig. 5.7 [28].

Each of the system components operates with entropy generation larger than zero (irreversible processes). System improvements (although within the confines of the imposed technologies) require fighting the irreversibilities and reducing entropy generation. Hence the entropy generation clearly represents a metric that must be considered on the path of process improvements.

To illustrate the approach to an evaluation of entropy generation (clearly not a property of a system but a metric of the process), let us select the heat exchanger heat transfer process as an example. Figure 5.8 depicts a two-fluid heat exchanger of Fig. 5.7 assumed to have counterflow flow arrangement.[38]

The sizing of such an exchanger for the given task requires implementation of mass and energy balancing (first law of thermodynamics) and utilizing the principles of heat transfer analysis [32]. This analysis leads to the determination of the heat exchanger effectiveness ε in terms of two parameters: (1) heat exchanger thermal size, $\mathrm{NTU} = \mathrm{UA}/(\dot{m}c_p)_{\min}$, and (2) heat capacity rate ratio, $C^* = (\dot{m}c_p)_{\min}/(\dot{m}c_p)_{\max}$ for a given flow arrangement, i.e.,

$$\varepsilon = \varepsilon(\mathrm{NTU}, C^*, \text{flow arrangement}). \tag{5.4}$$

The heat exchanger effectiveness (the figure of merit based on the first law of thermodynamics) takes values between 0 (nonexistent exchanger) and 1 (hypothetical, infinitely large counterflow exchanger). This metric, however, does not take into account the entropy generation manifested in a heat exchanger that is due to the heat transfer across the finite temperature differences (and other sources of irreversibility – neglected in this discussion for the sake of simplicity but without essential consequences for the delivery of the main message of this analysis). Entropy

[37] D.P. Sekulić, "The second law quality of energy transformation in a heat exchanger," *J. Heat Transfer, Trans. ASME* **112**, 295–300 (1990), see [32] as well.
[38] The "min" and "max" designations are assigned to fluids with smaller and larger heat capacity rates, respectively (regardless of the temperature level).

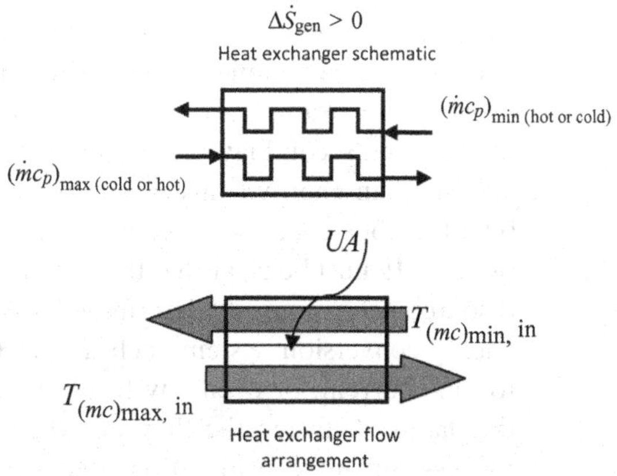

$$\Delta \dot{S}_{gen} > 0$$

Heat exchanger schematic

$(\dot{m}c_p)_{min\ (hot\ or\ cold)}$

$(\dot{m}c_p)_{max\ (cold\ or\ hot)}$

Figure 5.8. A heat exchanger schematic and flow arrangement (counterflow).

UA

$T_{(mc)min,\ in}$

$T_{(mc)max,\ in}$

Heat exchanger flow arrangement

balancing[39] must be implemented (that is, the consequences of both the first and second laws of thermodynamics must be considered). The resulting dimensionless entropy generation is [33]

$$\frac{\dot{s}_{gen}}{(\dot{m}c_p)_{max}} = C^* \ln\left(\frac{T_{out}}{T_{in}}\right)_{min} + \ln\left(\frac{T_{out}}{T_{in}}\right)_{max}$$

$$= C^* \ln\left[1 + \varepsilon\left(\frac{1}{\vartheta} - 1\right)\right] + \ln[1 + C^*\varepsilon(\vartheta - 1)]. \qquad (5.5)$$

In Eq. (5.5) ϑ represents the ratio of inlet temperatures, $T_{min,in}/T_{max,in} \neq 1$. It is apparent from Eq. (5.5) that if $\varepsilon = 1$ [see Eq. (5.4)] the entropy generation vanishes (i.e., for the infinitely large counterflow heat exchanger). That means that as long as the given technology uses a heat exchanger to transfer heat across the finite temperature differences, the entropy generation will be different from zero. This irreversibility cannot be eliminated (but can be reduced) and apparently other (transformational) technologies must be used to accomplish the task beyond the confines of the given solution (see Fig. 5.6, in which the heat transfer was delivered by a Carnot cycle operating system). It should be noted that in this kind of analysis, a single component should never be solely considered. The meaningful approach is implemented *only if the whole system* is optimized through entropy-generation minimization.[40]

One important additional aspect of this analysis has not yet been discussed. Namely, the rigorous calculation of physical limits of an ideal process operation for a separation process offers the minimum energy resources needed at the input into the process considered.[41] However, it does not consider the energy resources

---

[39] See Eq. (2.32) in Chap. 2. In a steady state, the left-hand side is equal to zero. The heat exchanger is adiabatic vs. the surroundings; therefore the entropy transfer terms are also diminished. Only two fluid streams enter or exit the system, so the entropy generation becomes equal to the total entropy change of the hot and cold fluids (i.e., of the fluid with larger and fluid with smaller heat capacity rates).

[40] See Chap. 15 on Exergoeconomics by Tsatsaronis in this book.

[41] A much more detailed discussion of a number of examples of the determination of minimum energy-resources use limits is presented in Chap. 6.

needed to generate that input within an associated energy subsystem; see Fig. 2.5, Chap. 2. This energy-resource amount, again in its physical limit, can be calculated from a thermodynamic analysis of the associated energy-conversion system point of view. For example, if that system is a power cycle, in an ideal reversible case, the first and second laws of thermodynamics require that the thermal energy input into such an energy-conversion closed system be $Q = W/(1 - T_0/T)$, i.e., larger by a Carnot's coefficient versus the required input into the considered separation process! It must be clear that this resource supply becomes larger than the projected resource use previously determined, even if one would be able to use a reversible energy-conversion system such as the Carnot cycle. Certainly real power cycles for any current, or even any future technology, would require more, in excess of the margin between the thermal efficiency of a given process versus an idealized process. In this consideration, the only limitation would be the selection of the energy-conversion system for the delivery of the required energy-resource quantity. If a different conversion process would be used, the corresponding assessment would take into account its physical limit of performance.

5.10 Conclusion

The selection of adequate metrics in assessing technologies for sustainable development assumes that a system considered is well defined. Any physical system prone to thermodynamic analysis must be well defined in the thermodynamic realm, and equally so a system considered in the sustainability realm must be characterized by adequate properties in the context of causal relations between the targeted state of sustainability and the selected metrics.

The changes of sustainability state may be successfully quantified by the selection of meaningful sustainability metrics. The current state of defining such metrics reflects the lack of mature sustainable development analysis approaches.

A thermodynamics metric, the entropy generation manifested within a system taken from a physical domain (technosphere), can be applied in an assessment of a technology used for performing a given task.

An analogous approach should be followed for a selection of all metrics (but based on respective, occasionally quite diverse, frameworks). The key message from this discussion is that the selection of metrics requires a rigorous approach and it must be founded on a well-established (in this case thermodynamics) framework.

5.11 Nomenclature

C^*	Heat capacity rate ratio, dimensionless
c	Specific heat, J/kg K
e	Specific energy, J/kg
E	Energy, J
$\mathscr{E}$	Exergy, J
M	Molecular mass, kg/kmol
$\dot{m}$	Mass flow rate, kg/s
NTU	Number of transfer units, $UA/(\dot{m}c_p)_{min}$
$\mathcal{R}$	Universal gas constant, J/mol K

R	Resource use, not specified
$\dot{S}$	Entropy rate, J/Ks
$\dot{S}_{gen}$	Entropy generation rate, J/Ks
s	Specific entropy, J/K kg
T	Temperature, K
t	Time, s
UA	Thermal size, W/K
y	Mol fraction, dimensionless
$\dot{W}$	Work rate, J/s $\equiv$ W
w	Specific work J/kg

Greek Letters

χ	Unit of a resource, say mass, kg
ε	Heat exchanger effectiveness, dimensionless
ϑ	Inlet temperature ratio, dimensionless

Superscripts

*	Signifies heat capacity rate ratio

Subscripts

0	Denoting environmental or reference state
Q	Signifies heat
in	Signifies inlet value
gen	Signifies entropy generation
H_2O	Signifies water
max	Signifies larger value
min	signifies smaller value
NaCl	Signifies NaCl
out	Signifies outlet value
p	At constant pressure
t	Time instant
used	Signifies a use in a process
W	Work

Acronyms

BOD	Biochemical oxygen demand
DESA	Department of Economic and Social Affairs
DWEER	Duol Work Exchanger Energy Recovery
EEA	European Environmental Agency
EF	Ecological footprint
EPI	Environmental performance index
ER	Energy Resources
FAO	Food and agriculture organization

GDP   Gross domestic product
HDI   Human development index
ICVS   International Crime Vietimization Survay
IPCC   Intergovernmental Panel on Climate Change
ISED   Indicator for sustainable energy development
ODP   Ozone depletion potential
ODS   Ozone depleting substances
OECD   Organization for Economic Co-Operation and Development
SDI   Sustainable development indicator
SNA   National Accounts of Japan
SS&T   Sustainable science and technology
SWRO   Seawater reverse osmosis
TPEC   Total primary energy consumption
UNCSD   United Nations Commission on Sustainable Development
UNDP   United Nations Development Programme
UNEP   United Nations Environmental Programme
UNESCO   United Nations Educational, Scientific and Cultural Organization
UNICRI   United Nations Interregional Crime and Justice Research Institute
UNFCCC   United Nations Framework Convention on Climate Change
WB   World Bank
WHO   World Health Organization
WRI   World Resources Institute
WSSSMR   Water supply and sanitation sector monitoring report

REFERENCES

[1] T. M. Parris and R. W. Kates, "Characterizing and measuring sustainable development," *Annu. Rev. Environ. Resour.* **28**, 13.1–13.28 (2003).

[2] S.A. Levin and W.C. Clark, Eds. "Toward a Science of Sustainability: Report from Toward a Science of Sustainability Conference, Airlie Center, Warenton, Virginia, November 29, 2009–December 2, 2009." CID Working Paper No. 196. Center for International Development at Harvard University. May 2010.

[3] S. Chu, Statement of Steven Chu, Secretary of Energy, before the Committee on Energy and Natural Resources, March 5, 2009, DOE. Available at http://www.energy.gov/news/print2009/6964.htm (accessed November 10, 2009).

[4] E.P. Gyftopoulos, "Building on the legacy of Professor Keenan. Entropy and intrinsic property of matter," in *Meeting the Entropy Challenge*, edited by G.P. Beretta, A.F. Ghoneim, and G.N. Hatsopoulos (American Institute of Physics, Melville, NY, 2008), pp. 124–140.

[5] B. Moldan and A.L. Dahl, "Challenges to sustainability indicators," in *Sustainability Indicators. A Scientific Assessment*, edited by T. Hak, B. Moldan, and A.L. Dahl (Scope 67, Island Press, Washington, D.C., Covelo, London, 2007), Chap. 1, pp. 1–24.

[6] Indicators of Sustainable Development: Framework and Methodologies, Commission on Sustainable Development, 9th Session, Background Paper No. 3, 2001; available at http://www.un.org/esa/sustdev/csd/csd9_indi_bp3.pdf.

[7] Human Development Indices, UNDP Report (2008); available at http://hdr.undp.org/en/media/HDI_2008_EN_Tables.pdf

[8] R. Prescott-Allen, R. *The Wellbeing of Nations: A Country by Country Index of Quality of Life and the Environment* (Island Press, Washington, D.C., 2001).

[9] *2008 Environmental Performance Index*, Yale Center for Environmental Law & Policy and CIESIN Columbia University; available at http://sedac.ciesin.columbia.edu/es/epi/downloads.html.

[10] M. Wackernagel, N. B. Schulz, D. Deumling, A. C. Linares, M. Jenkins, V. Kapos, C. Monfreda, J. Loh, N. Myers, R. Norgaard, and J. Randers, "Tracking the ecological overshoot of the human economy," *Proc. Natl. Acad. Sci. USA* **99**, 9266–9271 (2002); M. Wackernagel and W. Rees, *Our Ecological Footprint: Reducing Human Impact on the Earth* (New Society Publishers, Gabriola Island, British Columbia, Canada 1996).

[11] J. Talberth, C. Cobb, and N. Slattery, "The genuine progress indicator 2006. A tool for sustainable development, redifining progress." Available at http://www.rprogress.org/publications/2007/GPI202006.pdf.

[12] R. M. Solow, "An almost practical step toward sustainability," *Resources Policy* **19**, 162–172 (1993).

[13] D. C. Esty, M. A. Levy, C. H. Kim, A. de Sherbinin, T. Srebotnjak, and V. Mara, *Environmental Performance Index* (Yale Center for Environmental Law and Policy, New Haven, CT, 2008).

[14] Indicators of Sustainable Development: Framework and Methodologies, UN Commission on Sustainable Development, 9th Session, Background Paper No. 3 (DESA/DSD/2001/3, 2001), p. 2.

[15] "EEA core set of indicators – Guide," EEA Tech. Rep. 1/2005 (EEA, Luxembourg, 2005).

[16] "Measuring progress towards a more sustainable Europe," 2008 Monitoring Rep. (EUROSTAT, Luxembourg, 2008).

[17] "Indicators for sustainable energy development" (International Atomic Energy Agency and International Energy Agency).

[18] "Human development report 2009" (UNDP, New York, 2009).

[19] B. Ewing, A. Reed, S.M. Rizk, A. Galli, M. Wackernagel, and J. Kitzes. *Calculation Methodology for the National Footprint Accounts*, 2008 ed. (Global Footprint Network, Oakland, CA, 2008).

[20] J. Szargut, D.R. Morris, and F.R. Steward, *Exergy Analysis of Thermal, Chemical and Metallurgical Processes*, 1st ed. (Hemisphere, New York, 1988).

[21] N.U. Ukidwe and B.R. Bakshi, "Industrial and ecological cumulative exergy consumption of the United States via the 1997 input–output benchmark model," *Energy* **32**, 1560–1592 (2007).

[22] Gutowski, T., Sekulić, D.P., and Bakshi, B., "Preliminary thoughts on the application of thermodynamics to the development of sustainability criteria, ISSST 2009 – The International Symposium on Sustainable Systems and Technology, IEEE, Phoenix, AZ, May 18–20, CD Edition.

[23] E.P. Gyftopoulos and G. P. Beretta, *Thermodynamics: Foundations and Applications* (Macmillan, New York, 1991; Dover, Mineola, NY, 2005).

[24] Gleick, P.H., "Water resources," in *Encyclopedia of Climate and Weather* edited by S.H. Schneider (Oxford University Press, New York, 1996), vol. 2.

[25] S.S. Hatson, N.L. Barber, J.F. Kenney, K.S. Linsey, D.S. Lumia, and M.A. Maupin, "Estimated use of water in the US in 2000," U.S. Geological Survey Circular 1268, released March 2004, revised February 2005 (U.S. Department of the Interior, Reston, VA, 2005).

[26] UN Department of Economic and Social Affairs; http://esa.un.org/unpp/p2k0data.asp

[27] EIA, *Annual Energy Outlook 2006* (DOE/EIA-0383, Washington, D.C., 2006).

[28] Y.M. El-Sayed and A.J. Aplenc, "Application of the thermoeconomic approach to the analysis and optimization of a vapo-compression desalting system," *ASME J. Eng. Power* **92**, 17–26 (1970).

[29] R.B. Evans, G.L. Crellin, and M. Tribus, "Thermoeconomic considerations of sea water demineralization," in *Principles of Desalination*, 2nd ed., Part A, (Academic, New York, 1980), Chap. 1, pp. 1–54.

[30] Cerci, Y., "The minimum work requirement for distillation processes," *Exergy Intl. J.* **2**, 15–23 (2002).

[31] Y.A. Cengel, Y. Cerci, and B. Wood, "Second law analysis of separation processes of mixtures," *Proc. ASME Adv. Energy Syst. Div.* **39**, 537–543 (1999).

[32] R.K. Shah and D.P. Sekulić, *Fundamentals of Heat Exchanger Design* (Wiley, Hoboken, NJ, 2003).

[33] D.P. Sekulić, "The second law quality of energy transformation in a heat exchanger," *Int. J. Heat Mass Transfer* **33**, 295–300 (1990).

6 Thermodynamic Analysis of Resources Used in Manufacturing Processes

Timothy G. Gutowski and Dušan P. Sekulić

6.1 Introduction

The main purpose of manufacturing processes is to transform materials into useful products. In the course of these operations, energy resources are consumed and the usefulness of material resources is altered. Each of these effects can have significant consequences for the environment and for sustainable development, particularly when the processes are practiced on a very large scale. Thermodynamics is well suited to analyze the magnitude of these effects as well as the efficiency of the resource transformations. The framework developed here is based on the exergy analysis that is reviewed in the first two chapters of this book.[1] Also see [1–5]. The data for this study draw on previous work in the area of manufacturing process characterization, but also includes numerous measurements and estimates we have conducted. In all, we analyze 26 different manufacturing processes, often in many different instances for each process. The key process studies from the literature are as follows: for microelectronics, Murphy et al. [6], Williams et al. [7], Krishnan et al. [8], Zhang et al. [9], and Boyd et al. [10]; for nanomaterials processing, Isaacs et al. [11] and Khanna et al. [12]; for other manufacturing processes, Morrow et al. [13], Boustead [14, 15], Munoz and Sheng [16], and Mattis et al. [17]. Some of our own work includes Dahmus and Gutowski [18], Dalquist and Gutowski [19], Thiriez and Gutowski [20, 21], Baniszewski [22], Kurd [23], Cho [24], Kordonowy [25], Jones [26], Branham et al. [27, 28], and Gutowski et al. [29]. Several texts and overviews also provide useful process data [30–35] and additional manufacturing-process studies of ours include Sekulić [36], Jayasankar [37], Sekulić and Jayasankar [38], Bodapati [39] and Subramaniam and Sekulić [40].

6.2 Thermodynamic Framework

Manufacturing can often be modeled as a sequence of processes performed by open thermodynamic systems [27] as proposed by Gyftopoulos and Beretta for materials

---

[1] This chapter is based in large part on a paper that appeared in *Environmental Science and Technology* **43**, 1584–1590 (2009) titled "Thermodynamic analysis of resources used in manufacturing processes," by T.G. Gutowski, M.S. Branham, J.B. Dahmus, A.J. Jones, A. Thiriez, and D. Sekulić. New material has been added in several places, especially on the calculation of the minimum work for manufacturing processes.

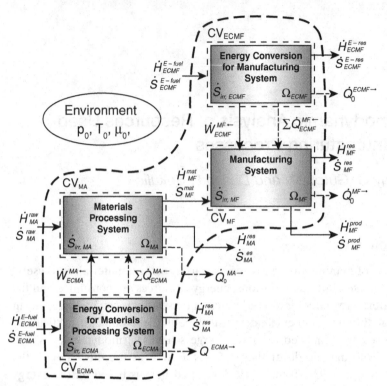

Figure 6.1. Diagram of a coupled manufacturing- and materials-processing system ([41], adapted from [1]).

processing [1]. Each stage in the process can have work and heat interactions, as well as materials flows. The useful output, primarily in the form of material flows of products and by-products from a given stage, can then be passed on to the next. Each step inevitably involves losses that are due to an inherent departure from reversible processes, hence generating entropy and a stream of waste materials and exergy losses (often misinterpreted as energy losses).[2]

Figure 6.1 depicts a generalized model of a manufacturing system [27, 41]. The manufacturing subsystem (Ω_{MF}) receives work W and heat Q from an energy-conversion subsystem (Ω_{ECMF}). The upstream input materials come from the materials-processing subsystem (Ω_{MA}), which also has an energy-conversion subsystem (Ω_{ECMA}). This network representation can be infinitely expanded to encompass ever more complex and detailed inputs and outputs [31, 32].

At each stage, the subsystems interact with the environment (at some reference pressure p_0, temperature T_0, and chemical composition, which is given by mole fractions x_i, $i \in (1, n)$, of n chemical compounds, characterized by chemical potentials $\mu_{i,0}$). The performance of these subsystems can then be described in thermodynamic terms by formulating mass, energy, and entropy balances. Beginning with the manufacturing subsystem Ω_{MF}, featuring the system's mass M_{MF}, energy E_{MF}, and entropy S_{MF}, we have the following three basic rate equations.[3]

---

[2] Note that the exergy losses involve not only exergy of the energy flows but also exergy of the material flows.

[3] For a discussion of mass, energy, entropy, and exergy balances see Chap. 2.

Mass Balance

$$\frac{dM_{MF}}{dt} = \left(\sum_{i=1} \dot{N}_{i,in} \tilde{M}_i\right)_{MF} - \left(\sum_{i=1} \dot{N}_{i,out} \tilde{M}_i\right)_{MF}, \quad (6.1)$$

where $\dot{N}_i$ is the amount of matter per unit time of the ith component entering or leaving the system and $\tilde{M}_i$ is the molar mass of that component.

Energy Balance

$$\frac{dE_{MF}}{dt} = \sum_k \dot{Q}_{ECMF,k}^{MF\leftarrow} - \dot{Q}_0^{MF\rightarrow} + \dot{W}_{ECMF}^{MF\leftarrow} + \dot{H}_{MF}^{mat} - \dot{H}_{MF}^{prod} - \dot{H}_{MF}^{res}, \quad (6.2)$$

where $\dot{Q}_{ECMF,k}^{MF}$ and $\dot{W}_{ECMF}^{MF}$ represent rates of energy interactions between the manufacturing subsystem (Ω_{MF}) and its energy-supplying subsystem (Ω_{ECMF}). The $\dot{H}$ terms signify the lumped sums of the enthalpy rates of all materials, products, and residue bulk flows into or out of the manufacturing system. Note that a heat interaction between Ω_{MF} and the environment, denoted by the subscript 0, is assumed to be out of the system (a "loss" into the surroundings) at the local temperature T_0.

Entropy Balance

$$\frac{dS_{MF}}{dt} = \sum_k \frac{\dot{Q}_{ECMF}^{MF\leftarrow}}{T_k} - \frac{\dot{Q}_0^{MF\rightarrow}}{T_0} + \dot{S}_{MF}^{mat} - \dot{S}_{MF}^{prod} - \dot{S}_{MF}^{res} + \dot{S}_{gen,MF}, \quad (6.3)$$

where $\dot{Q}^{MF}/T$ terms represent the entropy flows accompanying the heat transfer rates exchanged between the subsystem Ω_{MF} and energy-supplying subsystem (Ω_{ECMF}) and environment, respectively, and the $\dot{S}_i$ terms indicate the lumped sums of the entropy rates of all material flows. The term $S_{gen,MF}$ represents the entropy generation caused by irreversibilities generated within the manufacturing subsystem.

Assuming steady state and eliminating $\dot{Q}_0$ between Eqs. (6.2) and (6.3) yields an expression for the work rate requirement for the manufacturing process:

$$\dot{W}_{ECMF}^{MF\leftarrow} = \left[\left(\dot{H}_{MF}^{prod} + \dot{H}_{MF}^{res}\right) - \dot{H}_{MF}^{mat}\right] - T_0 \left[\left(\dot{S}_{MF}^{prod} + \dot{S}_{MF}^{res}\right) - \dot{S}_{MF}^{mat}\right]$$

$$- \sum_{k>0} \left(1 - \frac{T_0}{T_k}\right) \dot{Q}_{ECMF}^{MF\leftarrow} + T_0 \dot{S}_{gen,MF}. \quad (6.4)$$

The quantity $H - TS$ appears often in thermodynamic analysis and is referred to as the Gibbs free energy. In this case, a different quantity appears, $H - T_0 S$. The difference between this and the same quantity evaluated at the reference state (denoted by the subscript 0) is called exergy, $\dot{Ex} = (\dot{H} - T_0\dot{S}) - (\dot{H} - T_0\dot{S})_0$.[4] Exergy of a material flow represents the maximum amount of work that could be extracted from the flow considered as a separate system as it is reversibly brought to equilibrium with a well-defined environmental reference state. In general, the bulk-flow terms in Eq. (6.4) may include contributions that account for both the physical and chemical

---

[4] A detailed discussion of the physical meaning and a formal structure of this physical quantity is provided in Chaps. 1 and 2 of this book. In the text that follows, only the main features of exergy are discussed.

exergies, and hence $\dot{Ex} = \dot{Ex}^{ph} + \dot{Ex}^{ch}$, as well as kinetic and potential exergy (not considered in this discussion); see [2–5].

The physical exergy is that portion of the exergy that can be extracted from a system by bringing a system in a given state to the *restricted dead state* at a reference temperature and pressure (T_0, p_0). The chemical exergy contribution represents the additional available energy that can be extracted from the system at the restricted dead state by bringing the chemical potentials μ_i^* of a component $i \in (1, n)$ at that state (T_0, p_0) to equilibrium with its surroundings at the *ultimate dead state*, or just the *dead state* $(T_0, p_0, \mu_{i,0})$. In addition to requiring an equilibrium at the reference temperature and pressure, the definition of chemical exergies also requires an equilibrium at reference state with respect to a specified chemical composition. This reference state is typically taken to be (by convention) representative of the compounds in the Earth's upper crust, atmosphere, and oceans. In this chapter, exergy values are calculated with the Szargut reference environment [5]. Several updates and alternative references and environments are available, but they do not change the accuracy of this development.[5] See Appendix A of this book for an extensive list of standard chemical exergies taken from Szargut [42] and others.

Substituting and writing explicit terms for the expressions for physical and chemical exergy allows us to write the work rate as

$$\dot{W}_{ECMF}^{MF\leftarrow} = \left[\left(\dot{Ex}_{MF}^{prod,ph} + \dot{Ex}_{MF}^{res,ph}\right) - \dot{Ex}_{MF}^{mat,ph}\right] + \left(\sum_{i=1}^{n} e_{x_{i,0}}^{ch} \dot{N}_i\right)_{MF}^{prod} + \left(\sum_{i=1}^{n} e_{x_{i,0}}^{ch} \dot{N}_i\right)_{MF}^{res}$$
$$- \left(\sum_{i=1}^{n} e_{x_i}^{ch} \dot{N}_i\right)_{MF}^{mat} - \sum_{k>0}\left(1 - \frac{T_0}{T_k}\right) \dot{Q}_{ECMF}^{MF\leftarrow} + T_0 \dot{S}_{gen,MF}. \tag{6.5}$$

Using the same analysis for the system Ω_{ECMF} yields

$$\dot{W}_{ECMF}^{MF\leftarrow} = \left(\dot{Ex}_{ECMF}^{E-fuel,ph} - \dot{Ex}_{ECMF}^{E-res,ph}\right) + \left(\sum_{i=1}^{n} e_{x_{0,i}}^{ch} \dot{N}_i\right)_{ECMF}^{E-fuel} - \left(\sum_{i=1}^{n} e_{x_{0,i}}^{ch} \dot{N}_i\right)_{ECMF}^{E-res}$$
$$- \sum_{k>0}\left(1 - \frac{T_0}{T_k}\right) \dot{Q}_{ECMF}^{MF\leftarrow} - T_0 \dot{S}_{gen,ECMF}. \tag{6.6}$$

Here we have purposefully separated out the physical exergies, written as extensive quantities Ex, and the chemical exergies, where $e_{x_{i,0}}^{ch}$, represent the molar chemical exergies in the "restricted dead state" [2]. We do this to emphasize the generality of this framework and the significant differences between two very important applications. In resource accounting, as done in life-cycle assessment, the physical exergy terms are often not included. Hence the first bracketed term on the right-hand side of Eq. (6.5) becomes zero because the material flows enter and exit the manufacturing process at the restricted dead state.[6] However, many manufacturing processes

---

[5] Refer to Chap. 2 and Appendix A for a more detailed list of references.
[6] This is equivalent to the consideration of a system that would have an extended boundary so that material flows enter and leave the system at the state of environment. This brings all intermediate interactions into the considered system. Further discussion on this topic can be found in Section 6.5 on exergy efficiency in this chapter.

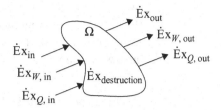

Figure 6.2. Components of an exergy balance for any arbitrary open thermodynamic system.

involve material flows with nonzero physical exergies at system boundaries [31, 32, 34]. To analyze these processes, and in particular to estimate the minimum work rate and exergy lost, these terms must be retained. This is typical for an engineering analysis of an energy system. Note that similar equations can also be derived for the systems Ω_{MA} and Ω_{ECMA}. Before proceeding, it is worth pointing out several important insights from these results. First, in both Eqs. (6.5) and (6.6) we see that the magnitude of the work input is included fully while the heat inputs are modified (reduced) by a Carnot factor $(1 - T_0/T_k)$. Hence, in exergy analysis, work and heat are not equivalent, as they are in first law analysis. Second, Eq. (6.5) provides the framework for estimating the minimum work input for any process, i.e., when irreversibilities are zero, $T_0 S_{gen} = 0$. The analytical statements formulated in Eqs. (6.5) and (6.6) feature all the energy interactions (including the energy carried by material streams) in terms of exergies – i.e., the available energy equivalents of all energy interactions. Such a balance may be written, in general, for an arbitrary open system Ω (including the one presented in Fig. 6.1] as follows (see Fig. 6.2):

$$\dot{Ex}_{in} + \dot{Ex}_{W,in} + \dot{Ex}_{Q,in} = \dot{Ex}_{out} + \dot{Ex}_{W,out} + \dot{Ex}_{Q,out} + \dot{Ex}_{destruction}. \qquad (6.7)$$

In Eq. (6.7), the exergy components (i.e., exergy modes) of the balance are as follows: (1) $\dot{Ex}_{in/out} = \dot{Ex}_{in/out}^{ph} + \dot{Ex}_{in/out}^{ch}$, (2) $\dot{Ex}_{W,in/out} = \dot{W}_{in/out}$, (3) $\dot{Ex}_{Q,in/out} = (1 - T_0/T) \dot{Q}_{in/out}$, and (4) $\dot{Ex}_{destruction} = T_0 \dot{S}_{gen}$. Work required beyond the minimum work, by definition, is lost. This represents exergy destroyed ($\dot{Ex}_{destruction}$).[7]

6.3 Estimating the Minimum Work for Materials Transformations in Manufacturing Processes

Materials transformations in manufacturing processes can be produced by energy or mass interactions in the form of either work inputs and outputs or heat inputs and outputs, respectively, or, in some cases, mass flow inputs involving chemical processes or mass flow inputs and outputs involving either auxiliary or waste material. These interactions are directly related to the resource needs and may be expressed in terms of exergy. The identification of inputs and outputs depends on how one draws the boundaries of a system exposed to the process. Here we adhere to the convention suggested in the system drawing in Fig. 6.1, that is, we identify the manufacturing- or materials-processing subsystem that must be provided with energy and material flows. Calculating the resources needs in terms of the minimum exergy input can be an important aid in identifying the efficiency of processes and components of

---

[7] The term "destruction" associated with systems change of state is often denoted as "loss." For the sake of clarity, we use the term "destruction" to signify internal irreversibilities and "loss" to mark external losses.

processes and thereby identifying opportunities for improvement.[8] Here we introduce the topic of estimating the minimum resources requirements for typical material transformations that take place in manufacturing processes. We also use these results to compare with the actual resources used.

Temperature and Pressure Changes for Open Materials-Processing Systems

Let us define as the system a material flow that must change the state by changing the temperature and pressure. Let us specify that this change should be made by a work interaction (an energy interaction characterized with no associated entropy flow). Starting from Eq. (6.5), consider the case of a continuous steady-state flow. Here there are no chemical reactions and correspondingly no changes in chemical exergy. There are no heat inputs. Further, we assume 100% yield so that the residual material stream is zero, and all of the input material is converted to product. To calculate the minimum work rate input, we set $T_0 \dot{S}_{gen} = 0$. This gives (with simplified notation),

$$\dot{W}_{min} = \dot{E}x^{prod} - \dot{E}x^{mat} = \dot{E}x_{out} - \dot{E}x_{in} = \dot{m}(e_{x,out} - e_{x,in}) = \dot{m}(b_{out} - b_{in}). \quad (6.8)$$

In Eq. (6.8), $e_{x,i} = h_i - h_{i,0} - T_0(s_i - s_{i,0})$ and $b_i = h_i - T_0 s_i$. This latter function is called the availability function.[9]

Writing the minimum work rate per unit of mass rate processed gives

$$w_{min} = \frac{\dot{W}_{min}}{\dot{m}} = b_{out} - b_{in}. \quad (6.9)$$

Determination of the specific minimum work reduces to the determination of availability functions at the given terminal ports of the open material-processing system, that is, determining the state properties h_{in}, h_{out}, and s_{in}, s_{out}, or, if expressed in terms of exergy change, determining additionally the enthalpy and entropy values for the same system at its dead state, i.e., when in equilibrium with the surroundings at (T_0, p_0).

In differential form,

$$dw_{min} = db = dh - T_0 ds. \quad (6.10)$$

(Note that we can write dw_{min} rather than δw_{min} because, whereas work is not a state variable, the minimum work, which is a function of other state and environmental variables, is a state property if T_0 = fixed).

The task of calculating the minimum work then comes down to integrating Eq. (6.10) between the terminal states of the materials process, for temperature and pressure dependencies for dh and ds for the given change of state. This calculation would be possible for a given system only after the related constitutive relationship is specified for the assumed substance models. We can do this by identifying real or assuming certain idealized behavior for the materials being processed and assuming a reversible change of state of the system exposed to bulk mass flow rates and work interaction only. For example, for a pure, simple compressible system and internally

---

[8] A more general interpretation of the importance of these extrema for transformational technologies development is discussed in Chap. 5.

[9] For details of the physical meaning of the availability function see Chaps. 1 and 2.

reversible process, we can write $Tds = dh - vdp$. Now, we can write for *incompressible substances*

$$dh = cdT + v\,dp, \tag{6.11}$$

$$ds = c\frac{dT}{T}. \tag{6.12}$$

Here c is the specific heat, v is the specific volume, and p is pressure.

Following a similar procedure for *ideal gases*, we find that these relationships are

$$dh = c_p dT, \tag{6.13}$$

$$ds = c_p\frac{dT}{T} - R\frac{dp}{p}. \tag{6.14}$$

Here c_p is the specific heat at the constant pressure of the ideal gas, $R = \overline{R}/M$, where $\overline{R} = 8.314$ J/mol K is the universal gas constant and M is the molar mass (molecular weight).

As an example, consider a process that increases the temperature of a material in preparation for molding. We are interested in the minimum electrical work rate required for causing this transformation. This could represent the minimum electrical power input needed for an electrical resistance heater, a device commonly used in manufacturing. Here the material remains in the condensed phase, and we assume the process takes place at atmospheric pressure. The system boundary crosses the electrical current leads, so the interaction does not involve heat transfer if no heat losses are assumed.

The substitution of Eqs. (6.11) and (6.12) into (6.10) and integration from T_0 to T yields

$$w_{min} = c(T - T_0) - T_0 c \ln \frac{T}{T_0}. \tag{6.15}$$

Reconsider now the same problem of energy-resource use in form of heat, not work. That means, consider the system boundary as being crossed by Joule energy delivered as heat across the system boundary.

To calculate the heat Q required for raising the temperature of our system from T_0 to T with no work ivolved, we should just use the energy balance as stated by the first law of thermodynamics. This heat input equals the enthalpy change in the steady-state process from the given state of equilibrium with the surroundings T_0 to the final state T, i.e.,

$$q_{min} = c(T - T_0). \tag{6.16}$$

This gives the heat transfer rate per unit of mass flow rate, which is larger than the minimum work. Note that c is assumed constant over the temperature range. This heat transfer rate is the minimum needed assuming that no heat losses are present.

Consider now a system exposed to phase change. The heat input needed is just the enthalpy of phase change (solid to liquid phase at the constant temperature) h_{sf} but the exergy value of the heat needed for phase change is less. This again is because the work can be converted to a full extent reversibly, but heat cannot. If

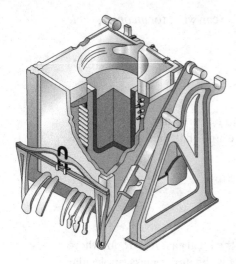

Figure 6.3. Diagram of an electric induction furnace for melting iron [26].

the enthalpy change during melting is h_{sf}, then because the liquid (f) and solid (s) phases are in equilibrium at T_m, i.e., $h_f - T_m s_f = h_s - T_m s_s$, it follows that $s_s - s_f = \Delta s = h_{sf}/T_m$, where $h_s - h_f = \Delta h = h_{sf}$. Rewriting Eq. (6.10) as $w_{\min} = \Delta h - T_0 \Delta s$ gives

$$w_{\min} = h_{sf}\left(1 - \frac{T_0}{T_v}\right). \tag{6.17}$$

The exergy needed for vaporization can be developed in a manner analogous to melting as

$$w_m = h_{fg}\left(1 - \frac{T_0}{T_b}\right). \tag{6.18}$$

EXAMPLE 6.1. HEEL ELECTRIC INDUCTION MELTING OF IRON

One way of preparing the melt for production in an iron foundry is to use a "heel" electric induction furnace (see Fig. 6.3). An electric induction furnace creates an electromagnetic field in the metal by virtue of an alternating current that runs through coils that are wound around a refractory lining. The effect is to melt and stir the metal. A metal heel is retained in the furnace to maintain the effect even while molten metal is periodically tapped from the furnace.

Here we model this process as an open system (including coils) in steady state at atmospheric pressure with a work input (electricity) that increases the temperature and then melts the input iron. For the sake of simplicity, we do not include in the system the material of the induction furnace (that is, we assume that the electrical work interacts with the processed material only). First we raise the temperature from $T_0 = 293$ K to $T_m = 1813$ K. We determine the minimum work required by using Eq. (6.15) and assuming a nominal value for c of 0.67 J/g K [43]. And then we assume an additional investment of work for melting the iron at a constant melt temperature T_m, using $h_{sf} = 272$ J/g and Eq. (6.17). This gives us an estimate of the minimum work as $w_{\min} = 889$ J/g, or 0.9 MJ/kg (melt). Electricity data for heel induction melters used in iron foundries range from 570 to 1000 kWh/tonne (melt) or 2.1 to 3.6 MJ/kg (melt), depending on operating parameters [26]. If we measure the efficiency of these melters in terms of w_{min}/w_{actual}, we get a range from 25% to 43%. (Efficiency measures are discussed in more detail later in this chapter and in Chaps. 1 and 2 of this book.) The lower value reflects a furnace run at a suboptimal production rate and inefficient input energy-resources conversion. For example, a heel furnace could be left on over a weekend or holiday without production to maintain the heel and avoid start-up. However, even the higher value (43%) may be considered as being lower than we might expect. For example, for the thermal efficiency defined differently using the enthalpy change as $c(T - T_0) + h_{sf}$ and using the enthalpy equivalent of the electricity input as the actual, we would get 62%. In general, using the minimum work in an efficiency comparison is a more stringent requirement than the more familiar so-called first law efficiencies such as the thermal efficiency given as the minimum enthalpy change required divided by the actual input. That aside, the reasons for the inefficiencies in the induction furnace include heat loss from the vessel and heat losses in the inductor, as well as inefficiencies in the interactions within the electromagnetic field and material. The first is controlled by the temperature gradient across the walls of the vessel, the second depends on load shifting between heating the metal and heating the coils, and the third depends on the energy conversion within the system. In most designs the heating coils are water-cooled copper tubes. The cooling of the copper tubes represents a dissipated heat and guarantees a temperature gradient across the vessel walls. One way to improve the performance in such an operation would be to use the dissipated heat to preheat the incoming material charge. This will be especially true in cold regions of the world, where the input material may be well below the standard value for T_0 and preheating with "waste heat" could make a significant improvement in efficiency. Second, also note that, although we have calculated the minimum work only up to T_m, actual operations will tend to raise the temperature higher than this.

EXAMPLE 6.2. PRODUCTION OF CARBON SINGLE-WALLED NANOTUBES (SWNTs) BY THE HIGH-PRESSURE CO PROCESS (H₁Pco)

Single-walled nanotubes (SWNTs) can be produced in a gas phase reaction with CO as the carbon source and an iron compound as a catalyst (see Fig. 6.4). For the reaction to proceed efficiently, both high temperatures ($\sim$1000 °C) and high pressures ($\sim$30 atm) are required. This process was developed by Richard Smalley's team at Rice University, who documented the process in sufficient detail so as to allow an estimate of the minimum work (electricity) rate to produce these materials [45].

Again we model the process as an ideal steady-state open system. There are no heat inputs or losses, and the temperature and pressure increases are performed reversibly. It can be shown that the change in chemical exergy for the process is small compared with the required change in physical flow exergy and can be ignored, primarily because the conversion rate is so low [46]. The simplified version of the process we analyze then comes down to the task of raising the temperature from 273 °K to 1273 °K and the pressure from 1 atm to 30 atm of a CO gas stream modeled as an ideal gas [45, 46] and determining the standard chemical energy of SWNTs [46–48].

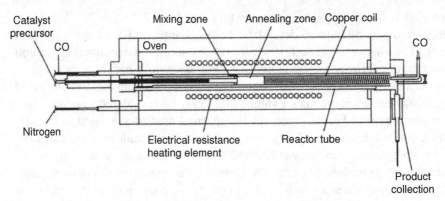

Figure 6.4. Apparatus for HiPco process [44].

Using Eqs. (6.10) (6.13), and (6.14) and integrating between T_0 and T and p_0 and p gives,

$$w_{\min} = c_p(T - T_0) - c_p T_0 \ln \frac{T}{T_0} + T_0 R \ln \frac{p}{p_0}. \qquad (6.19)$$

Now, by using the nominal value of $c_p = 1130$ J/kg K (this value ranges from 1040 at 293 K to 1227 J/kg K at 1250 K), we can calculate the work requirement to raise the temperature from 293 K to 1273 K as 0.65 MJ/kg CO and to raise the pressure of the CO gas from 1 to 30 atm as 0.64 MJ/kg CO. This yields a total exergy requirement of 1.29 MJ/kg CO. Very early versions of this process were performed open loop. The hot gas was exhausted and the SWNTs were collected and purified. At this stage the conversion rate was quite small, with the actual yield in terms of incoming CO to the SWNT output at 45,000:1. This results in an overall minimum exergy requirement of 58 GJ/kg SWNT. Even without any losses assumed, this high value makes SWNTs by this process among the most energy-intensive materials known.

Subsequent improvements, however, including the recycling of CO gas, have greatly improved the production yield for this process, resulting in current estimates of the actual electricity used to about 32 GJ/kg [46]. More details on the thermodynamic analysis of this process can be found in [46].

EXAMPLE 6.3. METAL CUTTING AND FORMING

A number of manufacturing processes shape the workpiece material primarily by plastic deformation. These would include machining and grinding processes as well as forming processes such as sheet metal stamping, rolling, extrusion, and forging. In all cases of deformation processing the plastic work is converted in large part to thermal energy and not recovered. Hence these processes are inherently irreversible. We can calculate a minimum work requirement from mechanics by using idealized material-behavior models and simplified loading configurations. Here we use but the simplest of model and give references to a number of texts on plasticity for a more detailed analysis.

Plastic work can be estimated from a knowledge of the applied stress and the resulting strain fields. For example, for any arbitrary loading, say in the x, y, z

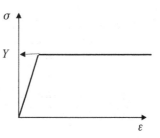

Figure 6.5. Stress–strain behavior for an idealized elastic–plastic material in a tension test.

reference frame, mechanics texts write an expression for a work increment per unit volume in terms of stress (σ and τ) and strain increments ($d\varepsilon$ and $d\gamma$) as

$$dw = \sigma_x d\varepsilon_x + \sigma_y d\varepsilon_y + \sigma_z d\varepsilon_y + \tau_{xy} d\gamma_{xy} + \tau_{xz} d\gamma_{xz} + \tau_{yz} d\gamma_{yz}. \qquad (6.20)$$

In principle, to obtain the plastic work of deformation we would integrate this equation over the incremental strains as the deformation evolved.

A good, relatively simple, manufacturing example of this would be the case of orthogonal machining of an ideal elastic–plastic material. The stress–strain behavior of an ideal elastic–plastic material is shown in Fig. 6.5 for a tension test.

At the critical stress Y the material yields and extends without strain hardening.

An idealized model for orthogonal machining can be built by use of the representation in Fig. 6.6. Here we conceive of this situation as a system with the material initially undeformed and then sheared in a narrow zone at shear angle ϕ because of a tool with rake angle α. The resultant plastic work per unit volume of material sheared, w, then is due to only one term in Eq. (6.19) as

$$\int dw = \int \tau d\gamma, \qquad (6.21)$$

where the dimensions of τ are for stress (force/area) and of γ are for strain (length/length).

The yield stress in shear can be obtained by transforming the tension test result as $\tau = y/2$. The shear strain can be estimated from a knowledge of the rake and

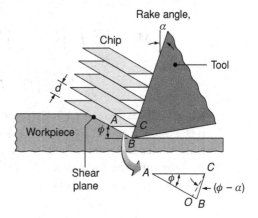

Figure 6.6 Idealized orthogonal machining, taken from [37].

Table 6.1. *Approximate range of exergy requirements in cutting operations at the drive motor of the machine tool*

Material	Specific exergy (kJ/kg)
Aluminum alloys	140–360
Cast irons	140–690
Copper alloys	160–360
High-temperature alloys	380–940
Magnesium alloys	170–340
Nickel alloys	570–800
Refractory alloys	290–880
Stainless steels	250–625
Steels	260–1150
Titanium alloys	440–1100

Note: Adapted from [32].

shear angles but generally is in the range of $2 \leq \gamma \leq 4$ for positive rake angles [49]. Using the value of $\gamma = 3$ gives

$$w_{\text{plastic}} \cong \frac{3}{2} Y. \tag{6.22}$$

The actual machining process involves significant friction at the tool–workpiece interface such that the actual work requirement is considerably more than the calculated plastic work [32, 49, 50]. This work is provided by the spindle motor of the machine tool and is tabulated for various workpiece materials as the so-called specific cutting energy u_s. Comparisons for various common workpieces and cutting conditions show that, as a rough approximation, $w_{\text{plastic}}/u_s \cong \frac{1}{2}$.

Typical values for u_s are given in Table 6.1, adapted from Kalpakjian and Schmid [32], assuming incompressible deformation.

In some cases, materials processing in manufacturing involves steps that may be characterized as a change of state of a closed system [36]. In such cases, no material flow takes place across the system boundary, i.e., the processing assumes a fixed quantity of matter exposed to a change of state. If we considers a noninsulated closed system, the change of physical exergy Ex in a given period of time t would be

$$\Delta \text{Ex} = \sum_{i=1}^{n} \int_{t_1}^{t_2} \left[\left(1 - \frac{T_0}{T_i} \right) \frac{dQ_i}{dt} \right] dt$$
$$- \int_{t_1}^{t_2} \left(\frac{dW}{dt} \right) dt + p_0 \int_{t_1}^{t_2} \left(\frac{dV}{dt} \right) dt - T_0 \int_{t_1}^{t_2} \left(\frac{dS_{\text{gen}}}{dt} \right) dt, \tag{6.23}$$

where $\Delta \text{Ex} = \text{Ex}(t = t_2) - \text{Ex}(t = t_1)$ represents the exergy difference and $\text{Ex} = [(U + E_k + E_p) - U_0] + p_0(V - V_0) - T_0(S - S_0)$ represents the exergy in a given state versus the reference state (T_0, p_0). Equation (6.23) is an integrated exergy-balance equation for a closed system in its rate form.[10] Let us consider a fixed mass of a material as a system exposed to an elastic and subsequently plastic deformation.

---

[10] Refer to Eq. (6.4) in this chapter and the material given in Chap. 2.

Table 6.2. *Exergy use (order of magnitude) for elastic–plastic deformation*

System (includes alloys)	Aluminum	Cooper	Magnesium	Steel	Titanium
Specific exergy use elastic (kJ/kg)	10^{-3}–10^{0}	10^{-3}–10^{-1}	10^{-1}–10^{0}	10^{-2}–10^{0}	10^{-1}–10^{0}
Specific exergy use (from zero load until fracture) (kJ/kg)	10^{1}–10^{2}	10^{1}–10^{2}	10^{0}–10^{1}	10^{-1}–10^{1}	$\sim 10^{1}$

The initial state (at time t_1) of a material is exposed to an action of forces that deforms the material in the presence of uniaxial tensile stresses only. The temperature of the system at which the heat dissipation reaches surroundings is assumed to be virtually the same as that of the environment. Furthermore, we assume that material density does not change; therefore, for the fixed mass, the volume of the system does not change either. The change of state during elastic deformation would be fully reversible, and the term signifying corresponding irreversibility would be equal to zero. So the exergy change during elastic deformation would be equal to the work of elastic deformation but it would be fully recoverable. In the case of further plastic deformation, we assume that the thermal state would be restored to the initial state, so the corresponding exergy change would be zero; hence the irreversible work invested in the process would be equal to the mechanical work of plastic deformation. As a result, only the work term in Eq. (6.23) would survive.

With the previously listed assumptions, the exergy change required for executing an elastic or plastic (or both) deformation would be proportional to the integral of the stress versus strain relationship for the given material, i.e., $w \sim \int \sigma \, d\varepsilon$ [51]. The total amount of energy resources used would be dependent on the type and the mass of material of the system and the ultimate strain achieved.

The results of calculations of exergy utilization are summarized in Table 6.2. The calculation was performed for actual stress–strain relationships of the selected materials.

In Fig. 6.7 a schematic representation of the exergy use for achieving either reversible or irreversible deformation is presented on a stress–strain diagram. Note that significant exergy use for achieving the required effect is mostly dissipated in the form of thermal energy virtually at the temperature of the environment under assumed conditions; therefore these exergy expenditures are lost. Note that elastic deformation is associated with reversible exergy use and plastic with irreversible [52].

Figure 6.7. Reversible (elastic) and irreversible (plastic) exergy uses.

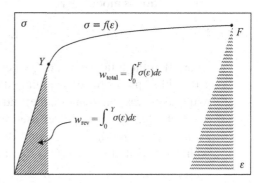

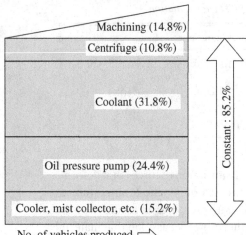

Figure 6.8. Electrical work rate used as a function of production rate for an automobile-production machining line [35].

No. of vehicles produced ⟹

Energy-Use Breakdown by Type

6.4 Electrical Work (Exergy) Used in Manufacturing Processes

Here we look at the actual electrical work used in manufacturing processes. In general, manufacturing processes are made up of a series of processing steps, which for high-production situations are usually automated. For some manufacturing processes, many steps can be integrated into a single piece of equipment. A modern milling machine, for example, can include a wide variety of functions, including work handling, lubrication, chip removal, tool changing, and tool-break detection, all in addition to the basic function of the machine tool, which is to cut metal by plastic deformation. The result is that these additional functions can often dominate energy-resources requirements at the machine. This is shown in Fig. 6.8 for an automotive machining line [29, 35]. In this case, the maximum energy-resources requirement for the actual machining in terms of electricity is only 14.8% of the total. Note that this energy rate represents an entity that is recognized in thermodynamics as a rate of work interaction. At lower production rates the machining contribution is even smaller. Other processes exhibit similar behavior. See, for example, data for microelectronics fabrication processes as provided by Murphy et al. [6]. Thiriez [20] and Thiriez and Gutowski [21] show the same effect for injection molding. In general, there is a significant energy resources requirement to start up and maintain the equipment in a "ready" position. Once in the "ready" position, there is then an additional requirement that is proportional to the quantity of material being processed. This situation is modeled as

$$\dot{W} = \dot{W}_0 + k\dot{m} \tag{6.24}$$

where $\dot{W}$ is the total work rate (power) used by the process equipment, in watts; $\dot{W}_0$ is the "idle" power for the equipment in the ready position, in watts; $\dot{m}$ is the rate of material processing in mass/time; and k is a constant, with units of joules/mass.

Note that the total power used by the process may alternatively be presented as the exergy rate that corresponds to the electrical work rate. Hence this equation is directly related to Eq. (6.5) for the work rate $\dot{W}$. Note that, with a model for the reversible work, we could directly calculate the lost exergy $T_0 \dot{S}_{\text{gen}}$ by comparing Eqs. (6.5) and (6.24).

The specific electrical work per unit of material processed w, in units of joules/mass, is then

$$w = \frac{\dot{W}_0}{\dot{m}} + k \qquad (6.25)$$

This corresponds to the specific or intensive work input (exergy) used by a manufacturing process. In general, the term $\dot{W}_0$ comes from the equipment features required to support the process, and k comes from the physics of the process. For example, for a cutting tool, $\dot{W}_0$ comes from the coolant pump, hydraulic pump, computer console, and other idling equipment, and k is the specific cutting work that is closely related to the workpiece hardness, the specifics of the cutting mechanics, and the spindle motor efficiency. For a thermal process, $\dot{W}_0$ comes from the power required for maintaining the processing environment at the proper temperature, and k is related to the incremental input required for raising the temperature of a unit of product; this is proportional to the material heat capacity, temperature increment, and the enthalpies of any phase changes that might take place.

We have observed that the electrical power requirements of many manufacturing processes are actually quite constrained, often in the range of 5–50 kW. This happens for several reasons related to electrical and design standards, process portability, and efficiency. On the other hand, when looking over many different manufacturing processes, we find that the process rates can vary by 10 orders of magnitude. This suggests that it might be possible to collapse the specific electrical work requirements for these processes versus process rate on a single log–log plot. We have done this, and in fact the data do collapse, as shown in Fig. 6.9 for 26 different manufacturing processes. What we see is that the data are essentially contained between four lines. The lower diagonal at 5 kW and the upper at 50 kW bound most of the data for the advanced machining processes and for the microprocesses and nanoprocesses. The horizontal lines are meant to indicate useful references for the physical constant k. The lower one at 1 MJ/kg is approximately equal to $c_{\text{ave}}(T_{\text{melt}} - T_{\text{room}}) + h_{sf}$ for either aluminium or iron. The work to plastically deform these metals, as in milling and machining represented by the so-called "specific energy," would lie just below this line (see Table 6.1). The upper horizontal line includes additional terms required for vaporizing these metals. Somewhat surprising, nearly all of the data we collected from a rather broad array of manufacturing processes, some of them with power requirements far exceeding 50 kW, are contained within these four lines. In the "diagonal region," the behavior is described by the first term on the right-hand side of Eq. (6.25). At about 10 kg/h there is a transition to a more constant work requirement, essentially between 1 and 10 MJ/kg. This group includes processes with very large power requirements. For example, the electric induction melters use between 0.5 and 5 MW, and the cupola uses approximately 28 MW power. Note that the cupola is powered by coke combustion and not electricity; hence the power was calculated based on the exergy difference between the fuel inputs and residue outputs at T_0, p_0 according to Eqs. (6.5) and (6.6). This difference includes any exergy losses during the process.

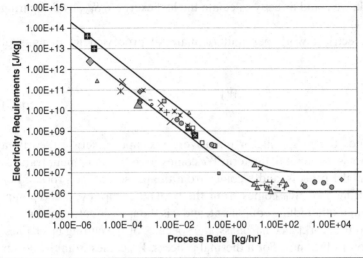

Figure 6.9. Work in the form of electricity used per unit of material processed for various manufacturing processes as a function of the rate of materials processing. EDM, electrical discharge machining; DMD, direct metal deposition; PCB, printed circuit board; CVD, chemical-vapor deposition; PECVD, plasma-enhanced CVD; SWNT, single walled nano tube.

The processes at the bottom right of the diagram in Fig. 6.9, between the horizontal lines, are the older, more conventional manufacturing processes, such as machining, injection molding and metal melting for casting. At the very top left of the diagram we see newer, more recently developed processes with very high values of electric work per unit of material processed. The thermal oxidative processes (shown for two different furnace configurations) can produce very thin layers of oxidized silicon for semiconductor devices. This process, which is carried out at elevated temperatures, is based on oxygen diffusing through an already oxidized layer and therefore is extremely slow [6]. The other process at the top (EDM drilling) can produce very fine, curved cooling channels in turbine blades by a spark discharge process [35]. Fortunately these processes currently do not process large quantities of material and therefore represent only a very small fraction of electricity used in the manufacturing sector.

In the central region of the figure are many of the manufacturing processes used in semiconductor manufacturing. These include sputtering, dry etching, and several variations on the chemical-vapor deposition (CVD) process. Although these are not the highest on the plot, some versions of these processes do process considerable amounts of materials. For example, the CVD process is an important step in the production of electronic-grade silicon (EGS) at about 1 GJ/kg. Worldwide production of EGS now exceeds 20,000 metric tons, resulting in the need for at least 20 PJ of electricity [31]. Notice also that recent results for carbon nanofibers are also in the same region [12]. These fibers are being proposed for large-scale use in nanofiber composites. Furthermore, carbon nanotubes and SWNTs generally lie well above the nanofibers – at least 1 order of magnitude [46] and possibly as much as 2 orders of magnitude or more [11, 53]. Hence it should not be thought that these very exergy-intensive processes operate on only small quantities of materials and therefore their total electricity usage is small. In fact, in some cases it is the opposite that is true.

When considering the data in Fig. 6.9, keep in mind that an individual process can move up and down the diagonal by a change in operating process rate. This happens, for example, when a milling machine is used for finish machining versus rough machining, or when a CVD process operates on a different number of wafers at a time.

Note also that the data in Fig. 6.9 may require further modification in order to agree with typical estimates of energy-resources consumption by manufacturing processes given in the life-cycle assessment literature. For example, the data for injection molding, given by Thiriez [20], averages about 3 MJ/kg. At a grid efficiency of 30% this yields a specific energy value of 10 MJ/kg. However, most injection molding operations include a variety of additional subprocesses, such as extrusion, compounding, and drying, all of which add substantially to the energy totals. If these additional pieces of equipment are also included, they result in a value for injection molding of about 20 MJ/kg, which agrees with the life cycle literature [14, 15, 20]. Additionally, the data in Fig. 6.9 do not include facility-level air handling and environmental conditioning, which for semiconductors can be substantial [28].

6.5 Exergy Efficiency of Manufacturing Processes

The figure of merit of resources use in terms of exergy in a process to which a system is exposed assumes two prerequisite analysis steps to be completed. The first is the definition of the system, and the second is writing the conservation or balance equation. As formulated earlier [see Eq. (6.7) and Fig. 6.2], the system Ω is defined quite generally and its exergy interactions marked. It is assumed that n material streams, $\dot{m}_{\text{in},i}$ (including auxiliary streams), each carrying total exergy rate $\dot{Ex}_i$, ($i = 1, n$), enter at the system inlets. In general, these streams may not be at the restricted dead state, so they may have both physical and chemical exergy components. Within the system the material streams participate in the process and are transformed into the product and the waste streams by the help of energy interactions, work $\dot{W}$, or in exergy terms $\dot{Ex}^W$; heat transfers $\dot{Q}$, or in exergy terms $\dot{Ex}^Q$ delivered at the temperature $T \neq T_0$, and heat transfer loss $\dot{Q}_0$ at T_0, or in exergy terms $\dot{Ex}^{Q_0} = 0$. The product material stream $\dot{m}^{\text{product}}$ carries, in general, both physical and chemical exergies, the total of which is $\dot{Ex}^{\text{product}}$, whereas m material streams include a waste or auxiliary material stream, $\dot{Ex}_j^{\text{waste}}$ ($j = 1, m$). The exergy balance for this system states that total exergy in $\dot{Ex}^{\text{in}}$ that includes material exergies, work exergy, and heat transfer exergies must be equal to the sum of exergy of the product, the exergy losses (which are due to the waste streams and heat loss at the temperature of the environment), and the exergy destruction within the system, $\Delta\dot{Ex}^{\text{destruct}}$. This situation is shown in Fig. 6.10. After Eq. (6.7) is applied, the steady-state balance becomes

$$\dot{Ex}^{\text{in}} = \sum_{i=1}^{n} \dot{Ex}_i + \dot{Ex}^W + \dot{Ex}^Q = \dot{Ex}^{\text{product}} + \sum_{j=1}^{m} \dot{Ex}_j^{\text{waste}} + \dot{Ex}^{Q_0} + \Delta\dot{Ex}^{\text{destruct}}. \quad (6.26)$$

From this balance equation, we define the figure of merit as a coefficient of resources-use performance (this coefficient should not be confused with a coefficient

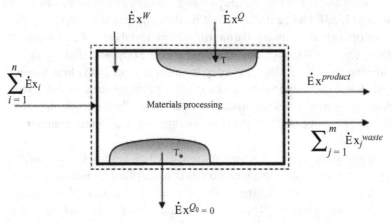

Figure 6.10. Materials-processing or manufacturing system.

of performance defined in thermodynamics for refrigeration cycles and heat pumps). We define this figure of merit as the ratio of the exergy rate value of the product [the first term in Eq. (6.26) on the right-hand side of the second equality, and the total exergy rate input into the system, the left-hand-side in Eq. (6.26)]. That is,

$$\eta_p = \frac{\dot{E}x^{product}}{\dot{E}x^{in}} = \frac{\dot{E}x^{in} - \left(\sum_{j=1}^{m} \dot{E}x_j^{waste} + \dot{E}x^{Q_0} + \Delta\dot{E}x^{destruct} \right)}{\dot{E}x^{in}}$$

$$= 1 - \frac{\sum_{j=x}^{m} \dot{E}x_j^{waste} + \Delta\dot{E}x^{destruct}}{\dot{E}x^{in}}. \tag{6.27}$$

Note that Eq. (6.27) takes two limiting values, i.e.,

$$\eta_p = \frac{\dot{E}x^{product}}{\dot{E}x^{in}}$$

$$= \begin{cases} 0 \ if \ \dot{E}x^{product} = 0 \\ \\ 1 \ if \ \sum_{j=1}^{m} \dot{E}x_j^{waste} + \Delta\dot{E}x^{destruct} = 0 \begin{cases} \sum_{j=1}^{m} \dot{E}x_j^{waste} = 0, \ \text{i.e., no waste} \\ \\ \Delta E x^{destruct} = 0, \ \text{i.e., rev. process} \end{cases} \end{cases} \tag{6.28}$$

Special Case: Resource Accounting

Consider the definition of the exergy efficiency of a manufacturing process as defined in Eqs. (6.27) and (6.28). This is identical to the so-called degree of perfection as given by Szargut et al. [5]. In these equations there is an exergy value for waste, $\sum \dot{E}x^{waste} > 0$. This represents an aggregated exergy equivalent of the material and possible other energy output streams. The materials streams may have both physical and chemical exergies. If this term is large, it represents an opportunity to reduce this value and thereby improve the efficiency of the process. It also could represent an opportunity to recapture these resources and to improve the performance of an extended system that would now include the original process plus the recapturing process. Practical examples of the first kind of efficiency improvement would be to reduce the cutting fluid used in a machining process or to insulate a furnace. Examples of the second kind would be to recycle the cutting fluid used in machining or to use waste enthalpy potential to preheat incoming materials.

In a special subset of exergy accounting, as distinct from traditional thermal engineering analysis, one can be concerned primarily with how much exergy resources are used and ultimately lost or recaptured to perform various manufacturing operations. In this context, the term "lost" may include an additional value system, not only thermodynamic based on exergy balancing. For example, consider a material stream of a residue of an intentionally destructive process that cannot be used regardless of the possibility of having nonzero exergy. To account for this kind of situation, we introduce an accounting scheme, which we refer to as *resource accounting*. This scheme is developed for the purpose of comparing alternative manufacturing processes. In this scheme, we would ask the question, "What normally happens to the waste material?" For example, if it is recaptured, as in recycling, we could reconsider the waste term as a product or a co-product. On the other hand, if it is disposed of in some way such that it becomes unavailable for use, it might be considered lost. This difference is important if we want to identify the resources lost as a consequence of a manufacturing process.

Often the most important form of physical exergy that is lost in the material output of a manufacturing process is the thermal component of it. In many cases, materials, both product and residues, exit a manufacturing process at an elevated temperature. However, for a large group of manufacturing processes, such as the kind analyzed here, this physical exergy is seldom captured. In almost every case it is dissipated to the environment. Hence, for resource accounting, the physical exergy is lost as the materials outputs are equilibrated with environmental conditions at T_0, p_0. In fact, for this accounting scheme, we generally consider that all materials enter and exit at their standard chemical exergy at T_0, p_0.

As far as chemical exergy is concerned, there are a number of processes that eventually recapture the residue materials and recycle them. A variety of machining processes such as turning and milling would fall into this category. At the same time, however, there is a large group of manufacturing processes that produces wastes that are not recaptured. In fact, in some cases waste materials may be destroyed purposefully to render them passive in the environment. In other cases material residues are directly dissipated to the environment, and in some cases are land filled where they may or may not be exploited in the future. In resource accounting we are concerned about these outcomes and want to include these outcomes in the analysis. In particular, we intend to address here those cases in which the output material exergy becomes unavailable and eventually destroyed beyond the manufacturing-process subsystem boundaries. To do this, we can imagine a postprocess subsystem similar to the subsystems drawn in Fig. 6.1. This new subsystem would use the residue or the waste output from the manufacturing process as an input and reduce it to its ultimate state. This postprocess then could be analyzed in a manner similar to the one we used to analyze the manufacturing process. After all, it is another version of a materials-transformation process. A simple example could help to illustrate this point. Consider the case of hot iron cuttings from a machining operation that are subsequently cooled and deposited to the ground and, because of natural action, eventually oxidize and become diluted in the soil at T_0, p_0. A second example could be a process that produces a waste stream of vaporized aluminum that oxidizes, cools, and is dispersed in the environment at T_0, p_0. These exergy losses are calculable. In both of these cases mentioned here, the waste exergy (aggregate physical and chemical) is reduced by over 99% as it goes though the postprocess subsystem we use to represent their usual treatment. And still further loses can be incurred with increasing dilution or possible reactions with the environment. Hence, by a variety of processes both industrial and natural, the residue or waste materials can ultimately be reduced to, or very near to, the ultimate dead state. Although there are indeed nuances to this description, there are so many manufacturing processes that produce waste materials that are not recaptured, are degraded in the environment, and are most likely lost for ever, that this needs to be addressed in order to account for resources lost in these processes. As a result, when analyzing a manufacturing process for resource accounting, we need to include a postprocess subsystem(s) combined with the manufacturing process to address the issue of the ultimate fate of the material wastes. Of course, for this combined analysis any additional inputs or outputs (or both) used for the postprocess subsystems must also be included in the analysis. However, for natural processes these inputs are often small in exergy value, if not zero. One of the most important is oxygen for oxidation. However, oxygen could enter the postprocess subsystem as standard average atmospheric air with close to zero exergy. Keep in mind that the materials themselves are not lost. Their mass is conserved. But their usefulness is destroyed.

This problem becomes particularly poignant for a special class of manufacturing processes that remove material from a workpiece in a destructive manner. This rather large class of manufacturing processes would include such processes as laser machining, electrical discharge machining, ion-beam machining, plasma- and chemical-etching processes, and still many more. In these cases of extended resource accounting, when the material system to be analyzed is necessarily focused on the material to be removed, it becomes difficult, if not impossible, to identify a product at the output. The goal of the process is to produce a hole where the material used to be. There is no suitable positive bulk exergy at the output to be identified as the product. Hence the resulting reduced balance equation is essentially $\dot{E}x^{in} \cong \dot{E}x^{lost}$, and it may become questionable to use the previous definition of efficiency. Note that in this interpretation we define as a system only the material to be removed and lost, that is, any possible material flow that passes through the process as not removed and lost is not included so that the preceding statement becomes valid. This may raise a question of an interaction of the considered process with this excluded reminder of the "unaffected" material flow. In other words, the material balance is preserved and residue just passes through the process setting; however, in principle, this residue may gain or lose certain exergy because of an interaction with the process exerted on the material to perform the destruction.

In spite of the fact that these destructive removal processes do not fit our traditional framework for the exergy analysis of manufacturing processes, there are several ways we might construct an efficiency metric to measure their exergetic use of resources. For example, we could identify a portion of the exergy lost as equal to the exergy of the removed material. This would constitute the goal of the process: to remove and reduce this exergy to zero.

We can form the balance equation for an idealized version of a destructive removal process in a manner similar to Eq. (6.26). In this case, the input flow exergy is confined to the material to be removed $\dot{E}x_{mr}$ and any other input material that is destroyed within the process or ultimately lost in the extended boundary version of the process $\dot{E}x_{mo}$. Other materials that transit through the system with their chemical exergy intact are not included. At the same time, we recognize that these very same materials may and often do interact with the process, most notably by a change in their physical exergy. But these very same effects are ultimately lost as the workpiece or auxiliary materials are equilibrated with the environmental at T_0, p_0, resulting in a loss. By this device, all inputs are ultimately lost. On the right-hand side of the balance equation, we differentiate only between the exergy lost, associated with the identified material to be removed $\dot{E}x_{mr}^{lost}$, and everything else, which is $\dot{E}x_{other}^{lost}$. Here we do not distinguish internal and external destruction, but simply label exergy that is ultimately lost as such. Hence the balance equation for this case would be

$$\dot{E}x^{in} = \dot{E}x_{mr} + \dot{E}x_{mo} + \dot{E}x^W + \dot{E}x^Q = \dot{E}x_{mr}^{lost} + \dot{E}x_{other}^{lost}. \qquad (6.29)$$

We form the exergy efficiency of destructive removal processes η_{DR} then as the ratio

$$\eta_{DR} = \frac{\dot{E}x_{mr}^{lost}}{\dot{E}x^{in}} = 1 - \frac{\dot{E}x_{other}^{lost}}{\dot{E}x^{in}}. \qquad (6.30)$$

This metric, η_{DR}, takes on a value of 1 when all losses are confined to the material to be removed. In practice, η_{DR} can be quite small (approaching zero, i.e., $< 10^{-3}$) for real processes. This is demonstrated in the next section.

6.6 Experimental Data for Exergetic Efficiency of Manufacturing Processes

The three processes considered here can all be found in Fig. 6.9. They include electric induction melting of iron [26] near the lower right-hand section of the figure, plasma-enhanced CVD (PECVD) of S_iO_2 [28], much higher on the figure, and dry etching of S_iO_2 [28] in between the two previous examples. In each case, we collect data on the input materials and electricity, convert them to exergy values, aggregate, and then compare them with the "product" or the "identified material to be removed" as given in Eqs. (6.27) and (6.30). The results are given in Tables 6.3–6.5. We can see immediately that the difference between the conventional process (melting) and the semiconductor processes is enormous, nearly 6 orders of magnitude. For the case of melting, we can see that the exergy inputs are dominated by the exergy of the working material (gray iron). For the semiconductor processes we see two interesting effects. First and foremost is the dominance of the electricity input. But, in addition, we see that the exergy of the auxiliary materials used in semiconductor manufacturing is also very large. In fact, for these two cases the exergy of the input materials alone is 2–4 orders of magnitude larger than the product output or the identified material to be removed. For example, in Table 6.4 for PECVD, we see that the exergy of

Table 6.3. *Exergy analysis of an electric induction melting furnace [26]*

Inputs	Mass (kg)	Exergy (MJ)
Input materials		
Scrap metallics	0.68	5.08
Cast iron remelt	0.30	2.51
Additives	0.05	1.13
Input energy		
Electricity		1.72
		10.43 (total in)
Useful output		
Gray iron melt	1.0	8.25
		8.25 (total out)
Degree of perfection (η_p)		**0.79**

the input cleaning gases alone is more than 4 orders of magnitude greater than the product output. Furthermore, these gases have to be treated to reduce their reactivity and possible attendant pollution. If this is done using combustion with methane, the exergy of the methane alone can exceed the electricity input [10, 29]. When still other manufacturing processes are analyzed, we find that, although the degree of perfection is generally in the range of 0.05–0.8 for conventional processes, the range for semiconductor processes is generally in the range of 10^{-4} to 10^{-6}. Note that this analysis uses only the direct inputs and outputs to the manufacturing system given as Ω_{MF} in Fig. 6.1. Hence the exergy cost of extraction and purifying the inputs, which would be captured in the system Ω_{MA} in Fig. 6.1, is not included in this analysis.

Before closing this section we would like to reemphasize that there are many ways in which one can construct a measure of efficiency. Each is done for a specific system and purpose, and one should be careful not to interpret the results in any context other than the intended.

Table 6.4. *Exergy analysis of a PECVD process for an undoped oxide layer [28]*

Inputs	Mass (g)	Mols	Specific chemical exergy (kJ/mol)	Exergy (kJ)
Input deposition gases				
N_2	276.3	9.86	0.69	6.80
SiH_4	8.57	0.267	1383.7	369.4
N_2O	440.6	10.01	106.9	1070.2
Input cleaning gases				
O_2	69.09	2.16	3.97	8.57
C_2F_6	298.0	2.16	962.4	2078.1
Input energy				
Electricity				50,516
				54,049 (total in)
Output				
Undoped SiO_2 Layer	1.555	2.59×10^{-2}	7.9	0.204
				0.204 (total out)
Degree of perfection (η_p)				$\mathbf{3.78 \times 10^{-6}}$

Table 6.5. *Exergy analysis of dry-etching process for SiO$_2$ [28]*

Inputs	Mass (g)	Mols	Specific chemical exergy (kJ/mol)	Exergy (kJ)
Input materials				
Ar	4.18	1.05×10^{-1}	11.69	1.22
CHF$_3$	0.378	5.4×10^{-3}	569.0	3.07
CF$_4$	0.389	4.42×10^{-3}	454.1	2.01
SiO$_2$	0.067	1.12×10^{-3}	7.90	8.86×10^{-3}
Input energy				
Electricity				5565
				5571 (total in)
Output				
Etched SiO$_2$	0.067	1.12×10^{-3}	7.90	8.86×10^{-3}
				8.86×10^{-3} (total out)
Exergetic efficiency of removal (η_{DR})				**1.59×10^{-6}**

6.7 Closing Comments

In this chapter we develop a thermodynamic framework for analyzing manufacturing processes. This can be used in several ways, but perhaps the most important are (1) for the identification of losses and inefficiencies in a process, which can be used to direct attention to potential areas of improvement, and (2) in the area of resource accounting, for example, as one would do in life-cycle assessment. In fact the principles presented in this chapter are not new, but the application of thermodynamics to manufacturing processes, as the ones discussed here, is indeed relatively new. We suspect that this area will receive more attention, particularly as concerns about energy and global warming rise.

Note that the information in Fig. 6.9 provides a kind of chronological tour through new manufacturing-process development. In general, new process development proceeds from the lower right to the upper left in that figure. For example, note that processes such as machining and casting date back to the beginning of the last century and long before, whereas the semiconductor processes were developed mostly after the invention of the transistor (1947), and the nanomaterials processes have come even more recently. The more modern processes can work to finer dimensions and smaller scales, but also work at lower rates, resulting in very large specific electrical work requirements. In addition to these trends in increased electricity use per unit of material processed, we also saw in our last three examples the increased use of high-exergy auxiliary materials in manufacturing. In general, the auxiliary materials are used in the manufacturing process, but do not get incorporated into the final product.

One should note that the systems considered in manufacturing processes differ vastly from the systems used in energy processes (where energy conversion and thus energy efficiency of a process are of ultimate importance). In all our cases, the material processed is considered a thermodynamic system, but the primary useful effect has been beyond the energy-resource-use scope. The product and its quality have marginalized considerations of energy resources (as well as the impact on the environment). With much more prominent concerns regarding resources use in the context of sustainable development, these issues must be reconsidered. An adoption of thermodynamics methods for such analysis may still pose a certain level of inadequacy when perceived in the same context as for energy systems (e.g., the dramatic difference in magnitudes – and meaning – of efficiencies), but we are convinced that these will find an increasing use, although possibly after additional streamlining of the metrics to better reflect the transdisciplinary aspects of resources utilization.

A particular example of the development of carbon nanotubes can be used to illustrate how this general trend in increased energy resources and exergy use for manufacturing has come to be. As shown earlier, carbon nanotubes are one of the most energy-intensive materials humankind has produced. Yet the cost of this energy turns out to be only a very small fraction of the price. For example, say the energy cost for making carbon nanotubes is of the order of 36 GJ of electricity per kilogram or 36 MJ/g. This is equal to 10 kWh/g. Now at 7 cents a kilowatt hour this yields a cost of 70 cents per gram. But carbon nanotubes can sell for around \$300/g. In other words, the electricity cost in this case is of the order of 0.2% of the price, and, according to a recent cost study, energy costs for all manufacturing processes for nanotubes result in about 1% of the cost [53]. It appears that new manufacturing processes can produce novel products with high demand, resulting in a value that far exceeds the energy-resources (electricity) cost. At the same time, however, because our current electricity supply comes primarily from fossil fuels, most of the environmental impacts associated with these materials (e.g., global warming, acidification, mercury emissions) are related to this use of electricity [53]. How can we reconcile this inconsistency? One comment would be that the current price for carbon nanotubes may well be inflated because of the rather substantial government funds for nanotechnology research worldwide. Another comment, of course, is that, from an environmental perspective, electricity from fossil fuels is significantly underpriced. That is, the environmental and health externalities associated with the use of fossil fuels are not included in the price of electricity.

These trends, in Fig. 6.9 and Tables 6.3–6.5, however, do not give the whole story for any given application. New manufacturing processes can improve and furthermore can provide benefits to society and even to the environment by providing longer life or lower energy (or both) required in the use phase of products. Furthermore, they may provide any number of performance benefits or valuable services that cannot be expressed only in energy–exergy terms. Nevertheless, the seemingly extravagant use of materials and energy resources by many newer manufacturing processes is alarming and needs to be addressed alongside claims of improved sustainability from products manufactured by these means.

6.8 REFERENCES

[1] E.P. Gyftopoulos and G.P. Beretta, *Thermodynamics: Foundations and Applications* (Dover, New York, 2005).
[2] A. Bejan, *Advanced Engineering Thermodynamics*, 3rd ed. John (Wiley, New York, 2006).
[3] J. de Swaan Aarons, H. van der Kooi, and K. Shankaranarayanan, *Efficiency and Sustainability in the Energy and Chemical Industries* (Marcel Dekker, New York, 2004).

[4] N. Sato, *Chemical Energy and Exergy – An Introduction to Chemical Thermo-dynamics for Engineers* (Elsevier, New York, 2004).

[5] J. Szargut, D.R. Morris, and F.R. Steward, *Exergy Analysis of Thermal Chemical and Metallurgical Processes* (Hemisphere and Springer-Verlag, New York, 1988).

[6] C.F. Murphy, G.A. Kenig, D. Allen, J.-P. Laurent, and D.E. Dyer, "Development of parametric material, energy, and emission inventories for wafer fabrication in the semiconductor industry," *Environ. Sci. Technol.* **37**, 5373–5382 (2003).

[7] E.D. Williams, R.U. Ayres, and M. Heller, "The 1.7 kilogram microchip: Energy and material use in the production of semiconductor devices," *Environ. Sci. Technol.* **36**, 5504–5510 (2002).

[8] N. Krishnan, S. Raoux, and D.A. Dornfield, "Quantifying the environmental footprint of semiconductor equipment using the environmental value systems analysis (EnV-S)," *IEEE Trans. Semicond. Manuf.* **17**, 554–561 (2004).

[9] T.W. Zhang, S. Boyd, A. Vijayaraghavan, and D. Dornfeld, "Energy use in nanoscale manufacturing," presented at the IEEE International Symposium on Electronics and the Environment, San Francisco, CA, May 8–11, 2006.

[10] S. Boyd, D. Dornfeld, and N. Krishnan, "Lifecycle inventory of a CMOS chip," presented at the IEEE International Symposium on Electronics and the Environment, San Francisco, CA, May 8–11, 2006.

[11] J.A. Issacs, A. Tanwani, and M.L. Healy, "Environmental assessment of SWNT production," presented at the IEEE International Symposium on Electronics and the Environment, San Francisco, CA, May 8–11, 2006.

[12] V. Khanna, B. Bakshi, and L. Lee James, "Carbon nanofiber production; life cycle energy consumption and environmental impacts," *J. Ind. Ecol.* **12**, 394–410 (2008).

[13] W.M. Morrow, H. Qi, I. Kim, J. Mazumder, and S.J. Skerlos, "Environmental aspects of laser based tool and die manufacturing," *J. Cleaner Prod.* **15**, 932–943 (2007).

[14] I. Boustead, *Eco-Profiles of the European Plastics Industry: PVC Conversion Processes* (APME, Brussels, Belgium, 2002). Available at <http://www.apme .org/dashboard/business_layer/template.asp?url=http://www.apme.org/media/ public_documents/   20021009_123742/EcoProfile_PVC_conversion_Oct2002 .pdf>. Accessed Feb. 2005.

[15] I. Boustead, *Eco-Profiles of the European Plastics Industry: Conversion Processes for Polyolefins* (APME, Brussels, Belgium, 2003). Available at <http://www.apme.org/dashboard/business_layer/template.asp?url=http:// www.apme.org/media/public_documents/20040610_153828/Polyolefins ConversionReport_Nov2003.pdf>. Accessed Feb. 25, 2005.

[16] A. Munoz and P. Sheng, "An analytical approach for determining the environmental impact of machining processes," *J. Mater. Process. Technol.* **53**, 736–758 (1995).

[17] J. Mattis, P. Sheng, W. DiScipio, and K. Leong, "A framework for analyzing energy efficient injection-molding die design," Tech. Rep. (University of California, Engineering Systems Research Center, Berkeley, CA, 1996).

[18] J. Dahmus and T. Gutowski, "An environmental analysis of machining," presented at the ASME International Mechanical Engineering Congress and RD&D Exposition, Anaheim, CA, Nov. 13–19, 2004.

[19] S. Dalquist and T. Gutowski, "Life cycle analysis of conventional manufacturing techniques: Sand casting," presented at the ASME International Mechanical Engineering Congress and RD&D Exposition, Anaheim, CA, Nov. 13–19, 2004.

[20] A. Thiriez, "An environmental analysis of injection molding," project for M.S. thesis (Department of Mechanical Engineering, MIT, Cambridge, MA, 2005).

[21] A. Thiriez and T. Gutowski, "An environmental analysis of injection mold-ing," presented at the IEEE International Symposium on Electronics and the Environment, San Francisco, CA, May 8–11, 2006.

[22] B. Baniszewski, "An environmental impact analysis of grinding," B.S. thesis (Department of Mechanical Engineering, MIT, Cambridge, MA, 2005).

[23] M. Kurd, "The material and energy flow through the abrasive waterjet machin-ing and recycling processes," B.S. thesis (Department of Mechanical Engineer-ing, MIT, Cambridge, MA, 2004).

[24] M. Cho, "Environmental constituents of electrical discharge machining," B.S. thesis (Department of Mechanical Engineering, MIT, Cambridge, MA, 2004).

[25] D.N. Kordonowy, "A power assessment of machining tools," B.S. thesis (Department of Mechanical Engineering, MIT, Cambridge, MA, 2001).

[26] A. Jones, "The industrial ecology of the iron casting industry," M.S. thesis (Department of Mechanical Engineering, MIT, Cambridge, MA, 2007).

[27] M. Branham, T. Gutowski, and D. Sekulić, "A thermodynamic framework for analyzing and improving manufacturing processes," presented at the IEEE International Symposium on Electronics and the Environment, San Francisco, CA, May 19–20, 2008.

[28] M. Branham, "Semiconductors and sustainability: Energy and materials use in the integrated circuit industry," M.S. thesis (Department of Mechanical Engi-neering, MIT, Cambridge, MA, 2008).

[29] T. Gutowski, J. Dahmus, M. Branham, and A. Jones, "A thermodynamic char-acterization of manufacturing processes," presented at the IEEE International Symposium on Electronics and the Environment, Orlando, FL, May 7–10, 2007.

[30] T. Gutowski, C. Murphy, D. Allen, D. Bauer, B. Bras, T. Piwonka, P. Sheng, J. Sutherland, D. Thurston, and E. Wolff, "Environmentally benign manufac-turing: Observations from Japan, Europe and the United States," J. Cleaner Prod., 13, 1–17 (2005).

[31] A. Luque and O. Lohne, Handbook of Photovoltaic Science and Engineering (Wiley, New York, 2003).

[32] S. Kalpakjian and S.R. Schmid, Manufacturing Engineering and Technology, 6th ed. (Prentice-Hall, Upper Saddle River, NJ, 2010).

[33] W.R. Morrow, H. Qi, I. Kim, J. Mazumder, and S.J. Skerlos, "Laser-based and conventional tool and die manufacturing: Comparison and environmental aspects," in Proceedings of Global Conference on Sustainable Product Devel-opment and Life Cycle Engineering (Uni-edition GmbH, Berlin, 2004).

[34] S. Wolf and R.N. Tauber, Silicon Processing for the VSLI Era, Vol. 1 of Process Technology (Lattice Press, Sunset Beach, CA, 1986).

[35] J.A. McGeough, Advance Methods of Machining (Chapman and Hall, New York, 1988).

[36] D. P Sekulić, "An entropy generation metric for non-energy systems assess-ments," Energy 34, 587–592 (2008).

[37] S. Jayasankar, "Exergy based method for sustainable energy utilization of a net shape manufacturing system, M.S. thesis (Mechanical Engineering Department, University of Kentucky, Lexington, KY, 2005).

[38] D.P. Sekulić and S. Jayasankar, "Advanced thermodynamics metrics for sus-tainability assessments of open engineering systems, Thermal Sci. 10, 125–140 (2006).

[39] V.S. Boddapati, "Exergy based metrics for assessment of manufacturing pro-cesses sustainability," M.S. thesis (Mechanical Engineering Department, Uni-versity of Kentucky, Lexington, KY, 2006).

[40] S. Subramaniam and D.P. Sekulić, "Balancing material and exergy flows for a PCB soldering process: Method and a case study, presented at the IEEE International Symposium on Sustainable Systems and Technology, Washing-ton, D.C., May 16–19, 2010.

[41] T.G. Gutowski, M.S. Branham, J.B. Dahmus, A.J. Jones, A. Thiriez, and D.P. Sekulić, "Thermodynamic analysis of resources used in manufacturing processes," *Environ. Sci. Technol.* **43**, 1584–1590 (2009).

[42] J. Szargut, *Egzergia, Poradnik Obliczania i Stosowania* [in Polish] (Widawnictwo Politechniki Slaskiej, Gliwice, Poland 2007).

[43] M.C. Flemings, *Solidification Processing*, McGraw-Hill Series in Materials Science and Engineering (McGraw-Hill, New York, 1974).

[44] K.R. Sharma, *Nanostructuring Operations* (McGraw-Hill, New York, 2010).

[45] M.J. Bronikowski, P.A. Willis, D.T. Colbert, K.A. Smith, and R.E. Smalley, "Gas-phase production of carbon single-walled nanotubes from carbon monoxide via the HiPco process: A parametric study," *J. Vac. Sci. Technol. A* **19**, 1800–1805 (2001).

[46] T.G. Gutowski, J.Y.H. Liow, and D.P. Sekulić, "Minimum exergy requirements for the manufacturing of carbon nanotubes," IEEE/ISSST (IEEE International Symposium on Sustainable Systems and Technology), Washington, D.C., May 16–19, 2010.

[47] J. Abrahamson, "The surface energies of graphite," *Carbon* **11**, 357–362 (1975).

[48] Q. Lu and R. Huang, "Nonlinear mechanics of single-atomic-layer graphene sheets," *Intl. J. Appl. Mech.* **1**, 443–467 (2009).

[49] N.H. Cook, *Manufacturing Analysis* (Addison-Wesley, Reading, MA, 1966).

[50] R. Hill, *The Mathematical Theory of Plasticity*, The Oxford Engineering Science Series (Oxford University Press, Oxford, 1950).

[51] S. Aceves-Saborio, J. Ranasinghe, and G. Reistad, "An extension to the irreversibility minimization analysis applied to heat exchangers," *J. Heat Transfer* **111**, 29–37 (1989).

[52] E.A. Avallone, T. Baumeister, and A. Sadegh, *Mark's Standard Handbook for Mechanical Engineers*, 11th ed. (McGraw-Hill, New York, 2007).

[53] M.L. Healy, L.J. Dahlben, and J.A. Isaacs, "Environmental assessment of single-walled carbon nanotube processes," *J. Ind. Ecol.* **12**, 376–393 (2008).

[54] P. Su, A. Gerlich, T.H. North, and G.J. Bendzsak, "Energy utilisation and generation during friction stir spot welding," *Sci. Technol. Welding Joining* **11**, 163–169 (2006).

7 Ultrapurity and Energy Use: Case Study of Semiconductor Manufacturing

Eric Williams, Nikhil Krishnan, and Sarah Boyd

7.1 Introduction

The notion that technological progress leads to reduced demand of materials and energy to manufacture products and deliver services is known as *dematerialization* [1]. The conventional conception of dematerialization views products and services as static, and from this perspective technological progress can but mitigate the impact per product produced. A demand for increased functionality and performance, however, induces changes in products. Automobiles, computers, and cell phones, for example, have become significantly more complex over the past two decades. A more complex design generally implies tighter tolerances in materials, parts, and manufacturing processes. Semiconductor, nanotechnology, and pharmaceutical manufacturing in particular require chemicals and processing environments that are much purer than traditional industries. Viewing this trend through the lens of thermodynamics, one can assert that the entropy of many products has been decreasing as a function of increasing sophistication. The second law of thermodynamics dictates that the entropy of an isolated system cannot decrease. A purified separation has lower entropy than a mixed one; thus purification implies interaction with the external world. In practice, this interaction is subjecting the system to processes that involve net inflows of energy, e.g., distillation. Purification in practice requires the input of energy, suggesting that increasing complexity should come at a cost of additional processing. This additional processing requires additional secondary energy and materials to attain the desired low-entropy form, a trend we call *secondary materialization* [2]. Secondary materialization implies that technological progress tends to increase energy and material use associated with products and is thus a counterforce to efficiency improvements attributed to dematerialization. Thermodynamic considerations suggest that secondary materialization will occur to some degree: The minimum work required for extracting a target material from a mixture increases as the target material becomes more dilute (see Subsection 7.3.1). In practice, however, it has yet to be established how energy demand scales with increasing purity requirements. In addition, process efficiency tends to improve with technological progress, and it is not yet clear if and when increased energy demand driven by secondary materialization is relevant to the scale of energy savings driven by efficiency improvements.

Our intent in this chapter is to study secondary materialization in the case of semiconductor manufacturing. Technological progress in the semiconductor industry has been rapid and sustained, with a steady drive toward smaller feature size and more complex designs, resulting in steadily improving performances of devices. Moore's law, first formulated by Intel cofounder Gordon E. Moore, quantifies this progress: The number of transistors that can be fabricated on an integrated circuit increases exponentially, doubling approximately every 18 months [3]. This rate of pace has been maintained for several decades. In semiconductor fabrication, technology advances are very rapid, with swift decreases in transistor sizes and corresponding increases in the number of transistors in a semiconductor device – such as an integrated circuit or memory chip. The state of a semiconductor technology is referred to as a technology "node" and corresponds to the smallest transistor gate length that can be fabricated in a repeated pattern. In 1990 the typical transistor size was 1 μm; in 2010 it is 0.045 μm. The technology to realize a smaller feature size also requires purer processing materials and environments. The performance of modern chips depends on ultrapurity in materials used and the processing environment; semiconductor manufacturing has the most stringent purity standards of any industry. All input materials, from silicon wafers to chemicals to water, must be purified to impurity levels in the parts in 10^9 (parts per billion, or ppb) and processing done in special clean rooms with fewer than 10 particles greater than 0.1 μm in size per cubic meter. Semiconductor-grade ammonia, for example, is 99.999%–99.9995% pure, compared with industrial grades, which run in the 90%–99% range.

Purity standards in semiconductor manufacturing lead to a surprisingly high intensity of energy and materials use. The total weight of secondary fossil fuel and chemical inputs consumed in the supply chain to manufacture a single 2-g 32-MB dynamic random-access memory (DRAM) chip are at least 1200 g and 72 g, respectively [2]. To express this in conventional energy units, the energy needed to make a 32-MB DRAM chip is at least 55 megajoules (MJ). Secondary materials (fossil fuels and chemicals) used in production total 630 times the mass of the final product, indicating that the environmental weight of semiconductors far exceeds their small size. This intensity of use is orders of magnitude larger than that for "traditional" manufactured goods; for an automobile or refrigerator, the ratio of fossil fuels embodied in production to the weight of the final product is 1–2. The explanation offered for this high-energy intensity is the need to produce low-entropy materials and processing environments. The production of silicon wafers is some 160 times more energy intensive per kilogram than that of the usual industrial grade [2]. Purifying water and nitrogen accounts for 5% and 7% of semiconductor-fabrication-facility electricity use, respectively [4]. Clean-room operation (ventilation, heating, cooling) accounts for 30%–46% of electricity consumption in fabrication facilities [4].

The effects of secondary materialization should be most prominent in semiconductor manufacturing. Energy and materials use associated with purity dominate current manufacturing, and there have been continuing trends toward stricter purity standards over time. One expects that the degree of secondary materialization in semiconductor manufacturing has been increasing over time. In this chapter we undertake theoretical and empirical analyses of the effect of increasing purity standards in semiconductor manufacturing on energy use. In Section 7.2 we review the purity standards for materials and environments used in semiconductor manufacturing and the technologies used to attain these standards. In Section 7.3 we analyze the thermodynamics literature relating to purification and examine its implications for energy use in attaining ultrapurity. In Section 7.4 we undertake empirical analysis of trends in energy used for semiconductor materials and processing environments: ultrapure water, gases, and clean rooms. In Section 7.5 we discuss implications of these results and suggest future work.

7.2 Ultrapurity Requirements in Semiconductor Manufacturing and Technologies Used to Attain Them

Attaining ultrapurity is an advanced form of separation, so we begin with an overview of separation processes. Separation is fundamental in the functioning of both biological and industrial systems. Separation processes are also energy intensive, accounting for 22% of on-site industrial energy use in the United States [5]. Separation processes exploit differences in physical or chemical properties of constituents in a mixture, such as particle size, density, solubility, or reactivity. The main separation technologies used in attaining high purities related to semiconductor manufacturing are distillation, adsorption, ion exchange, filtration/membrane, and crystallization.

Distillation is a thermally driven process based on the boiling points of constituents. It takes advantage of differences in vapor–liquid equilibrium to effect a separation of miscible components. Distillation is well known as an energy-intensive process, accounting for around 7% of energy consumption of the U.S. chemical industry as a whole [6]. Cryogenic distillation involves the fractional separation of gases by freezing point, typically in a series of multiple stages of increasing purity.

Adsorption involves a gas or liquid solute accumulating on the surface of a solid or a liquid (adsorbent), forming a film of molecules or atoms on the adsorbate. The adsorbate must be periodically purged or regenerated by high pressure or temperature cycling. Less energy intensive than distillation, the main energy use is due to losses in regenerating the adsorbate, generally achieved through heating (direct heat loss) or pressure (losses that are due to pumping).

Ion exchange is an exchange of ions between two electrolytes or between an electrolyte solution and a complex. In most cases the term is used to denote purification, separation, and decontamination of aqueous and other ion-containing solutions with solid polymeric or mineral-based ion exchangers. Membrane-separation processes are mass transfer limited, whereas distillation, adsorption and ion-exchange processes are equilibrium dependent.

In ultrafiltration and reverse osmosis, a filter/membrane has the property of selectively allowing some constituents to pass through but not others. An external potential such as pressure or an electric field is applied to move material through the filter/membrane.

Crystallization is a solid–liquid separation technique. In some cases a precipitate is crystallized from an aqueous solution but in other cases a liquid melt (often a metal) is purified by crystallizing the melt such that impurities are left behind.

Semiconductor manufacturing requires ultrapure silicon wafers, water, chemicals and gases, and clean-room environments. We review evolving purity requirements and technologies used to attain this purity.

Table 7.1. *200-mm wafer specifications [7]*

Property	Specification
Diameter tolerance	±0.25 mm
Crystal orientation	<100> ±1 degree
Resistivity	2.7–4 Ω cm
Resitivity gradient	10%
Oxygen	25–29 ppm
Oxygen gradient	5%
Carbon	0.3 ppm
Heavy-metal impurities	<1 ppb

7.2.1 Silicon Wafers

Silicon wafers are extremely pure, smooth, thin disks of monocrystalline silicon. They are a key element in the production of microchips, providing the base on top of which integrated circuits are laid. To achieve good yields in microcircuit fabrication, silicon wafers must satisfy very strict requirements for size, flatness, crystal structure, and impurity levels. Table 7.1 shows a set of specifications for a typical 200-mm wafer [7].

Conditions on the resitivity and oxygen gradients reflect requirements that the properties of the wafer be spatially uniform. Wafers are produced in a number of discrete diameters: 100, 125, 150, 200, and 300 mm. Wafer thickness ranges from 0.5 to 0.75 mm, with 0.725 mm being typical. There has been a continuing shift to larger radii because of favorable economies of scale at the semiconductor production stage. A larger radius means more chips in one fabrication batch and less wasted space at the boundary. 150 and 200 mm were the standard sizes, and the industry is currently in transition to 300-mm wafers, though this process has been impeded by the recent slump in the global semiconductor industry.

The chain of processes yielding wafers starting from raw quartz is technologically advanced. To recap the chain of processes [2], a simplified flow of the transformations involved is

$$SiO_2 \xrightarrow{C} Si \xrightarrow{Cl_2} HSiCl_3 \xrightarrow{H_2} \text{hyper-pure Si} (+HCl) \rightarrow \text{single-crystal Si} \rightarrow \text{Si wafers}.$$

The starting point is the reduction of quartz (typically 99% SiO_2) with some carbon source such as coal or charcoal in an electric furnace. The resulting "raw" silicon is typically 98.5%–99.0% pure and must be purified to meet the demands of semiconductor fabrication. Powdered industrial-grade silicon is reacted with chlorine to yield trichlorosilane ($HSiCl_3$) [and silicon tetrachloride ($SiCl_4$)] that is purified by distillation [8]. The resulting $HSiCl_3$ is at least 99.9% pure with metallic impurities in the several ppb [9]. The next step is to convert $HSiCl_3$ back to silicon. In the most commonly used Siemens process, $HSiCl_3$ is reacted by chemical-vapor deposition with hydrogen to yield pure elemental silicon that is 99.9999% pure (metals < 0.4 ppb) [9]. This hyperpure silicon is referred to as polysilicon in the industry. Molten polysilicon is further purified by crystallization and drawn into single-crystal ingots typical of the Czochralski method. Polysilicon is first melted at around 1400 °C in a fused-silica crucible, surrounded by an inert atmosphere of pure argon.

Table 7.2. *Energy use and silicon losses in the chain of processes to produce silicon wafers from quartz [2]*

Stage	Electrical energy input /kg silicon out (kWh)	Silicon (%)
Quartz + carbon → silicon	13	90
Silicon → trichlorosilane	50	90
Trichlorosilane → polysilicon	250	42
Polysilicon → single-crystal ingot	250	50
Single-crystal ingot → silicon wafer	240	56
Process chain to produce wafer	2130	9.5

Note: kWh, kilowatt hour.

The melt is cooled to a precise temperature, at which point a single-crystal seed is dipped in the melt and then pulled out slowly while rotating. Melted silicon crystallizes onto the seed, and an ingot of pure silicon slowly forms. The diameter of the rod is determined by the temperature and rotation speed. The monocrystalline ingots are sliced into wafers by special saws and are cleaned and polished to a mirror finish by means of chemical mechanical polishing, which involves use of special slurries, acids, and ultrapure water [7].

In Section 7.4 we analyze trends in energy use to purify water and gases for the semiconductor but, because of the lack of available data, we do not examine similar trends for silicon wafers. We do, however, review results of a static snapshot of energy use of the chain of processes involved in wafer manufacturing in the late 1990s [2], shown in Table 7.2. Silicon yield refers to the fraction of silicon in inputs embodied in the output used for the next stage. Significant silicon losses along the chain suggest that 9.4 kg of raw silicon are needed per kg of final wafer, increasing the total energy demand to yield wafers. The main result is that 2130 kilowatt hours (kWh) per kilogram is used in the production chain for silicon wafers, some 160 times the amount used to produce crude silicon. Energy consumption in the purification is thus much more important than in the preparation of the starting crude material.

7.2.2 Ultrapure Water

The semiconductor manufacturing process also requires large amounts of high-purity water. Generally each etching or cleaning step is followed by rinsing with water, and throughout the entire fabrication process, the wafer may spend a total of several hours in water-rinse systems [7]. A typical 6-in. (15.24-cm) wafer fabrication plant processing 40,000 wafers reportedly consumes 2–3 millions of gallons per day, which corresponds to 18–27 L/cm^2 of silicon [10]. Water is generally purified on site in order to remove contaminants such as dissolved minerals, particulates, bacteria, organics, dissolved gases, and silica. A typical purification system takes municipal water with impurity levels in the parts per hundred (pph) or parts per thousand (ppt) to few ppb [11]. The purification system is complex, including application of reverse osmosis (RO), a vacuum degasifer to remove gases such as O_2 and CO_2, ion-exchange treatment, (UV) ultraviolet treatment to remove organics, and filtration [7]. Ultrapure water (UPW) is alternatively referred to as DI (deionized) water or RO/DI water.

Table 7.3. *Excerpt from UPW specifications for the semiconductor industry [12]*

Technology node	250 nm	180 nm	130 nm	90 nm	65 nm
Particles (cts/L) 0.1–0.2 μm	350	250	100	100	100
Total silica (ppb)	3	2	1	0.5	0.5
Aluminum (ppt)	10	5	3	3	1

The technical guidelines for UPW are extremely complex, with >50 specified criteria, and include details on allowable concentrations of ionic species, metals, and resistivity [12]. It can be seen that UPW specifications are becoming increasingly stringent with reduced sizes in the technology nodes (Table 7.3). (The term "technology node" refers to the DRAM half-pitch and is used as common shorthand in industry to describe the wafer processing technologies of a given year or device generation.)

The evolving energy use associated with attained higher standards for UPW is analyzed in Subsection 7.4.2.

7.2.3 Ultrapure Chemicals and Gases

High-purity chemicals are used in every semiconductor process step. These materials range in volume composition from 99.99997% to 99.999999%, or 0.3- to 0.001-ppm allowed contaminants. Among the high-purity process materials, those used in largest quantity are the "bulk gases" [2, 13]. Nitrogen, oxygen, helium, hydrogen, and argon are used as carriers or for dilution of active materials or as inert gases to maintain purity in the process chamber between process steps. Distillation, adsorption, and membrane-permeation methods may be used in the purification of semiconductor-grade bulk gases.

Nitrogen, oxygen, and argon are all present in the atmosphere at concentrations that allow separation from air to be the preferable means of production. Pressure-swing adsorption or cryogenic distillation is used to separate nitrogen, oxygen, and argon from air. In practice, cryogenic distiller systems also employ some adsorption systems to remove major impurities such as water before freezing. The products of distillation are purified further before use as semiconductor process materials. For example, a semiconductor fab typically purchases standard-grade nitrogen that has been cryogenically distilled and purify it to semiconductor grade through adsorption with a reduced metal catalyst. Oxygen would similarly be purified from a standard-grade by catalytic adsorption of carbon monoxide, carbon dioxide, hydrogen, water and methane impurities. Standard-grade argon would be purified at high temperature over an adsorption bed composed of metal alloys.

The dominant means of producing crude hydrogen is steam methane reforming [14]. Semiconductor-grade hydrogen is typically purified from the crude state by adsorption over metal alloys at a high temperature.

Crude helium of 50%–70% purity is a by-product of natural-gas extraction. This gaseous mixture is purified using first by cryogenic distillation and then by adsorption to achieve purities required in semiconductor processing. Adsorbents include metal alloys containing titanium and vanadium.

Table 7.4. *Particulate concentration standards for different clean-room classes [15]*

U.S. clean-room class	"Not to exceed concentrations" particles per ft^3 by particle diameter (μm)			
	0.1	0.2	0.3	0.5
1	35	75	3	1
10	350	75	30	10
100	–	750	300	100

Trends in energy use to produce semiconductor process gases are analyzed in Subsection 7.4.3.

7.2.4 Clean-Room Environments

Particles on the scale of the smallest feature size of a microelectronic device cause contamination during fabrication. Ordinary air contains dust, bacteria, and other particles ranging from tens to ten thousands of nanometers in size. Thus, not only must semiconductor processing materials such as wafers, chemicals, and glassware be ultrapure, the air in semiconductor fabrication facilities must be kept free of particles. This is done by processing in special clean rooms, in which air is continuously circulated through special filters. The bunny suits featured in Intel ads keep human hair and skin from workers from contaminating chips. Clean rooms are also carefully controlled for, in addition to particles, temperature and humidity; the latter is particularly important for avoiding damages to circuits through sparking.

Clean rooms are rated by class, which places upper limits on the concentrations of particles of different sizes. Table 7.4 summarizes these specifications according to U.S. standards for clean rooms.

Along with the trend of increasing material purity needed when semiconductors are fabricated with smaller feature sizes, the standards for clean rooms likewise become more stringent. Table 7.5 shows typical feature sizes, wafer diameters, and clean-room classes for different generations of central processing units (CPUs). There are exceptions to the progression illustrated in this table of increasing clean-room class with each subsequent technology generation; our intent is to roughly characterize trends.

The basic configuration for a clean room is a space with high-efficiency particulate air (HEPA) and ultralow-penetration air (ULPA) filters set into the ceiling. For lower classes, the entire ceiling is emplaced with filters. Fans run air through the

Table 7.5. *Trends in feature size, wafer diameter, and cleanroom class*

Feature size (nm)	Wafer diameter (mm)	U.S. clean-room class	ISO clean-room class	Representative CPU
<90	300	1	3	Pentium VI (Northwood on)
500–130	200	10	4	Pentium II–III
1000–500	150	100	5	486-Pentium I

Note: ISO, International Organization for Standardization.

filters, changing the air at a rate increasing with decreasing class: 200 times/h for class 6, 400 times/h for class 5, and 600 times/h for 4. In addition, heating, cooling, and humidification equipment keep this air within specified temperature and humidity ranges.

Trends in energy use in clean rooms are analyzed in Subsection 7.4.4.

7.3 Theory: Thermodynamics and Purification

Before examining empirical trends in energy requirements for ultrapurification in Section 7.4, here we review how thermodynamics informs the relationship between energy use and purity. Thermodynamics is the primary physical theory used in understanding separation processes and is applied routinely in their modeling and optimization. In this section we focus on the fundamental relationship between energy use and purity as opposed to modeling of particular separation processes. The central question posed is this: How do general thermodynamic considerations constrain the behavior of the minimum work needed to achieve ultrapurity? There appears to be little previous analysis of this question. Here we simply apply the well-known entropy of mixing formula to ultrapurification and review previous literature.

7.3.1 State-Variable Approach – Entropy of Mixing

The entropy of mixing describes a fundamental thermodynamic aspect of purification: Because of the increased number of possible configurations, mixing two separated substances results in a quantifiable increase in entropy. To recap the standard derivation of the entropy of mixing from first principles [16], first recall the fundamental definition of entropy:

$$S = k \ln W, \tag{7.1}$$

where k is Boltzmann's constant and W is the number of states available to a system. Consider a binary mixture with N_1 particles of type 1 and N_2 particles of type 2. The total number of particles is $N = N_1 + N_2$, and the number of ways to positionally arrange them is given by the permutations of N objects, given that N_1 are alike and N_2 are alike:

$$W = N!/N_1!N_2!. \tag{7.2}$$

With Stirling's approximation used for the factorial function for large N, $\ln N! \sim N \ln N - N$, the entropy of mixing becomes

$$S = -k(N_1 \ln N_1/N + N_2 \ln N_2/N). \tag{7.3}$$

The next step is to use this formula to determine the minimum work needed for purification. A purification process is modeled using the entropy of mixing in Fig. 7.1. An initial sample with N particles and J impurities is separated into waste and purified portions. The number of particles in waste and purified portions is denoted by N_w and N_p, respectively, and is subject to the constraint $N = N_w + N_p$. The number of impurities in the waste and purified portions is denoted by J_w and J_p, respectively, and is subject to the constraint $J = J_w + J_p$. A "higher-purity" result of the process is reflected by a smaller value of J_p/N_p.

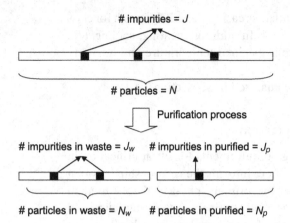

Figure 7.1. Purification modeled by entropy of mixing.

The entropy of mixing before purification is

$$S_{\text{before}}/k = -(N-J)\ln(1-l/N) - J\ln(J/N). \qquad (7.4)$$

The entropy of mixing after purification is the sum of the separate entropies of waste and purified components, respectively:

$$S_{\text{after}}/k = -(N_w - J_w)\ln(1 - J_w/N_w) - J_w\ln(J_w/N_w)$$
$$- (N_p - J_p)\ln(1 - J_p/N_p) - J_p\ln(J_p/N_p). \qquad (7.5)$$

The minimum work required for purification is

$$\text{minimum work} = -T(S_{\text{after}} - S_{\text{before}}). \qquad (7.6)$$

Discussed in detail in Chap. 4, this relation is based on the idea that entropy "lost" in the subsystem (the substance being purified) is achieved through an input of work from an external system (the purification process) such that entropy of the total system does not decrease. These expressions can be used to determine how the minimum work for separation this is due to entropy of mixing changes as one moves toward ultrapurity. To do this, take the limit of (7.6) as the impurity level in the purified fraction, J_p/N_p, approaches zero. To ensure the limiting process makes sense, we first take the thermodynamic limit, work/N with $N \to \infty$, of Eq. (7.6). We define f as the fraction of starting particle number in the waste fraction and the following variations for concentrations:

$$f = N_w/N, \; x = J/N, \; x_w = J_w/N_w, \; x_p = J_p/N_p. \qquad (7.7)$$

The thermodynamic limit of (7.6) is thus

$$\text{minimum work per particle} = -kT\,[(1-x)\ln(1-x) + x\ln x$$
$$- f(1-x_w)\ln(1-x_w) - f x_w \ln(x_w)$$
$$- (1-f)(1-x_p)\ln(1-x_p) - (1-f)x_p \ln x_p]. \qquad (7.8)$$

Next, taking the limit of (7.8) in the case of perfect purity, we obtain

$$\lim_{x_p \to 0} \text{minimum work per particle} = -kT[(1-x)\ln(1-x) + x\ln x$$
$$- f(1-f_w)\ln(1-x_w) - f x_w \ln(x_w)]. \qquad (7.9)$$

The work per particle required for separation approaches a constant that is a function of the starting concentration of impurities (x) and characteristics of the waste portion (f, x_w). Note that in Eq. (4.15) in Chap. 4 the result was that the minimum work needed to mine a mineral from ore increases with decreasing ore grade. These two results are consistent because the limit of extracting a resource from declining ore grades is qualitatively different from extracting a perfect pure subsample from an impure starting point. For mining, the conditions were $x_p = 0$ and $x_w = 1$; both are states after separation that have zero entropy. Let the concentration of substance to be mined be $x' = (1 - x) = N_p/N$. The work per particle to extract the desired mineral from declining ore grade corresponds to

$$\lim_{x' \to 0} \text{work}/N_p \sim kT \ln (x'),\tag{7.10}$$

which goes to infinity, as in Chap. 4.

7.3.2 Process Approach

Analyzing changes in entropy is a state-variable approach. An alternative approach is to use thermodynamics to model the separation process. To quote Sciamanna and Prausnitz [17],

> The conventional wisdom is that it is not possible to achieve perfect purity by absorption, extraction, or other diffusion operations. In any standard separation device, the mole fraction x_p can be made arbitrarily small but not zero. The basis of this assertion is that perfect purity cannot be obtained in a single process step.

To justify this, note that, by definition, purification processes must allow particle transfer between the waste and purified fractions, i.e., purification opens some form of sieve between the two portions in Fig. 7.1. The purification processes creates a chemical potential that preferentially pushes impurities in the purified fraction to the waste side. The "diffusive" chemical potential based on the entropy of mixing is [17]

$$\mu = \text{constant} + kT \ln x_p.\tag{7.11}$$

As $x_p \to 0$, this chemical potential goes to negative infinity. Assuming that any given purification process will have a chemical potential limit beyond which it cannot purify, this implies that an additional process step is needed to purify beyond this limit.

In this framework the approach to ultrapurity is a repeated application of purification, the minimum work for which can be described by

$$\text{minimum work} = \sum W(x_j \to x_{j+1}),\tag{7.12}$$

where $W(x_j \to x_{j+1})$ is the energy needed to purify from level x_j to x_{j+1} and the sum is over an number of individual process steps. The convergence of the minimum energy use for ultrapurification thus is determined by the behavior of $W(x_j \to x_{j+1})$ as $x_j \to 0$. Based on minimum work formula (7.9) from the entropy of mixing, the minimum energy will approach infinity because a constant energy value for each step is required.

If the minimum work $W(x_j \to x_{j+1})$ were to approach zero, $x_j \to 0$, the series in Eq. (7.12) could converge. In the late 1980s, two articles were published arguing that the chemical potential does not go to infinity in the dilute limit [17, 18]. In other words the authors dispute that Eq. (7.9) is the correct limit of (7.6) in the case of ultrapurity. The basis of the argument is the observation that Stirling's approximation to the factorial function holds only for large numbers of particles and that by definition ultrapurity is the case of small numbers of impurities. The two articles postulate through a simple entropy-based theoretical analysis that utilizing highly dispersed systems (with small phase sizes) could in principle be purified with a finite energy requirement. Two analyses following these papers disputed the results of Reis and Sciamanna and Prausnitz, asserting that droplet size does not affect the traditional result that chemical potential tends toward negative infinity in the ultrapurity limit [19, 20]. Also note that the empirical purities used in semiconductor manufacturing discussed in Section 7.2 are in parts per billion. This implies that the numbers of impurities are well within the domain of application of Stirling's law. Thus relationships between energy and ultrapurity, at least with respect to what ultrapurity means today, are not driven by failures in Stirling's law. Since this exchange, the question of ultrapurity apparently has not been considered in the journal literature. We believe this to be a worthwhile and relevant area that deserves further study.

7.4 Empirical Studies of Ultrapurity

In this section we explore the relationship between energy and ultrapurity from an empirical perspective. On the surface the approach appears straightforward: Collect data on the energy consumption of purification processes as a function of different purity standards as described in Section 7.2 and compute the trends. There are, however, practical and conceptual complications. On the practical side, technologies in semiconductor material and manufacturing industries are still evolving, resulting in firms being reluctant to release data compared with many other industries. The conceptual complication is direct energy use of purification processes may not sufficiently describe energy use: The energy overhead to produce equipment, replace filters, and purchase chemicals could be substantial.

To address these complications, we use a method known as hybrid life-cycle assessment (LCA), subsequently introduced in Subsection 7.4.1. In Subsections 7.4.2–7.4.4 we apply LCA techniques to analyze energy-use trends for ultrapure water, gases, and cleanrooms respectively.

7.4.1 Life-Cycle Assessment Methodologies

LCA is a quantitative method designed to assess the environmental impacts of a product or service, including relevant phases of the entire life cycle, from mining, materials production, assembly, and distribution, to use and disposal. LCA can be conceptually divided into inventory and assessment phases. In the inventory phase a model is built that describes the materials and energy use and emissions over the life cycle. This set of material and energy quantities is termed the life-cycle inventory (LCI). In the assessment phase these quantities are interpreted and often combined to estimate impacts on one or more environmental issues. In our analysis we focus on the LCI. The three main methods for estimating life-cycle inventories of material and energy used are process–sum, economic input–output (EIO), and their combination, hybrid analysis.

The *process-sum method*, on which most existing LCIs are based, connotes both a calculation method and a type of data or normalization used [21]. The method starts with a process network diagram, for which materials input–output data have been collected for each element in the network. The flows between processes are usually described in material terms, e.g., kilograms of emissions per unit mass of product output. The sources of data are often facility based, though sometimes they reflect industry or even national averages. The net materials use and emissions associated with a unit output of product or service being studied are obtained by use of a linear increment of materials flow associated with each process.

Economic input–output life-cycle assessment (EIOLCA) is based on Wassily Leontief's formulation [22] of an economy as a matrix describing economic trans-actions between sectors. The core of the model is the input–output matrix, usually denoted by Z_{ij}, which describes the economic purchases and sales between economic sectors. This matrix, being a national aggregate of results of (confidential) firm-level surveys, is normally formulated by a government agency, such as the U.S. Bureau of Economic Analysis. The most detailed tables divide an economy into 400–500 sectors. Although originally formulated to address economic questions, the model can also be supplemented with environmental information to estimate supply-chain materials use and emissions for products. This method has been used since the 1970s to estimate the net energy cost of products and facilities and more recently was expanded to cover a broad variety of emissions and impacts [23–25]. The basic for-mula used to calculate the net materials use or emissions associated with a unit of economic output for economic sectors is

$$E_{SC} = E_D (1 - A)^{-1}, \tag{7.13}$$

The result, E_{SC}, is the vector of sector-level supply-chain energy-use intensities (MJ/\$). A is the requirements matrix built from the transaction matrix ($A_{mn} = Z_{mn}$/total economic output of sector n). E_D represents direct energy use of a sector and is constructed from national (or sometimes process-level) information by LCA practitioners. In the United States, researchers at Carnegie Mellon have developed and maintained a public-use model based on the 491-sector benchmark U.S. input–output tables [25]. The LCI result for energy use for manufacturing the target product is found by

1. identifying its representative sector in the input–output table,
2. calculating E_{SC} for the materials or emissions of interest, and
3. multiplying by the producer price of the product (or consumer price, depending on the formulation of the input–output model).

Process-sum LCA inevitably excludes some processes in the supply chain for which materials input–output data are unavailable, leading to a truncation error. EIOLCA is a coarse-grain model that often combines many different processes into economic sectors, leading to an aggregation error. *Hybrid LCA* combines the process-sum model with EIOLCA [26–28] with the goal of reducing the cutoff error in the former and the aggregation error in the latter [29]. The term hybrid generically refers to any method combining process and EIOLCA: There are a number of approaches to achieve this. The simplest is the additive hybrid in which economic data are identified as covering processes for which materials data are unavailable and associated with sectors in an EIO model [30]. An economic-balance hybrid calculates the value added covered in a materials-process model, subtracts this from the total price, and estimates the impacts associated with the remaining value by using EIOLCA [28]. A mixed-unit hybrid model constructs a matrix with both physical and economic quantities [31].

As cost, but not materials, input–output data are identified for purification processes; the hybrid model used in this analysis is the additive method. Costs are based on a cost-of-ownership (CoO) model. CoO is also known as total cost of ownership (TCO) and is a useful tool to estimate total costs related to equipment or production activities. CoO models include all of the direct and indirect costs related to owning and operating equipment, including, in addition to equipment costs, costs related to materials, energy, labor, facilities, and other allocated overhead. For example, whereas the capital cost of an UPW purification system includes only its purchase price, the CoO includes the capital cost as well as the costs of electricity, consumable materials, and maintenance. The concepts of CoO and TCO have been around since the mid-20th century, and a more detailed explanation can be found in finance and accounting references [32]. Detailed guidelines for CoO analysis are also available in the semiconductor industry [33]. To understand the life-cycle impacts associated with a process, a CoO model can be combined with EIOLCA. The energy to manufacture a product is given by

$$E_{\text{production}} = \sum_i C_i E_{\text{SC},i}, \tag{7.14}$$

where the subscript i refers to the ith line item in the CoO model, C_i ($) refers to the cost of line item i, and $E_{\text{SC},i}$ (MJ/$) refers to the corresponding supply-chain energy intensity from EIOLCA [25].

7.4.2 Ultrapure Water

In this section we analyze the life-cycle energy and secondary materials requirements to clean water to different degrees of purity. Although producing UPW is a complex and multistage process, we simplify analysis by considering two principal steps – a bulk purification step using ion exchange (IX) or reverse osmosis (RO), and a final polishing step using activated carbon and ultraviolet (UV) light [34]. We examine the energy requirements with increasing purity by comparing life-cycle energy for the bulk purification of city water, with the final polishing, with UPW quality at current technology nodes (180–90 nm).

Incoming water to facilities, or industrial city water (ICW), can vary significantly in quality depending on local water sources and treatment systems. We assume concentration levels of 80–480 ppm of dissolved solids in ICW [34]. Because water quality is determined by the purity levels of a number of different species (Table 7.3) [12], we adopt a simplified approach and use the concentration of silica as a proxy for the bulk purification step (silica is difficult to remove). After initial treatment

Table 7.6. *Purification cost inputs [34]*

Purification stage	Bulk purification		Final polish		
Technology	IX	IX	RO	RO	UV
ICW water quality (ppm)	80	480	80	480	–
Cost inputs ($/1000 gal)					
Electricity (pumps)	0.06	0.07	0.62	0.63	1.3
Electricity (heating)	–	0.02	–	–	–
Sulfuric acid	0.07	0.04	0.01	0.08	–
Caustic	0.16	0.63	–	–	–
Lime	0.01	0.08	–	–	–
Antiscalant	–	–	0.04	0.09	–
Resin replacement	0.05	0.18	–	–	–
Membrane replacement	–	–	0.21	0.21	–
Maintenance	0.19	0.26	0.31	0.31	–

Note: IX = ion exchange, RO = reverse osmosis, UV = ultraviolet, ICW = industrial city water.

through IX or RO, ICW is purified to approximately 0.01-ppm silica levels. After final treatment to UPW quality, silica concentrations are reduced to about 0.001 ppm at the 130-nm node (see Table 7.3).

CoO modeling is used to determine energy needs in UPW purification. We examine purification primary energy requirements under two key bulk purification routes (IX and RO), but we also consider two incoming ICW water purity levels for each route (80 and 480 ppm). The main input in the final polishing step is assumed to be related to energy use in the UV lamps. Key cost inputs are summarized in Table 7.6 [34], and the corresponding EIOLCA sectors are summarized in Table 7.7.

Using this model, we now examine how the incremental primary purification energy requirements change at different purity levels for two different technologies (IX and RO). The results are shown in Figs. 7.2 and 7.3. The empirical trend is clear: purification energy requirements increase dramatically with higher purity. Figure 7.2 shows results for IX: Purifying from 80 or 480 ppm to 0.01 ppm (a purification of 4–5 orders of magnitude) requires 18–45 kJ/gal of water purified. However, purifying from 0.01 to 0.001 ppm (an additional order of magnitude) requires significantly (~4 to 9 times as much) more energy (160 kJ/gal).

For the RO case, Fig. 7.3 shows that achieving higher purity levels requires increasing energy, although the difference is less marked than in the IX case. Bulk purification with reverse osmosis from 80 or 480 ppm to 0.01 ppm uses significantly more primary energy than IX: 80 kJ/gal for RO versus 18–45 kJ/gal for IX. The

Table 7.7. *EIOLCA sectors and primary energy intensities [25]*

Purification input	EIOLCA sector	E_{SC} (MJ/$)
Electricity	Power generation and supply	121
Sulfuric acid, caustic, antiscalant	Other basic inorganic chemical manufacturing	36.1
Lime	Lime manufacturing	54.9
Resin, membrane	Plastics material and resin manufacturing	22.7
Maintenance	Commercial machinery repair and maintenance	3.82

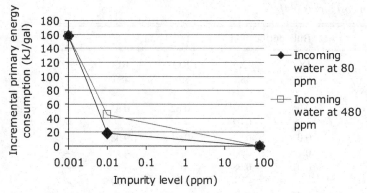

Figure 7.2. Primary energy requirements for water increase with increasing impurity level: Bulk purification is performed with a three-bed IX system.

energy required for the final step of purifying from 0.01 to 0.001 ppm (an additional order of magnitude) is similar for both processes (160 kJ/gal).

Although the bulk purification stage uses significantly more primary energy with RO (80 kJ/kg) than with IX (18–45 kJ/kg), the RO energy requirements are much less sensitive to variations in incoming-water quality. This was expected because water with high concentrations of dissolved solids will exhaust IX beds rapidly, requiring more frequent regenerations and higher costs and primary energy requirements. In regions with highly impure ICW, it may therefore make sense to use RO over IX for UPW purification – from both a CoO and primary energy perspective.

The results of this empirical analysis indicate that, for UPW systems, energy requirements dramatically increase with increasing purity. Bulk purification from 80–480 ppm to 0.01 ppm requires less energy than 0.01 ppm to ppb levels. With more stringent purity specifications related to future semiconductor technology nodes, we expect that energy consumption related to UPW production may increase in the future. Finally, with UPW consumption growing across a wide range of high-technology industries beyond semiconductor manufacturing (such as nanotechnology, biotech, and pharmaceutical sectors), we suspect that the primary energy requirements of UPW production and more generally for other process inputs could emerge as key issues of concern.

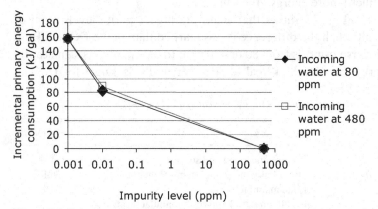

Figure 7.3. Primary energy requirements for water increase with increasing purity: Bulk purification is performed with a double-pass RO system.

Table 7.8. *Gas purification system cost of ownership model*

Parameter	Amount	Unit
System lifetime	8	year lifetime
	59,520	h/lifetime
System capacity	1000	L/min
	3.57×10^9	L/lifetime of system
Power	15	kW (regenerating)
	2	kW (flow)
	10	% time regenerating
	3.30	kW (average over lifetime)
	196,416	kW h/system life
	0.1	$/kW h
	$ 19,640	$/system life
Metal catalyst	90	$/g
	20	g/system
	$ 1800	$/system
Machine parts / equipment	$ 9500	$/system
Maintenance	100	$/hour
	38.4	h/lifetime
	$ 3840	$/system life
Total cost of ownership	$ 34,780	

7.4.3 High-Purity Gases

The energy demand of chemicals purification is examined in this subsection, looking specifically at high-purity gases that are the materials used in the largest volume in wafer fabrication. The adsorption processes used to purify elemental gases to semiconductor grade do not have a representative sector in the EIO model, so we model by using a CoO model and EIOLCA, as we did in the previous section. Table 7.8 shows the CoO model assumptions for a hypothetical adsorption gas purification system that was developed based on specifications of available gas purification equipment as well as cost estimates and technical guidance from a senior engineer at a semiconductor process equipment manufacturer.

The next task is to identify sectors that correspond to these cost elements in the EIO table. Our selections are shown in Table 7.9 along with respective values of EIOLCA-defined supply-chain energy [Eq. (7.14)]. The energy consumption per dollar for each cost item is determined with the EIOLCA model [25] results for that item's economic sector.

Because each elemental gas requires processing to a different degree to achieve a given purity, the value contributed by each economic sector in the CoO model is described as a dollar value per cost of purification. (The cost of purification for each gas is defined here as the price difference between the standard purity and semiconductor grade of that gas.) Results of the analysis for argon, oxygen, nitrogen, and hydrogen are shown in Fig. 7.4. Data points between 10^5- and 10^6-ppm contaminants represent the ambient concentration of each species in the atmosphere. Points between 10 and 1000 ppm correspond to industrial- or standard-grade purity and points near 0.1 ppm represent semiconductor-grade gases. As in the case of

Table 7.9. *Cost items and their energy intensities [25] for adsorption purification*

Cost item	Cost inputs ($ cost/$ purified)	Economic sector in EIOLCA model	E_{SC} (MJ/$)
Electricity (pumps)	0.31	Power generation and supply (sector # 221100)	121
Electricity (heating)	0.26	Power generation and supply (sector # 221100)	121
Metal catalyst	0.05	Primary nonferrous metal, except copper and aluminum (sector # 331419)	17.7
Capital equipment	0.27	(All other) industrial machinery manufacturing (sector # 333298)	6.73
Maintenance	0.11	Commercial machinery repair and maintenance (sector # 811300)	3.82

UPW, the energy intensity of the purification process for each of these elemental gases increases with increasing purity.

7.4.4 Clean-Room Energy Use

Along with purity requirements for UPW and gases, clean-room standards have also become increasingly stringent (see Table 7.4). As before, we seek to characterize how stricter standards affect energy use. Because of lack of available data on CoO, the analysis of clean rooms considers only direct energy use. Publicly available data on direct energy use in clean rooms are also scarce; the primary source is the High Performance Buildings for the High-Tech Industry group at Lawrence Berkeley National Labs. This group has carried out analyses since the 1990s to characterize and improve energy efficiency in clean rooms. Table 7.10 shows a selection of data normalized to total electricity use per square foot of clean room [35, 36]. To construct

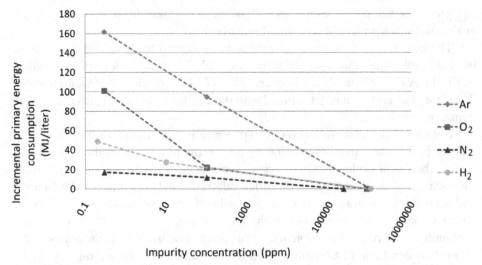

Figure 7.4. Energy intensity of bulk gas purification.

Table 7.10. *Electricity use per area in different classes of clean rooms [35, 36]*

Clean-room class (U.S. standard)	Total electricity use (annual kW h/ft^2)	Fan electricity use (annual kW h /ft^2)
1	not available	350
10	500	263
100	280	116

this table we selected typical figures, but it is important to note that there is substantial variation in clean-room energy use within the same class. One reason for this variation is that class is a design standard: Implementation in practice exceeds the standard to varying degrees. In addition, the semiconductor manufacturing equipment put in clean rooms affects their clean-room design and, in turn, energy use. Energy use uniformly increases with higher cleanliness standards. Fans account for 30%–50% of total energy use.

Energy per unit area of clean room does not necessarily correlate to energy per product manufactured. The question of the appropriate unit of analysis for clean-room energy is more complex than for water and gas because the resulting product, integrated circuits, is not usefully characterized by the mass of output [37]. In line with previous analyses, the unit area of a wafer processed is chosen as a functional unit [38] with the understanding that this is an imperfect measure because the functionality of microcircuits evolves over time. Table 7.11 shows anonymous industry data on trends in clean-room areas needed per number in the second column, converted to area of wafers processed in the third column. Significantly, more clean-room area is needed as wafer diameter increases; this is likely driven by a need for more processing space, given the increased complexity of processing. On the other hand, larger wafers imply that substantially more area is processed and clean-room area per wafer area tends to decrease.

Energy Use Trends Per Wafer Processed

Following the logic of the introduction with other factors held constant, we would expect the energy use associated with clean rooms to increase along with stricter requirements on air quality. We test this hypothesis by assessing electricity used in

Table 7.11. *Clean-room area per wafer and per wafer area processed in semiconductor fabrication facilities*

Wafer diameter (mm)	Clean-room area / # wafers processed (m^2 per 1000 wafers per week)	Clean-room area/ area wafer processed (cm^2/cm^2)
150	1400	1.5
200	1910	1.2
300	2000	.54

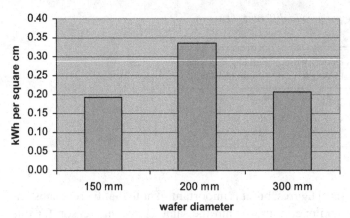

Figure 7.5. Clean-room fan electricity use per area of silicon processed as a function of wafer diameter.

clean room per area of wafer processed for 150-, 200-, and 300-mm wafer fabrication per this equation:

$$\frac{\text{clean-room electricity}}{\text{wafer processed}} = \frac{\text{clean-room electricity}}{\text{area}} \frac{\text{area}}{\text{wafer processed}}. \quad (7.15)$$

The first term in (7.15) is characterized by use of the correlation between 150-, 200m-, and 300-mm wafer processing with class 100, 10, and 1 clean rooms, respectively, and drawing on the existing literature on electricity use per area for clean rooms from Table 7.10. Given the lack of data for total energy for class 1 clean rooms, we use fan electricity use as a proxy to indicate trends. We estimate the second term by using Table 7.11.

Figure 7.5 shows the results of the analysis: Clean-room energy use increases in the transition from 150-mm to 200-mm wafers, but then decreases for 300-mm wafers. The latter decrease indicates that the purification-induced increases in energy use are weaker than other factors. The explanation is likely that the need for increasing purity has induced the shift in the use of mini-environments in 300-mm facilities, qualitatively distinct from previous clean rooms. A mini-environment is a tool-scale clean room that workers do not enter. The shift to 300-mm-based facilities has been accompanied by dramatic growth in the use of mini-environments. There are two main driving factors. One is that it becomes more difficult to protect the product from the worker for higher environmental purities. The second reason is that increasing the purity of clean rooms also drives up the cost of energy to maintain the environment, making the savings accrued to the smaller area of a mini-environment more relevant.

The clean-room example illustrates a case in which increasing purity demands at first induced increases in the energy use but, as these increases led to increased technological challenges and costs, caused an industry to shift to a different technological path.

7.5 Discussion

Our results indicate that for UPW and bulk gas purification there is a clear trend toward increasing life-cycle energy consumption at higher purity. In clean-room environments, the trend toward increasing energy use is countered through the shift toward mini-environments from uniformly purified large-scale clean rooms. These results provide confirmation that secondary materialization is present and relevant for semiconductor manufacturing. This in turn implies that a life-cycle perspective is crucial in assessing environmental implication in semiconductor manufacturing, because by definition facility-level studies do not account for secondary use of energy and materials. The semiconductor industry discourse on energy efficiency, in contrast, continues to focus on facility-level issues.

Our analysis suggests that the energy overhead to realize pure materials and environments for semiconductors will increase in the future unless alternative technology paths, such as mini-environments, are developed. Our purely empirical analysis leaves many important questions unanswered. How rapidly will the secondary energy demand associated with purity increase in the future? What are the trajectories of energy uses for different purification technologies, such as distillation and reverse osmosis, as functions of increasing purity? Is it possible to forecast the relationship between energy and purity to provide useful input into decisions on research and development of new technologies? Addressing these previous questions will require a combination of theoretical and empirical work.

REFERENCES

[1] C. J. Cleveland and M. Ruth, "Indicators of dematerialization and the materials intensity of use," *J. Ind. Ecol.* **2**, 15–50 (1999).
[2] E. D. Williams, R. U. Ayres, and M. Heller, "The 1.7 kilogram microchip: Energy and material use in the production of semiconductor devices," *Environ. Sci. Technol.* **36**, 5504–5510 (2002).
[3] G. E. Moore, "Cramming more components onto integrated circuits," *Electronics* **38**, 114–117 (1965).
[4] LBNL, "Energy efficiency in semiconductor cleanrooms: A technical perspective" (Lawrence Berkeley National Laboratory, CA, 2002), available at http://ateam.lbl.gov/cleanroom/technical.html.
[5] ORNL, *Materials for Separation Technologies: Energy and Emission Reduction Opportunities* (Oak Ridge National Laboratories, Oak Ridge, TN, 2005).
[6] O. Oppenheimer and E. Sorenson, "Comparative energy consumption in batch and continuous distillation," *Comput. Chem. Eng.* **21**, S529–S534 (1997).
[7] P. Van Zant, *Microchip Processing*, 5th ed. (McGraw-Hill, New York, 2004).
[8] M. Howe-Grant and J. Kroschwitz, in *Kirk-Othmer Encyclopedia of Chemical Technology* (Wiley, New York, 1997), Vol. 22, pp. 1–154.
[9] K. A. Jackson, *Materials Science and Technology*, Vol. 16 of Processing of Semiconductors Series (VCH Press, Weinheim, Germany, 1996).
[10] L. Peters, "Ultrapure water: Rewards of recycling," *Semiconduct. Intl.* **21**, 71–76 (1998).
[11] J. Genova and F. Shadman, *Environ. Prog.* **16**, XX–XX (1994).
[12] *Ultrapure Water Monitoring Guidelines*, Rev. 2.0, Air Liquide, 2009. Available at http://www.balazs.com/jmen/PDFDocs/Guidelines/ UPW_Guidelines_rev2.0.pdf.
[13] N. Krishnan, S. Boyd, A. Somani, S. Raoux, D. Clark, and D. Dornfeld, "A hybrid life cycle inventory for semiconductor manufacturing," *Environ. Sci. Technol.* **42**, 3069–3075 (2008).
[14] W. H. Scholz, "Processes for industrial production of hydrogen and associated environmental effects," *Gas Sep. Purif.* **7**, 131–139 (1993).

[15] B. Bhatia, *A Basic Design Guide for Cleanroom applications*, 2007. Available at http://www.pdhonline.org/courses/m143/m143.htm.

[16] L. Reichl, *A Modern Course in Statistical Physics* (University of Texas Press, Austin, 1984).

[17] S. F. Sciamanna and J. M. Prausnitz, "Thermodynamics of highly dilute solutions and the quest for ultrapurity," *AIChE J.* **33**, 1315–1321 (1987).

[18] J. C. R. Reis, "Phase equilibria and thermodynamic conditions for the obtention of pure materials," *J. Phys. Chem.* **90**, 6078–6080 (1986).

[19] L. Kubic, "Fluctuations in highly dilute solutions," *AIChE J.* **34**, 1581–1583 (1988).

[20] A. H. Harvey, "On the irrelevance of phase size in purification," *Comments J. Phys. Chem.* **92**, 6477–6478 (1988).

[21] H. Baumann and A. M. Tillman, *The Hitch Hikers Guide to LCA* (Studentlitteratur AB, Lund, Sweden, 2004).

[22] W. Leontief, "Environmental repercussions and the economic structure: An input–output approach," *Rev. Econ. Statist.* **52**, 262–271 (1970).

[23] C. Bullard and R. Herendeen, "The energy cost of goods and services," *Energy Policy* **3**(4), 268–278 (1975).

[24] C. T. Hendrickson, A. Horvath, S. Joshi, and L. B. Lave, "Economic input–output models for environmental life-cycle assessment," *Environ. Sci. Technol.* **32**, 184A (1998).

[25] Economic Input-Output Life Cycle Assessment (EIO-LCA), Carnegie Mellon University, Available at: http://www.eiolca.net.

[26] B. C. W. Engelenburg, T. F. M. van Rossum, K. Blok, and K. Vringer, "Calculating the energy requirements of household purchases," *Energy Policy* **22**, 648–656 (1994).

[27] S. Suh, M. Lenzen, G. J. Treloar, H. Hondo, A. Horvath, G. Huppes, O. Jolliet, U. Klann, W. Krewitt, Y. Moriguchi, J. Munksgaard, and G. Norris, "System boundary selection in life-cycle inventories using hybrid approaches," *Environ. Sci. Technol.* **38**, 657–664 (2004).

[28] E. Williams, "Energy intensity of computer manufacturing: Hybrid analysis combining process and economic input–output methods," *Environ. Sci. Technol.* **38**, 6166–6174 (2004).

[29] E. Williams, C. Weber, and T. Hawkins, "Hybrid approach to managing uncertainty in life cycle inventories," *J. Ind. Ecol.* **15**, 928–944 (2009).

[30] C. Bullard, P. Pennter, and D. Pilati, "Net energy analysis: Handbook for combining process and input–output analysis," *Resources Energy* **1**, 267–313 (1978).

[31] T. Hawkins, C. Hendrickson, C. Higgins, H. S. Matthews, and S. Suh, "A mixed-unit input-output model for environmental life-cycle assessment and material flow analysis," *Environ. Sci, & Tech.* **41**, 1024–1031 (2007).

[32] P. Baily, D. Farmer, D. Jessop, and D. Jones, *Purchasing, Principles and Management*, 9th ed. (Financial Times Press, Upper Saddle River, NJ, 2005).

[33] SEMI, *Guide To Calculate Cost Of Ownership (COO) Metrics For Semiconductor Manufacturing Equipment*, SEMI Standard E35–0307 (Semiconductor Equipment and Materials International, San Jose, CA, 2007).

[34] M. C. Lancaster, *Ultrapure Water – The Real Cost* (Rose Associates, Los Altos, CA, 1996).

[35] T. Tengfang Xu and W. Tschudi, "Energy Performance of Cleanroom Environmental Systems," LBNL Doc. 793761 (Lawrence Berkeley National Laboratory, 2001).

[36] T. Xu, "Characterization of minienvironments in a cleanroom: Assessing energy performance and its implications," *Building Environ.* **42**, 2993–3000 (2007).

[37] L. Deng and E. Williams, "Functionality versus 'Typical Product' measures of energy efficiency: Case study of semiconductor manufacturing," in press, *J. Ind. Ecology*, DOI: 10.1111/j.1530-9290.2010.00306.x (2011).

[38] S. Boyd, A. Horvath, and D. Dornfeld, "Life-cycle energy demand and global warming potential of computational logic," *Environ. Sci. Technol.* **43**, 7303–7309 (2009).

8 Energy Resources and Use: The Present Situation, Possible Sustainable Paths to the Future, and the Thermodynamic Perspective

Noam Lior

8.1 Energy Resources and Use Summary (Year 2008)

The status of energy resources and use in 2008 is briefly summarized in this section, with some elaboration to follow.

8.1.1 Current Energy Resources and Consumption

The current energy-resources and consumption situation has generally worsened relative to that at the end of 2006:

- A major concern (or opportunity?): The price of oil was growing very rapidly, from $28 in 2003 to $38/barrel in 2005 and occasionally to above $80 in 2006 and peaking at $147 in 2008, but then precipitously dropping to $40 by the end of 2008.
- The peak price is 1–2 orders of magnitude higher than the cost of extraction, possibly meaning that financial speculation is overwhelming supply and demand and all technical improvements.
- In 2007, world primary energy-resources use rose by 2.4%, with the increase rate slightly dropping (Fig. 8.1), but is likely to rise again soon, as the large developing countries in Asia keep improving their standard of living; China's rose by 7.7% (lowest since 2002), India's by 6.8%, and the United States' by 1.6%, Japan's dropped by 0.9%, and EU's dropped by 2.2% (EU is the European Union).
- The reserves-to-production (R/P) ratio remains rather constant: ~40 for oil, ~60 for gas, and 200+ for coal, and mostly rising (Figs. 8.2 and 8.3)! There are probably sufficient oil and gas for this century and coal for two or more.
- Tar sands and oil shales are becoming more attractive and available in quantities probably exceeding those of oil and gas.
- Nuclear power produces ~16% of world electricity; the number of reactors is increasing very slightly; public perception is improving, new government initiatives started, but the same problems remain.
- Renewable energy resources can satisfy ~2 orders of magnitude more than the world energy demand, but negative impacts are not inconsequential. Wind

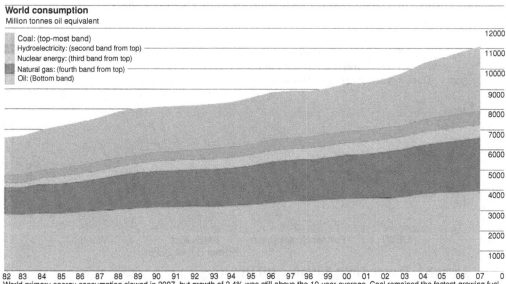

Figure 8.1. World primary energy consumption 1981–2007 [1].

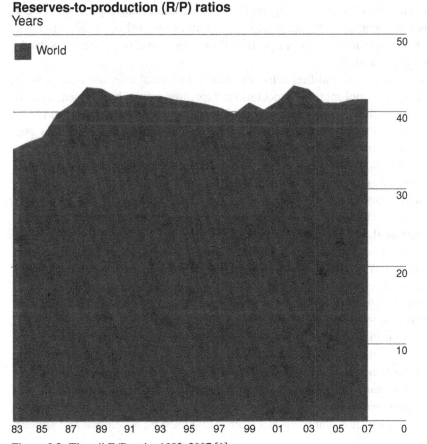

Figure 8.2. The oil R/P ratio, 1982–2007 [1].

Fossil fuel reserves-to-production (R/P) ratios at end 2007
Years

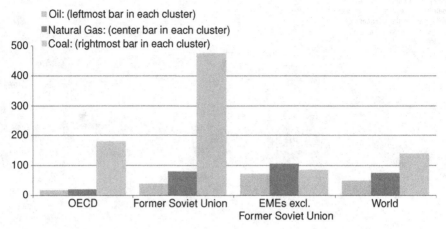

Figure 8.3. The fossil fuels R/P ratio, 2007 [1].

and solar photovoltaics (PVs) are experiencing an exponential growth as costs decrease; interest is renewed in solar-thermal power.

- Strong subsidies for converting food to fuel are increasingly proving to be a mistake, helping to triple the price of foods and reduce their availability, and raise water consumption, all as predicted by some experts before the subsidy program was started.
- Although hydrogen and fuel cells continue to be valuable in the energy portfolio, they have not met the expectations expressed by the huge research-and-development (R&D) investments made by many governments. This could have been foreseen by more careful early analysis, and some of the moneys and valuable scientists' time could have been spent better.
- The plug-in electric or hybrid car seems to be the preferred route to private transportation. The development of traffic management, roads, and public transit is at least as important.
- Costing of energy resources remains inequitable, as it does not include subsidies, environmental, and other consequences.
- The development of renewable energy resources, and of all energy systems for that matter, is dominated by the highly controlled, cost-unrelated, highly fluctuating, and unpredictable conventional energy-resources prices.
- Fuel and energy consumption in general must be significantly constrained, with due attention to the prevention of the rebound effects.
- The "Living Planet Index" is estimated to have declined since 1970 by about 30% and the "Ecological Footprint" increased by 70% in the same period: We seem to be *running out of environment much faster than out of resources.*
- *It is highly inadvisable, and unlikely, that energy resourcing, conversion, and consumption will continue to be developed unsustainably.*
- Sustainability is only emerging as a science and must be developed and applied urgently.

8.1.2 Future Power Generation

- The most imminent challenge is that the expected demand for electricity would require during the coming two decades the installation of as much power-generation capacity as was installed during the entire 20th century.
- Although the plug-in hybrid electric car and electric-driven public transportation seem to be the most promising ways toward energy-efficient transport, this would further raise the demand for electricity in a most significant way, perhaps doubling it.
- To mitigate associated negative effects of such a massive increase, it would increasingly have to be done sustainably.
- Because of its abundance in the most energy-consuming countries such as China, the United States, parts of Europe, India, and Australia, coal is likely to be increasingly the main basic fuel for these plants, partially after conversion to gaseous or even liquid fuels, with the reduced-emissions IGCC (integrated gasification combined-cycle) plant receiving major attention.
- The combined-cycle power-generation plants are the most desirable, having efficiencies of up to about 60% even at present, less emission than other plants when using natural gas, and a reasonable cost that would keep decreasing as the technology advances further.
- Despite the unresolved problems of waste storage, proliferation risk, and, to some extent, safety, nuclear power plants are likely to be constructed at least for special needs, such as in countries that have much better access to uranium than to fossil fuels and if carbon emissions become costly. The amount of uranium-235 in the world is insufficient for massive long-term deployment of nuclear power generation, which can change only if breeder reactors are used, but that technology is not safe and mature enough and is not likely to be in the next couple of decades.
- Wind power generation will be deployed rapidly and massively, but will be limited to regions where wind is economically available and will be limited by the extent and quality of the electricity distribution grid.
- PV power generation will continue increasing in efficiency, decreasing in price, and being employed in many niche applications, but being three to five times more expensive now than other power-generation methods, and also limited by the extent and quality of the electricity distribution grid, and even by availability of materials, it may not reach parity in the coming decade.
- Improvements and technological advances in the distribution and storage of electric power will continue and should be advanced much faster.
- The investments in energy R&D appear to be much too low, less than half a percent of the monetary value of the energy-resources use, to meet future needs.

8.2 Introduction

This chapter is a brief summary of the state of current energy resources and use, their limitations and consequences, and possible paths to the future, including energy research funding trends, especially in the United States. The data are taken from many sources, including the latest (June 2008) energy statistics annual report of

British Petroleum (BP) for 2007 [1],[1] the excellent websites of the U.S. Department of Energy (USDOE) [2], its Energy Information Administration [3], Office of Budget [4], Office of Energy Conservation and Renewable Energy [5], Office of Fossil Energy [6], and the National Renewable Energy Laboratory [7], the Energy Research website of the EU [8], the International Energy Agency (IEA) [9], and the International Atomic Energy Agency [10]. The analysis, interpretation, and comments are entirely the author's and do not represent any institutional or government views. Reviews of a similar nature were published by the author for the situation in 2002 and 2008 [11, 12] and in 2009 [13] to update this very dynamic field; this chapter uses the latter data.

A decline in energy research experienced during the 1980s was somewhat arrested toward the end of the 1990s, primarily because of increasing concerns about global warming from energy-related combustion. This has invigorated R&D in efficiency improvement, use of energy sources that do not produce CO_2, and in methods for CO_2 separation and sequestration. The interest in energy has received another important boost in the last couple of years, driven by the exponentially rising energy consumption by the highly populated countries of China and India, accompanied by the heightening tensions with many of the oil- and gas-producing countries. Interest in the energy issue and support for energy R&D are now rising rapidly, abetted by concerns about energy-resources security. The EU and Japan appear at present to have and to be able to afford the most forward-looking and extensive programs, partially because they do not have to bear the enormous relatively recent defense expenses that the United States does.

8.3 Sustainable Energy Development

8.3.1 The Motivation for Sustainable Development

Energy development is increasingly dominated by major global concerns of overpopulation, pollution, deforestation, biodiversity loss, and global climate deterioration. For example, more than 20% of the Arctic ice cap melted away between 1979 and 2003 [14], the Living Planet Index, a metric that measures trends in the Earth's biological diversity, is estimated to have declined since 1970 by about 30%, and the Ecological Footprint (defined in [15], extended in [16]), which is the area of biologically productive land and water needed to provide ecological resources and services including land on which to build and land to absorb CO_2 released by burning fossil fuels, increased by 70% in the same period [17].[2] These trends are clearly unsustainable and alarming.

Obviously, energy-resources consumption increases with population size, but not in a linear way: A new population from developing countries typically requires more energy resources per capita than their parents did. Although the rate of population increase had been dropping since the 2.2%/year peak in 1962 to 1.2%/year

---

[1] Although British Petroleum (BP) has published the *Annual Statistical Review of World Energy* for 57 years without significant challenges and serves most frequently as the source of the proved fuel reserves data, the accuracy is unknown and is subject to large errors.

[2] Although there is an ongoing argument about the proper definition of the Living Planet Index and the Ecological Footprint metrics, the general alarming trends appear to be correct.

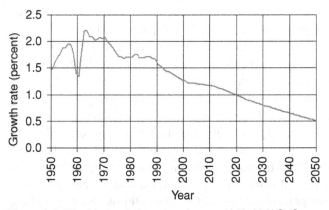

Figure 8.4. World population growth rates 1950–2050 [18].

currently (Fig. 8.4) [18], the increase from the current 6.7 billion to the projected 9.6 billion in 2050 is 43%. The projections are obviously in some doubt, especially if the most populous countries, such as China and India, do not continue or start family size control. It would be impossible to achieve sustainable development if population size is not seriously addressed.

To prevent disastrous global consequences, it would increasingly be impossible to engage in large-scale energy-related activities without ensuring their sustainability, even for developing countries in which there is a perceived priority of energy development and use and power generation over their impact on the environment, society, and indeed on the energy sources themselves. Although having various definitions [19–21], here the original broad one is simply given, that sustainable activities mean that they meet the current needs without destroying the ability of future generations to meet theirs, with a balance among economic, social, and environmental needs.

8.3.2 Sustainability Analysis

Clearly the quantification of a project's sustainability metrics (indicators) is the first step in sustainable development, design, and monitoring, but it is very difficult because these are large very complex systems that have technical, ecological, economic, and societal components [19]. It is of vital importance to have, where available, or to develop, where not, agreeable and unambiguous definitions of all the needed metrics. Unfortunately, even the more technical and economic metrics are not always well defined and internationally agreed on yet. For example, it was discussed in some detail in [22] the definitions of and differences among energy, exergy, second law, and economic efficiencies, energy criteria considering environmental effects, and embodied energy payback. Exergy and emergy were proposed by several authors to serve, arguably, as metrics for environmental and even social aspects [23–25]. Entropy generation was proposed as a metric for the sustainability of energy-unrelated materials processing for manufacturing focused on product quality conditions rather than on energy consumption [26].

All the needed metrics must obviously satisfy the laws and other facts of nature. Observation of the laws of nature, such as the second law of thermodynamics, not only avoid wrong metrics, but also provide effective guidance for improvements.

In that context, it is increasingly recognized and included in practically all textbooks on thermodynamics and energy systems design [27, 28] that exergy[3] (or second law) analysis must be added to conventional energy accounting analysis during the conception, analysis, development, and design of such systems [29]. Only exergy analysis can identify the specific irreversibilities and is uniquely required for providing the guidance needed in this process. The benefits of exergy analysis are clearly demonstrated in such an analysis of a simple Rankine cycle described in [30]. *Energy* analysis indicates that the major energy loss, 70%, is due to the heat rejected in the condenser. Examination of the *exergy* analysis chart shows, however, that this large energy loss amounts to only 3% of the fuel exergy, and even complete elimination of the condenser heat rejection (if it were at all feasible) would increase the cycle efficiency by up to only three percentage points. Of course, this is because the heat rejected in the condenser is at a low temperature, only slightly elevated above that of the ambient, and thus has commensurately little potential to perform work despite its large energy. Another very significant difference between the energy and exergy analyses is the fact that the exergy analysis identifies the major losses, 69%, to be in the boiler, because of the combustion and gas-to-steam/water heat transfer processes, whereas the energy analysis associates no loss with these processes. Finally, the exergy analysis attributes much less loss to the stack gas than the energy analysis does. The example shows it is only exergy analysis that can correctly identify and evaluate the losses (irreversibilities) that diminish the ability of processes to perform useful work.

An example of how proper thermodynamic analysis and process experience can reduce irreversibility and energy consumption in chemical technology processes is shown in [31]. Thermodynamic reversibility requires that all process driving forces, such as temperature, pressure, and chemical potential differences, be zero at all points and times. Thus the theoretical thermodynamically reversible chemical process must proceed along an equilibrium line that is in chemical equilibrium at each point of a reactor. Accordingly, the driving force for the process must be zero throughout the entire process, not just at the end. Such a theoretical process results in the production of the maximal amount of useful work or in the consumption of the minimal amount of work. Unfortunately, a reversible chemical process operates at an infinitesimally slow rate and requires an infinitely large plant. To operate a chemical process in finite time and at finite cost, it is necessary to have finite driving forces, i.e., to expend some thermodynamic availability (exergy) and, as a result, to consume energy resources. The goal of the process designer is to expend this thermodynamic availability wisely while achieving the technological goals of

---

[3] Exergy (a) is defined as the measure of a system's potential to perform useful work between any given state and the so-called "dead state," (subscript 0 in the following equation) at which the system can undergo no further spontaneous processes:

$$a = h - h_0 - T_0(s - s_0),$$

where h is the enthalpy and s is the entropy. Because enthalpy is the measure of the energy in flow systems, the preceding equation shows clearly that the portion of the energy h that cannot be converted to useful work is the product $T_0 s$. Some general references on exergy are M. J. Moran, *Availability Analysis* (Prentice-Hall, Englewood Cliffs, NJ, 1982); J. Szargut, *Exergy Method* (WIT Press, Southampton, Boston, 2005); M. J. Moran and H. N. Shapiro, *Fundamentals of Engineering Thermodynamics*, 6th ed. (Wiley, Hoboken, NJ, 2008).

the process. Too large a driving force expends more exergy than is necessary and wastes energy resources, whereas too small a driving force requires excessive capital investment. In particular, the designer should avoid an apparatus that has too large a driving force in one part and too small a driving force in another part. In such a case, both energy resources and capital are wasted. The study ends with twelve second law of thermodynamics "commandments" for reducing entropy generation and energy consumption, up from three proposed 47 years earlier in a paper by Denbigh [32].

Consideration of other facts of nature, such as reliable data on resources availability and accessibility, allows sustainable development planning that takes into account use of a resource for both single or multiple demands and interrelations among the uses of different types of resources. Cogent examples are the discussion of "peak oil," availability of water for exploiting tar sands and oil shales and for many other purposes, and the possible competition over lithium use between batteries for electric vehicles and fusion power generation (if either achieves massive use). Very important is the inverse relationship that often exists in processes between the consumption of paid energy and resources. Two fundamental examples were previously discussed: (1) Increasing energy efficiency by a closer approach to the process thermodynamic reversibility requires a decrease of driving forces and the associated, usually inevitable, increase in equipment materials, and (2) increasing use of renewable energy that is typically available only with very low fluxes thus requires large areas of energy collection. A third example is the use of "waste" energy (such as "waste" or rejected heat), which is also of low exergy potential and thus requires large amounts of equipment, such as heat exchangers. An attempt to start the discussion on quantifying the depletion or resources and of the associated complexities is included in [33].

As an example of somewhat less used/known metrics is the concept of energy embodied[4] in the production of a plant, in the materials produced by it, and in the materials and labor needed for its operation and for the distribution of its material products to the customer. Increasingly used in ecoconscious design of buildings, it is, however, a very important metric in sustainable development in general. Such a criterion answers, for example, the commonly posed question on the length of time that it takes for an energy-conversion system to generate the energy originally required for its manufacturing and operation. Significantly, it also provides valuable guidance about the importance of the manufacturing, ultimate disposal, and recycling aspects of a product. Furthermore, careful consideration of embodied

---

[4] Some references about embodied energy are D. B. Reister, "The energy embodied in goods," *Energy* **3**, 499–305 (1978); K. Nishimura, H. Hondo, and Y. Uchiyama, "Derivation of energy-embodiment functions to estimate the embodied energy from the material content," *Energy* **21**, 1247–1256 (1996); C. Atkinson, S. Hobbs, J. West, and S. Edwards, "Life cycle embodied energy and carbon dioxide emissions in buildings," *Ind. Environ.* **2**, 29–31 (1996); W. R awson, *Embodied Energy of Building Materials, Environment Design Guide* (Royal Australian Institute of Architects, Manuka, Australia, 1996); B. V. V. Reddy and K. S. Jagadish, "Embodied energy of common and alternative building materials and technologies," *Energy Buildings* **35**, 129–37 (2003); *Manufacturing Energy Consumption Survey* (USDOE Energy Information Administration), available at http://www.eia.doe.gov/emeu/mecs/contents.html (accessed Oct. 31, 2009); and X. Yan, "Energy demand and greenhouse gas emissions during the production of a passenger car in China," *Energy Convers. Manage.* **50**, 2964–2966 (2009).

energy is of vital importance in renewable energy development, because renewable energy sources typically use, as previously discussed, significantly larger amounts of material per unit useful energy output than conventional fossil and nuclear fuel plants.

Life-cycle analysis (LCA), the investigation and valuation of the environmental (and often economic and social) impacts of a given product or service caused or necessitated by its existence, is a commonly used tool in sustainability analysis. In its full form, when addressing environmental, economic, and social impacts, it is a cornerstone of sustainable design and development: It systematizes a comprehensive consideration or analysis of all conceivable impacts of a project, process, or product, and its results serve as an objective function, allowing quantitative comparison between alternatives and optimization, based on the chosen sustainability metrics. It is noteworthy, however, that LCA is not equivalent to sustainability analysis but can be a component of it.

The LCA time period may be cradle-to-gate (gate being the exit of the plant that makes the product), cradle-to-grave, cradle-to-cradle (includes recycling), or any period chosen by the LCA performer as long as it is clearly defined. The spatial extent (boundaries) depends on legislation or choice. Because of its increasing use, the LCA procedures were defined by the ISO 14000 environmental management standards.

It must also be recognized from the start that LCAs, just as sustainability analysis in general, are subject to serious uncertainties (cf. [34, 35]), because the future is hard to predict ("The art of prophecy is very difficult, especially with respect to the future"), because the extent of the space of interest and its content or purpose may change with time, and the life-cycle impact may vary with time because of legislation, discovery of new information, changes in attitudes, population, events, etc. These inevitable uncertainties and the difficulties in evaluating them make the value of absolute quantitative LCA outcomes meaningless, but the process and methodology by themselves are very valuable in learning about the object of the LCA and about areas that need better information, and ways that it affects the sustainability pillars of environment, economics, and social impact. It is also useful for considering alternative approaches if all the inputs and scenarios are the same and reasonable. Expectations from LCA outcomes should be constrained by recognition of the uncertainties, and uncertainties can be reduced by less-ambitious LCA goals. Qualitative uncertainty analysis in LCA will improve its value, and the LCA community should develop a better understanding of the importance of uncertainty and variability and develop protocols for reliably characterizing, propagating, and analyzing uncertainty in LCA.

Once all the relevant metrics for a sustainability analysis are determined, they need to be aggregated with sensible weighting factors, the objective function for the system optimization must be determined, and then the an optimal solution must be found. This modeling and solution are also very difficult because the problems are dynamic, multiscale, and in many parts nondeterministic, and the data are difficult to collect, so better knowledge and tools are needed. Achieving sustainability requires a new generation of engineers and scientists who are trained to adopt a holistic view of processes as embedded in larger systems. Useful work to develop sustainability science is under way but much remains to be done.

8.4 Future Power Generation

8.4.1 The Technologies

From the 18 billion kW h of electricity generated worldwide in 2006, about 66% is produced from fossil fuel, 17% from hydropower, 15% from nuclear fuel, and the remaining 2% from geothermal, wind, solar, wood, and wastes. Coal provides 62% of the fossil fuel electric power generation, gas 29%, and oil 9%. Practically all of the coal- and oil-fired electricity generation is by Rankine-type steam power plants, and some of the gas-fired plants use combustion gas turbines. A small but increasing fraction of power generation is by combined-cycle systems, using a topping gas turbine system and bottoming steam turbine one. Such plants have an efficiency approaching 60%, 35% higher than that of regular cycles, at a competitive capital cost. Nuclear power plants generate electricity by means of steam turbine Rankine-type cycles, with an efficiency of about 33%. It is noteworthy that this efficiency is much lower than those of fossil fuel power plants because of the lower top temperature in the nuclear power plants and proportionally increases the amount of waste heat discharge to the environment. Large hydropower plants operate at efficiencies approaching 90% and large wind power plants below 30%.

8.4.2 The Future Power-Generation Problem and Likely Solution Trends

The most eminent problem is that expected demand for electricity would require during the coming two decades the installation of as much power-generation capacity as was installed during the entire 20th century [3]. This translates to the stunning number of one 1000-MW power station brought on line every 3.5 days over the next 20 years, on average!

To mitigate associated negative effects of such a massive increase, it would increasingly have to be done sustainably. The first step is clearly energy conservation, a less wasteful, wiser, and more modest use of electricity.

Because of its abundance in the most energy-consuming countries such as China, the United States, parts of Europe and India, and Australia, coal is likely to be increasingly the main basic fuel for these plants, partially after conversion to gaseous or even liquid fuels. Compared with other energy sources, coal-fueled power plants also produce the cheapest electricity. The extensive use of coal will increase the need for more stringent mining and emissions controls and other ecological and social problems associated with a coal economy. The reduced-emissions IGCC plants, increasingly with CO_2 separation, are thus likely to be receiving major attention. Using fossil fuels, the combined cycle plants are the most desirable, having efficiencies of up to about 60% even at present, less emission than other plants when using natural gas, and a reasonable cost that would keep decreasing as the technology advances further.

Despite the unresolved problems of waste storage, proliferation risk, and, to some extent, safety, nuclear power plants are likely to be constructed at least for special needs, such as countries that have much better access to uranium than to fossil fuels. Furthermore, if carbon emissions are made expensive enough, nuclear power plants would become more viable. At the same time, the amount of uranium-235 in the world is insufficient for satisfying the world energy demand by nuclear energy, a situation that can change only if breeder or natural uranium or thorium reactors are used. The technology for breeders is not, however, safe and mature enough, and is not likely to be in the next couple of decades. The use of breeders and natural uranium reactors also produces plutonium, with the associated safety and proliferation problems. The latter problem, as well as that of nuclear waste storage, cans be alleviated if transmutation technology is developed to break down the long-half-life actinides to shorter-half-life elements. Thorium-based reactors are under development, but many problems have to be overcome before commercial units could be built.

The economic competiveness of all renewable energy power-generation plants depends of course on the cost of the fuel used by fossil or nuclear power plants. Wildly fluctuating and unpredictable oil and gas prices make reliable planning of renewable, or even nuclear, power generation nearly impossible.

Wind power generation is typically competitive when oil prices are around $60/barrel, currently has a respectable worldwide capacity of about 94,000 MWe ($\sim$2.5% of the world electric-generation capacity of about 4 million MWe), and will be deployed rapidly and massively, but it will be limited to regions where wind is economically available and limited by the extent and quality of the electricity distribution grid. PV power generation is estimated to be marginally competitive at an oil price above $150/barrel and will continue increasing in efficiency and decreasing in price. Its widespread use is also is limited by the extent and quality of the electricity distribution grid, and even by the availability of materials. It may not reach parity in the coming decade. Hybrid solar-thermal power plants that use solar heat at a lower temperature and the fossil fuel for raising the temperature of the working fluid prior to its inlet to a turbine, of the type described in [36–38], are becoming competitive. The time dependency of wind and solar power introduces major problems that could be resolved by use of energy storage (expensive and often unavailable when hydro or compressed air storage are considered) or grid storage.

Hydroelectric power provides most of the $\sim$6% contribution of renewable energy to the total energy supply and shows steady but slow growth. Perhaps the most remarkable event is the addition of 18.2 GWe with Three-Gorges Dam in China. The hydroresources are becoming more limited, and the construction of such projects poses various environmental, social, and security problems; this dam, for example, created an upstream lake of 600 km, displacing millions of people. It is also of importance to note that hydroelectric projects in warm-climate vegetated regions cause significant release of CO_2 and methane.

Biomass energy has the very important benefits of contribution to the security of fuel supply, lower greenhouse-gas emissions, and support for agriculture; there are also some important concerns and obstacles. These include the fact that bioenergy production and policies have mostly not been based on a broad cost-and-benefit analysis at multiple scales and for the entire production chain, which is particularly true for bioenergy's impact on agriculture. For example, although many publications extol the advantages of converting corn or other crops to ethanol, many of these analyses are flawed, at least in that they do not consider the entire system and cycle [39]. Furthermore, there is strong concern about the effects on food production and cost: Over the past couple of years, corn prices in the United States have doubled despite record crops because of its rapidly increasing use for ethanol production. Filling the 25-gal ($0.094 m^3$) tank of an SUV with pure ethanol requires over 450 lb (204.5 kg) of corn – which contains enough calories to feed one person for a year.

Cellulosic source ethanol may be better but final proof is absent, and conversion demonstrations have only started. There is also a significant interest and effort in producing butanol, which is a much better and more transportable fuel than ethanol, and in biodiesel fuels.

It is noteworthy that the biofuels well-to-wheel greenhouse-gas abatement potential is not as certain and high as may be thought: less than 20% for corn ethanol, but over 90% for sugar cane based [39]. Furthermore, some recent results have shown that growing plants release methane [40], which has a greenhouse-gas potential at least 20-fold that of CO_2.

IEA analyses and projections for biomass uptake by 2030 at competitive costs are 15 to 150 EJ/yr [9, 39]. The proposed research needed for this major progress in using biomass [41] includes development of (1) "new" biomass, by means of improved land use, waste utilization, and crop management, together with modified processing methods; (2) new methods of cultivating and harvesting aquatic organisms; (3) genomics and transgenic plants (e.g., to engineer plants and microorganisms that would yield novel polymers or to maximize carbon for high-energy content); (4) new processes, such as enzymatic conversion of corn carbohydrates to polylactic acid (PLA) and other polymers, and combinations of photosynthetic processes with special enzymes to create solid structures that would intercept sunlight and fix carbon into energy-rich materials; (5) improved use of traditional biomass (lignin and cellulosics) by more efficient gasification, enzymatic conversion of lignocellulosic biomass to ethanol; and (6) cultivation of hybrid rapidly growing plants (e.g., poplar or willow, switchgrass).

Improvements and technological advances in the distribution and storage of electric power must and will continue. These are needed for accommodating varying demands with electricity generated by nonrenewable conventional fuels, and even more important when renewable intermittent sources such as solar and wind are used. Also, the development of superconductors that would become commercial and affordable must continue, as they have great potential in increasing electric system efficiency and allowing economical longer-distance transmission, say, from energy-rich to energy-needy regions.

8.4.3 Fuel Cells and Hydrogen

The very active development of fuel cells, encouraged by the governments of practically all industrialized nations, is ongoing, primarily aimed at using hydrogen fuel in transportation, but also for large stationary power-generation units. It seems that this major effort has peaked by now, because various important technical issues must be resolved before fuel cells attain significant market penetration and the cost must be reduced by an order of magnitude. Conducting vigorous R&D is reasonable, but has to be balanced against equally important support needed for improved internal and external combustion engines that have in some cases already attained efficiency higher than those of fuel cells at much lower costs.

Hydrogen derived from coal is stated to be the USDOE's primary goal in the fuels program, with a primary objective of developing modules for co-producing hydrogen from coal at prices competitive with crude oil equivalents when integrated with advanced coal power systems (cf. [42]). The development of hydrogen as an energy carrier is also very active by other industrialized countries. Despite its advantages in producing near-zero harmful emissions *in the process of its conversion to power* and the declared plans for its development, the general opinion of the scientific community in this field is that widespread use of hydrogen as a fuel in the foreseeable future appears to be doubtful because of the high energy demand and emissions in its production, and issues of safety, storage, and distribution.

8.4.4 Micropower Systems (cf. [43, 44])

There is increasing interest in the construction and use of very small, of the order of 1000 μm, power-generation systems for various applications, ranging from the military to the medical. Such systems include miniaturized thermal power cycles and direct-energy-conversion systems, including fuel cells [45], mostly intended to replace batteries as much longer operation and low weight and volume devices. Because the power produced by such a device is of the order of milliwatts at best, it does not at first glance appear that they will be used to produce a significant fraction of the overall power demand. At the same time one cannot help but note that use in very large numbers can create significant worldwide capacity. For example, the many very low-capacity computers that are increasingly being used in just about any electrical device, including cars and home appliances, constitute by now a computing capacity far exceeding the total capacity of the existing personal, workstation, and mainframe computers, and the total power produced by batteries of various types is of the order of magnitude of the total electric power generation.

Micropower generators pose very interesting research, development, and construction challenges, many related to the very complex flow, transport, and thermodynamic phenomena. The extraordinary benefits of micropower generators in many known and yet unknown applications make the challenges associated with their development very worthwhile.

8.4.5 Further-Future Paths: Fusion and Power From Space

8.4.5.1 Fusion

The major appeal of this process for power generation is that its fuel is composed of rather abundant elements, deuterium that is plentifully available in ordinary water (a liter of water would thus have an energy content of 300 L of gasoline) and tritium that can be produced by combining the fusion neutron with the abundant lithium. The radiation from the process is very low and short lived, but the environmental problems are not negligible. Thus fusion has the potential to be a very abundant and relatively clean source of energy, with minimal global-warming emissions. The biggest problem, not yet solved after more than 50 years of research, is to create a fusion reaction that continuously produces more energy than it consumes. Past predictions of success and commercialization repeatedly had a 25-year target, and those have increased to about 35 years based on the ambitious multinational

International Thermonuclear Experimental Reactor (ITER) program that is constructing a 500-MW fusion test facility in Cadarache, France [46].

8.4.4.2 Electricity from Space? (cf. [47, 48])

Power can be produced in space for terrestrial use by use of a number of energy sources, including solar, nuclear, and chemical. The generated power can be transmitted back to Earth by a number of ways, including transmission by microwaves or laser beams or on-site manufacturing of easily transportable fuels for electrochemical or combustive energy conversion.

This is a very complex method, but in view of the rising demand for energy, the diminishing fuel and available terrestrial area for power plant siting, and the alarmingly increasing environmental effects of power generation, the use of space for power generation seems to be rather promising and perhaps inevitable in the long term: (1) It allows the highest energy-conversion efficiency, provides the best heat sink, allows maximal source use if solar energy is the source, and relieves the Earth from the penalties of power generation; and (2) it is technologically feasible, and both the costs of launching payloads into space and those of energy transmission are declining because of other uses for space transportation, dominantly communications.

The technology for such systems is in principle available, and the major obstacle is the exorbitantly high cost, which under current conditions requires the reduction of all costs by orders of magnitude; for example, space transportation costs by at least a 100-fold to less than $200/kg into orbit, for competitiveness.

It is noteworthy that any comparative economical analysis must be conducted on an equitable basis: here specifically including all of the costs of power generation including those of the environmental effects, resource depletion, and embodied energy. Other issues also need to be resolved, some of a general nature, such as environmental effects and security and legal aspects, and some system specific, such as safety of nuclear power plants and the realization of higher energy-conversion and transmission efficiencies.

Perhaps most interesting is the change of paradigm that space power presents: Earth becomes less of an isolated, closed system. National and international work on this subject should be invigorated so that humankind will continue having the energy it needs for its happiness and, indeed, survival.

8.5 Some Recent Energy R&D Budgets and Trends

The information presented here must be prefaced with a statement that examination of governmental and institutional aims and budgets is very difficult, in part because of duplication and overlap of programs and frequent changes across them, and all the numbers given here are thus not always precise. Perhaps a very cogent introduction is the fact that the average government annual expenditure for renewable energy research for all nations is less than $1 per person [39].

The total USDOE budget[5] dedicated to energy R&D was requested to increase in 2009 by about 4%, to about 3.9 B$, and perhaps more than 1 B$ in basic energy

---

[5] These numbers very likely changed with the advent of the new U.S. administration in 2009.

sciences (out of the 4.7 B$ USDOE Office of Science after its 19% increase, which funds also several other areas that are not directly related to energy), for a total of about 5 B$.

The U.S. Department of Energy's National Renewable Energy Laboratory (NREL) had its budget increased by 80% during 2007, to $378 million.

Out of the USDOE energy R&D part, the programs of energy efficiency and renewable energy, fossil energy, and nuclear energy are about equally budgeted, nuclear having a slight lead. This is a significant change in the apportioning compared with the situation in 2006, when energy efficiency and renewables had about half of the budget, with the other two areas a quarter each.

The most important budget changes are as follows:

- 19% ($748 million) increase in the DOE's *Science* programs (nuclear physics including major facilities, materials, nanoscience, hydrogen, advanced computing).
- 27% decrease in the *Energy Conservation and Renewable Energy* program [with gains in biomass (+37%), wind, geothermal, and building technologies; drop of 31% in hydrogen].
- 23% increase in the *Fossil Energy* program to $1.13 billion, which includes
  - $648 million for coal carbon capture-and-storage (CCS) research, including $149 million for sequestration. (It is noteworthy that despite the U.S. administration's refusal to sign the Kyoto 1997 protocol, its stated goal is to reduce greenhouse-gas intensity by 18% by 2012),
  - zero for petroleum and natural gas
- 84% increase (to $344 million) for the Strategic Petroleum Reserve (capacity expansion from 727 million barrels to 1.0 billion barrels beginning in fiscal year 2008 and later to 1.5 billion barrels).
- $1.65 billion in investment tax credits will accelerate commercial deployment of technologies central to carbon capture and storage.
- Nuclear energy, $1.4 billion, up 37%, including
  - $302 million to begin investments in the Global Nuclear Energy Partnership (GNEP), enable an expansion of nuclear power in the United States and around the world, promote nuclear nonproliferation goals, and help resolve nuclear waste disposal issues,
  - $495 million for permanent geologic storage site for nuclear waste at Yucca Mountain, Nevada, planned for 2017 pending many difficult obstacles,
  - 3.5-fold increase (over 2007), $214 million for ITER (fusion) (this is uncertain as of the time of writing).
  - 5% increase (to $110.6 million) for the Energy Information Administration to improve energy data and analysis programs.

These numbers are rough because there are research areas in the basic sciences that apply across energy-source categories, and there are separately very large budgets that are dedicated to high-energy physics and to the maintenance of large experimental facilities in the national laboratories.

Japan's energy R&D program is above 2.5 B$ (three quarters of which are for fission and fusion).

Table 8.1. *A qualitative assessment of promising research directions and their current U.S. government funding trend (valid for the beginning of 2008)*

Direction	Potential	Foreseen improvement	Time scale (years)	2009 government funding trend
Conservation	★★★+	50% of use	ongoing	☹☹
Transportation	★★★+	50% of use; 120 g CO_2/km by 2012	3–20	☹
Biomass	★★+	30% U.S. energy	4–40	☺☺☺
Wind	★★★	2.5 c/kW h,15%	1–15	☺
Solar PV	★★★+	Competitive price	6+	☺☺
Solar thermal	★★	Competitive price	5+	☹
Geothermal (deep)	★★	Competitiveness	20	☹☹
Hydrogen	★★	Affordable transport fuel	15	☺
Fossil fuel power	★★	65%–75% efficiency, ~0 emissions	6–15	☺☺
Oil and gas	★+	Exploration, recovery, transportation	3–15	☹☹☹
Coal	★+	Exploration, recovery, transportation, conversion	7	☺
Global warming/CO^2	★★	0 CO^2	10–15	☺☺
Fuel cells	★+	60%+; price	9	☹
Superconductivity	★★★	Order of magnitude	30+	☹☹
Nuclear fission	★	Manageable wastes, no proliferation	9	☺☺
Nuclear fusion	★★★	Feasibility	25+	☺
Micropower	★★★	Cost, market penetration	7+	☺☺
Space power	★★★+?	Competitiveness	50+	☹☹

Note: ☺, increased; ☹, decreased.

The EU (which is the largest importer and second–largest consumer of energy in the world) 7th Framework Programme (2007–2013) had a 50% increase in the energy area (energy, environment, transport) over the 6th program and is about 1.75 B\$ plus 0.84 B\$ for the nuclear research in Euratom. Some of the goals for the year 2020 include a 20% reduction of energy use, a 20% share to renewables, and all new coal power plants being of the CCS type. It is noteworthy that individual European countries also have their own energy R&D budgets that in total exceed that of the EU.

Table 8.1 summarizes the author's view of the promising energy R&D areas, their potential, foreseen improvements and their time scale, and last year's trends in government funding.

8.6 Possible Paths to the Future

The first step in any path to the future is wiser use of the energy resources, also referred to as conservation. This would include elimination of obvious waste, higher energy-conversion efficiency, substitution for lower-energy-intensity products and processes, recycling, and more energy-modest lifestyles.

The omnificent politician, publisher, and scientist Benjamin Franklin (who also founded the University of Pennsylvania in 1740), a believer in conservation and frugality, wrote "a penny saved is a penny earned." In the energy area in general, and in power generation in particular, one could safely say that "a Joule saved is worth significantly more than a Joule earned": it takes significantly more than 1 J of energy to generate 1 J of power. This is amplified severalfold when one considers the resources and environmental impact associated with the construction and operation of a power plant or even a vehicular engine. It is clear therefore that the first priority in meeting the challenges of the 21st century is energy conservation, but not implemented in a way that would deprive large fractions of humanity of the basic comforts of life. Indeed, as one of the drafters and signers of the U.S. Constitution, Franklin believed in facilitating people's "pursuit of happiness" and practiced it himself whenever he could. Such pursuit is made very difficult, or impossible, for a population living under energy-conservation measures that it considers to be harsh. Indeed, a lifestyle of health and sustainability study conducted in the United States in 2008 by the Natural Marketing Institute (www.nmisolutions.comlclohas.html) found that there are very few consumers (5%–10%) who are willing to accept higher cost or lesser performance of a product that has environmental benefits. The majority felt that, although environmental issues are important, they are not willing to make sacrifices [49].

The pursuit of more efficient and less polluting transportation must include not only vehicular improvements (with preference for the plug-in electric or hybrid car) but also traffic management, significant development of efficient public transit, and redesign of cities.

Buildings are the biggest single contributor, ~45%, to world energy and greenhouse-gas emissions. An excellent and practically attainable way to reduce this problem is the design and retrofit of buildings such that they consume less energy (including embodied energy) over their lifetime, with and without incorporation of renewable-energy sources, and further with an extension to "eco-efficient" buildings that not only reduce their negative environmental impact but also help heal and improve the environment. A broader method is to design residential communities in a way that reduces both indirect use of energy and emissions by reducing the need for transportation and resources by the residents.

At least for this century, more efficient and less polluting use of fossil fuels, as well as better and cleaner exploration and extraction of such fuels, is continuing to be pursued. Important steps must also be taken to prevent energy-efficiency "rebound," the frequent outcome in which higher efficiency and lower costs lead to increased consumption (cf. [50, 51]).

It appears that the massive use of nuclear fission power would be stymied unless permanent and economical solutions to nuclear waste, such as element transmutation, can be attained. Nuclear fusion power could produce a very satisfactory long-term solution, but is still rather far from being achieved. R&D and implementation of renewable energy must continue vigorously, with the most promising technologies being solar PVs, wind, and, to some extent, biomass. Very deep drilling, or generally access, technologies for reaching the enormous renewable geothermal heat resources should be pursued.

R&D to develop commercial superconductors would reduce energy losses significantly, but will take some decades at least. Space power generation for terrestrial use must be explored as a long-term solution.

The inequitable costing of energy resources and their conversion must stop, by governments and industry assigning a true value based on all short- and long-term externalities. In-depth scenario studies are necessary for quantitative forecasting of the best ways to spend government research moneys, but qualitatively, and based on the current knowledge and situation, they should develop effective commercial ways for attaining the just-described objectives.

Sustainability is only emerging as a science and must be developed and applied urgently to provide analysis and evaluation tools. It is of immediate importance because energy conversion and use are associated with major environmental, economical, and social impacts, and all large energy projects should therefore be designed and implemented sustainably.

The critical problems that energy development poses and the possible paths to the future create at the same time great opportunities for respected solutions by the engineering and scientific communities that promote new and expanded creativity, higher employment, and higher job satisfaction. It also offers special prospects for small enterprises and nations that are not hampered by the inertia inherent in larger organizations.

A frequent major obstacle is the political system needed to support rapid and effective movement along the new paths and to plan beyond its tenure, which often prefers solutions that are primarily supportive of its own survival: popular support for sensible paths should be sought and people should be educated to diminish this obstacle.

Many of the innovative solutions require very long periods of time. It is of vital importance to start intensively now, so we won't be too late.

REFERENCES

[1] British Petroleum, *Statistical Review of World Energy 2008*, available at http://www.bp.com/productlanding.do?categoryId=6929&contentId=7044622 (accessed July 31, 2008).

[2] U.S. Department of Energy, http://www.energy.gov/ (accessed July 31, 2008).

[3] U.S. Department of Energy, Energy Information Administration, http://www.eia.doe.gov/ (accessed July 31, 2008).

[4] U.S. Department of Energy, Office of Management, Budget, and Evaluation (MBE), Office of Budget, http://www.cfo.doe.gov/crorg/cf30.htm (accessed July 31, 2008).

[5] U.S. Department of Energy, Energy Efficiency and Renewable Energy, http://www.eere.energy.gov/ (accessed July 31, 2008).

[6] U.S. Department of Energy, Office of Fossil Energy, http://www.fossil.energy.gov (accessed July 31, 2008).

[7] National Renewable Energy Laboratory, http://www.nrel.gov/ (accessed July 31, 2008).

[8] The European Commission website on Energy Research, http://europa.eu.int/comm/research/energy/index_en.html (accessed July 31, 2008).

[9] International Energy Agency, http://www.iea.org/ (accessed July 31, 2008).

[10] International Atomic Energy Agency, http://www.iaea.org/ and http://www.iaea.org/programmes/a2/index.html (accessed July 31, 2008).

[11] N. Lior, "The state and perspectives of research in the energy field," invited keynote paper in *Advances in Energy Studies: Reconsidering the Importance of Energy*, Proceedings of the Third Biennial Workshop, edited by S. Ulgati (2002), pp. 351–364.

[12] N. Lior, "Energy resources and use: The present situation and possible paths to the future," invited keynote lecture for PRES 06 (Ninth Conference on "Process Integration, Modeling and Optimisation for Energy Saving and Pollution Reduction"), joint with CHISA 2006 (17th International Congress of Chemical and Process Engineering), Praha, Czech Republic, August 2006; in revised form in *Energy* **33**, 842–857 (2008).

[13] N. Lior, "Energy resources and use: The present situation and possible sustainable paths to the future," invited keynote presentation for SET 2008, the Seventh Conference on Sustainable Energy Technologies, Seoul, Korea, Aug. 24–27, 2008. (Also in the proceedings published by Korea Institute of Ecological Architecture and Environment, Seoul, Korea, 2009, Vol. 1, pp. 55–67).

[14] NASA, http://www.nasa.gov/centers/goddard/news/topstory/2003/1023esuice .html (accessed July 31, 2008).

[15] M. Wackernagel and W. Rees, *Our Ecological Footprints* (New Society Publishers, Gabriola Island, British Columbia, Canada, 1996).

[16] H. Nguyen and R. Yamamoto, "Modification of ecological footprint evaluation method to include non-renewable resource consumption using thermodynamic approach," *Resources Conserv. Recycl.* **51**, 870–884 (2007).

[17] WWF (World Wildlife Federation; World Wide Fund For Nature), The Living Planet, http://www.panda.org/news_facts/publications/living_planet_ report/lp_2006/index.cfm (accessed July 31, 2008).

[18] U.S. Census Bureau, http://www.census.gov/ipc/www/idb/worldpopinfo.html (accessed July 31, 2008).

[19] N. Lior, "About sustainability metrics for energy development," invited keynote presentation at the 6th Biennial International Workshop, "Advances in Energy Studies," Graz, Austria, June 29–July 2, 2008 (Graz University of Technology Publication, Graz, Austria, 2008, pp. 390–401).

[20] S. K. Sikdar, P. Glavič, and R. Jain, editors, *Technological Choices for Sustainability* (Springer, Berlin, 2004).

[21] C. Böhringer and P. E. P. Jochem, "Measuring the immeasurable – A survey of sustainability indices," *Ecol. Econ.* **63**, 1–8 (2007).

[22] N. Lior and N. Zhang, "Energy, exergy, and second law performance criteria," *Energy* **32**, 281–296 (2007).

[23] H. S. Yi, J. L. Hau, N. U. Ukidwe, and B. R. Bakshi, "Hierarchical thermodynamic metrics for evaluating the environmental sustainability of industrial processes" *Environ. Prog.* **23**, 302–314 (2004).

[24] S. E. Jørgensen and S. N. Nielsen, "Application of exergy as thermodynamic indicator in ecology," *Energy* **32**, 673–685 (2007).

[25] J. Dewulf, H. Van Langenhove, B. Muys, S. Bruers, B. R. Bakshi, G. F. Grubb, D. M. Paulus, and E. Sciubba, "Exergy: Its potential and limitations in environmental science and technology," *Environ. Sci. Technol.* **42**, 2221–2232 (2008).

[26] D. P. Sekulić, "An entropy generation metric for non-energy systems assessments," *Energy* **34**, 587–592 (2009).

[27] A. Bejan, G. Tsatsaronis, and M. Moran, *Thermal Design and Optimization* (Wiley, New York, 1996).

[28] J. Szargut, *Exergy Method: Technical and Ecological Applications* (WIT Press, Southampton, UK, 2005).

[29] N. Lior, "Thoughts about future power generation systems and the role of exergy analysis in their development," *Energy Conversion Manage. J.* **43**, 1187–1198 (2002).

[30] N. Lior, "The second law of thermodynamics and entropy," in *Handbook of Engineering*, edited by E. Dorf (CRC Press, Boca Raton, FL, 1996), Chap. 44, pp. 462–478.

[31] I. L. Leites, D.A. Sama, and N. Lior, "The theory and practice of energy saving in the chemical industry: Some methods for reducing thermodynamic irreversibility in chemical technology processes," *Energy* **28**, 55–97 (2003), with Corrigendum, *Energy* **29**, 301–304 (2004).

[32] K. G. Denbigh, "The second-law efficiency of chemical processes," *Chem. Eng. Sci.* **6**, 1–9 (1956).

[33] S. Lems, H. J. Van Der Kooi, and J. de Swaan Arons, "The sustainability of resource utilization," *Green Chem.* **4**, 308–313 (2002).

[34] R. Heijungs and M. A. J. Huijbregts, "A review of approaches to treat uncertainty in LCA," in *Complexity and Integrated Resources Management. Transactions of the 2nd Biennial Meeting of the International Environmental Modelling and Software Society*, edited by C. Pahl-Wostl, S. Schmidt, A. E. Rizzoli, and A. J. Jakeman Proceedins of the 2nd Blennial Meeting of iEMSs, Complexity and integrated resources management, 14–17 June 2004, Osnabrück, Germany, iEMSs, Orlando, Flo.

[35] S. M. Lloyd and R. Ries, "Characterizing, propagating, and analyzing uncertainty in life-cycle assessment – a survey of quantitative approaches," *J. Ind. Ecol.* **11**, 161–169 (2007).

[36] N. Lior, "Solar energy and the steam rankine cycle for driving and assisting heat pumps in heating and cooling modes," *Energy Convers.* **16**, 111–123 (1977).

[37] K. Koai, N. Lior, and H. Yeh, "Performance analysis of a solar-powered/fuel-assisted Rankine cycle with a novel 30hp turbine," *Solar Energy* **32**, 753–764 (1984).

[38] J. Dersch, M. Geyer, U. Herrmann, S. A. Jones, B. Kelly, R. Kistner, W. Ortmanns, R. Pitz-Paal, and H. Price, "Trough integration in power plants – A study on the performance and economy of integrated solar combined cycle systems," *Energy* **29**, 947–959 (2004).

[39] R. K Dixon and R. E. Sims, "IEA secretariat perspectives on bioenergy," presentation at the International Renewable Energy Alliance meeting "100% Renewable Energy Mix Scenarios for Africa and Asia," Hon. Peter Rae, Chair, United Nations, New York, May 7, 2007. Data from the International Energy Agency, http://www.iea.org/RDD/ (accessed July 31, 2008).

[40] F. Keppler, J. G. Hamilton, M. Brass, and T. Rockmann, "Methane emissions from terrestrial plants under aerobic conditions," *Nature (London)* **439**, 187–191 (2006).

[41] U.S. Department of Energy, "The biobased materials and bioenergy vision," draft, July 18, 2001. Available at http://www1.eere.energy.gov/biomass/pdfs/final_2006_vision.pdf-1355.5KB (accessed July 31, 2008).

[42] U.S. Department of Energy, Office of Fossil Energy Vision 21 program, http://www.fossil.energy.gov/programs/powersystems/vision21/index.html (accessed July 31, 2008).

[43] A. Majumdar and C. L. Tien, "Micro power devices," *Microscale Thermophys. Eng.* **2**, 67–69 (1998).

[44] A. Kribus, "Thermal integral micro-generation systems for solar and conventional use," *ASME J. Solar Energy Eng.* **124**, 180–197 (2002).

[45] J. Holladay, E. O. Jones, M. Phelps, and J. Hu, "High-efficiency microscale power using a fuel processor and fuel cell," in *Components and Applications for Industry Automobiles, Aerospace and Communication*, edited by H. Helvajian, S. W. Janson, and F. Laermer, Proc. SPIE **4559**, 148–156 (2002).

[46] U.S. Department of Energy, Department of Science, http://www.science.doe.gov/ News_Information/News_Room/2006/ITER/index.htm (accessed July 31, 2008).

[47] P. E. Glaser, F. P. Davidson, and K. I. Csigi, editors. *Solar Power Satellites, the Emerging Energy Option* (Ellis Horwood, New York, 1993).

[48] N. Lior, "Power from space," *Energy Convers. Manage. J.* **42**, 1769–1805 (2001).

[49] L. Sauers and S. Mitra, "Sustainability innovation in the consumer products industry," *Chem. Eng. Prog.* **105**, 36–40 (2009).

[50] H. Herring, "Energy efficiency – A critical view," *Energy* **31**, 10–20 (2006).

[51] S. Sorrell and J. Dimitropoulos, "The rebound effect: Microeconomic definitions, limitations and extensions," *Ecol. Econ.* **65**, 636–649 (2008).

PART III

LIFE-CYCLE ASSESSMENTS
AND METRICS

9 Using Thermodynamics and Statistics to Improve the Quality of Life-Cycle Inventory Data

Bhavik R. Bakshi, Hangjoon Kim, and Prem K. Goel

9.1 Introduction

Methods that consider the life cycle of alternative products are popular for evaluating their environmental aspects. Such methods, including *life cycle assessment* (LCA), net energy analysis, and exergetic LCA, rely on *life-cycle inventory* (LCI) data. These data are usually compiled from various sources, including industrial measurements, government databases, and fundamental knowledge, and include information about resource use and emissions. The reliability of the results of these assessment methods depends largely on the quality of the inventory data. Like all measured data, LCI data are also subject to various kinds of *errors*. The need for addressing *uncertainty* in LCA has been identified over the years by many researchers, and several papers have discussed their characteristics and sources. The types of uncertainties include those that are due to data inaccuracy, data gaps, model uncertainty, spatial and temporal variability, and mistakes [1–3]. Many methods have been explored for understanding the effect of such uncertainties, as summarized recently in [3, 4]. The nature and extent of uncertainties in LCA is such that formal methods for dealing with all of them are truly challenging to find.

This chapter focuses on reducing data uncertainty in the LCI. Such uncertainty is due to process variability and failures such as instrument malfunction, poor sampling, or mistranscription of data. In general, these errors may be divided into two categories: random and gross errors. *Random errors* are uncorrelated and Gaussian, whereas *gross errors* are non-Gaussian and include outliers. *Process engineering* research has resulted in methods for reducing the contribution of such errors in data obtained from physical and chemical processes, based on the insight that the true values of the measured variables must satisfy thermodynamic laws. The approach of data rectification combines thermodynamic constraints with statistical knowledge about the measurements to enhance data quality and estimate missing values by exploiting redundant information.

Such an approach may also be used for rectifying LCI data, but such data pose some unique challenges that are often not encountered or are not addressed in process engineering. In general, LCI rectification is much more challenging because of the broad scope of the life cycle, diversity of data, and many missing variables. Some of the characteristics of LCI data are described in the rest of this section.

Table 9.1. *LCI of an ammonia process from SimaPro (Demo version 6.0, ammonia LCI by BUWAL, 250 projects)*

Components	Input (kg)	Output (kg)
Natural gas	464.0	0.0
Water	922.0	0.0
Methane	0.0	7.1
Carbon monoxide	0.0	2.5×10^{-2}
Carbon dioxide	0.0	4.4×10^{-1}
Nonmethane VOC	0.0	9.3×10^{-1}
Nitrogen dioxide	0.0	0.3
Sulfur dioxide	0.0	0.1
Carbon dioxide (co-product)	0.0	1155.9
Ammonia (co-product)	0.0	1000.0

Inconsistency. Table 9.1 shows some typical LCI data, in this case from the bundesamt für umwelt, wald and landschaft (BUWAL) database included in SimaPro [5]. Table 9.2 shows the material balance of relevant atoms. The standard composition of natural gas is used for the calculation of atomic balances. The large inconsistency in the LCI data is quite clear from this table. In this case, ammonia is produced by reacting nitrogen and hydrogen; the latter is supplied by the reforming reaction of natural gas, and the former gas is usually supplied from air. However, nitrogen is missing in the original LCI data shown in Table 9.1, and is the main cause of the large inconsistency in total mass balance. The inconsistencies in carbon, hydrogen, oxygen, and sulfur may be due to incorrect measurements of input and output flows or the compilation of LCI data from multiple sources. Such inconsistencies in LCI data are quite common [6–8]. Purveyors of LCI data may argue that the focus of LCA is on emissions and their impact, and these data are reliable. If so, the proposed rectification approach will not make them worse.

Data at Multiple Scales. LCI data are often available at different spatial scales. For example, data about specific processes may be available from the relevant industry at the equipment scale, whereas more aggregate data about typical processes may be available as the industrial average at the process scale. Even more aggregate are data at the economy scale that are often available in government databases. Hybrid LCA methods aim to combine data from such sources. Their rectification

Table 9.2. *Inconsistency of atomic balances for LCI of an ammonium nitrate process*

Atoms	Input (kg)	Output (kg)
Carbon	340.5	321.7
Hydrogen	213.9	179.5
Oxygen	823.2	841.0
Nitrogen	8.4	822.5
Sulfur	1.2×10^{-1}	5.0×10^{-3}
Total	1386.1	2164.7

poses unique challenges because conservation equations and other constraints across scales would also need to be satisfied. Such challenges are not addressed in process data rectification [8].

Chemical Species and Reaction. In addition to satisfying basic conservation laws, data from chemical processes must also satisfy stoichiometric and reaction constraints. As shown in [7], the quality of rectified data improves with more information about the processes in the life cycle. Such information is not readily available in most LCI databases at the process scale or at the economy scale. Obtaining such information can be tedious and time consuming and represents another challenge that is not encountered in traditional process data rectification.

9.2 Background on Data Rectification

Data rectification relies on thermodynamic laws and statistical knowledge to "make data right." The measurements are assumed to contain two types of additive errors: random and systematic. Random errors are often caused by process variability and sensor errors and are inherent in most measured data. They are usually assumed to follow a normal distribution with mean zero and be independent of other errors. In contrast, systematic errors (or gross errors) behave in a nonrandom manner and are often considered as outliers. The size of a gross error is often larger than that of a random error and its effect on other measurements can be significant. Therefore it is essential to carefully identify and correct these errors. Gross errors may also appear in measurement data because of incomplete or incorrect constraints. This situation is not common in process data rectification but often happens in LCI rectification. To compensate for such errors, obtaining appropriate constraints in the form of fundamental models becomes important. Data rectification is composed of data reconciliation and *gross error compensation*. Data reconciliation assumes no gross errors in the measurements and solves the optimization problem by using constraints to reduce the random error. Gross error compensation corrects the variables with gross errors.

9.2.1 Data Reconciliation

Generally, data reconciliation can be formulated as follows:

$$y = x + \varepsilon, \quad \varepsilon \sim N(0, \Sigma_\varepsilon),$$

where y is the vector of process measurements, x is the vector of true values of the measurements, and ε is the vector of random errors. To obtain the estimate of x, we solve the following least-squares optimization problem.

$$\min_{\hat{x}, \hat{\delta}} (y - \hat{x})^T \Sigma_\varepsilon^{-1} (y - \hat{x})$$

$$\text{s.t.} \quad D\hat{x} = 0. \tag{9.1}$$

The analytical solution of Eq. (9.1) is given by

$$\hat{x} = y - \Sigma_\varepsilon D^T \Phi^{-1} Dy, \tag{9.2}$$

where $\Phi = D\Sigma_\varepsilon D^T$, $D = PA$, P is a projection matrix to represent the constraints in the form shown in Eq. (9.1), and A is an incidence matrix that captures the network structure. The covariance matrix of rectified values $\Sigma_{\hat{x}}$ is given by Eq. (9.3), provided that the model has no gross error:

$$\Sigma_{\hat{x}} = \Sigma_\varepsilon - \Sigma_\varepsilon D^T \Phi^{-1} D\Sigma_\varepsilon. \tag{9.3}$$

This equation may be used to estimate the error in the reconciled data. Note that it does not require the LCI data but relies on only its statistical knowledge and available constraints.

9.2.2 Gross Error Detection

In the presence of gross errors, the following steps are needed [9]. The presence of gross errors needs to be *detected*, their location among variables *identified*, their magnitude *estimated*, and their presence *compensated*. Each step has received considerable attention, and the methods relevant to LCI rectification are summarized in this section.

Many statistical approaches are introduced to detect whether a gross error exists: the chi-square test [10], the nodal test [10, 11], the measurement test [11, 12], the global test [13], generalized likelihood ratios (GLRs) [14], Bonferroni tests [15], and principal component tests (PCTs) [16]. For example, the global test uses the statistic that follows the chi-square distribution:

$$\gamma = (Dy)^T \Phi^{-1}(Dy), \tag{9.4}$$

where the base model $y = x + \varepsilon$, $\varepsilon \sim N(0, \Sigma)$ s.t. $D\hat{x} = 0$. The hypotheses of this test are stated as follows:

$$H_0: E(\hat{\varepsilon}) = 0 \quad \text{vs.} \quad H_1: E(\hat{\varepsilon}) \neq 0,$$

where $\hat{\varepsilon} = y - \hat{x}$ and $\hat{x}$ is an estimated value of x. Under the H_0 hypothesis, the test statistic γ follows chi distribution with v degrees of freedom, where v is the rank of D. If the statistic γ is larger than $\chi^2_{1-\alpha,v}$, the critical value at the chosen α level of significance, then the null hypothesis is rejected and we conclude that there is a gross error.

Another method for detecting gross error is the measurement test proposed by Mah and Tamhane [11] and modified by Crowe [17]. This test considers the extent to which the imposed constraints are satisfied by applying statistical tests to each constraint residual. This test relies on the transformed residual vector [18], $r^* = \Sigma_\varepsilon^{-1}(y - \hat{y}) = D'\Phi^{-1}Dy$. Under the hypothesis H that there are no gross errors present,

$$E(r^*) = 0 \quad \text{and} \quad \text{cov}(r^*) = D^T \Phi^{-1} D, \tag{9.5}$$

the presence of a gross error in the ith measurement can be detected by the statistic

$$|z_i| = \frac{|r_i^*|}{\sqrt{(D^T \Phi D)_{ii}}} = \frac{|d_i^T \Phi Dy|}{\sqrt{d_i^T \Phi d_i}}. \tag{9.6}$$

Here, d_i is the ith column of D. A gross error is considered to be present in the ith measurement iff $|z_i| > k$, where k may be chosen to reflect a desired confidence, as suggested in [19].

9.2.3 Gross Error Compensation

If the model has gross errors, measured values are modified with gross errors as

$$y = x + \varepsilon + \mathbf{B}_{rm}\delta. \tag{9.7}$$

Here, δ is the vector of gross errors and $\mathbf{B}_{rm}$ is a matrix whose binary elements represent the existence of gross errors. The linear steady-state *optimization problem* can be formulated as

$$\min_{\hat{x},\hat{\delta}} \left(y - \hat{x} - B_{rm}\hat{\delta}\right)^T \Sigma_\varepsilon^{-1} \left(y - \hat{x} - B_{rm}\hat{\delta}\right),$$

$$\text{s.t.} \quad D\hat{x} = 0, \tag{9.8}$$

where $\hat{x}$ and $\hat{\delta}$ are the rectified estimates of x and δ, respectively. The analytical solution of Eq. (9.8) is given by

$$\hat{\delta} = \left[Q^T \Phi^{-1} Q\right]^{-1} Q^T \Phi^{-1} Dy, \tag{9.9}$$

$$\hat{x} = y - \Sigma_\varepsilon D^T \Phi^{-1}(Dy - Q\hat{\delta}) - B_{rm}\hat{\delta}, \tag{9.10}$$

where $Q = DB_{rm}$, $\Phi = D\Sigma_\varepsilon D^T$, and Σ_y is a covariance matrix of measured values.

Various modifications and enhancements to this basic gross error detection, identification, and compensation method have been suggested. Among these, the simultaneous estimation of gross errors (SEGE) approach of Sanchez et al. [20] handles multiple gross errors. It uses the measurement test to identify suspected gross errors. A simultaneous detection and estimation strategy is followed to estimate and compensate for the errors.

If the location of gross errors is correctly identified, then the data rectification approach possesses some attractive theoretical properties: the estimated or rectified values and the estimated gross errors may be proved to be unbiased. This has motivated much research on these topics. Satisfying these properties is also essential for methods that can guide the rectification of LCI data, as described in the next section.

9.3 Approach for LCI Data Rectification

As mentioned in Section 9.1, rectification of LCI data poses some unique challenges compared with the rectification of process data. This section describes a general approach that guides the procurement of new LCI data according to the results of rectification. The flow chart in Fig. 9.1 shows the proposed procedure to enhance LCI data by means of rectification. This procedure starts with the available LCI data and corresponding error information. Domain-specific constraints such as material balances, if available, should also be considered at this stage.

The second step is to test whether the data have gross errors. If no gross errors are detected, the data are reconciled with the available information. Otherwise, the

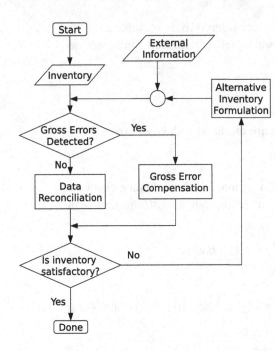

Figure 9.1. Flow chart for rectification of LCI data.

gross errors are compensated by means of the selected rectification method. Next, the enhanced LCI data after rectification are evaluated by comparing the error variances of the measured and rectified values. Generally, more useful information decreases the uncertainty of the estimator. This implies that LCI data rectified with better information are likely to be more reliable than those rectified with less information. If all gross errors have been correctly detected, it has been proved that the proposed rectification approach cannot worsen the quality of the rectified data. Indicators such as the *relative error covariance metric*, defined as

$$R_{m,i} = 1 - \frac{\sigma^2_{m+1,i}}{\sigma^2_{m,i}}, \qquad (9.11)$$

may be utilized to quantify the improvement in data quality. In Eq. (9.11), the subscript m represents the current LCI model and $m + 1$ represents the model with new information. If the finer-scale data are refinements of the coarser-scale data, then it can be proved that the metric defined in Eq. (9.11) will not decrease and indicates the improvement.

Figure 9.2 shows the nested and multiscale nature of many LCI data. The rectification approach has been extended to improve the quality of such data. In this figure, consider that the LCA is desired for a product from the node $V2$. This may represent a process-model-based LCI, which is commonly available as the average or typical industrial data representing many such processes in the selected geographical region. Data at a finer and coarser scale than that of $V2$ may also be available, as depicted in Fig. 9.2. At the finer scale, the detailed flow sheet for the selected process may be available, including flow rates of intermediate, raw material and product streams, measured at the equipment scale. At the coarser scale, the node $V2$ is included in an economic sector, $S2$, which also has corresponding physical flow data representing

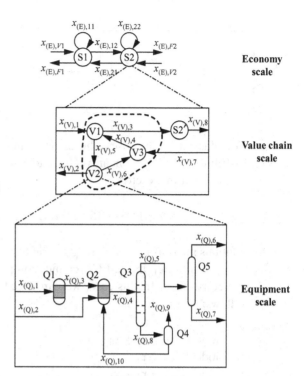

Figure 9.2. Conceptual diagram showing the characteristics of multiscale data [8].

raw material use and emissions. The data at this economy scale usually include the detailed data at the value chain and equipment scales.

In such a problem, it is common to have multiple values of resource flows and emissions from each scale. For example, the resource flows $x_{(Q),1}$ and $x_{(Q),2}$ are available at the equipment scale. In addition, independent values for the same variables are also likely to be available at the value chain scale for $V2$, aggregated as a single variable, $x_{(V),5}$. Similarly, the flow of products, $x_{(Q),6}$ and $x_{(Q),7}$, are at the equipment scale, as well as the value chain scale, which are $x_{(V),2}$ and $x_{(V),6}$. The measured variables of $x_{(V),1}, x_{(V),2}, x_{(V),7}$, and $x_{(V),8}$ at the value chain scale correspond to the measured variables of $x_{(E),12}, x_{(E),21}, x_{(E),V2}$ and $x_{(E),F2}$, respectively. Data from such a multiscale network may also be rectified by means of the proposed approach, but additional constraints representing the relationship between variables across scales need to be imposed. For example, the sum of $x_{(V),3}, x_{(V),4}, x_{(V),5}$, and $x_{(V),6}$ should be equal to $x_{(E),22}$. Details about this approach are available in [8], and the results of application to practical examples are described in the next section.

9.4 Illustrative Examples

9.4.1 Example of Reconciliation: Data Without Gross Error

We simulate the flow sheet, originally used by Rosenberg et al. [21] (see Fig. 9.3). It consists of a recycle system with four units. The parameter values of each flow are

$$x = [5.00 \quad 15.00 \quad 15.00 \quad 5.00 \quad 10.00 \quad 5.00 \quad 5.00]^T.$$

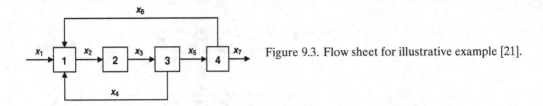

Figure 9.3. Flow sheet for illustrative example [21].

For illustration purposes, the measurement vector is generated from the normal distribution with x as mean vector and 5% of x as the standard deviation (SD). We assume that this vector,

$$y = [4.60 \quad 15.82 \quad 15.15 \quad 5.11 \quad 10.09 \quad 4.66 \quad 5.30]^T,$$

contains measurements corresponding to each underlying parameter value.

The first step in the LCI rectification procedure is to collect information about the constraints. In this case, we can specify the relation among the flow rates as follows:

node 1 : $x_1 + x_4 + x_6 = x_2$
node 2 : $x_2 = x_3$
node 3 : $x_3 = x_4 + x_5$
node 4 : $x_5 = x_6 + x_7$

Using these equations, we formulate the constraints matrix as

$$D = \begin{bmatrix} 1 & -1 & 0 & 1 & 0 & 1 & 0 \\ 0 & 1 & -1 & 0 & 0 & 0 & 0 \\ 0 & 0 & 1 & -1 & -1 & 0 & 0 \\ 0 & 0 & 0 & 0 & 1 & -1 & -1 \end{bmatrix}. \quad (9.12)$$

The second step is to test whether the model has gross errors. Using Eq. (9.4), we calculate the global test value as $\gamma = 6.22$. Comparing with $\chi^2_{(4)}(\cdot)$, we can conclude that there are no gross errors under the confidence level of 95%. Without any gross errors, we get the reconciled values and their estimated variances by using Eqs. (9.1) and (9.3).

Figure 9.4 shows the result of reconciliation. From the first graph, we know that measurements are adjusted to obtain reconciled values that are closer to the parameters or true values. The result shows that the random errors in the measurement are reduced by the constraints in (9.12). The second graph shows that the uncertainties in the measurements decrease by reconciliation. The ratio in the second graph is calculated as $R = $ SD/measurement (or reconciled value). Because we assume that the SD of measurements is 5% of the measured value, the ratios of measurements are indicated as one bar whose value is 0.05. From the graph, we know that the SDs of each variable become smaller after reconciliation. In variables x_2 and x_3, the ratio decreases to 0.02. This result shows that reconciliation brings measurements closer to the true values and decreases their uncertainties.

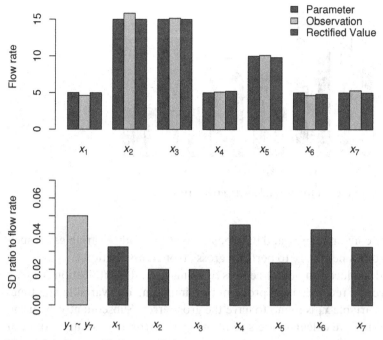

Figure 9.4. Reconciliation result for example without gross errors.

9.4.2 Example of Rectification: Data With Gross Error

Now we assume that the observed value in measurement x_4 is 3.81 instead of 5.11, which was assumed in the previous subsection. This situation causes significant inconsistency of mass balance around the measurement x_4. The global test is applied to detect the presence of the gross error. From the data, we can get $\gamma = 6.22$. Comparing with $\chi^2_{(4)}(\cdot)$, we find that the statistic is much higher so we can conclude that gross errors exist in the model.

If we disregard the existence of gross errors, the reconciled values are far from the true values, as seen in Fig. 9.5 for the estimators of x_2, x_3, and x_4. Moreover, the reconciled x_2 and x_3 have larger errors than their original measurements. This is due

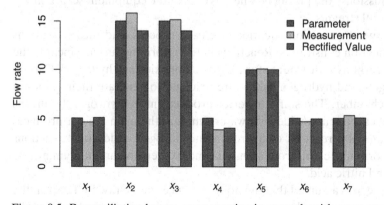

Figure 9.5. Reconciliation but no compensation in example with gross error.

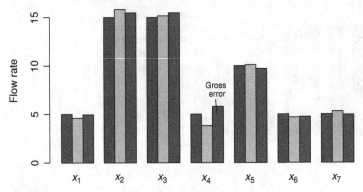

Figure 9.6. Rectification result for example with gross error.

to the gross error contained in x_4 and its propagation to the other variables. In such a situation it becomes necessary to perform gross error compensation.

First, we find the location of gross errors by using a serial identification strategy in which we solve the reconciliation problem by eliminating one variable at a time. In this example, variable x_4 is found to have the gross error. Subsequently, by using Eqs. (9.9) and (9.10), we calculate the size of the gross error and estimates of each variable. After setting the gross error location as

$$B = [0 \quad 0 \quad 0 \quad 1 \quad 0 \quad 0 \quad 0],$$

we calculate the gross error $\delta = -1.97$. Figure 9.6 shows the rectification result and size of the gross error. Compared to the Figure 9.5, each rectified value gets closer to the true value.

9.4.3 Practical Example: Ammonium Nitrate Process

This example, taken from [8], solves a multiscale data rectification problem for an ammonium nitrate process and its life cycle. The scales in this process are similar to those depicted in Fig. 9.2 and include information about the *ammonium nitrate process* at the equipment scale in the form of engineering data and at the value chain scale in the form of LCI. At the value chain scale the inventory data include only raw materials, emissions, and products, whereas data at the equipment scale contain detailed flows for all streams.

Table 9.3 shows the reaction equations for each process, and these equations are used as rectification constraints. Reactions in the reforming furnace include the reaction of hydrocarbons with water to form carbon monoxide and hydrogen. Carbon monoxide, nitrogen, and hydrogen sulfide are oxidized to generate their oxides in the combustion chamber. The shift converter produces more hydrogen by means of the reaction of carbon monoxide with water, whereas the main reactor produces ammonia by means of a reaction of hydrogen and nitrogen. Additional reactions in this life cycle form nitric acid from ammonia and oxygen and ammonium nitrate from ammonia and nitric acid.

In this example, it is assumed that the noise variances in the raw LCI data at the value chain and equipment scales are available, and their values, are between 5%

Table 9.3. *Reaction equations for each process*

Reforming furnace	$CH_4 + H_2O \longrightarrow CO + 3\,H_2$
	$C_2H_2 + 2\,H_2O \longrightarrow 2\,CO + 5\,H_2$
	$C_3H_8 + 3\,H_2O \longrightarrow 3\,CO + 7\,H_2$
	$C_4H_{10} + 4\,H_2O \longrightarrow 4\,CO + 9\,H_2$
	$C_5H_{12} + 5\,H_2O \longrightarrow 5\,CO + 11\,H_2$
Combustion chamber	$CO + \frac{1}{2}\,O_2 \longrightarrow CO_2$
	$H_2S + \frac{3}{2}\,O_2 \longrightarrow H_2O + H_2O + SO_2$
	$N_2 + 2\,O_2 \longrightarrow 2\,NO_2$
Shift converter	$CO + H_2O \longrightarrow CO_2 + H_2$
Reactor	$N_2 + 3\,H_2 \longrightarrow 2\,NH_3$

and 10%. Figure 9.7 shows the difference between the rectified and observed values. The rectified amount of natural gas is smaller than the observed value, whereas for oxygen the trend is reversed. In this example, it is not possible to compare the results with the true values because the latter are unknown. However, the second graph in Fig. 9.7 shows that the SDs of the rectified variables are significantly smaller than

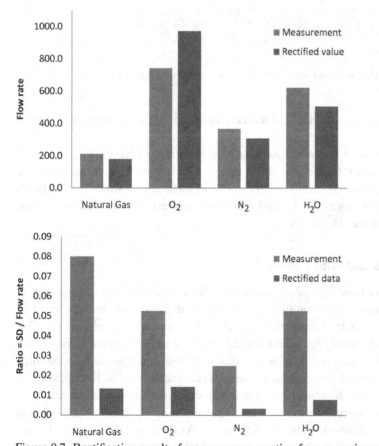

Figure 9.7. Rectification result of resource consumption for ammonium nitrate life cycle.

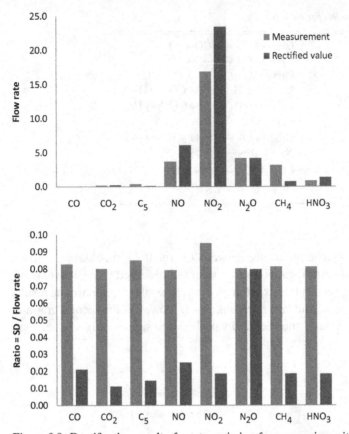

Figure 9.8. Rectification result of waste emission for ammonium nitrate life cycle.

those of the measured variables and indicate considerable improvement, which is due to rectification.

The LCA results of waste emission and their variance ratios are represented in Fig. 9.8. They also indicate that the rectified data are more accurate than the observed value because SD ratios of the rectified data are so small. The ratio of measurement ranges from 0.08 to 0.09, but most of the ratios in rectified values, except N_2O, are smaller than 0.03.

9.5 Conclusions and Future Work

This chapter shows how thermodynamics may be used to enhance the quality of LCI data. The approach relies on the fact that thermodynamic laws must be satisfied and rectifies LCI data by minimizing the error between measured and rectified data with statistical knowledge about the data subject to constraints derived from the laws of thermodynamics. This work uses only the first law, as is commonly done to rectify measured data from the chemical process industries. Rectification of LCI data poses unique challenges because of the multiscale nature of the life-cycle data, lack of uncertainty information, and unavailability of detailed information about the processes, such as reaction information. As demonstrated in this chapter through illustrative examples based on simulated and real data, the proposed LCI rectification can enhance the quality of LCI data and provide indicators of data quality. As shown by the last example, this approach may rectify data at multiple scales, which is a common situation in LCA.

For the LCI data rectification approach to become a mainstream tool for enhancing LCI data, several practical challenges need to be overcome. The rectification approach relies on knowledge about uncertainty in the LCI data, but such information is often not available. Although newer LCI databases are providing uncertainty information, such information is still not available for many products, especially for emerging technologies. The model at the economy scale that is most commonly used is based on economic input–output data. This is usually available as a deterministic model, and may be used in a manner analogous to the deterministic process model. Errors may be considered in the emissions and resource intensities for each sector or the prices used to obtain the final demand vector (or both). Additional research is needed to extend the proposed approach to deal with such practical considerations. In addition to the first law, the second law may also be used to rectify data, but the constraint is less convenient because of its being an inequality. As shown in [7], the second law may be used to check the validity of LCI data by ensuring that exergy is lost in each subsystem and life cycle.

REFERENCES

[1] A. E. Björklund, "Survey of approaches to improve reliability in LCA," *Intl. J. Life Cycle Assess.* **7**, 64–72 (2002).

[2] M. A. Huijbregts, W. Gilijamse, AD M. J. Ragas, and L. Reijnders, "Evaluating uncertainty in environmental life-cycle assessment. A case study comparing two insulation options for a Dutch one-family dwelling," *Environ. Sci. Technol.* **37**, 2600–2608 (2003).

[3] S. M. Lloyd and R. Ries, "Characterizing, propagating, and analyzing uncertainty in life-cycle assessment – A survey of quantitative approaches," *J. Ind. Ecol.* **11**, 161–179 (2007).

[4] J. Reap, F. Roman, S. Duncan, and B. Bras. "A survey of unresolved problems in life cycle assessment," *Intl. J. Life Cycle Assess.* **13**, 374–388 (2008).

[5] SimaPro, available at www.pre.nl.

[6] R. U. Ayres, "Life-cycle analysis – A critique," *Resource Conserv. Recycl.* **14**, 199–223 (1995).

[7] J. L. Hau, H.-S. Yi, and B. R. Bakshi, "Enhancing life cycle inventories via reconciliation with the laws of thermodynamics," *J. Ind. Ecol.* **11**, 1–21 (2007).

[8] H. S. Yi and B. R. Bakshi, "Rectification of multiscale data with application to life cycle inventories," *AIChE J.* **53**, 876–890 (2007).

[9] S. Narasimhan and C. Jordache, *Data Reconciliation and Gross Error Detection* (Gulf Publishing, Houston, TX 2000).

[10] P. Reilly and R. Carpani, "Application of statistical theory of adjustments to material balances," in *Proceedings of the 13th Canadian Chemical Engineering Conference*, Montreal, Canada (1963).

[11] R. S. H. Mah and A. C. Tamhane, "Detection of gross errors in process data," *AIChE J.* **28**, 828–830 (1982).

[12] C. Crowe, "Data reconciliation progress and challenges," *J. Process Control* **6**, 89–98 (1996).

[13] C. Iordache, R. S. H. Mah, and A. C. Tamhane, "Performance studies of the measurement test for detecting gross errors in process data," *AIChE J.* **31**, 1187–1201 (1985).

[14] S. Narasimhan and R. S. H. Mah, "Generalized likelihood ratio method for gross error identification," *AIChE J.* **33**, 1514–1521 (1987).

[15] D. K. Rollins and J. F. Davis, "Unbiased estimation of gross errors in process measurements," *AIChE J.* **38**, 563–572 (1992).

[16] H. Tong and C. M. Crowe, "Detection of gross errors in data reconciliation by principal component analysis," *AIChE J.* **41**, 1712–1722 (1995).

[17] C. M. Crowe, Y. A. G. Campos, and A. Hrymak. "Reconciliation of process flow rates by matrix projection. I: Linear case," *AIChE J.* **29**, 881–888 (1983).

[18] A. C. Tamhane, "A note on the use of residuals for detecting an outliers in linear regression," *Biometrika* **69**, 488–499 (1982).

[19] A. C. Tamhane and R. S. H. Mah, "Data reconciliation and gross error detection in chemical process network," *Technometrics* **27**, 409–422 (1985).

[20] M. C. Sanchez, J. A. Romagnoli, Q. Jiang, and M. Bagajewicz, "Simultaneous estimation of biases and leaks in process plants," *Comput. Chem. Eng.* **23**, 841–857 (1999).

[21] J. Rosenberg, R. S. H. Mah, and C. Iordache, "Evaluation of schemes for detecting and identifying gross errors in process data," *Ind. Eng. Chem. Res.* **26**, 555–564 (1987).

Developing Sustainable Technology: Metrics From Thermodynamics

Geert Van der Vorst, Jo Dewulf, and Herman Van Langenhove

10.1 Introduction

Several definitions for sustainability have been proposed, one of which is widely accepted. This definition was formulated by the Brundtland Commission in 1987, stating that sustainable development is a development that fulfills the needs of present generations without compromising the possibilities of fulfillment of the needs of future generations [1].

Technology is essential in sustainable development because it provides the products to fulfill the needs of the society. However, for delivering these products, the technosphere has to extract resources from the ecosphere, thereby diminishing the pool of resources available for next generations. Furthermore, emissions from the technosphere and from society are sent back to the ecosphere, causing several environmental problems such as acidification and global warming [2]. For these reasons, it is important that sustainable technologies be developed, resulting in a neverending delivery of products to society and involving less impact on the environment.

Stating that new technologies are more sustainable has to be argued by metrics [3–6]. Such metrics are sought by academics (e.g., Oldenburg University, Delft University of Technology, and Ghent University [4, 5, 7] as well as by industry (e.g., GlaxoSmithKline [8], Janssen Pharmaceutica [9], and Shell [10]). Although sustainability metrics should cover the economic, social, and environmental concerns of sustainability, matching the "Triple Bottom Line" of Elkington [11], only the environmental concern is discussed here.

In this chapter, four approaches to quantify the environmental sustainability of technology by means of exergy analysis are presented. These four levels are illustrated in Fig. 10.1 and are explained thoroughly in the four following sections.

At the first level (boundary 1 in Fig. 10.1), exergy analysis is performed on a single technology whereby the sustainability is determined based on the exergetic efficiency and the occurring irreversibilities of the processes in the technology. This approach is explained in Section 10.2, Exergy Analysis of Technologies. Several technologies as identified in level 1 can be combined into an industrial metabolism in which the waste of one technology is the resource for other technologies. This concept is called industrial ecology and is covered in the second level (boundary 2 in Fig. 10.1). A

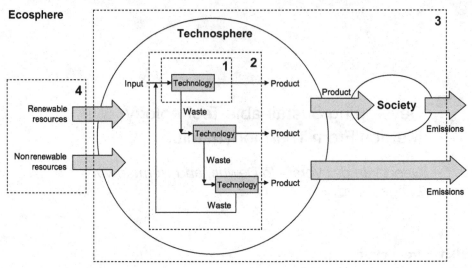

Figure 10.1. Material exchange among ecosphere, technosphere, and society combined with the four approaches for the sustainability assessment of technology.

thorough explanation follows in Section 10.3, Assessment of Technology Integration Into the Industrial Ecology System. For this approach, the overall exergetic efficiency of the metabolism is taken into account.

Expanding the system boundaries outside the technosphere brings us into the third level (boundary 3 in Fig. 10.1) and in the field of life-cycle analysis (LCA). Level 3, explained in Section 10.4, covers the exergetic LCA (ELCA) and the zero-ELCA. The amount of resources extracted from the environment and the amount of emissions emitted into the environment are the input data for such a sustainability assessment. The fourth level (boundary 4 in Fig. 10.1) focuses on the renewability of the resource consumption. This is discussed in Section 10.5, Assessment of the Renewability of the Resource Intake.

By means of these four approaches, it should become clear that the quantitative assessment of environmental sustainability of technologies has two issues: first, the optimization of the industrial system and the efficiency improvement of the process chain. This is covered in the first and second levels. Second, optimizing the resource extraction and considering the renewability of these resources, which is covered in the third and fourth levels. This is all possible by use of metrics based on thermodynamics, more specifically, by use of the exergy concept. In this chapter the emphasis is not on the exergy calculations, but on the exergy concepts at the four levels. More details on the exergy calculations as well as more examples concerning the different levels can be found in the other chapters or in Dewulf et al. [12] and Sciubba and Wall [13].

10.2 Exergy Analysis of Technologies

The most common application of exergy is probably the analysis of a technology consisting of thermal, mechanical, or chemical processes and trying to optimize these processes [14, 15]. Most of the publications from the beginning of the 1970s and several books [16–21] deal with these optimization procedures [12, 13].

Table 10.1. *Exergy destruction and loss in a simple steam power plant [27]*

Description	Type	Exergy destruction or exergy unrecovered (% of the fuel exergy)
Steam-generator combustion	Destruction	30
Steam-generator heat transfer	Destruction	30
Turbine	Destruction	3
Condensor	Destruction	5
Pump	Destruction	Negligible
Stack gas	Loss	3
Cooling water	Loss	2

These publications on exergy analysis of a technology focus, on the one hand, on exergy flows, which are the inputs, the products, and the waste as illustrated in Fig. 10.1, and, on the other hand, on the loss of exergy through the specific process units. This information is then used to identify opportunities for improving process efficiency by modifying the parts of a process with the highest loss of exergy. Through this, the thermodynamic degree of perfection of the different processes can be increased [12].

The use of process efficiencies, however, can cause a lot of confusion when not properly defined [22]. The debate about the definitions of efficiency converged to three fundamental definitions, given in Sciubba and Wall [13]. However, because of the multiple definitions, confusion can be avoided only by defining the used efficiency indicator for each exergy analysis seperately. In this chapter the efficiency indicator defined in Dewulf and Van Langenhove [23] is used as illustration. Here it is stated that efficiency is the ratio between all useful outputs (products) and the required input (resources), all quantified in exergy values. With respect to the inputs, both extracted virgin resources and reused waste materials have to be taken into account. The efficiency metric η is thus defined as $\eta = \frac{Ex_{Prod}}{Ex_{Extr} + Ex_{Reused}}$ where Ex_{Prod} is the useful exergetic output of the process and Ex_{Extr} and Ex_{Reused} are the exergy inputs of the process based on extracted virgin resources and reused materials, respectively. This efficiency metric is dimensionless (input and output in the same units) and is scaled between 0 (unsustainable) and 1 (sustainable). It has to be mentioned that all process inputs that originate from virgin resources have to be traced back in the analysis. Intermediates have to be analyzed with respect to the virgin inputs they require, and these inputs are to be included within the production process chain block. This concept matches the cumulative exergy consumption (CExC) approach of Cornelissen [22] and Szargut et al. [17] and the cumulative exergy demand (CExD) [24], which are being similar to the cumulative energy consumption [25] and the cumulative energy demand [26].

The use of exergy analysis for process optimization is illustrated with an example from Dewulf et al. [12] and Paulus [27]: The advantages of exergy-over-energy analyses are demonstrated by considering a simple natural-gas-fired steam Rankine cycle, operating at 6000 kPa with "typical" values for stack-gas temperature, cooling-water temperature rise, etc. Carrying out the modeling, followed by writing exergy balances on each component, yields the results in Table 10.1. The "ratio" column gives the percent of fuel exergy either destroyed in a component or exhausted into the environment. The results contrast with a traditional energy analysis. Here, energy is conserved, so all inefficiencies are attributed to energy losses. Thus, in this plant, all losses would be attributed to the stack gas (2%) and cooling water (6%). Because this plant has only a single product, an energy analysis thus suggests to improve the process efficiency by reducing the energy losses associated with the cooling water and the stack gas. This is clearly not true, and an exergy analysis clearly points at the true sources of inefficiency.

The major process inefficiencies occur here in the steam generator, where nearly 60% of the fuel exergy is destroyed. This is due to (a) heat transfer from very high-temperature combustion products to the relatively low-temperature boiling water, and (b) the combustion process itself. The next largest source of inefficiency is the condenser, although the 5% destruction and 2% loss (cooling water) are an order of magnitude smaller than those of the steam generator. The exergy destruction in the turbine and the exergy loss with the stack gas each account for about 3% of the total fuel input. The exergy destruction in the pump is negligible. Improving the process efficiency thus clearly starts with the improvement of the steam generator.

This is just one of the many examples of exergy analysis used for the optimization of processes and industrial systems. More examples and references can be found in reviews of Dewulf et al. [12] and Sciubba and Wall [13]. Although the comparison of an exergy analysis with an energy analysis is not repeated in the following sections, the advantages, as mentioned here, remain valid on the three following levels.

10.3 Assessment of Technology Integration Into the Industrial Ecology System

If technology strives for stronger sustainability, it is vital that resources be used with maximum efficiency. This can first be accomplished by improving the efficiency of the used processes, as indicated in the previous section. Another method is keeping materials as long as possible in the technosphere and in society as useful products [28, 29]. This means that, instead of getting resources from the environment or sending waste back to the environment, it is possible to keep these transfers inside the technosphere. Waste from one process should be considered as a possible resource for other processes. This concept is called industrial ecology, and the group of technologies on which the concept is performed is called an industrial metabolism. The idea of industrial ecology can be compared with natural ecological systems, where different base types are found. First, there are linear flows through systems with an unlimited resource input and waste output. Second, there are quasi-cyclic systems with energy input, a limited resource input, and a waste output. Third, there are cyclic systems with energy input and no resource input or waste output [30]. To make technology more sustainable, linear flow through material systems should be converted as much as possible in cyclic systems [31]. This matches the main principle of industrial ecology, which states that individual firms have to be connected into industrial ecosystems [32]. This goal can be achieved by closing loops through reuse and recycling and by defining all wastes as potential products or resources and seeking markets for them [28].

Based on the advantages of exergy analysis used for assessing sustainability of technological processes, it is here shown that exergy can also be used as a tool for assessing sustainability of industrial metabolisms. This is illustrated by the exergy analysis of different solid-waste-treatment systems of polyethylene (PE) [33]. The solid-waste streams of PE are studied as described by Finnveden et al. [34]. Three treatment options are investigated: recycling, incineration with co-generation of heat and electricity, and disposal without any energy or material recovery. Process data on PE are taken from Sundqvist et al. [35]. First, an exergy analysis on the different waste-treatment methods as single technologies is performed. Resources to transform PE waste are all calculated in cumulative exergy terms. Incineration of 1-kg PE requires electricity, transport, and chemicals with a total added CExC (ACExC) of 1.23 MJ. The output (the conversion products) leads to 6.25 MJ of heat and 14.61 MJ of electricity per kilogram of PE incinerated, all in exergy terms ($Ex_{ConvProd}$). Recycling PE requires a higher amount of virgin resources, being 7.19 MJ ACExC per kilogram of PE waste. The higher input simultaneously results in a higher output of conversion products ($Ex_{ConvProd}$) per kilogram of PE waste, being 27.9 MJ PE, 1.69 MJ heat, and 3.95 MJ of electricity. The third option for PE is disposal for which no extra exergy input is taken into account and no extra output is delivered. So it can be concluded that recycling requires a higher additional input (ACExC) than incineration, but the exergy output embodied in conversion products ($Ex_{ConvProd}$) is also higher. The best utilization of resources for the treatment of 1 kg PE waste is thus the recycling technology, reminding us that the method from Section 10.2 is used.

From the industrial ecology perspective, however, waste treatment cannot be seen as an isolated set of processes. Recycling makes it possible to generate new PE out of the used PE. This means, on one hand, that less virgin resources will be extracted for primary PE production, but, on the other hand, other virgin resources are necessary for the recycling conversion process. In this example, incineration is not able to deliver PE as recycling can do, so incineration implicitly demands the maintenance of PE production starting from virgin resources. This makes it clear that waste treatment has to be placed in a broader perspective than was done in the first analysis. This is illustrated by Fig. 10.2.

In this case, the incineration metabolism requires exergy for virgin PE production ($CExC_{PrimProd} = 86.0$ MJ) and the incineration process itself ($CExC_{ConvProd} = 47.8$ MJ with 46.5 MJ from PE waste and 1.3 MJ as ACExC). The metabolism delivers 1 kg of PE (= 46.5 MJ exergy) and 20.8 MJ exergy of H + EL (where H is heat and EL is electricity). To produce the same output, the recycling metabolism of PE requires less input for the primary process (34.4 MJ), because 27.9 MJ PE is produced by recycling. However, the recycling process delivers only 5.6 MJ H + EL, so additional virgin H + EL of 15 MJ exergy is needed, which costs 30.2 MJ ($CExC_{AddProd}$) of virgin resources. Total CExC data necessary to deliver 1 kg of PE and 20.8 MJ H + EL as well as the overall efficiency for the three treatment methods are given in Table 10.2.

It can be concluded that the recycling metabolism is the most efficient set of processes, followed by disposal and incineration.

However, with respect to the overall metabolic efficiency and sustainability, waste inputs have to be distinguished from virgin inputs. Essentially, reused wastes are not inputs but are part of internal loops. Recycling requires a total virgin input of 71.9 MJ, so the output–virgin input ratio R is 0.936. In this view, incineration is

Table 10.2. *Total CExC and the overall efficiency of three treatment methods for 1 kg PE waste*

Method	Total CExC (MJ)	Overall efficiency η (%)
Incineration	133.7	0.503
Recycling	118.4	0.568
Disposal	127.4	0.528

the second-best option, with $R = 0.772$, followed by disposal ($R = 0.528$). Although disposal has a set of technological operations that are more efficient than those of incineration, it is the worst option for industrial metabolism because of its high input of virgin resources.

In conclusion, we have shown that exergy analysis of industrial metabolisms can be a suitable tool in sustainable resource management. Together with tools focusing

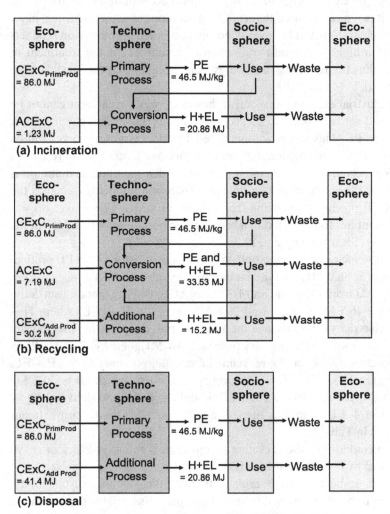

Figure 10.2. (a) Incineration metabolism of PE, (b) recycling metabolism of PE, (c) disposal metabolism of PE.

on overall net emissions and on social and economic impacts, it can contribute to a more quantitative approach of the sustainability of industrial metabolisms.

10.4 Exergetic Life-Cycle Assessment

To determine the thermodynamic perfection of a system, not only processes that occur within the system should be considered, as was done in Sections 10.2 and 10.3. Also, the interactions between energy and mass flows outside the system's boundaries should be taken into account. In this section, the irreversibilities in the whole life cycle of a product are considered, allowing us to assess and evaluate the degree of thermodynamic perfection of the production processes of the complete process chain. Cornelissen [22] developed a method called the exergetic life-cycle assessment (ELCA), which is based on the life-cycle approach combined with exergy analysis. The LCA method has four major steps: defining goal and scope, data collection, impact assessment, and the interpretation of results [36]. ELCA uses the same framework as LCA, but the only criterion in step three (impact assessment) is now the exergy loss during the whole life cycle [37]. ELCA can show where in the process natural resources are lost, which makes it possible to optimize processes and reduce the loss of natural resources more efficient [22]. The environmental impact of a product is, however, due not only to the total resource consumption. Emissions coming from the technosphere as well as from society also have their effect on the environment. The effect of these emissions can be determined with other LCA impact assessment methods like CML 2001 or Eco-Indicator 95 [38]. However, ELCA can also be extended to zero-ELCA to include the abatement exergy of emissions. This is the exergy consumption required for the abatement of waste streams. In this way, all environmental problems associated with emissions can also be taken into account with the use of exergy calculations and not only depletion of natural resources [22].

As an example of the use of ELCA on consumer products, the ELCA and the zero-ELCA of a porcelain mug and a disposable polystyrene (PS) cup are included [22]. Much discussion has taken place on the question of which cup is preferable from an environmental point of view: the disposable PS cup or the porcelain mug. An ELCA study was performed to give an answer to this question. The functional unit is taken to be the use of 3000 cups. A repeated use of 3000 times is taken for the porcelain mug, which is the average lifetime. The PS cups, which weigh 4 g, are used only once. The PS cups consist of about 49% high-impact PS and 49% general-purpose PS. An amount of 2% of TiO_2 is added for the coloring of the cup. The transport of the different materials is taken into account in this analysis; the packing material of the PS cups and porcelain mugs is not taken into account. PS is created by the polymerization of styrene and the disposable PS cups are formed by a foil-extrusion process. The electricity use for the foil-extrusion process is 4 MJ/kg of cups. After usage the cups end up in a municipal waste flow, of which 48% is incinerated and 52% is dumped at a waste site. The porcelain mug is made up of a mixture of minerals brought in the desired form and subsequently heated in an industrial furnace to 900 °C. Glazing is added to the mug, and the porcelain mug is finished with a heat treatment at 1400 °C. The mass for the porcelain mug is 250 g. For the production of 1 kg of porcelain mugs, 1 m^3 of natural gas is used. After each

Table 10.3. *Main incoming flows of the two cycles in kilograms for the use of 3000 cups*

Components	PS cup	Porcelain mug
Coal	0.378	4.700
Crude oil	23.600	
Natural gas	0.685	7.478
Oxygen	63.961	34.615
Phosphate		0.045
Total	88.624	46.793

use, the porcelain mug is washed in an industrial dishwasher, in which a detergent containing 30% of phosphates is used. The electricity use is 45 kJ and detergent use is 0.25 g for each time the mug is washed [39]. After usage, this mug ends up in a municipal waste flow that is dumped at a waste site.

For the ELCA, the black-box approach was used. For the sake of simplicity, a simplified mass balance was used, which takes into account only the main flows, as can be seen in Tables 10.3 and 10.4. Other flows with a small mass flow or low exergy content, like clay, are neglected.

The exergy of the incoming natural resources was determined. The exergies of the emissions, except one, were considered to be negligible. It is assumed that these emissions diffuse into the environment and dissipate. Only the PS flow that ends up at the waste site was taken into account, because this plastic waste forms an exergy reservoir that might be used in the future. The life-cycle irreversibility for the 3000 PS cups is 817 MJ, whereas the life-cycle irreversibility for the porcelain mug is 442 MJ, a factor of 1.85 less, as can be seen in Figures 10.3 and 10.4. The main causes of irreversibility for the PS cups are the production of PS and the burning of the waste in a waste-incineration plant. The main irreversibilities in the case of the porcelain mug take place in the electricity generation and the dishwasher.

Table 10.4. *Main outgoing flows of the two life cycles in kilograms for the use of 3000 cups*

Components	PS cup	Porcelain mug
CO_2	55.100	29.600
H_2O	26.520	14.773
SO_2	0.428	0.066
PS waste	6.115	
NO	0.304	0.172
N_2	0.020	1.701
Ash	0.032	0.395
CO	0.023	0.021
C_xH_y	0.081	0.023
CH_4		0.043
Phosphate		0.045
Total	88.624	46.793

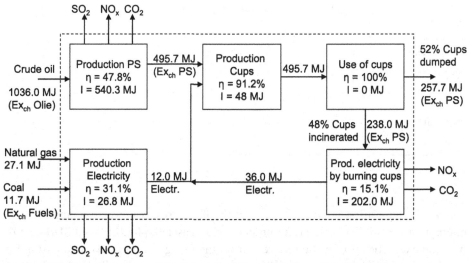

Figure 10.3. Exergy diagram of the life cycle of the 3000 disposable PS cups.

To make a zero-ELCA of the two products, the production processes have to be transformed into zero-exergy emission processes. The four major contributors to the pollution of these two products are SO_2, NO_x, CO_2, and phosphate. These substances together form 95% of the environmental impact of the emissions caused by the life cycle of the two products according to Eco-Indicator 95 [22]. The abatement exergy for these emissions was determined. For SO_2 the abatement exergy was found to be 57 MJ/kg based on 90% SO_2 removal with limestone in a flue-gas desulfurization (FGD) unit in a coal-fired power plant. The limestone is converted into gypsum in the FGD. For NO_x, the abatement exergy was found to be 16 MJ/kg, based on an 80% removal in a $DeNO_x$ unit in a coal-fired power plant. In this unit the NO_x react with NH_3 to N_2 and H_2O. It is well known that these abatement techniques are now applied everywhere in Western Europe. For CO_2, a figure of 3 MJ/kg is calculated for

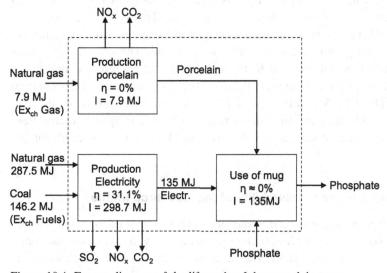

Figure 10.4. Exergy diagram of the life cycle of the porcelain mug.

Table 10.5. *Abatement exergy of the main harmful emissions (in megajoules)*

Components	PS cup	Porcelain mug
CO_2	148.8	79.9
SO_2	22.0	3.4
NO_x	3.9	2.2
Phosphate		0.8
Total	174.7	86.3

the case of separation of 90% CO_2 out of the flue gases, compression, and storage, in empty gas fields according to Göttlicher and Pruschek [40]. In this 3 MJ/kg CO_2, only the exergetic cost for the recovery from the flue gases and the compression for storage is included. The complete life-cycle assessment of actually performing this abatement was not included, making of this 3 MJ/kg CO_2 an underestimation of the real resource consumption for abatement. Finally, for phosphate a figure of 18 MJ/kg was found for 99% removal [41]. The total abatement exergy for each emission is given in Table 10.5.

The zero-exergy-emission life-cycle irreversibility for the 3000 PS cups is 992 MJ. This is an increase of 21% compared with the situation without emission reduction. The zero-exergy-emission life-cycle irreversibility of the porcelain mug is 528 MJ. This is an increase of 20% compared with the situation without emission reduction and a factor of 1.88 less than zero-exergy-emission life-cycle irreversibility for the 3000 PS cups.

It can be concluded that, by including the environmental effect of depletion of natural resources, the ELCA is a valuable addition to the LCA. From the ELCA it can be concluded that the disposable PS cups make two times more use of the exergy reservoir of natural resources than the porcelain mug. From this section it is again clear that, when an exergy analysis is made, the energetic resources as well as the nonenergetic resources are taken into account. This is not the case in an energy analysis. More examples of the advantages of exergy as a life-cycle impact indicator above energy can be found in Bösch et al. [24]. The zero-ELCA shows that only a relatively small increase in the use of natural resources is needed to avoid 80%–95% of the harmful emissions and reduce the associated environmental effects greatly. The ELCA and the zero-ELCA are objective criteria, taking into account the irreversibilities and the efforts required for reducing the environmental effects on the resource side and with zero-ELCA as well on the side of the emissions [22].

10.5 Assessment of the Renewability of the Resource Intake

As seen in the previous section, technology also interacts with the natural environment. Two basic streams can be identified. First, on their way to the production chain, natural resources pass the boundary between the environment and the technosphere. The origin of these resources can be renewable or nonrenewable. A resource can be

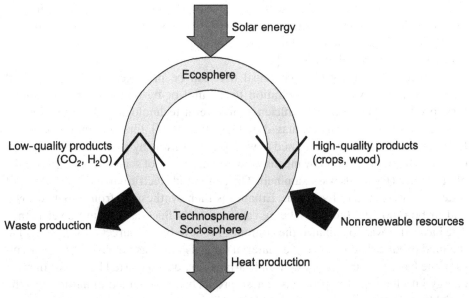

Figure 10.5. Solar-driven closed-cycle condition for a sustainable ecosphere–technosphere interaction.

called renewable when the consumption rate is equal to the production or regeneration rate in the natural environment. Fossil fuels, for example, are consumed much faster than they are generated, which makes them nonrenewable resources.

Second, emissions are brought into the natural ecosystem. These emissions are formed during the production or use phase and also come from the disposal of products. These emissions contribute to two opposite phenomena in the natural environment. On one hand, they are known to disrupt the natural system, ending up in global warming, stratospheric ozone depletion, eutrophication, etc., which is mentioned in the previous section. On the other hand, they bring the elements like carbon and oxygen back into the natural ecosystem. By doing so, new renewable resources can be produced by photosynthesis [2].

When one aims at the assessment of the sustainability of technology, the nature of virgin resources is also one of the key issues and can play a major role in achieving environmental sustainability [2]. Because resources from nonrenewable origin make it impossible to close material cycles, as indicated in Fig. 10.5, renewable resources should be preferred when striving for sustainability.

Making use of renewable resources means closing the loop shown in Fig. 10.5, hereby avoiding the need to use nonrenewable and avoiding emitting waste products. The ultimate sustainable situation is the one in which the black arrows from Fig. 10.5 are avoided and the closed cycle is driven only by solar energy.

To quantify the renewability of resources, exergy is used. More specifically, a renewability indicator α based on thermodynamics was proposed by Dewulf and Van Langenhove [23] and Dewulf et al. [4], as $\alpha = (Q_{Prod}/Q_{Extr})$, where Q_{Prod} is the production rate of resources in the environment and Q_{Extr} is the extraction rate of virgin resources. If renewable resources are used, i.e., if the production rate in the ecosystem equals and compensates for the consumption rate, then $\alpha = 1$. For nonrenewable resources such as fossil resources and mineral ores, their production rate is negligible compared with the consumption rate so that α tends to 0 [23].

By means of a case about ethanol production, it is illustrated how exergy analysis copes with this renewability issue [2].

Ethanol is a widely used chemical that can be produced by cracking fossil resources into ethylene and hydration to ethanol or by fermentation of agricultural products. The exergetic efficiency of several technological pathways for the production of ethanol was evaluated by Dewulf et al. [4]. First, ethanol production from fossil resources was examined as well as the pathway by means of fermentation of wheat. Next, production by means of hydrogen by use of solar energy captured by photovoltaic (PV) cells was examined. The generated electric current is used to split water into oxygen and hydrogen. Ethanol is then synthesized from this hydrogen and carbon dioxide, the latter being captured from flue gases from power plants. The basic idea was to enhance the content of renewable resources (solar energy) in the final product and to close the material cycle as much as possible. As in agriculture, the basic resources are water and carbon dioxide, converted by means of solar energy into high-quality products. These products are degraded in nature into the initial low-quality products, CO_2 and H_2O, closing the material cycle.

The wheat fermentation pathway to ethanol is mainly based on solar energy. However, the pathway also requires nonrenewable resources such as mineral resources, fossil fuels for the production of nutrients and pesticides, and power for the use of machinery and equipment. Furthermore, this way of production also delivers wheat straw (from agriculture), gluten, and cake (both from fermentation), all of which are useful products. The resources required in the agricultural step must be attributed in a proportional way to ethanol and to the other useful products. Also, the PV-based route requires substantial nonrenewable inputs mainly for the construction of the PV cells and to operate the chemical synthesis of ethanol from carbon dioxide and hydrogen.

From the exergy analysis, it turned out that the fossil-resources-based route was the most efficient, with a cumulative exergy requirement of 60.13 MJ to deliver 1 kg ethanol. After an exergy-based allocation of the by-products in the bio-route, it was found that a nonsolar input of 7.6 MJ is necessary to produce 1 kg of ethanol out of wheat. This value is almost negligible compared with the required solar irradiation for the photosynthesis process (4240 MJ). The route through PV-generated electricity and electrolysis of water to produce hydrogen for the synthesis of 1 kg ethanol requires only 338 MJ of solar irradiation but 26.4 MJ of nonsolar resources are required.

This case shows that CExC as such cannot be the sole indicator to analyze the sustainability of the three options. In other words, not only efficiency should be taken into account, but the nature of the resources as well. The two exergy-based parameters to assess the sustainability used in this case are the efficiency η and renewability α of resources. Results for this case are presented in Table 10.6.

With respect to efficiency, fossil-resources-based technology proved to be better than the PV route and the bio-route. If renewability is considered, the results are the other way around. This illustration shows that the PV route is largely based on renewables, but far more efficient than the bio-route, the latter being limited by the efficiency of the photosynthesis process. In conclusion, this example shows that

Table 10.6. *Exergy-based parameters to assess the sustainability of three ethanol-producing methods*

parameter	Overall efficiency η (%)	Renewability α (%)
Fossil resource based	49	0.02
PV route	7	91.1
Bio-route	0.7	99.8

exergy analysis is able to take into account quantitatively the role of renewables in technology development based on a scientifically sound basis [2].

10.6 Conclusion

In this chapter it is shown that the exergy approach, based on the fundamental laws of thermodynamics, is applicable as sustainability metric in the field of environmental technology. The exergy approach makes it possible to compare different technologies based on the same unit: joule of exergy. The comparison can be made based on the exergetic efficiency and the exergetic losses that appear while processes are carried out. Further on, it was shown that exergy analysis can be of high value quantitatively, indicating the closing of loops in industrial ecology. This means keeping waste flows into the technosphere and using it as a resource for other processes. It has also been shown that exergy can serve as a basis for LCA in assessing the sustainability. The exergy approach allows judging different technology options on their CExC. The major difficulty in ELCA, however, is the treatment of the negative timeless effects of emissions. Therefore, it has been proposed that the exergy required for the abatement of the emissions prior to their release into the ecosphere should be taken into account in a zero-ELCA, so that emissions do not longer affect the ecosphere and human health.

At last, it was illustrated that exergy allows describing the use of resources in terms of renewables and nonrenewables. Also the efficiency of the employment of resources throughout the production chain can be quantitatively described. Sustainable societies and ecosystems must maintain the potential of the sources they use to perform work. If a society consumes the exergy resources at a faster rate than they are renewed, it will not be sustainable.

It can be concluded that the assessment of the sustainability of technological processes can been performed in a quantitative way, based on the basic laws of thermodynamics. However, it is clear that a sustainable society needs more than only a sustainable technosphere. Next to the technological condition of sustainability, also social, economical, and political aspects affect the fulfillment of the needs of the current generation without endangering the needs of the future generations.

10.7 List of Symbols

α	Renewability indicator
η	Efficiency
ACExC	Added cumulative exergy

CExC	Cumulative exergy consumption
$CExC_{PrimProd}$	CExC of the primary products
$CExC_{AddProd}$	CExC of the added products
$CExC_{ConvProd}$	CExC of the conversion products
EL	Electricity
Ex	Exergy
Ex_{ch}	Chemical exergy
Ex_{prod}	Exergy of the product
Ex_{extr}	Exergy of the extracted resources
$Ex_{re\text{-}used}$	Exergy of the reused waste materials
$Ex_{Conv.\ Prod}$	Exergy of the converted products
H	Heat
R	Output over virgin input ratio
Q	Production/Extraction Rate

REFERENCES

[1] G. H. Brundtland, *Our Common Future. World Commission on Environment and Development (WCED)* (Oxford University Press, 1987).

[2] J. Dewulf and H. Van Langenhove, "Exergy," in *Renewables-Based Technology: Sustainability Assessment*, edited by J. Dewulf and H. Van Langenhove (Wiley, Chichester, UK, 2006).

[3] R. Darton, "Sustainable development and energy: Predicting the future," presented at the *15th International Conference of Chemical and Process Engineering* (Praha, Czech Republic, 25–29 August 2002).

[4] J. Dewulf, H. Van Langenhove, J. Mulder, M. M. D. Van Den Berg, H. J. Van Der Kooi, and J. D. Arons, "Illustrations towards quantifying the sustainability of technology," *Green Chem.* **2**, 108–114 (2002).

[5] S. Lems, H. J. Van Der Kooi, and J. de Swaan Arons, "The sustainability of resource utilization," *Green Chem.* **4**, 308–313 (2002).

[6] N. Winterton, "Twelve more green, chemistry principles," *Green Chem.* **3**, G73–G75 (2001).

[7] M. Eissen and J. R. O. Metzger, "Environmental performance metrics for daily use in synthetic chemistry," *Chem. Eur. J.* **8**, 3580–3585 (2002).

[8] P. Smith, "How green is my process? A practical guide to green metrics," in *Proceedings of the Conference on Green Chemistry of Sustainable Products and Processes* (University of Wales Swansea, Wales, April 3–6, 2001).

[9] J. Liessens, "Sustainability performance assessment: A sustainable development pilot project at Janssen pharmaceutica," presented at the *Euro Environment Conference* (Aalborg, Denmark, 2002).

[10] J. P. Lange, "Sustainable development: Efficiency and recycling in chemicals manufacturing," *Green Chem.* **4**, 546–550 (2002).

[11] J. Elkington, *Cannibals With Forks: The Triple Bottom Line of the 21st Century Business* (Capstone, Oxford, 1997).

[12] J. Dewulf, H. Van Langenhove, B. Muys, S. Bruers, R. Bakshi, G. Grubb, R. A. Gaggioli, D. M. Paulus, and E. Sciubba, "Exergy: Its potential and limitations in environmental science and technology," *Environ. Sci. Technol.* **42**, 2221–2232 (2008).

[13] E. Sciubba and G. Wall, "A brief commented history of exergy from the beginnings to 2004," *Intl. J. Thermodyn.* **10**, 1–26 (2007).

[14] A. Rucker and G. Gruhn, "Exergetic criteria in process optimisation and process synthesis – Opportunities and limitations," *Comput. Chem. Eng.* **23** (Suppl. 1) S109–S112 (1999).

[15] M. Sorin, A. Hammache, and O. Diallo, "Exergy based approach for process synthesis," *Energy* **25**, 105–129 (2000).

[16] T. J. Kotas, *The Exergy Method of the Thermal Plant Analysis* (Krieger, Malabar, FL, 1985).

[17] J. Szargut, D. R. Morris, and F. R. Steward, *Exergy Analysis of Thermal, Chemical and Metallurgical Processes* (Hemisphere, Berlin, 1988).

[18] J. de Swaan Aarons, H. Van Der Kooi, and K. Sankaranarayanan, *Efficiency and Sustainability in the Energy and Chemical Industries* (Marcel Dekker, New York, 2004).

[19] N. Sato, *Chemical Energy and Exergy: An Introduction to Chemical Thermodynamics for Engineers* (Elsevier, Amsterdam, 2004).

[20] A. Bejan, G. Tsatsaronis, and M. Moran, *Thermal Design and Optimization* (Wiley, New York, 1996).

[21] J. E. Ahern, *The Exergy Method of Energy Systems Analysis* (Wiley, New York, 1980).

[22] R. L. Cornelissen, "Thermodynamics and sustainable development: The use of exergy analysis and the reduction of irreversibility," in *Proefschrift ter verkrijging van de graad van doctor* (Technische Universiteit Delft, Delft, The Netherlands, 1997).

[23] J. Dewulf and H. Van Langenhove, "Integrating industrial ecology principles into a set of environmental sustainability indicators for technology assessment," *Resources Conserv. Recycl.* **43**, 419–432 (2005).

[24] M. E. Bösch, S. Hellweg, M. A. J. Huijbregts, and R. Frischknecht, "Applying cumulative exergy demand (CExD) indicators to the ecoinvent database," *Intl. J. Life Cycle Assess.* **12**, 181–190 (2007).

[25] H. Schaefer, "Cumulative energy consumption of products – methods for determinations – problems of evaluation," *Brennstoff-Warme-Kraft* **34**, 337–344 (1982).

[26] VDI, *Cumulative Energy Demand – Terms, Definitions, Methods of Calculation* (VDI, Düsseldorf, Germany, 1997).

[27] D. M. J. Paulus, *Energy Engineering Class Notes* (Technische Universität Berlin, Berlin, Germany, 2004).

[28] E. A. Lowe, J. L. Warren, and S. R. Moran, *Discovering Industrial Ecology – An Executive Briefing and Sourcebook* (Battelle, Columbus, OH, 1997).

[29] T. Graedel and B. R. Allenby, *Design for Environment* (Prentice-Hall, Englewood Cliffs, NJ, 1996).

[30] J. Dewulf and H. Van Langenhove, "Thermodynamic optimization of the life cycle of plastics by exergy analysis," *Intl. J. Energy Res.* **28**, 969–976 (2004).

[31] T. E. Graedel, B. R. Allenby, and M. Sharfman, "Industrial ecology," *Acad. Manage. Rev.* **20**, 1090–1094 (1995).

[32] J. Dewulf and H. Van Langenhove, "Quantitative assessment of solid waste treatment systems in the industrial ecology perspective by exergy analysis," *Environ. Sci. Pollution Res.* **9**, 267–273 (2002).

[33] G. Finnveden, J. Johansson, P. Lind, and A. Moberg, "Life cycle assessments of energy from solid waste," Fms Rep. (Stockholm University Press, Stockholm, 2000).

[34] J. O. Sundqvist, A. Baky, A. Björkland, M. Carlsson, O. Eriksson, B. Frostell, J. Granath, and L. Tyselius, "Systemanalys av energiutnyttjande fran avfall – utvardering av energi, miljö och ekonomi," *IVL Rep. 1379* Swedish (Environmental Research Institute, Stockholm, 2000).

[35] ISO, "Environmental management – Life cycle assessment – Principles and framework," ISO14040.

[36] M. Gong and G. Wall, "On exergy and sustainable development – Part 2: Indicators and methods," *Exergy Inl. J.* **1**, 217–233 (2001).

[37] Ecoinvent Centre. Ecoinvent data v1.2, final reports ecoinvent 2000, No 1–15 (Swiss Centre for Life Cycle Inventories, Düssendorf, Switzerland, 2005).

[38] J. Van Eijk, J. W. Nieuwenhuis, C. W. Post, and D. J. H. Zeeuw, "Re-usable versus disposable?" *Productenbeleid 1992/11* (Dutch Ministry of Housing, Infrastructure and Environment, Zoetermeer, The Netherlands, 1992).

[39] G. Göttlicher and R. Pruschek, "Comparison of CO_2 removal systems for fossil-fuelled power plant processes," in *Proceedings of the Third International Conference on Carbon Dioxide Removal* (MIT University Press, Cambridge, MA, 1996).

[40] STOWA, *The cleaning of municipal waste water in the context of sustainable development for pollution control* (Hageman Verpakkers B.V., Zoetermeer, The Netherlands, 1996).

11 Entropy Production and Resource Consumption in Life-Cycle Assessments[*]

Stefan Gößling-Reisemann

11.1 Measuring Resource Consumption

This book is all about thermodynamics and resources, the meaning of resources, their interpretation in thermodynamics, their consumption, and the path to a sustainable way of managing resources. It has been clarified in other chapters that the consumption of resources is to be understood as a process of diminishing their usability. This is the basis for this chapter: understanding resource consumption as a loss of *potential utility* of these resources. The concept of potential utility is explained later in this chapter in more detail. For now it should suffice to note that consumption is only partly measurable in physical terms. We speak of consumption when the states of matter and energy change in a way that makes them less usable for human needs. Thus the general meaning of consumption has an anthropogenic notion, which cannot be evaluated in objective physical terms only. In some cases, this change in the usability of material or energy flows is accompanied by a proportional change in one or more of their physical properties, as, for example, when the temperature of a heat flow decreases. In these cases, measuring consumption is straightforward. In other cases, however, the changes in the material and energy flows are of a different nature with possibly no way of direct measurement. We might still be able to find changes in the physical, chemical, or biological properties of the flows that are responsible for the loss in usability, but the relation between consumption and the change in these properties is much more complicated. A generalization of consumption and the derivation of a consistent measure for these cases is much more difficult, if not impossible. Examples of this kind of consumption are the addition of toxic substances to drinking water (in which the actual consumption, or loss of usability of the resource, depends on toxicity thresholds), the discoloring of fabric (in which the consumption is more or less defined by markets and demand), or the disruption of structure in structurally used materials (like paper or construction wood).

A preliminary conclusion from the preceding paragraph would be that measurement of consumption on a general level is impossible. Still, there have been many attempts at measuring consumption, and many of them have been applied very successfully. In general, it is quite straightforward to measure resource consumption

[*] Some parts of this chapter were published as an article in the Journal of Industrial Ecology, volume 12, number 1, pages 10–25, © Yale University.

when (a) there is only one (type of) resource involved and (b) the transformation of this resource is either very extensive or irrelevant. As an example, when the fuel consumption of a car is concerned, the measure for consumption is obvious: the volume (or energy content) of the fuel, measured in liters (or megajoules, respectively). Here, only one resource is relevant, and the transformation of that resource is very extensive: The fuel is transformed into heat at ambient temperature, water vapor, and CO_2 (and further emissions). All of these products are of virtually no direct use to humans, and thus the transformation can be considered maximal. When the transformation is minimal, as for example in the case of water flowing through a hydropower station, we generally do not speak of consumption at all. One physical property of the flow of water has been changed drastically (the potential energy), but the general usability of the water is not altered very much. The real challenge for measuring resource consumption arises when several resources are transformed to varying degrees. When fossil fuels are transformed into plastics, water and sugar are transformed into soft drinks, or metal ores are transformed into different pure metals, slag, and sulphuric acid, how much of these resources have then been consumed? I believe there can be only approximative answers, and I believe that different questions concerning resource consumption demand different measures. Nevertheless, because thermodynamics provides a framework for analyzing the transformations of matter and energy, it seems most logical to look for measures in this field of science. Consequently this is reflected in the different approaches to measuring resource consumption that have been developed so far.

11.1.1 Throughput versus Consumption

Consuming resources means transforming them. The term *resource consumption* already refers to such transformations, as can be seen, for example, from its dictionary definition:

> Consumption (noun): The utilization of economic goods in the satisfaction of wants or in the process of production resulting chiefly in their destruction, deterioration, or transformation [1].

For our purposes, I will subsequently try to define resource consumption more precisely. Still, it is instructive to highlight the conceptual differences between resource use and resource consumption, two commonly used terms, which sometimes get confused. Whenever a material or energy flow enters a production system, is then either transformed or not and then either leaves the system or is stored within, one can speak of the material or energy being used. The material or energy, i.e., the resource, was used to produce a product (or service). If this flow is transformed inside the system, either by changing its quantity or quality, then, and only then, it has (partly or wholly) been consumed. Thus use and consumption have related but different meanings. The term use is broader in its meaning, which implies a certain vagueness when one tries to measure resource use. As an example, consider the use of one ton of water in three different scenarios: a hydropower station, in the cooling tower of a fuel-fired power plant, and in a cleaning process. In each case, we could imagine the same amount of water entering the process and finally leaving it again. If we then ask for the respective resource use, in each case the answer would be the same in all three cases: one ton. Unfortunately, this answer, though formally correct, is totally blind to the actual quality of the process, i.e., the way in which the water is transformed. In the hydropower example, the potential energy of the water is decreased (and transformed into another energy form); in the cooling-tower example, the heat contents and possibly the phase state of the water have changed; and in the cleaning process, the "water" in fact becomes a complex mixture of many different materials. The water flowing from the hydropower station can still be used in many different processes, as it is chemically and physically undistinguishable from freshwater (except for its potential energy). This is not the case for the water outputs from the cooling plant and the cleaning process. Their potential utility has decreased (quite markedly in the cleaning example). If we asked for the consumption of water (instead of the use), the answer should really contain a reference to this decrease in potential utility. For the cleaning process, this would include the change in composition (taking the polluting substances into account); for the hydropower example, this would include the generation of heat at ambient temperature to account for these losses (this requires the closing of the energy balance). However, a measure based on only mass, volume, or energy will not be able to map this change correctly. It will measure only the actual throughput of matter and energy. As subsequently argued, entropy production might serve as a proxy for measuring real consumption covering the physical aspect of transformations.

For completeness, there is another concept that needs to be mentioned in this context: resource depletion. This term usually refers to the decrease in quantity of a naturally occurring resource and is thus more limited in its scope than resource use or resource consumption. Whereas use and consumption can apply to any form of material or energy, depletion applies only to the materials found in nature. This is not to say that the measurement of resource depletion is irrelevant! On the contrary: With diminishing reserves of fossil fuels and minerals, the depletion of these resources is an important factor for the sustainability of our economy. However, measuring resource depletion means focusing on the input side only, neglecting the transformations following the extraction. Thus depletion needs to be accompanied by other measures that allow for a more detailed look into these transformations, yielding insight into the causes for depletion and starting points for optimisation.

11.1.2 Resource Consumption and Entropy Production

One of the qualitative aspects of matter and energy transformations is what I call the potential utility. The potential utility P can be thought of as the size of the set of all possible uses of a given amount of resources; see Fig. 11.1.

When a physical resource is consumed, its potential utility is decreased. As an example, let's take the production of paper from wood, water, and energy. At first sight, a stack of paper might seem much more useful than the materials and energy it was made from, but in fact the potential utility has decreased. Starting with wood, water, and energy (plus the right tools), we could produce all sorts of products: a table plus lots of extra water, a wooden swimming basin (filled with water), cardboard boxes, a boat, and some extra drinking-water rations, or, of course, a stack of paper. Each of these products could in turn be used in many different fashions, whereas the stack of paper can be used only as such: a stack of paper. The set of possible uses has decreased by the transformation of the materials and energy into paper. The size of this set would then correspond to the potential utility of the resources. It

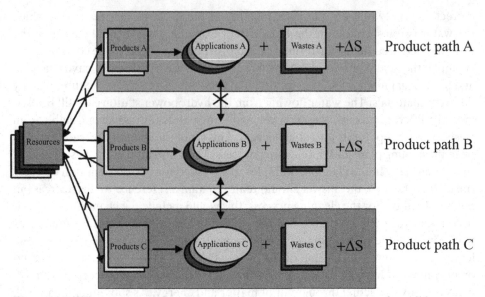

Figure 11.1. When resources are consumed in a process, the actual set of possible uses or applications decreases. The size of this set can be viewed as the potential utility of these resources.

should be noted, though, that the *actual* usefulness of the raw materials, or products, respectively, is very much dependent on the preferences, abilities, and circumstances of the respective user. A sheet of paper is much less useful to someone who cannot write. Nevertheless, the set of potential uses is much clearer defined. It will probably not be possible to precisely determine the size of this set, because it is also dependent on human ingenuity and technical possibilities. Still, it should be obvious that, by choosing one production path, a certain subset of potential uses will no longer be available. It would certainly be desirable to measure the actual *loss* of potential utility and use this as a measure for resource consumption. Unfortunately, this seems to be impossible, for the same reasons that limit our knowledge of the actual size of the set of potential uses. However, there is something in the transformations that reflects the loss of potential utility on a physical level: the irreversibility of the process. Irreversibility in simple terms means "the impossibility of a process to run in a complete cycle without changing its environment" [2]. This definition can be quantified by the introduction of entropy: An open system undergoes an irreversible transformation if the combined entropy of the system and the environment increases. The second law of thermodynamics states that the combined entropy can only remain constant or increase. Thus the reversible transformation corresponds to a constant combined entropy. We have to bear in mind that these statements are valid only for the combination of the system plus its environment. The open system itself can have increasing, constant, or decreasing entropy. If we denote the entropy of the system by S_{sys}, we can express the system's entropy change as

$$\Delta S_{sys} = \Delta_e S_{sys} + \Delta_i S_{sys}, \tag{11.1}$$

where $\Delta_e S_{sys}$ is the entropy exchange between system and the environment and $\Delta_i S_{sys}$ is the entropy produced within the system. The entropy exchanged with the environment does not increase the overall entropy (of system and environment), because the decrease of the system's entropy (by export) is the increase of the environment's entropy (by import) and vice versa. Thus the irreversibility of the transformations taking place inside the system is measured by the internally produced entropy $\Delta_i S_{sys}$. The second law of thermodynamics simply states that $\Delta_i S_{sys} \geq 0$, where the equal sign applies to only the (unrealistic) case of reversibility. Coming back to the discussion of consumption, potential utility, and irreversibility, I argue that the more irreversible a process, the more potential utility is lost. The basis for this argument is the fact that irreversibility describes for thermodynamic systems what the loss of potential utility describes for resources: the loss of potential pathways for development. Because entropy change measures irreversibility, I propose that the change in potential utility ΔP (i.e., the consumption) that occurs inside a system, caused by internal processes, is a negative and monotonically decreasing function of the entropy change ΔS caused by these processes. Because the entropy change caused by these processes is equal to the internal entropy production $\Delta_i S_{sys}$, we found a physical approximation to the loss of potential utility: entropy production.

It is important to note that the entropy production always has to be analyzed for the whole system under investigation, including all flows into and out of the system. Entropy production does not measure the consumption of only one resource. Instead, all resources are considered together and the overall consumption of resources is assessed. If, for example, we analyze a water reservoir with respect to its potential energy, we have to make sure we close the energy balance when the system undergoes transformations to account for all resource losses. If we lower the elevation of this reservoir by some means, we would of course lose potential energy stored in the water (or, more precisely, stored in the gravitational field). Naively, we could argue that the entropy of the system would not change under this transformation and thus entropy analysis would not detect any resource consumption. However, if we close the energy balance, we must account for the lost potential energy somehow. It is transferred to some other part of the system (e.g., by lifting a corresponding weight to a higher elevation), it is transformed into heat and transferred to the surroundings, or it is transformed into another form of energy (e.g., electricity). Whichever is the case, the analysis is complete only if all parts of the system are taken account of: the reservoir, the weight, or the interface between system and surroundings. Then the entropy production of the transformation tells us exactly how much of the resource, which is the potential energy in this case, has been lost. In the case of a lifted weight there is no loss (apart from possible losses that are due to friction, which become visible in the balance as heat transfers to the surroundings). In the case of electricity production, the losses depend on the efficiency of the conversion. Again, these losses will appear at the interface of system and environment as heat flows (at probably ambient temperature). We can also see this from Eq. (11.1): The overall entropy change (of system and environment) caused by processes inside the system is $\Delta_i S_{sys}$. The question that has to be answered correctly is this: What is *inside* the system, i.e., where do we draw the boundaries? If we define them in such a way that all losses occur outside the system, e.g., by defining the lifted weight to be part of the environment, we would of course have no irreversibilities inside the system, and all losses (which are due to friction when lifting the weight) would occur in the environment. However, setting the system boundaries in such a way would be nonsensical when the goal of such an analysis is to measure resource consumption. When the system boundaries are set adequately, such that they include all relevant transformations of matter and energy, the term $\Delta_i S_{sys}$ in Eq. (11.1) correctly describes the irreversibility of the resource transformation.

We have seen how entropy change is related to resource consumption, but what about the absolute value of the entropy of a system? The absolute values of the entropy of a system have no direct meaning with regard to resource evaluation. Two identical water reservoirs at different elevations have the same entropy, even though the stored energy within (and thus their potential utility) is different. Thus entropy content cannot be used for evaluating the absolute utility of a resource, but entropy *production* accounts for changes in utility. Exergy is usually used for the purpose of estimating a resource's physical potential for doing useful work. The quantification of this potential, however, is based on the assumption of reversible processes, i.e., it measures the physical work that could ideally, i.e., reversibly, be extracted from this resource. Because real-life processes are always irreversible and the minimum irreversibility is determined by many factors (including economic, social, technological, and environmental factors), the absolute exergy value is of only limited significance. More about this in the next section.

11.1.3 Related Concepts: Entropy Production and Exergy Loss

The relevance of entropy (production) in the discussion of resource consumption has a history that really dates back to the beginning of thermodynamics. After all, Sadi Carnot's main motivation was to find an expression for the efficiency of steam engines, i.e., how much work could be produced from a given amount of steam (the resource in this case) (see [2]). Clausius later developed his ideas on the basis of the Carnot cycle and introduced the term entropy to describe the lost work potential (see [2]), which is a concept already close to the resource consumption as understood in this chapter. Yet, to my knowledge, entropy analysis has not yet been used to assess the overall resource consumption of general industrial (or economic) processes.

On the other hand, there are quite a few authors who have used second law analysis in an attempt to assess the *economic* consequences of resource consumption, the *efficiency* of industrial processes, or the *environmental impacts* of industrial processes. Essentially, they all use either one of the two concepts: entropy or exergy. Georgescu-Roegen argued [3], in 1971, that the inevitable entropy increase described by the second law of thermodynamics will lead to the final standstill of all economic activities.[1] He claimed that the true resource is "low entropy." Notwithstanding Georgescu-Roegen's enormous impact on economics, this interpretation is somewhat naive, because it is not the low entropy of a material flow that defines its resource character, as I have laid out in the previous chapter. However, his observation that economic activities produce entropy, that their inputs have a lower entropy than their outputs (if we include stocks and wastes in the outputs), and that thus

---

[1] His reasoning was based on the idea that there exists a fourth law of thermodynamics describing the dissipation of matter alone. He concluded that matter could never be completely recycled and thus, eventually, the economy would run out of resources. There is no evidence that such a fourth law exists; it is not in agreement with thermodynamics.

the consumption of resources is linked to entropy production is without a doubt correct. Bejan introduced entropy production (he uses the term entropy generation) as a means to thermodynamically assess and optimize the losses connected with heat transfer, fluid flow, and mass transfer irreversibilities [4]. The "entropy-generation minimization" (EGM) approach of Bejan is basically a way to optimize the design of systems and to assess their efficiency. It requires knowledge of the geometrical setup of machines and of the thermodynamic properties of the internal flows of matter and energy inside the analyzed system (or model). It should be obvious that increased thermodynamic (second law) efficiency of any device implies a reduction of resource consumption attributable to this device. However, to fully assess the effects of optimizations at the process level, one has to carry out a life-cycle-wide analysis. Whereas Bejan's EGM method aims at improving the design of systems at a very detailed level, Georgescu-Roegen's argumentation aims at the consideration of entropy production at a very aggregate level, namely, the whole economy. The approach presented here is situated in between these two extremes, at the level of production and consumption systems that span several plants and regions. It is not so much aimed at optimizing the design of specific systems, but at assessing the resource consumption accompanying the economic production and consumption of goods. As mentioned earlier, the most promising approach so far to measuring resource consumption on an aggregate level has been exergy (loss) analysis. The history and the spectrum of application of the exergy concept are well explained in the introductory chapter and elsewhere in this book, and I will not go into the details here. Exergy is defined as the useful energy stored in resources and intermediate process streams, and thus an important application of exergy analysis is the technical and economic optimization of energy conversion technologies; see, for example, the long-standing series of ECOS conferences (e.g., [5]). In this field, it was undoubtedly most successful. But apart from optimization, exergy, or more precisely exergy loss, was also proposed as a measure for resource consumption. Szargut, for example, introduced exergy as a concept to account for "ecological costs," defined as the cumulated depletion of nonrenewable exergy resources (see [6]). He excludes renewable streams from the analysis, based on the assumption that the consumption of these does not come with relevant costs to the environment. As long as the depletion rates stay below regeneration rates, this is of course justified. The exergy of natural resources, in this context, is understood as a common measure of their quality. For resources being consumed as energy carriers, this bears significance, because their exergy content describes the potential useful energy that could be extracted from them. For nonenergy resources, however, exergy as a quality measure is questionable, because their exergy content is probably never utilized. This is especially true for end-of-life materials that serve as secondary resources. For example, when copper and iron scrap are compared, their specific exergies differ by a factor of 3. But how does that correspond to their quality? One could argue that this difference in exergy value reflects the necessary expenditure when these metals are manufactured from the reference environment. But in the case in which they are manufactured from scrap, what is the significance in the different exergy values? At least we can say this much: When the specific exergy of a material stream changes, it loses some of its quality, in the sense that it is less available to further processing. If further processing is desired, new exergy has to be added to the process. For natural resources a difference in specific exergy could thus mean a difference in energy requirements for further processing. Sulfidic copper ore (concentrate) is a good example: It has a fairly high specific exergy, such that the first pyrometallurgical step is almost energetically autonomous. After that, the remaining material flows (copper matte and slag) do not contain enough exergy to drive further refining steps on their own: Exergy has to be added. Iron ore (containing Fe_3O_4 and Fe_2O_3), on the other hand, has a rather low specific exergy, and it needs large amounts of exergy for its metallurgy. In this respect it makes sense to compare the two exergy values (of copper ore and iron ore) as they represent the "dowry," provided by nature, of the two metal ores. But what is the relevance of this information? In the absence of alternative ores, this information is purposeful only in the sense that one should take good care in making use of the given exergy. This again is better be done by analyzing the exergy losses within a process. In summary, the absolute values of exergy seem not to have an importance of their own. The interesting question is this: How do we best manage the material and energy flows so as to minimize exergy losses?

In this line of thought, Ayres has long been an advocate of using exergy loss as a measure for resource consumption [7, 8]. His work builds mainly on the ground-breaking work by Szargut [9]. In addition, Ayres proposed exergy as a measure for resource quality [10] and as a measure of pollution [11]. The latter notion is rejected by other authors (especially Conelly and Koshland and Szargut). Conelly and Koshland [12, 13] consider exergy loss of natural resources as the ultimate measure for assessing the burden placed on the environment. Their reasoning is not based on the notion of exergy as a measure of the "value" of a resource (they explicitly reject this interpretation), but rather on the fact that resource extraction is always accompanied by environmental damage.[2] Thus, to minimize this damage, the extraction of resources must be minimized. They propose three strategies for optimizing industrial systems toward this goal: efficiency, recycling, and use of renewable resources. Indicators for all three of these strategies are derived from analyzing the exergy flows within and through the respective system. In this manner they identify exergy removal (i.e., the irreversible loss of exergy) as a measure for resource consumption. They also take into account the possibility of "upgrading" a resource by adding exergy to it, enabling the distinction between consumption and depletion: A resource that is consumed and then "refilled" with exergy is not depleted. Depletion takes place only when a resource's lost exergy is not renewed.[3]

In the framework of life-cycle assessment (LCA), exergy loss has also been used as a measure for resource consumption, e.g., by Cornelissen [15], Cornelissen and Hirs [16], and Ayres et al. [7]. In these analyses, exergy loss or exergy destruction (or both) is used to describe the consumption of renewable and nonrenewable resources during the life cycle of a product. It is basically an extension of Szargut's notion of cumulative exergy consumption, which includes all exergy consumption from resource extraction to final product [17]. Dewulf and colleagues have used the

---

[2] Szargut, in one of his later articles, also dismisses the interpretation of exergy as a measure of a resource's value [14].

[3] There seems to be a problem with material resources, however. When a material is "consumed," i.e., part of its exergy is removed, it often also changes its physical and chemical form (burning fossil fuels might be a good example). In this case, the resource is actually gone. So how can one renew its exergy content?

cumulative exergy approach to analyze different waste-treatment systems [18, 19] and to define an exergy-based measure of the sustainability of technology [20]. The life-cycle exergy analysis can also be used to measure depletion by distinguishing between renewable and nonrenewable resources. If only nonrenewable inputs to the life cycle are counted, the results (in terms of lost exergy) reflect the depletion of natural resources.

Whereas exergy is defined as to measure the availability of energy, entropy rather measures the nonavailability. The main conceptual difference, however, lies in the respective reference states. In general, energy is always defined only up to an arbitrary constant. For exergy this constant is chosen such that the system under investigation has zero exergy when it is in thermodynamic equilibrium with the so-called reference environment. This equilibrium state is also referred to as the dead state. For entropy the dead state for each substance (and for each system) is the state at absolute-zero temperature. This "reference state," however, is not arbitrary. It is rather an empirical fact that the entropy of a system is zero at absolute-zero temperature.

Following from the preceding discussion, the definition of exergy brings with it five conceptual issues that limit the explanative power of the concept and that made me choose entropy instead as a measure for resource consumption:

1. The reference environment is a basic prerequisite for the interpretation of exergy as available energy. The reference environment is usually chosen as to approximate the actual natural environment as an end point for all transformations of materials and energy. The assumption that the system under investigation can be brought into equilibrium with the reference environment assumes that the reference environment is itself in equilibrium. This is not the case for the natural environment, which makes the choice for a specific reference environment challenging. The natural environment is rather to be described as being far from equilibrium with large spatial and temporal variations of thermodynamic properties. This would (at least) call for an adjustment of the exergy reference environment with regard to space and time, which is usually not done in exergy analysis. As an extreme example, the exergy calculations for processes in interstellar space would have to be done on the basis of a reference environment very different from the one currently adopted. As a less extreme example, consider the reference state for water. In its fluid state, there exist several forms to be found in the natural environment (which should be the basis for a reference environment): sweetwater in lakes and rivers, groundwater, and seawater. Each form is in constant transformation into one of the other forms and into gaseous water in the atmosphere. This makes the definition of a reference state for liquid water ambiguous. These ambiguities can lead to definitions of reference environments that make the interpretation of exergy as "the maximum useful work obtainable from a flow" no longer valid. See, for example, [21] for a discussion of different choices of reference environments.
2. The interpretation of exergy as available energy suggests that the exergy of material and energy flows is, in principle, available to be used by humans. This could be true only if the calculation of exergy would take into account the actual environment the potential user of the material and energy flows is operating in, which is usually not the case (see previous point). In addition, this interpretation

quietly assumes that the limits for the technical utilization for resource flows are reversible processes, which is far from the truth and therefore highly theoretical. It has to be noted, though, that this goes for any attempt to evaluate the thermodynamic perfection of industrial processes without taking into account the intrinsically irreversible nature of all processes. Thus exergy is a measure for the theoretically available energy of matter and energy flows. How much of this energy could practically be used depends very much on the technical, spatial, and temporal limitations of the process in question. A lot of scientific work has been done on this subject in the fields of entropy generation minimisation (see, e.g., [4]) and finite-time thermodynamics (see, e.g., [22]), which both approach this problem from a process perspective, not from a substance perspective. In conclusion, the interpretation of exergy as available energy can be misleading, because the "availability" of a substance depends very much on the context of its uses and the available processes.

3. On another note, the interpretation of exergy as a measure of a substance's energetic value or the inherent driving force makes sense only for substances that are somehow used to drive a process. In view of the preceding example of the exergy of pure Cu and Fe, the absolute exergy value of these substances has relevance only when these substances are used energetically (which includes, e.g., driving a chemical reaction). The meaning of their exergy content is then the amount of useful energy that can be "harvested" from these substances. For substances that are used nonenergetically (e.g., by providing structure or by conducting heat and electricity), their exergy content has no practical relevance. For substances or materials that are used in a cascading way, the relevance of their exergy content might, however, come into play in a later-use phase.

4. In most exergy analyses it is not the absolute exergy content of material and energy flows that is analyzed, but rather the differences between input and output. The exergy loss is then interpreted as a measure for the actual resource consumption. Under these circumstances the definition of a reference environment becomes irrelevant because it cancels out in the calculation. On the other hand, the loss of exergy is directly proportional to the accompanying production of entropy (in many cases the the Gouy–Stodola equation provides the correct relation: $T_0 \, \Delta S = \Delta B$). Thus, by use of entropy production instead of exergy loss, the definition of a reference environment, with all the mentioned difficulties, can be avoided.

5. The exergy content of a material is sometimes interpreted as the minimum amount of energy needed for producing the material in question from the average environmental composition. The scope of this interpretation is limited because real-life material deposits have characteristics far from the average composition of the environment, which is why we call them deposits in the first place. Hence the minimum energy needed to produce refined metals, minerals, and other materials is not at all linked to their exergy content. Nevertheless, the fact that resource deposits are not in equilibrium with their environment highlights the fact that these deposits are to be viewed as some form of "natural inheritance." The earth as a far-from-equilibrium system has provided us with sources of highly concentrated or highly structured materials, which humans learned to utilize in order to derive different services from them (fuel services,

structure, conduction, optical activity, etc.). The exergy of natural deposits can thus be seen as a measure for the energetic capital endowed to us by the evolutionary processes in earth's past.

11.2 Resource Consumption in Life-Cycle Assessment

As I mentioned in the preceding section, exergy has been employed as a concept to account for resource consumption in LCA. Nevertheless, thermodynamics-based measures are the exception, and resource consumption in general is not having a lot of weight in the current LCA practice. Because resource consumption is mainly an economic problem, and LCA is primarily assessing ecological impacts, this is plausible. However, with dwindling resources and ore grades becoming increasingly smaller, resource consumption may very soon become an ecological problem, too. That is, when the material and energy expenditures for getting at the resources become large enough to constitute an important factor in the overall assessment. As a precautionary measure, even from an ecological viewpoint, a careful management of resources thus seems advisable. Additionally, LCA is a very well-suited tool for implementing a measure for resource consumption, because it requires a detailed and extensive data pool, which is also a prerequisite for correctly assessing resource consumption.

Whether part of LCA methodology or not, resource consumption is undoubtedly an important aspect of society's metabolism, and low resource consumption is also undoubtedly a key factor for a sustainable way of living. Thus the sustainable engineer's toolbox should contain an appropriate measure. The most promising approach to adequately describe resource consumption so far has been exergy analysis, as described with a great deal of detail in other chapters in this book. An approach that has gotten far less attention, which is also based on the second law of thermodynamics and is presented here as an alternative, is entropy-production analysis. But let us first turn toward the importance of analyzing resource consumption in LCA and other methods of technology assessment.

Material and energy flows are at the heart of any analysis of the interactions between the anthroposphere and its environment. Different methodologies were developed to assess these flows and the ecological impacts associated with regions, processes, and products. The methodologies are different in their objective and their scope, but they share a common feature: They are built around the material and energy exchange with the environment in terms of a quantitative material flow model of the respective system or region. Some of the methods are product oriented [e.g., LCA, material intensity per service unit (MIPS)], and others focus on spatial or social units [e.g., material flow analysis (MFA), substance flow analysis (SFA), and ecological footprint (EF)]. The assessment in any of these cases is usually based on a detailed inventory, comprising all relevant flows of matter and energy exchanged with the environment. Which flows are considered relevant, however, and which ones are not, depends very much on the methodology used and the analyst applying it. LCA has generally the most complete list of flows, which is one of the reasons for making it a good candidate for implementing thermodynamics-based measures for resource consumption. If impacts on the environment are considered (as is the case with LCA), these impacts are calculated from the quantitative material flow model by applying impact categorization and impact factors to the flows. The results are then interpreted and can be compared with the results for other (equivalent) products and processes.[4] The idea behind these approaches based on material flow models is that the total effect on the environment associated with a process can be approximated by measuring and analyzing the material and energy flows between the environment and the process or region. In other words, the industrial metabolism is measured and analyzed at its interface with the environment. At first glance, this approach seems to be valid. There are limitations, however, and most of the current methodologies have their blind spots. A shortcoming that is worthwhile being discussed in this context is the fact that the just-mentioned methodologies in their current form are mainly dealing with flows *into* and *out of* the technosphere, but pay little attention to the ways these flows are dealt with *within* the technosphere [23]. Although it is true that many of the effects on nature can best be described by measuring the inputs and outputs, the actual solution to environmental problems is most often found within the technosphere (cf. [24]). In the same way as the effects on the environment can be quantified only when there exist impact models linking environmental impacts to matter (and energy) flows, we also need models describing the internal transformations of matter and energy to assess the efficiency (and sustainability) of the technosphere. Thus, if we want to avoid environmental problems, we should ask ourselves this: "How can we better manage our internal flows of material and energy, how can we make better use of them?" One of the immediate effects will then be the decrease of inputs from and outputs to the environment. The question remains of how we can assess the changes in the matter and energy flows along their path through the technosphere. This sets the stage for the development of a measure for the transformations of matter and energy taking place inside the technosphere (or inside a given process). Looking at the transformations of matter and energy will focus the attention of ecological assessment on the quality of flow transformations in addition to the exchanged quantities.

11.2.1 Established Measures for Resource Consumption

To motivate the introduction of an entropy-based measure for resource consumption in LCA, I would now like to discuss the way in which consumption is evaluated in the aforementioned methods (LCA, MIPS, MFA, SFA, and EF).

Evaluation in these methods occurs in different ways, but usually without reference to transformations of materials or energy. The LCA methodology speaks of the "depletion of abiotic (and biotic) resources" [25], which is to be distinguished from resource consumption, but is sometimes still used synonymously. Although depletion can be interpreted as *a decrease in the available amount*, for the purpose of my argument, consumption is to be understood as *decreasing the (overall) utility* (see the section on resource consumption). In LCA studies, resource depletion or consumption is often analyzed only marginally. In the minimal case, only the amount of fossil fuels entering the life cycle of a product is considered. Indicators for the depletion of basically all natural resources are described in the literature [25]; these

---

[4] This, of course, requires the detailed definition of the target system and its boundaries in order to make the process or product alternatives comparable.

are based on ultimate reserves and current deaccumulation rates. The numerical values of this indicator of course depend on our current knowledge of the reserves of the corresponding resource. In view of the approach taken in this chapter, the more prominent disadvantage of the reserves or deaccumulation approach is that it is considering only the extraction of resources from the environment (and correspondingly their input to the economic system), but neglecting the transformations these resources are subject to within the economic system.

Another approach to resource depletion is given by Müller-Wenk and operationalized in the Eco-Indicator 99 method [26]. It is not based on the questions of how long resources last and how much they are decreased by current practices, but instead evaluates the effects on future generations by examining the future additional investment (in energy terms) that is due to the extraction of resources in the present time [26, 27]. This approach has been further refined by including the functionality of the resources, the ultimate quality limit to deliver this functionality, and the backup technology needed to deliver the functionality when the limit is crossed [28]. This approach does not take into account resource consumption in the sense of a transformation of resources. It is rather based on a resource's actual utility in terms of the functions it performs (as opposed to the potential utility, as previously explained). A problem with this approach is that the actual utility of a resource is dependent on the current state of technology and knowledge, and thus the question of whether a resource can perform a certain function or not (in the latter case it would have reached its ultimate quality limit) can be answered for only the present time.

Another assessment method within the LCA methodology based on the second law of thermodynamics is exergy analysis, as explained in this chapter and elsewhere in this book. It can be seen as an extension to the conventional LCA methodology. This approach takes into account all resource consumption in the form of lost exergy (see, e.g., Ayres et al. [8]; Cornelissen [16]; Dewulf et al. [18]; Dewulf and van Langenhove [19, 20]; Finnveden and Östlund [29]).

In the MIPS approach, resource consumption (which is also referred to as resource use) is not distinguished from material movement [30]. Whenever a certain mass of material is moved in the course of a production or consumption process, it is considered to be a material input to this process and is counted as material consumption. This also applies to materials not entering the actual process (e.g., overburden from mining minerals), which are sometimes termed hidden flows. There is no reference made to the physical or chemical changes this material is subjected to in the process (apart from the fact that it is changing its positional parameters).

MFA treats resource consumption very much like the MIPS concept, but is usually applied on the national or regional level (as opposed to the product or service level). The parameter most closely related to resource consumption in this approach is direct material consumption (DMC), consisting of the domestic extraction of fossil, mineral, and biotic materials plus imported materials and minus the exports [31]. In essence, DMC describes the mass of the materials entering into and remaining in the economy or region (as stocks). Although the term consumption is used explicitly, no reference is made to the change in physical, chemical, economic, or other parameters describing the material flows. MFA also considers hidden flows, but they do not enter the definition of DMC.

SFA, which is sometimes seen as a submethod of MFA, focuses on one or a few substances and analyzes their mass flows into, through, and out of a regional or economic unit (for methodology and case studies, see [32]). Transformations are covered to a certain extent when the flows of elements (e.g., N, P, Cu) are analyzed with regard to the chemical compounds they appear in. Consumption, however, is not the main aspect of SFA. It is mainly used for tracking environmentally or economically relevant substances in order to optimize their economic use or increase recycling rates. Consumption, as it is understood in this article, is not addressed; the focus is on throughput and stocks.

The EF approach determines the area necessary to sustainably support a given region (city, nation) by starting out with the national or regional consumption (production plus imports minus export of goods) [33]. Here also consumption is understood in the sense of materials entering the region, and no reference is made to their state or any respective transformations. Consumption is thus seen as the throughput of materials, and its relevance lies in the fact that these materials need to be created (and CO_2 sequestered) by bioproductive land.

In conclusion, most of the standard tools for measuring the metabolism of the technosphere refer to resource use, depletion, or consumption. Only LCA, however, has been equipped with a measure for assessing the actual consumption of resources (exergy loss); all others are only input or throughput oriented.

11.2.2 Scope and Limitations of a Thermodynamic Assessment of Resource Consumption in Life-Cycle Assessment

The inclusion of thermodynamic assessment in LCA comes with a few challenges (see next paragraph), so the LCA practitioner has to decide whether the effort is worth it or not. In general, this depends on the goal of the LCA study. When the study has a focus on only a few environmental implications (for example, only global warming and ozone depletion), then the assessment of resource consumption, or materials transformation in general, is of no direct concern. If, however, the study aims at highlighting the sustainability of the product system more generally, resource consumption should be part of the analysis, thus warranting the application of a thermodynamic assessment. This is especially true if one of the goals is optimization of resource consumption. Without a proper method to assess the *overall* resource consumption (not only focusing on a few selected resources), this optimization will lead to erroneous conclusions. The thermodynamic assessment of resource consumption then guarantees that shifting consumption from one type of resource to another gets recognized properly and thus burden shifting can be avoided. Additionally, the more detailed look at the product system will lead to the detection of fundamental technological limits, which can induce innovation and further development of alternative technologies.

When applying thermodynamic assessments, one should be aware of the challenges and limitations coming with it. One of the main challenges for LCA in general is its data-collection intensity. This intensity results from the breadth of the environmental flows covered (elementary flows), as well as from the depth of the analysis (from cradle to grave). It is not uncommon to have thousands of processes making up the life cycle of a product. For each process in a product system, the environmentally relevant inputs and outputs have to be determined and quantified. This includes the processes' resource consumption and direct emissions as well as the materials and energy flows upstream of the process with their own emissions and resource consumption. It is obvious that this can lead to a product life-cycle description encompassing many thousand different material streams and emissions. For a thermodynamic assessment, these material streams and emissions have to be characterized regarding their thermodynamic parameters, further increasing the data requirements. Depending on the product system under investigation and the system boundaries chosen, the data requirements for a thermodynamic assessment change. However, the minimum requirements include the chemical composition of all material streams and the temperature of all heat flows. When the state of the materials crossing the system boundaries deviates markedly from the standard conditions, defined by ambient temperature and pressure, these parameters have to be known too. The latter parameters are, for example, required when the analysis focuses on subprocesses within the product system, where intermediate products and materials are exchanged with neighboring processes. In summary, although the number of data points needed to analyze the environmental implications of a product life cycle goes into the thousands, this number might easily double or grow even further for an added thermodynamic analysis. Fortunately, material specific data can in many cases be found in thermodynamic databases. However, these cost-intensive databases usually require thermodynamic expertise in order to obtain scientifically sound results.

The application of thermodynamic assessments in LCA is thus limited regarding two main aspects: (a) the already high data intensity grows markedly and (b) the expertise necessary for conducting such an assessment goes beyond that of a general LCA practitioner. The benefits are given by (a) an overall assessment of resource consumption, fully taking resource shifting into account, and (b) an in-depth look at the causes for resource consumption, delivering starting points for process optimization.

11.3 Entropy Production in LCA

11.3.1 Entropy Production for Steady-State Processes

As previously mentioned, the basis for an analysis of entropy production in a process is a detailed material and energy balance, including their thermodynamic parameters (cf. [34]). We can derive an expression for the entropy production by starting with Eq. (11.1) and making a few assumptions on the system boundaries and flows. For infinitesimal changes, Eq. (11.1) becomes

$$dS_{sys} = d_e S_{sys} + d_i S_{sys} \quad \text{or} \quad dS_{sys} - d_e S_{sys} = d_i S_{sys}. \tag{11.2}$$

The quantity we are interested in is $d_i S_{sys}$, because the internally produced entropy is our approximation for resource consumption. We can measure $d_e S_{sys}$ if we know all flows of energy and matter into or out of the system at hand, because the exchanged entropy is always the entropy "of something" that is exchanged with the environment, be it materials, heat, or radiation. The system entropy change dS_{sys} is harder to determine because it would require detailed knowledge of the system's material composition, its temperature gradients, the heat stored within, and so on. For a general system, this knowledge is practically unavailable. Nevertheless, many technical processes leave the actual system unchanged, i.e., $dS_{sys} = 0$. These are called steady-state processes, because the system itself does not change its internal state. Under this assumption, Eq. (11.2) reduces to

$$d_i S = \left(\underbrace{\sum_l \frac{e_{q,l}^{out}}{T_{q,l}} - \sum_l \frac{e_{q,l}^{in}}{T_{q,l}}}_{\text{Heat balance}} + \underbrace{\sum_j \frac{4}{3} \frac{e_{s,j}^{out}}{T_{s,j}^{out}} - \sum_j \frac{4}{3} \frac{e_{s,j}^{in}}{T_{s,j}^{in}}}_{\text{Radiation balance}} + \underbrace{\sum_k s_k m_k^{out} - \sum_k s_k m_k^{in}}_{\text{Material balance}} \right) dt,$$

(11.3)

where $d_i S$ is the infinitesimal entropy production within the investigated steady-state system during an infinitesimal time interval dt, e_q denotes (conductive) heat flow rates at temperature T_q, e_s denotes radiative energy flow rates with radiation temperature T_s, and m_k denotes mass flow rates (of material flow k) with specific entropy s_k. The specific entropies s_k are calculated from the specific (or molar) entropies of components of material flow k and the mass (molar) composition, including a mixing term. The corresponding formula is

$$s_k = \frac{1}{m_k} \left\{ \sum_j n_j^k y_j^k s_j' - R \sum_j n_j^k y_j^k \ln y_j^k \right\},$$

(11.4)

where R is the gas constant, s_j' is the molar entropy of the component j (of material flow k), y_j^k is the molar fraction of component j in material flow k, and n_k is the molar amount of component j. The components' specific (molar) entropies s_j' are either taken from standard reference sources [35–38] or calculated from

$$s_j'(p, T) = s_j'(p, T_0) + R \int_{T_0}^{T} \frac{c_p^j(T')}{T'} dT',$$

(11.5)

where $c_p(T)$ is the isobaric molar heat capacity of the component at temperature T. Equation (11.3) can be integrated over a time interval Δt to yield the entropy production $\Delta_i S$ associated with a certain amount of product produced in this time interval. This is especially useful for processes that do not leave the system in a steady state continually, but in which the system returns to a previous state after a certain function has been performed (as for example in batch processes). Because almost all processes relevant for LCA are either steady-state or batch processes, the assumptions leading to Eq. (11.3) are almost always valid.

11.3.2 Data Requirements for Calculating Entropy Production

LCA studies are usually developed around a more or less complete material and energy balance, the set of the m_k in Eq. (11.3). However, the completeness of the material balance strongly depends on the goal and scope definition of the LCA study. Flows that are exchanged with the environment (*elementary flows* in LCA language), but that have no environmental effect, are usually not considered in the material balance. On the input side this typically applies to air, whereas on the output side water vapor, cooling water, or heat radiation is often neglected in the balance. This can create problems for the entropy calculation when these flows are thermodynamically relevant for the process. A solution can usually be found by applying balance equations, like mass balance, energy balance, or elementary balance. With the help of these balances, the missing material flows and their thermodynamic parameters can be determined.

The material-properties data for the entropy-balance calculation can be found in thermodynamic databases[5] to a very high degree of accuracy. The LCA data itself, however, is usually not available at the level of detail needed for a rigorous calculation. This applies especially to the chemical composition, the temperature, and pressure of the material flows, and for the quantity and temperature level of heat flows and radiation flows. To remedy this situation, the material and energy flow model of the product system under investigation have to be improved. One can do this by measuring the missing parameters directly (if at all possible), by obtaining an expert opinion, or by finding average values in the technical literature describing the processes. The latter solution is in most cases the only approach available, because direct access to the processes is usually not possible for an LCA practitioner.

11.3.3 Implementation of Entropy Calculation in LCA

When the set of material flows is thus completed and specified to the necessary level of detail, the calculation of the entropy balance can commence by means of Eq. (11.3). We can perform the calculation in different ways, by using thermodynamic or thermochemistry software tools, spreadsheets, or LCA software. Thermodynamic software tools have the advantage that they come with a built-in interface to a thermodynamic database. The database itself might be included as well. This eases the actual calculation, because the thermodynamic properties of the components can easily be looked up and thus the thermodynamic flows can easily be calculated. Thermodynamic software can also perform some of the auxiliary calculations to complete the material and energy balance (like mass balancing, stoichiometric calculations, and heat balance). The challenge for using thermodynamic software in LCA, however, lies in the fact that the thermodynamic software is usually badly equipped to perform the environmental impact calculations necessary for an LCA. The impact factors of thousands of substances for many impacts would have to be added to the database. Additionally, some of the impacts in LCA are nonmaterial in nature, like noise, land-use change, and biodiversity loss. The storage and manipulation of these impacts are difficult to implement in a materially oriented thermodynamic database. There might be ways to accomplish this, but the effort would be enormous. As a result, under the usual time constraints, the life-cycle impact assessment can reasonably be performed only by LCA software. Thus, if thermodynamic software were to

---

[5] In principle, the entropy of a material is also available from direct measurement (by means of its heat capacity). However, for standard LCA applications, the efforts of this approach would be out of proportion in every respect.

be used for the thermodynamic calculations and LCA software for the environmental impact calculations, the same energy and material flow models would have to be constructed two times.

LCA software, on the other hand, is usually well equipped to handle all kinds of flows, including nonphysical flows, and all kinds of material parameters. In Umberto©, one of the leading LCA tools, generic "materials" (a placeholder term for exchanged quantities within the flow network) can have an arbitrary number of properties, including thermodynamic properties. The software does not know anything about these properties or how to calculate meaningful results from these, but they can be accessed by use of an interface to the built-in component objects model (COM). This interface is available through several scripting languages (Visual Basic Script, Java Script, Python). Scripts written in any of these languages can be used to perform the actual calculations, store data in the material database, retrieve data from external databases, add material properties, or perform any other data manipulation. What the scripting engine lacks, in comparison with thermodynamic software, is a simple way for checking the consistency of the calculated results. Although in thermodynamic software there are several ways of cross-checking the results (mass balance, energy balance, elementary balance), all of these checks have to be implemented individually in the scripted LCA software. This is cumbersome, but worth the effort, because it saves the researcher from even more cumbersome script debugging. Needless to say, in a simple spreadsheet calculation the situation would be worse.

Between thermodynamic software and LCA software there seems to be a gap that has to be closed in order to do any meaningful and efficient thermodynamic calculations. An ideal solution would be an exchange of data on flows and properties between the two expert tools. This seems possible, in principle. It requires, however, suitable interfaces on both ends of the exchange and access to a suitable thermodynamic software package.

After all the options are weighed, it seems appropriate to implement the thermodynamic calculations inside the LCA software by extending the materials database with thermodynamic properties from several thermodynamics databases and to perform the actual calculations by using scripts within the LCA software. An example of this approach is given in the next subsection. Currently this implementation serves as a proof of concept only, showing that thermodynamic calculations are feasible and readily implemented in LCA software. The next step would be to establish a working exchange of data between thermodynamic software and LCA software. This is left for future research.

11.3.4 Implementation of Entropy Calculation in LCA Software (Umberto)

As previously explained, the calculation of the entropy balance requires access to thermodynamic properties of (possibly many) chemical components. This access can be achieved by storing the thermodynamic data (temporally or persistently) in the material database of the LCA software. In Umberto, material properties can be used for that. Material properties can take any form, from physical properties (like molar mass) to free-style text descriptions (like hazardous-waste specifications). The actual calculation of the entropy production is then carried out inside the LCA software from user-defined scripts that access these material properties. For storing the composition of mixtures, there are two options: as another material property in the materials database or by defining respective "mixing arrays" (basically a lookup table) in the calculation script. Each approach comes with specific trade-offs. Here the latter approach is used for higher flexibility when material compositions are changed. Each material flow and energy flow in the LCA model also has specific temperatures, which are defined in a "temperature array" within the script. For calculating the entropy of mixtures, the script has to look up the properties of the individual components and calculate the entropy (including a mixing term) according to Eqs. (11.3) and (11.4). Because the equations require molar fractions for calculating, e.g., the mixing entropy, and the flow data in LCA models is usually given in mass terms, a conversion has to be added to the script. This requires the molar mass of components to be added as a material property too. After all material (and energy) flow entropies have been calculated, the overall balance and thus the entropy production can be computed.

For an example of how the preceding process can be implemented, let us look at a model (in Umberto) of a specific process network for copper production from concentrated ore. The main inputs to this process are copper ore concentrate, oxygen-enriched air, flux material (mainly sand), and energy carriers (fuel oil, natural gas, electrical energy). The main outputs are copper cathodes (the actual product), slag, off-gas, and waste heat. In addition to the metallurgical processes, the upstream generation of electricity has been added to the thermodynamic analysis in Umberto. Other upstream processes (as part of the LCA) are analyzed regarding their entropy production (see [39]), but not by the Umberto model. The mass flows of the processes analyzed here are shown in Fig. 11.2.

The compositions of the material flow mixtures (concentrate, air, fuel oil, natural gas) are taken from literature sources (cf. [34]) and then defined as an array of mass ratios in a Python script. The mixtures and pure components of the flow model must further be specified regarding their temperature and phase state. This is also done for the whole model in the script, which calculates the entropy balance; see Fig. 11.3.

The mixture and the components are stored as materials in the Umberto materials database. In addition, mixtures get the property IsMixture and components get their thermodynamic properties added. The molar heat capacity and thus the specific entropy of the components are temperature dependent. This requires the thermodynamic properties in the material database also to be given for the widest possible temperature range. Whenever available, this has been implemented by use of the Shomate equation, as used by the *NIST Chemistry WebBook* [36]. This equation parameterizes the heat capacity (and by integration the entropy and enthalpy) for a wide range of temperatures:

$$C_p^0 = A + Bt + Ct^2 + Dt^3 + E/t^2, \text{ with } t = T/1000, \qquad (11.5)$$

$$H^0 - H_{298.15}^0 = At + Bt^2/2 + Ct^3/3 + Dt^4/4 - E/t + F - H, \qquad (11.6)$$

$$S^0 = A \ln(t) + Bt + Ct^2/2 + Dt^3/3 - E/2t^2 + G. \qquad (11.7)$$

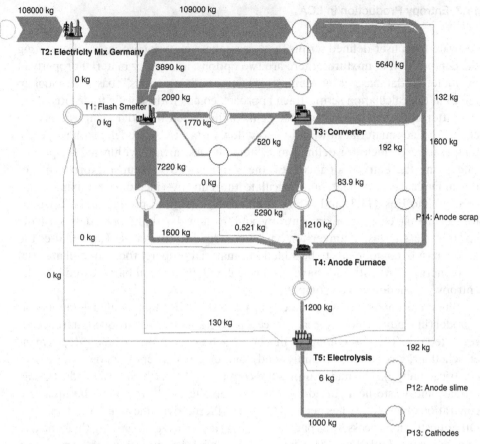

108000 kg

109000 kg

T2: Electricity Mix Germany

3890 kg

5640 kg

0 kg

2090 kg

132 kg

T1: Flash Smelter

0 kg

1770 kg

520 kg

T3: Converter

192 kg

1600 kg

7220 kg

0 kg

83.9 kg

5290 kg

0.521 kg

1210 kg

P14: Anode scrap

0 kg

1600 kg

0 kg

T4: Anode Furnace

1200 kg

130 kg

192 kg

T5: Electrolysis

6 kg

P12: Anode slime

1000 kg

P13: Cathodes

Figure 11.2. Mass flows (aggregated sums) of the analyzed process network (copper production from ore concentrate), including one upstream process: electricity production. Other upstream processes (e.g., ore mining and beneficiation) are not shown.

```
Script Editor - Entropie-Berechnung (Python)

Script | Options | Log |

112     # Extensions müssen mit angegeben werden, sofern sie existieren: "Material,Ext". Ansonsten nicht!
113
114     MixArray[1] = {"Concentrate":{"CuFeS2":0.8442, "SiO2":0.074, "C":0.023, "Al2O3":0.0224, "ZnS":0.0212,
115                   "PbS":0.0035, "Water,L":0.003, "As2S3":0.003, "Ni":0.0005, "CaO":0.0052},
116                   "Flux":{"SiO2":0.95, "Al2O3":0.03, "Fe3O4-Low":0.02},
117                   "Air":{"N2":0.7542, "O2":0.232, "Ar":0.0138}}
118
119     MixArray[2] = {"Matte,L":{"FeS":0.2045, "Cu2S":0.746, "CuO":0.0185, "FeO":0.0121, "ZnO":0.0125,
120                   "PbO,L":0.0054, "Ni":0.001},
121                   "O-Slag,L":{"Cu2S":0.0251, "SiO2":0.094, "Fe2SiO4":0.7132, "As2S3":0.0044, "Fe3O4-High":0.1
122                   "Al2O3":0.0408, "CaO":0.0102, "ZnO":0.0306}}
123
124     MixArray[3] = {"Offgas":{"N2":0.5047, "SO2":0.4023, "CO2":0.0577, "Water,G":0.0266, "Ar":0.0087}}
125
126     # erzeuge Temperatur-Array für Pfeile mit Key = Arrow.ID
127     # und der Temperatur für jedes Material (nicht nur Mischungen).
128     # Materialien ohne T sollen später Tref zugewiesen bekommen
129     # Element mit Key=0 bleibt leer (enthält Eintrag (0: 0))
130     # Braucht man das?
131
132     T = dict([(k,0) for k in range(ArrowCount+1)])
133
134     T[1] = {}
135     T[2] = {"Matte,L":1460, "O-Slag,L":1500 }
136     T[3] = {"Offgas":1570}
137
138
139 def Main():
140     global Prj, Sce, Net, ArrowCount, Period
141     Tref = 298.15 # K
142     R = 8.314472 # J/(K mol)
143

97: 1          Insert    Python
```

Figure 11.3. Excerpt from the Python script for calculating the entropy balance for the flash smelter example. The array MixArray stores the composition of each flow, and the array T holds the temperatures.

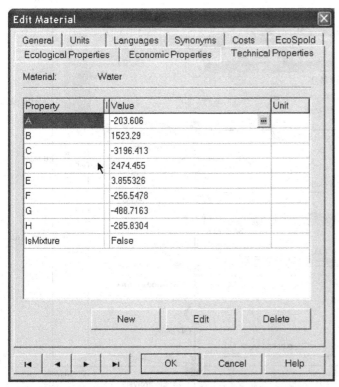

Figure 11.4. Material properties for water. The properties A–H are parameters for the Sho-mate equation, a parameterized form of the (temperature-dependent) molar heat capacity. The value "False" for "IsMixture" indicates that water is treated as a pure component.

The Shomate parameters A–H are stored as properties in the material database; see Fig. 11.4.

   When the entropy-balancing script is executed, it checks for all relevant material properties (compositions, thermodynamic data, temperature) and, when missing, either stops executing or sets appropriate defaults (in this case, warnings are issued to allow debugging). The calculated entropy values for each flow are written back to the model, appearing as additional material flows (a "material" in Umberto can virtually be anything from energy to computer parts). In this way, entropy flows can be visualized with the internal graphics tools. It is then easy to recognize the processes in the network with the largest contribution to entropy production (i.e., resource consumption). This enables a quick judgment on the process steps for which resource efficiency might be improved. For a Sankey-style visualization of the entropy flows, see Fig. 11.5.

   The flows have to complemented by an overview of the entropy production (i.e., the entropy balance in each node) in order to draw conclusions regarding the respective resource consumption. The respective numbers are given in Table 11.1.

11.3.5 Interpretation of Entropy Production

It is interesting to analyze where the entropy production comes from. Intuitively, one would assume the conversion of final energy to play an important part (i.e., burning

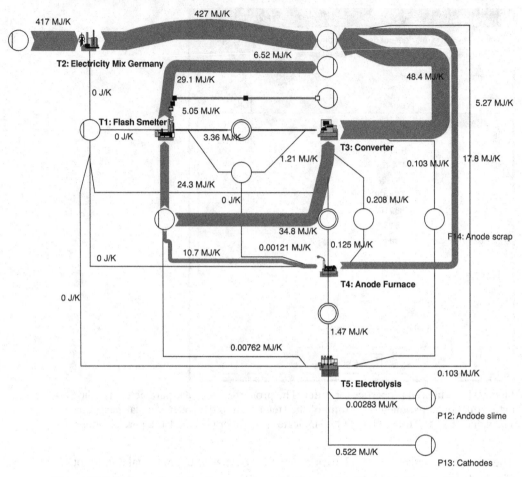

Figure 11.5. Entropy flows for the copper production model. There is a limit on the maximum width of the arrows, so that the two largest entropy flows (input and output of the electricity production) are not scaled correctly.

Table 11.1. *Entropy production of the main processes for producing 1 metric ton of primary copper from sulfidic copper ores*

Process	Entropy production (MJ/k)
Mining and beneficiation	37
Metallurgy	
Flash smelter	19
Converter	21
Anode furnace	8
Electrolysis	4
Electricity production	10
Sum metallurgy	62
Sum total	99

Note: The entropy production of the mining and beneficiation processes does not include any upstream processes (e.g., diesel production). For the metallurgical part, the main upstream process (electricity production) is included.

of fossil fuels and consuming electrical energy). One can calculate the contribution from this conversion when assuming a complete oxidation of the fossil fuels and when assuming that the electrical energy is fully transformed into heat at ambient temperature. For the mining and beneficiation stages, final energy conversion contributes approximately 36.1 MJ/K, which is about 97.7% of the entropy production of this stage. For the metallurgical part, the contribution from final energy is 38 MJ/K, or 61%, respectively. Thus, in the latter case, around 40% of the entropy production and thus the physical resource consumption stem from material transformations alone (i.e., mixing, structural changes, phase transitions, chemical reactions, etc.). It is therefore evident that energy or material accounting alone does not suffice to reveal the complete picture of resource consumption.

Another noteworthy fact is that some of the material flows are not really consumed, for example, the overburden. Notwithstanding the environmental impacts of relocating large amounts of soil and rock, this should really not be considered under the terms of physical resource consumption. Therefore using second law concepts (as entropy production) for measuring resource consumption helps to clarify the concept of consumption in general by giving a reference as to what constitutes consumption (in physical terms) and what does not. If entropy production (or, equivalently, exergy destruction) is not included in the assessment of consumption, the thermodynamic core of the assessment is missing and consumption as a concept is ill-defined. Concepts based solely on mass flows would indeed consider the moved overburden as being consumed. This interpretation of consumption is thus not in agreement with thermodynamics.

As an example for the necessity of thermodynamic analysis of material and energy flows, let us look at the entropy-production contribution from the mixing of off-gas streams in the metallurgical part of copper making. The major SO_2 emissions come from the flash smelter and the converter process. Although the flash smelter off-gas is mainly a direct reaction product of the reaction, the converter process leaks in quite a large amount of extra ambient air. This results in a large entropy production that is due to the mixing of SO_2-rich reaction off-gas with ambient air. As a secondary effect, the heat contained in the reaction off-gas is now "diluted" to a larger volume of final off-gas, thus losing in "quality." Both effects can be measured in terms of additional entropy production, resulting in an extra 6 MJ/K per ton of refined copper cathodes. This reveals some substantial inefficiencies of the converter process (which are in part due to the batch-wise operation). It should be noted though that the dilution of the converter off-gas also serves a technical purpose: It is, in part, required for the subsequent processes (off-gas dedusting, sulphuric acid plant). Nevertheless, this should not cloud the fact that there are inefficiencies that might be avoided by a different technology. These inefficiencies are usually not discovered through conventional energy and material flow analyses; they become visible only when the second law of thermodynamics, i.e., entropy, is taken into account.

What do we learn from looking at the contributions of the different processes to the overall resource consumption? First, we see that resource consumption has many different causes, and thus the conservation of resources must be approached from many different angles. Although the conversion of final energy carriers dominates resource consumption in the mining and beneficiation stage, the metallurgical stage has a broader variety of causes for resource consumption. Regarding the optimization of resource consumption, this would imply a focus on energy carriers for the mining stage, e.g., increasing the efficiency of the ore crushers in the beneficiation process. For the metallurgical stage, mechanisms other than the conversion of energy carriers play an important part for resource consumption. Important contributions come from mixing off-gas streams with ambient air (especially in the converter process), transferring stored heat to the surroundings (e.g., open-air cooling of slag and copper anodes), and expulsion of waste heat to the surroundings (by radiation and convection, especially in the flash smelting process). Some of the resource consumption can be avoided by the implementation of advanced technology (e.g., heat recuperation with steam production for the flash smelter, decreasing ambient air influx by more efficient suction hoods). Nevertheless, some of the main causes for resource consumption are predetermined by the pyrometallurgical route. Once the chemically stored energy in the concentrate has been released (in the form of heat), an unavoidable amount of entropy is produced, and the necessity to dispose of the heat will inevitably produce more entropy (because it will finally be transferred to the surroundings at ambient temperature). A process that circumvents this high-temperature route is the solvent extraction – electrowinning (SX/EW) process to produce refined copper, in which copper is leached from the ore by use of H_2SO_4 and then concentrated with the help of an extractant. In the final step, pure copper is won from the concentrated copper solution in an electrolytic process (electrowinning). Although the SX/EW process will probably have less direct entropy production (less final energy needed for mining and beneficiation, no conversion of chemical energy to heat), it does need a fairly large amount of electrical energy in the electrowinning process, which will offset this advantage to some degree. Nevertheless, total final energy demand is lower (see [40]). On the material side, the SX/EW process needs large amounts of water, which will be partly consumed (as explained for the beneficiation stage) and will add significantly to the entropy production. It is thus not a priori clear whether the SX/EW process can produce copper with less resource consumption than the pyrometallurgical route; a detailed entropy analysis is necessary.

11.3.6 Entropy-Based Efficiency of Processes

Using the results from an entropy analysis, one can ask for further metrics describing the processes. From an engineering viewpoint, the thermodynamic efficiency of the processes seems worth further analysis.

With entropy production used as a measure for resource consumption, the preceding results can be interpreted as a measure of the resource efficiency:

resource efficiency = (entropy production)/(output of functional unit).

If the functional unit consists of only one material stream and *output of functional unit* is measured in mass units, we can use the preceding definition to compare different systems for producing the same amount of material (e.g., 1 ton of copper). The entropy analysis from the previous section yields a resource efficiency for copper production of 99 MJ/K per ton of copper. If the functional unit is more complex, however, the definition per mass is of little value. Most metallurgical processes, especially recycling processes, produce a whole variety of materials from a complex input stream. Even the copper production process from the preceding discussion really produces a mixture of materials: copper, gold, silver, zinc, molybdenum, sulfuric acid, construction material (slag), and others. The functions these processes perform can thus not simply be reduced to one single mass indicator, and the definition of resource efficiency must be extended to allow for more than one functional unit.[6] Luckily, many metallurgical processes produce pure material streams as their output so that their function can be described as an increase in concentration of the produced metals, or a decrease in mixing. The decrease in mixing can be evaluated on a per-metal basis and expressed by means of the partial entropy of mixing (see [41, 42, 43]). The overall functional unit is then the decrease in mixing entropy aggregated over all metals.

REFERENCES

[1] Merriam-Webster, *Definition of "consumption."* Available at http://www .merriam-webster.com/dictionary/consumption (accessed Dec. 7, 2008, from Merriam-Webster Online Dictionary).

[2] D. Kondepudi and I. Prigogine, *Modern Thermodynamics: From Heat Engines to Dissipative Structures* (Wiley, Chichester, UK, 1998).

[3] N. Georgescu-Roegen, *The Entropy Law and the Economic Process* (Harvard University Press, Cambridge, MA, 1971).

[4] A. Bejan, *Entropy Generation Minimization: The Method of Thermodynamic Optimization of Finite-Size Systems and Finite-Time Processes*, Vol. 2 of Advanced Topics in Mechanical Engineering Series (CRC Press, Boca Raton, FL, 1996).

[5] M. Ishida, *Selected Papers From the Proceedings of Efficiency, Costs, Optimization, Simulation and Environmental Aspects of Energy Systems (ECOS '99)*, Energy Conversion and Management, Special issue, 43.2002, 9/12 (Pergamon, Oxford, 2002).

[6] J. Szargut, A. Ziebik, and W. Stanek, "Depletion of the non-renewable natural exergy resources as a measure of the ecological cost," *Energy Convers. Manage.* **43**, 1149–1164 (2002).

[7] R. U. Ayres, L. W. Ayres, and K. Martinás, "Eco-thermodynamics: Exergy and life cycle analysis," INSEAD R&D Working Papers, 96/04/EPS (INSEAD, Fontainebleau, France, 1996).

[8] R. U. Ayres, L. W. Ayres, and A. Masini, "An application of exergy accounting to five basic metal industries," in *Eco-Efficiency in Industry and Science: Sustainable Metals Management. Securing Our Future – Steps Towards a Closed Loop Economy*, edited by A. von Gleich, R. U. Ayres, and S. Gößling-Reisemann (Springer, Dordrecht, The Netherlands, 2006), pp. 141–194.

[9] J. Szargut, D. R. Morris, and F. R. Steward, *Exergy Analysis of Thermal, Chemical, and Metallurgical Processes* (Hemisphere, New York, 1988).

[10] R. U. Ayres, "Eco-thermodynamics: Economics and the second law," *Ecol. Econ.* **26**, 189–210 (1998).

[11] R. U. Ayres and K. Martinás, "Waste potential entropy: The ultimate ecotoxic?," INSEAD R&D Working Papers, 94/36/EPS (EPS, Fontainebleau, 1994).

---

[6] In LCAs this problem would be solved by allocation rules, attributing the resource consumption to each functional unit separately based on allocation factors (mass, energy, economic value, etc.). This section, however, is about process efficiency, so all functions the process delivers should be taken into account simultaneously.

[12] L. Connelly and C. P. Koshland, "Exergy and industrial ecology – Part 1: An exergy-based definition of consumption and a thermodynamic interpretation of ecosystem evolution," *Exergy* 1, 146–165 (2001).

[13] L. Connelly and C. P. Koshland, "Exergy and industrial ecology – Part 2: A non-dimensional analysis of means to reduce resource depletion," *Exergy* 1, 234–255 (2001).

[14] J. Szargut, "Exergy in the thermal systems analysis," in *High Technology,* Vol. 69 of Thermodynamic optimization of complex energy systems (Proceedings of the NATO Advanced Study Institute on Thermodynamics and the Optimization of Complex Energy Systems, Neptun, Romania, July 13–24, 1998), edited by A. Bejan and E. Mamut (Kluwer Academic, Boston, 1999).

[15] R. L. Cornelissen, "Thermodynamics and sustainable development – The use of exergy analysis and the reduction of irreversibility," Ph.D dissertation (University of Twente, Twente, The Netherlands, 1997).

[16] R. L. Cornelissen and G. Hirs Gerard, "The value of exergetic life cycle assessment besides LCA," *Energy Convers. Manage.* 43, 1417–1424 (2002).

[17] J. Szargut and D. R. Morris, "Cumulative exergy consumption and cumulative degree of perfection of chemical processes," *Intl. J. Energy Resources* 11, 245–261 (1987); retrieved Dec. 8, 2008, from http://doi.wiley.com/10.1002/er.4440110207.

[18] J. Dewulf, H. van Langenhove, and J. Dirckx, "Exergy analysis in the assessment of the sustainability of waste gas treatment systems," *Sci. Total Environ.* 273, 41–52 (2001).

[19] J. Dewulf and H. van Langenhove, "Quantitative assessment of solid waste treatment systems in the industrial ecology perspective by exergy analysis," *Environ. Sci. Technol.* 36, 1130–1135 (2002).

[20] J. Dewulf and H. van Langenhove, "Assessment of the sustainability of technology by means of a thermodynamically based life cycle analysis," *Environ. Sci. Pollut. Res. Intl.* 9, 267–273 (2002).

[21] I. Dincer and Y. A. Cengel, "Energy, entropy and exergy concepts and their roles in thermal engineering," *Entropy* 3, 116–149 (2001).

[22] B. Andresen, "Finite-time thermodynamics," Ph.D. dissertation (University of Copenhagen, Copenhagen, Denmark, 1984).

[23] S. Gößling-Reisemann and A. von Gleich, "Ressourcen, Kreislaufwirtschaft und Entropie am Beispiel der Metalle," in G. Hösel, B. Bilitewski, W. Schenkel, and H. Schnurer (Eds.): Müllhandbuch (Berlin: Erich Schmidt Verlag, Lfg. 6/08, 2008), pp. 1–27.

[24] A. von Gleich, "Outlines of a sustainable metals industry," in *Eco-Efficiency in Industry and Science: Sustainable Metals Management. Securing Our Future – Steps Towards a Closed Loop Economy*, edited by A. von Gleich, R. U. Ayres, and S. Gößling-Reisemann (Springer, Dordrecht, The Netherlands, 2006), pp. 3–39.

[25] J. B. Guinée and E. Lindeijer, *Handbook on Life Cycle Assessment: Operational Guide to the ISO Standards*, Vol. 7 of Eco-Efficiency in Industry and Science series (Kluwer, Dordrecht, The Netherlands, 2002).

[26] R. Müller-Wenk, "Depletion of abiotic resources weighted on the base of 'virtual' impacts of lower grade deposits used in future, *IWÖ-Diskussionspapier,* 57 (Institut für Wirtschaft und Ökologie, Universität St. Gallen, Switzerland, 1998).

[27] M. Goedkoop and R. Spriensma, "The Eco-Indicator 99 – A damage oriented method for life cycle assesment," *Method. Rep.*, Amersfoot, The Netherlands (2001) (available at http:\\www.pre.nl).

[28] M. Stewart and B. P. Weidema, "A consistent framework for assessing the impacts from resource use – A focus on resource functionality," *Intl. J. Life Cycle Assess.* 10, 240–247 (2005).

[29] G. Finnveden and P. Östlund, "Exergies of natural resources in life-cycle assessment and other applications," *Energy Oxford* **22**, 923–932 (1997).

[30] M. Ritthoff, H. Rohn, and C. Liedtke, *Calculating MIPS: Resource productivity of products and services. Wuppertal spezial, 27.* (Wuppertal Institute for Climate, Environment and Energy, Wuppertal, Germany, 2002).

[31] E. Matthews, *The Weight of Nations: Material Outflows From Industrial Economies* (World Resources Institute, Washington, D.C., 2000).

[32] R. U. Ayres and U. E. Simonis, *Industrial Metabolism: Restructuring for Sustainable Development* (United Nations University Press, Tokyo, New York, 1994).

[33] M. Wackernagel and W. E. Rees, *Our Ecological Footprint: Reducing Human Impact on the Earth*, Vol. 9 of The New Catalyst Bioregional Series (New Society Publishers, Gabriola Island, British Columbia, Canada, 1998).

[34] S. Gößling, "Entropy production as a measure for resource use: Method development and application to metallurgical processes," Ph.D. dissertation (University of Hamburg, Hamburg, Germany, 2001), available http://www.sub.uni-hamburg.de/opus/volltexte/2004/1182/.

[35] D. R. Lide, *CRC Handbook of Chemistry and Physics: A Ready-Reference Book of Chemical and Physical Data*, 77th ed. (CRC Press, Boca Raton, FL, 1996).

[36] P. Linstrom and W. Mallard, editors, *NIST Chemistry WebBook* (*http://webbook.nist.gov*): *NIST Standard Reference Database Number 69, June 2005.* (National Institute of Standards and Technology, Gaithersburg, MD, 2005).

[37] Centre for Research in Computational Thermochemistry, "F*A*C*T pure substances inorganic database – REACTION-Web" (2008), available from École Polytechnique de Montreál, http://www.crct.polymtl.ca/fact/.

[38] I. Barin, *Thermochemical Data of Pure Substances* (VCH, Weinheim, 1989).

[39] S. Gößling-Reisemann, "What is resource consumption and how can it be measured?: Application of entropy analysis to copper production," *J. Ind. Ecol.* **12**, 570–582 (2008).

[40] P. Maldonado, S. Alvarado, and I. Jaques, "The Chilean copper industry and energy-related greenhouse gases emissions," in *Energy and Technology: Sustaining World Development Into the Next Millennium; 17th Congress of the World Energy Council* (World Energy Council, London, 1998).

[41] S. Gößling-Reisemann, "Thermodynamic costs and benefits of recycling," presented at the Eco-X 2007 Conference, Vienna, May 9–11, 2007.

[42] S. Gößling-Reisemann, "Entropy analysis of metal production and recycling," *Manage. Environ. Qual.* **19**, 487 (2008).

[43] S. Gößling-Reisemann, A. von Gleich, V. Knobloch, B. Cebulla, "Bewertungsmaßstäbe für metallische Stoffströme: von Kritikalität bis Entropie," in: F. Beckenbach, A.I. Urban (Eds.): *Methoden der Stoffstromanalyse – Konzepte, agentenbasierte Modellierung und Ökobilanz* (Metropolis Verlag, Marburg, Germany, 2011).

12 Exergy and Material Flow in Industrial and Ecological Systems

Nandan U. Ukidwe and Bhavik R. Bakshi

12.1 Introduction

Ecological resources constitute the basic support system for all activity on Earth. These resources include products such as air, water, minerals, and crude oil, and services such as carbon sequestration and pollution dissipation [1–4]. However, traditional methods in engineering and economics often fail to account for the contribution of ecosystems despite their obvious importance. The focus of these methods tends to be on short-term economic goals, whereas long-term sustainability issues get shortchanged. Such ignorance of ecosystems is widely believed to be one of the root causes behind a significant and alarming deterioration of global ecological resources [5–8].

Several methods have been developed to address the shortcomings of existing methods and to make them ecologically more conscious [9]. Among these, the preference-based methods use *human valuation* to account for ecosystem resources [3, 10, 11]. These methods use either a single monetary unit to readily compare economic and ecological contributions or multicriteria decision making to address trade-offs between indicators in completely different units. However, preference-based methods do not necessitate compliance with basic biophysical laws that all systems must satisfy and require knowledge about the role of ecological products and services that is often inadequate or unavailable.

Biophysical methods comply with the basic scientific laws such as the conservation of mass and energy and the universal degradation of energy quality (second law). For instance, *material flow analysis* (MFA) determines the material basis of national economies [12–14]. These analyses consider natural resources such as metals, water, and organic soil that can be readily quantified in material units. However, MFA does not consider quality differences between different materials and proposes to add, say, 1 ton of coal to 1 ton of water, two vastly different substances. MFA also completely fails to account for natural resources that cannot be expressed in material units. This includes energetic resources such as sunlight, wind, and deep earth heat. Last, MFA ignores the contribution of ecosystems in making natural resources available for human consumption and consequently is of limited use in environmentally conscious decision making. Methods such as *net energy analysis* (NEA) and full-fuel-cycle Analysis (FFCA) are improvements over MFA studies, as they consider material and energy interactions [15, 16]. However, such methods also fail to consider the ecosystem contribution in making natural resources available as well as quality differences between different energy streams. As a result, energy-based methods consider 1 J of sunlight equivalent to 1 J of electricity when the two are vastly different in their available energy or exergy content. Furthermore, although mass and energy are conserved according to the first law of thermodynamics, exergy is lost in all thermodynamically irreversible activities according to the second law. Therefore exergy, not mass or energy, represents the truly limiting resource on Earth that is of greater relevance to any discussion on sustainability.

Exergy analysis has been traditionally used for identifying sources of inefficiency in manufacturing processes and equipment [17]. Many extensions to larger scales have also been developed. *Cumulative exergy consumption* (CEC) considers the exergy consumed in industrial processes and their supply chains. *Exergetic life-cycle assessment* (LCA) [18] is conceptually similar to CEC analysis, but also considers exergy consumed in the demand chain. Both approaches, however, focus on only a few, most important processes in the supply and demand chains. Because a life cycle is usually a large and complicated network of processes, including only the most important ones can entail a large truncation error [19]. Thermodynamic analysis of nations has also been popular because of its potential for providing useful insight into the efficiency of different countries, constituent economic sectors, and specific technologies [20, 21]. Many of these efforts have resulted in exergy efficiencies of specific economic sectors based on accounting for the direct and indirect CECs and the exergy of products. However, these efforts focus on a small number of economic sectors, such as energy, transportation, waste, and manufacturing, and lack the extent of detail commonly available in *economic input–output* (EIO) models. These methods also fail to consider aspects such as emissions and their impact and the contribution of human labor and capital that are significant to industrial operations. Recent exceptions are the use of Sciubba's *extended exergy accounting* (EEA) to account for the contribution of labor to the Norwegian economy [22] and exploration of the relationship between the exergy of emissions and their impact [23, 24], and calculating the exergy of abatement for different pollutants [25]. Although exergy-based methods show a marked improvement over MFA or NEA, they still fail to account for the ecosystem contribution to natural resources. Such a contribution comes in the form of ecological functions necessary to produce, transport, concentrate, and transform various natural resources so that they can be consumed by the economic activity. For instance, freshwater is made available by the global hydrological cycle and top organic soil crucial for agriculture is made available by a complex sedimentary cycle. These ecosystem functions are vital to the sustenance of all economic activity but are largely ignored in the aforementioned material, energyor exergy-based approaches.

Emergy analysis [4] stands out as being the most comprehensive biophysical approach for quantifying the contribution of ecosystems to economic activity. It does consider the contribution of global biogeochemical cycles to natural resources. Unfortunately, emergy analysis is often misunderstood and faces quantitative and algebraic challenges, and its broad claims about ecological and economic systems have been quite controversial [26–28]. Besides, emergy analysis is usually applied to the most important processes, usually in a short supply chain, with the contribution of inputs from the economy captured by means of coarse economy-wide emergy-to-money ratios. Recently, the theoretical underpinnings of emergy analysis were bolstered by a demonstration of their similarity to CEC analysis by Hau and Bakshi. The resultant methodology is called *ecological cumulative exergy consumption* (ECEC) analysis. ECEC analysis extends *industrial CEC* (ICEC) analysis to include exergy losses in the industrial as well as ecological stages of a production chain. It is equivalent to emergy under conditions of an identical system boundary, allocation rule, and approach for combining global exergy inputs. However, the controversial aspects of emergy, such as the use of prehistorical emergy, *maximum empower principle*, and substitution for *economic valuation*, are not relevant to ECEC because it relies on only the thermodynamics of direct inputs of ecosystems to economic sectors and only on a current flow of energy. The partitioning or allocation of emergy between multiple outputs is another controversial aspect of emergy. ECEC takes a view of allocation that is different from that of emergy analysis and, like LCA, treats it as a subjective decision. However, this view shows that the approach used in emergy analysis for allocating renewable and nonrenewable resources does make sense and can be easily justified. Further theoretical details and illustrative examples of ECEC analysis are beyond the scope of this chapter, but can be found in the introductory chapter of this book.

Thermodynamic input–output analysis (TIOA) builds on ECEC analysis by providing it with a formal algorithm to evaluate flows in linear static networks. TIOA combines existing approaches from life-cycle analysis, exergy engineering, emergy analysis, and EIO analysis to formulate such an algorithm. TIOA has many unique features that distinguish it from other contemporary thermodynamic methods and their application to nations:

- TIOA combines exergy analysis of industrial systems with the ability of emergy analysis and systems ecology to account for ecosystems and input–output analysis to consider direct and indirect effects in networks.
- TIOA acknowledges the economic network and provides industry-specific results. Such results are more accurate in appreciating the heterogeneous nature of economy than a single aggregate metric for the entire economy [4, 29].
- TIOA can accommodate a wide variety of ecological products and services, human resources, and the impact of emissions on human and ecosystem health, making it a holistic approach. This approach is used for ICEC and ECEC analyses in this chapter, but can be readily modified to other methods such as NEA and extended exergy analysis.

In the past, TIOA was applied to study contribution of ecological resources to a 91-sector 1992 U.S. economy [30]. Such an analysis, though better than a completely aggregate analysis of the entire economy, can still be improved by use of more detailed models of the U.S. economic system. For instance, whereas the 91-sector 1992 model aggregates all agricultural activity into a single sector, namely the sector of other agricultural products (SIC 2: SIC stands for Standard Industrial Classification), the 1997 benchmark model separates agricultural activity into 10 subsectors (NAICS 1111A0–1119B0: NAICS stands for the North American Industrial Classification System). Naturally the 1997 benchmark model is more detailed than the

91-sector 1992 model and likely to provide more accurate results. Recent benchmark models representing today's economy are being compiled by the Bureau of Economic Analysis.

This chapter applies TIOA to study exergy flows in the 1997 U.S. economy. The analysis is performed with and without accounting for the contribution of ecosystems. When the ecosystem contribution is excluded, the results are analogous to the ICEC in the economy at the scale of individual industry sectors. When the ecosystem contribution is included, the results represent the ECEC at the scale of individual industry sectors. The total ECEC requirement captures the thermodynamic basis of industrial operations and is analogous to the concept of ecological cost or environmental footprint expressed in exergy units. The analysis also calculates ECEC–money ratio to juxtapose the thermodynamic basis of an industrial operation with corresponding monetary activity and captures the discrepancy between thermodynamic work and the willingness of people to pay for a good or service. Such a discrepancy is believed to be the root cause behind the lack of internalization of ecological resources into classical economics [31, 32]. Industry-specific ECEC–money ratios quantify the magnitude of such discrepancy and are potentially useful for macroeconomic policy decisions such as determination of proecological taxes.

The rest of this chapter is organized as follows. Section 12.2 introduces key ideas in EIO analysis and TIOA. Section 12.3 discusses data requirements and sources for applying TIOA to the 488-sector 1997 benchmark model of the U.S. economy. Section 12.4 presents exergy flows in the economic network arising on account of specific natural and human resource inputs and impact of emissions on human health. Section 12.5 discusses aggregate metrics including total ICEC and ECEC requirements and ICEC–money and ECEC–money ratios for individual industry sectors. These aggregate metrics are obtained by combining the results presented in Section 12.4. Section 12.5 discusses the implication of ECEC–money ratios to address macroscale sustainability issues such as international trade and corporate restructuring. Section 12.5 also discusses performance metrics such as yield ratio, environmental loading ratio, and a sustainability index based on resource consumption. Section 12.6 discusses application of ICEC–money and ECEC–money ratios in hybrid thermodynamic LCA (thermoLCA) and illustrates the unique insights provided by thermoLCA by comparing six alternative electricity-generation systems.

12.2 Background

12.2.1 Economic Input–Output Analysis

Input–output analysis is a general equilibrium model that describes interactions between different units of a network [33, 34]. It can be applied to any network as long as interactions between network units are linear. Figure 12.1 shows one such network containing three units, S_1, S_2, and S_3. This network may represent process equipment such as a reactor, a distillation column, and a heat exchanger in chemical process design; photosynthetic plants, herbivores, and carnivores in ecosystems; or three industry sectors in an economy. The network shown in Fig. 12.1 can also be conveniently represented in the form of a transaction table, shown in Table 12.1. Here, X_{ij} represents the magnitude of the transaction from unit i to unit j. *Final demand* (FD) F_i represents output from the ith unit that does not go to any other

Table 12.1. *Transaction table for the system shown in Fig. 12.1*

Output–input	S_1	S_2	S_3	Intermediate output (IO)	Final demand (FD)	Total output (TO)
S_1	X_{11}	X_{12}	X_{13}	O_1	F_1	X_1
S_2	X_{21}	X_{22}	X_{23}	O_2	F_2	X_2
S_3	X_{31}	X_{32}	X_{33}	O_3	F_3	X_3
Intermediate input (II)	I_1	I_2	I_3			
Value added (VA)	V_1	V_2	V_3			
Total input (TI)	X_1	X_2	X_3			

unit of the network. Similarly, *value added* (VA) V_i represents the input to unit i that does not come from any other unit of the network. When applied to an economic network, final demand represents the sale of goods and services to consumers, government establishments, etc., whereas value added represents employee compensation, indirect business taxes, and property-type income [34]. Input–output analysis establishes a balance on each network unit. This is represented by

$$\left(\sum_{i=1}^{n} X_{ik} \right) + V_k = I_k + V_k = \left(\sum_{i=1}^{n} X_{ki} \right) + F_k = O_k + F_k = X_k \quad \text{for } k = 1, \ldots, n.$$

$$(12.1)$$

If transaction coefficients are defined as $D_{ik} = X_{ik}/X_k$ and $D_{vk} = V_k/X_k$, where $X_k = \sum_{i=1}^{n} X_{ik} + V_k$, Eq. (12.1) can be rewritten as

$$\left(\sum_{i=1}^{n} D_{ik} \right) X_k + D_{vk} X_k = \left(\sum_{i=1}^{n} D_{ki} X_i \right) + F_k \quad \text{for } k = 1, \ldots, n. \quad (12.2)$$

Because

$$\left(\sum_{i=1}^{n} D_{ik} \right) + D_{vk} = 1 \quad \text{for } k = 1, \ldots, n, \quad (12.3)$$

Eq. (12.2) can be written as

$$X_k = \left(\sum_{i=1}^{n} D_{ki} X_i \right) + F_k \quad \text{for } k = 1, \ldots, n, \quad (12.4)$$

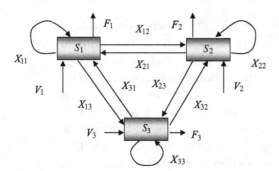

Figure 12.1. Network structure of a typical input–output system.

or, alternatively in matrix form as,

$$\mathbf{x} = \mathbf{D}\mathbf{x} + \mathbf{f}, \tag{12.5}$$

$$\mathbf{x} = (\mathbf{I} - \mathbf{D})^{-1}\mathbf{f}, \tag{12.6}$$

if

$$\mathbf{T} = (\mathbf{I} - \mathbf{D})^{-1} \approx \mathbf{I} + \mathbf{D} + \mathbf{D}^2 + \cdots + \mathbf{D}^n \quad \text{for large } n, \tag{12.7}$$

$$\mathbf{x} = \mathbf{T}\mathbf{f}. \tag{12.8}$$

In input–output literature, $\mathbf{D}$ is called the direct-requirements matrix and $\mathbf{T}$ is called the total-requirements matrix. The direct-requirements matrix captures only the first-order interactions whereas the total-requirements matrix captures all higher-order interactions. For example, in an economic network an increase in demand for automobiles entails a corresponding increase in direct inputs such as steel and plastic. A change in direct inputs has a further cascading effect throughout the economy. The additional production of steel requires an additional consumption of electricity. The additional generation of electricity requires an additional consumption of coal and so on. Direct inputs such as steel and plastic represent first-order interactions whereas indirect inputs such as electricity and coal represent second- and third-order interactions, respectively. A network comprising n units can have infinite-order interactions.

If transaction coefficients are defined as $\gamma_{ki} = X_{ki}/X_k$ and $\gamma_{kf} = F_k/X_k$, where $X_k = \sum_{i=1}^{n} X_{ki} + F_k$, Eq. (12.1) can be rewritten as

$$\left(\sum_{k=1}^{n} \gamma_{ik} X_i \right) + V_k = \left(\sum_{i=1}^{n} \gamma_{ki} X_k \right) + \gamma_{kf} X_k \quad \text{for } i = 1, \ldots, n. \tag{12.9}$$

Because

$$\left(\sum_{i=1}^{n} \gamma_{ki} \right) + \gamma_{kf} = 1, \tag{12.10}$$

Eq. (12.9) can be written as

$$\left(\sum_{i=1}^{n} \gamma_{ik} X_i \right) + V_k = X_k \quad \text{for } k = 1, \ldots, n, \tag{12.11}$$

or alternatively in matrix form as

$$\mathbf{x} = \gamma^{\mathsf{T}}\mathbf{x} + \mathbf{v}, \tag{12.12}$$

$$\mathbf{x} = (\mathbf{I} - \gamma^{\mathsf{T}})^{-1}\mathbf{v}. \tag{12.13}$$

Equation (12.13) is used in the TIOA discussed in the next subsection.

Since its development to study monetary interdependencies in the economic system [33], input–output analysis has also been widely used to address several environmental issues pertinent to industrial ecology [35, 36]. For instance, Leontief et al. [37] explored the integration of material flows of 26 nonfuel materials in conventional input–output models. Ayres developed the material process–product model to address questions on the boundary between traditional economic criteria such as cost and prices and material processing [38–40]. Duchin's structural

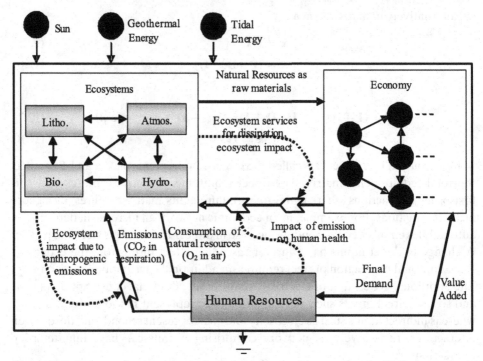

Figure 12.2. Integrated economic–ecological–human resource system (solid lines represent tangible interactions and dotted lines represent intangible interactions occurring as a consequence of emissions) [30].

economics approach combined the physical interconnectedness in the economic system with corresponding representation of costs and prices [41]. Input–output analysis has also been applied to study energy flows in the economic system [42–44]. A NEA or FFCA establishes a correlation between embodied energy intensities of economic goods and services and their economic prices [15, 16, 45]. *Economic input–output life-cycle assessment* (EIO-LCA) [46, 47] applies input–output analysis to overcome the problem of system boundaries in LCA. With the national economy as the analysis boundary, EIO-LCA calculates the economy-wide total environmental burden associated with an exogenous final demand. Although EIO-LCA successfully calculates total emissions over the entire life cycle of an industrial sector, it does not calculate their impacts on human or ecosystem health. Moreover, it gives results in disparate units, such as tons of CO_2 equivalents and tons of SO_2 equivalents. As a result, a comparison across different impact categories, in this case total global warming potential and total acidification potential, cannot be done without arbitrary human valuation.

12.2.2 Methodology for Thermodynamic Input–Output Analysis

TIOA recognizes the network structure of the integrated Economic–Ecological–Social (EES) system shown in Fig. 12.2. Such a system is an open thermodynamic system with energy inputs from the three fundamental sources of energy, namely sunlight, geothermal heat, and tidal or gravitation forces. The fourth fundamental source, namely nuclear energy, is not considered as it does not appear naturally in ecosystems. In addition, internal energy storages such as petroleum reservoirs, coal stocks, and metallic and nonmetallic mineral deposits are considered in the proposed approach. Material may also enter the EES system in the form of national imports and exit in the form of national exports. Imports and exports, however, are not considered in this analysis as their inclusion would require knowledge about the global economy that is beyond the scope of this analysis.

TIOA focuses on the economic system, which is divided into smaller functional units called industry sectors. In the United States, this task is accomplished by the Bureau of Economic Analysis, which defines industry sectors according to the Standard Industrial Classification (SIC) or the North American Industrial Classification System (NAICS) codes. *Ecological systems*, on the other hand, are divided into four conceptual ecospheres that encompass land (lithosphere), water (hydrosphere), air (atmosphere), and living flora and fauna (biosphere). Such a classification assists categorization of vast numbers of ecological resources into smaller groups and is by no means critical to the applicability of TIOA. Any other user-defined classification scheme would also work as long as renewable and nonrenewable resources are distinguished.

Figure 12.2 shows interactions among economic, ecological, and social systems. Interactions represented by solid lines arise on account of resource consumption and emissions, whereas those represented by dotted lines are intangible interactions, indicating the impact of emissions on human and ecosystem health. For instance, the dotted arrow between the economy and the ecosystems represents ecological services required for dissipating industrial emissions and their impact on ecosystem health. The solid arrow from the ecosystems to the economy, on the other hand, represents tangible interactions that include consumption of ecological resources as raw materials by the economic activity.

The network structure of the economic system and monetary interactions between industry sectors are typically well known and are captured in EIO analysis [33, 34]. This information is used in allocating cumulative exergy flows between industry sectors in ICEC analysis. Conversely, the network structure of an ecological system need not be completely known as the ECEC analysis can deal with partially known ecological networks by use of appropriate allocation rules [48]. ICEC and ECEC analyses provide a common unit to compare economic and ecological resources, as any system, economic or ecological, can be considered as a single network of energy flows [4]. In ICEC analysis this common unit represents the exergy content of natural resources whereas in ECEC analysis this common unit represents cumulative exergy content of natural resources.

The emphasis of this methodology is not on predicting how a complex, holarchic, and chaotic system such as the EES system would evolve under the influence of external energy sources [49], but is on analyzing available resource consumption and emissions data to understand how different industry sectors rely on ecosystems for their operations. In other words, TIOA does not attempt to forecast emergent, nonlinear, nonequilibrium, and self-organizing properties of the EES system, but assumes that these properties are manifested in the measured material and energy flows. The algorithm of TIOA can be summarized in the form of the following three tasks:

1. *Identify and quantify ecological and human resource inputs to the economic system.* Ecological inputs include ecosystem products, such as crude oil, metallic and nonmetallic minerals, and atmospheric nitrogen, and ecosystem services, such as wind and fertile soil. Human resources include employment of labor for economic activities. Emissions and their impact on human and ecosystem health are also included.
2. *Calculate the CEC of ecological inputs* by using *transformity* values from systems ecology for ECEC or exergy of inputs (unit transformities) for ICEC. For ECEC, these inputs are classified as additive or nonadditive to avoid double counting and to be consistent with the network algebra rules used in emergy analysis [4, 48]. In general, nonrenewable resources are additive, whereas renewable resources are nonadditive. For an ICEC, such a distinction is not necessary as the question of double counting does not arise.
3. *Allocate direct inputs to economic sectors by use of input–output data and the network algebra of ECEC analysis* [33]. The network algebra of ECEC analysis is based on a *static* input–output representation of the economic system. Dynamic versions of input–output analysis that consider temporal changes in the economic network are also available and are currently being explored. Also, use of monetary data for allocation is not a limitation of the approach, but is rather caused by lack of comprehensive material or energy accounts of interindustry interactions.

12.3 Data Sources for the 1997 U.S. Industry Benchmark Model

This section describes the resources considered in this analysis, along with their data sources. All required data were obtained from nonproprietary public-domain databases.

12.3.1 Transformities

The ECEC of ecological and human resources was determined by means of their transformity values [4, 50, 51]. Transformities can be viewed as reciprocals of global exergetic efficiencies of ecological resources. Consequently they enable calculation of total exergy consumption in the economic and ecological stages of a production chain. Transformities, as used in this analysis, are not subject to the controversial aspects of Odum's work, such as the maximum empower principle, emergy theory of value, or energy consumption over geological time scales. Transformities used in this analysis correspond to the 1996 base of 9.44×10^{24} sej/yr (sej/yr indicates solar equivalent joules per year) [4]. Furthermore, TIOA uses only transformities of direct inputs from nature, and derived transformities of economic goods and services are not required. ICEC analysis does not consider ecological stages of a production chain and consequently assumes cumulative exergy of direct natural-resource inputs to be equal to their standard exergy values. This is tantamount to assuming a uniform transformity of 1 for all natural resources.

12.3.2 *Ecosystem Services*

This analysis focuses only on supply-based services, as their contribution can be quantified independently of human valuation. Supply-based services are always accompanied by corresponding material and energy flows, and hence can be readily included in TIOA. As shown in Table 12.2, this analysis considers sunlight for 24-h photosynthesis, fertile soil and wind, and geopotential and hydropotential for electricity generation. Other supply-based services such as those involved in pollination, carbon sequestration, and dissipation of pollutant streams can also be included in TIOA, but would entail a more thorough understanding of their geobiochemical mechanisms. Unlike supply-based services, the value-based services cannot be measured with biophysical principles only. Examples of value-based services include those for recreational and cultural purposes. They depend on how people perceive them and are dealt with in environmental economics literature [3, 11].

12.3.3 Ecosystem Products

Ecosystem products refer to the natural raw materials consumed by economic activities. These raw materials are extracted by basic infrastructure activities such as mineral mining, coal mining, petroleum and natural-gas (ng) extraction, and logging and timber tract harvesting. Table 12.3 lists the ecosystem products considered in this analysis, their flows in the 1997 U.S. economic model, the industry sectors that receive their direct inputs, and corresponding data sources. The inputs to agricultural activities were adjusted by use of data about the number of farms and their average size during 1992 and 1997. Currently work is underway to expand the scope further by including material and energy flow information about additional ecosystem products.

12.3.4 Human Resources

Industry sectors consume human resources in the form of labor. The amount of human resources consumed is a function of number of individuals employed and their skill level. In this chapter, the average annual payroll is chosen as a measure of the quality of labor. Data about the number of people employed and their average annual payroll are available from the U.S. Department of Labor's Bureau of Labor Statistics [60]. In this analysis, human resources are considered to be exogenous to the economic model representing interindustry interactions. Therefore, in the absence of a single input–output model integrating industry sectors and social sectors, interactions between economy and human resources need to be considered independently. This is done through the use of transformity of unskilled labor, obtained from Odum [4], and calculated as the ratio of the total emergy budget to the total population of the United States. Odum assumes that the total emergy input to the U.S. economy is passed on to human resources by means of the final demand, which represents sale of economic goods and services to consumers, and consumers, in turn, feed the emergy flow back to the economy by means of the value added, which includes employment of labor.

Table 12.2. *Data for ecosystem service inputs to 1997 U.S. economy*

Ecosystem service	Sector receiving direct input and corresponding NAICS code	Energy or material flow (F)	Data source for F	ICEC flow (J/yr)	Transformity (τ) (sej/J)	Data source for τ	ECEC flow ($C = F \times \tau$) (sej/yr)
Sunlight for photosynthesis	Agricultural sectors (NAICS 1111A0–1119B0)	2.23×10^{22} J/yr[a]	[52, 53]	2.23×10^{22}	1	[4]	2.23×10^{22}
	Forest nurseries, forest products and timber tracts (NAICS 113A00)	1.19×10^{22} J/yr	[52, 53]	1.19×10^{22}	1	[4]	1.19×10^{22}
Hydropotential for power generation	Power generation and supply (NAICS 221100)	1.28×10^{18} J/yr	[54]	1.28×10^{18}	27,764	[4]	3.55×10^{22}
Geothermal heat for power generation	Power generation and supply (NAICS 221100)	5.3×10^{16} J/yr	[54]	5.3×10^{16}	6055	[4]	3.21×10^{20}
Wind energy for power generation	Power generation and supply (NAICS 221100)	1.18×10^{16} J/yr	[54]	1.18×10^{16}	1496	[4]	1.77×10^{19}
Soil erosion	Agricultural sectors (NAICS 1111A0–1119B0)	34.49×10^{8} ton/yr	[12, 13]	3.12×10^{18}	4.43×10^{4}	[51]	1.38×10^{23}[b]
	Construction sectors (NAICS 230110–230250)	35.65×10^{8} ton/yr	[12, 13]	3.22×10^{18}	4.43×10^{4}	[51]	1.43×10^{23}

[a] Sunlight for photosynthesis: $(2.26 \times 10^{22}$ J/yr, 1993 flux) × $(1.91 \times 10^{6}$, farms in 1997) × (487 acres, average size of farm in 1997) / $(1.93 \times 10^{6}$, farms in 1993) / (491 acres, average size of farm in 1993) = 2.23×10^{22} J/yr, 1997, flux.

[b] $(34.49 \times 10^{8}$ ton/yr, topsoil loss) × (4% organics in soil) × (5.4 Kcal/g, energy content of organic soil) × (4186 J/Kcal) × $(4.43 \times 10^{4}$ sej/J) = 1.38×10^{23} sej/yr; transformity adjusted to 1996 base of 9.44×10^{24} sej/yr.

Table 12.3. *Data for ecosystem product inputs to 1997 U.S. economy*

Resource considered in this analysis	Industry sector receiving direct input and its NAICS code	Material or energy flow (F)	Data source for F	ICEC flow (J/yr)	Transformity (τ)	Data source for τ	ECEC flow $c = \tau \times F$ (sej/yr)
Lithosphere							
Crude petroleum field production	Oil and gas extraction (NAICS 211000)	1.06×10^{19} J/yr[a]	[55]	1.06×10^{19}	53,000 sej/J	[4]	5.61×10^{23}
Natural Gas	Oil and gas extraction (NAICS 211000)	18.9 MMCuF/yr	[56]	1.99×10^{19}	48,000 sej/J	[4]	9.58×10^{23}
Iron-ore mining	Iron ore mining (NAICS 212210)	202 MMT/yr	[57]	2.08×10^{16}[b]	1×10^9 sej/g	[4]	2.02×10^{23}
Copper mining	Copper, nickel, lead, and zinc mining (NAICS 212230)	342 MMT/yr[c]	[57]	2.80×10^{16}[d]	1×10^9 sej/g	[4]	3.42×10^{23}
Gold mining	Gold, silver, and other metal mining (NAICS 2122A0)	217 MMT/yr[e]	[57]	5.63×10^{16}[f]	1×10^9 sej/g	[4]	2.17×10^{23}
Crushed Stone	Stone mining and quarrying (NAICS 212310)	1390 MMT/yr	[57]	1.83×10^{17}[g]	1×10^9 sej/g	[4]	1.39×10^{24}
Sand	Sand, gravel, clay, and refractory mining (NAICS 212310)	961 MMT/yr	[57]	1.27×10^{17}[h]	1×10^9 sej/g	[4]	9.61×10^{23}
Raw coal, excluding overburden	Coal mining (SIC 212100)	988 MMT/yr	[57]	5.73×10^{19}[i]	1×10^9 sej/g	[4]	9.88×10^{23}

[a] $(4.803 \times 10^6$ barrels, on-shore production/day$) \times (30$ days/month$) \times (12$ months/yr$) \times (6.12 \times 10^9$ J/barrel$) = 1.06 \times 10^{19}$ J/yr.

[b] Iron ore mining: Mass flow rate (F) = 202 MMT/yr; exergy of Fe_2O_3 (b) = 103 J/g [17]; ICEC flow = $F \times b = 2.08 \times 10^{16}$ J/yr.

[c] Copper mining: $(2.07$ MMT/yr, 1997 mine production$) \times (297$ MMT/yr, 1993, domestic ores input$) / (1.79$ MMT/yr, 1993 mine production$) = (342$ MMT/yr, 1997 domestic ores input$)$; assuming flotation, concentration, smelting, and refining technologies and ratio of domestic to imported concentrates unchanged between 1993 and 1997.

[d] Copper mining: 342 MMT/yr domestic ores input (F); exergy of CuO (b) = 82J/g [17]; ICEC flow = $F \times b = 2.80 \times 10^{16}$ J/yr.

[e] Gold mining: $(325$ tons/yr, 1997 gold production$) \times (221$ MMT/yr 1993, domestic gangue$) / (331$ tons/yr, 1993 gold production$) = (217$ MMT/yr, 1997 domestic gangue$)$; assuming flotation, concentration, smelting, and refining technologies and ratio of domestic to imported concentrates unchanged between 1993 and 1997.

[f] Gold mining: 217 MMT/yr, domestic ores input (F); exergy of Au_2O_3 (b) = $(114.7$ KJ/mol$)/(441.93$g/mol of Au_2O_3) [17]; ICEC flow = $F \times b = 5.63 \times 10^{16}$ J/yr.

[g] Crushed stone: Mass flow rate (F) = 1390 MMT/yr; exergy of SiO_2 (b) = 132J/g [17]; ICEC flow = $F \times b = 1.83 \times 10^{17}$ J/yr.

[h] Sand: Mass flow rate (F) = 961 MMT/yr; exergy of SiO_2 (b) = 132 J/g [17]; ICEC flow = $F \times b = 1.27 \times 10^{17}$ J/yr.

[i] Raw coal excluding overburden: Mass flow rate (F) = 878 MMT/yr; exergy of coal (b) = 29,000 J/g [17]; ICEC flow = $F \times b = 2.56 \times 10^{19}$ J/yr.

Table 12.3 (continued)

Resource considered in this analysis	Industry sector receiving direct input and its NAICS code	Material or energy flow (F)	Data source for F	ICEC flow (J/yr)	Transformity (τ)	Data source for τ	ECEC flow $c = \tau \times F$ (sej/yr)
Nitrogen from mineralization	Agricultural sectors (NAICS 1111A0–1119B0)	2.96 MMT/yr[j]	[58]	1.15×10^{15}[k]	4.19×10^9 sej/g	[4]	1.24×10^{22}
Phosphorous from mineralization	Agricultural sectors (NAICS 1111A0–1119B0)	1.97 MMT/yr[l]	[58]	9.75×10^{14}[m]	2×10^9 sej/g[n]	[4]	3.94×10^{21}
N deposition from atmosphere[o]	Agricultural sectors (NAICS 1111A0–1119B0)	1.97 MMT/yr[p]	[58]	7.76×10^{14}[q]	4.19×10^9 sej/g	[4]	8.25×10^{21}
Return of decomposing detritus to agricultural soil	Agricultural sectors (NAICS 1111A0–1119B0)	−433 MMT/yr[r]	[58]	-8.77×10^{18}[s]	2.24×10^8 sej/g of residue[t]	[4]	-9.70×10^{22}

(continued)

j Nitrogen from mineralization: (3 MMT/yr 1993, Nitrogen flux from mineralization) × (1.91 × 10⁶, farms in 1997) × (487 acres, average size of farm in 1997) / (1.93 × 10⁶, farms in 1993) / (491 acres, average size of farm in 1993) = 2.96 MMT/yr, 1997 nitrogen flux from mineralization.

k Nitrogen from mineralization: Mass flow rate (F) = 2.96 MMT/yr; exergy of $Mg(NO_3)_2$ (b) = 387J/g [17]; ICEC flow = $F \times b$ = 1.15×10^{15} J/yr.

l Phosphorous from mineralization: (2 MMT/yr, 1993 phosphorous flux from mineralization) × (1.91 × 10⁶, farms in 1997) × (487 acres, average size of farm in 1997) / (1.93 × 10⁶, farms in 1993) / (491 acres, average size of farm in 1993) = 1.97 MMT/yr, 1997 phosphorous flux from mineralization.

m Phosphorous from mineralization: Mass flow rate (F) = 1.97 MMT/yr; exergy of $Mg_3(PO_4)_2$ (b) = 495 J/g [17]; ICEC flow = $F \times b$ = 9.75×10^{14} J/yr.

n (4.6×10^8 sej/g of P_2O_5) × (1 g of P_2O_5/0.23g of P) = 2×10^9 sej/g of P.

o N deposition from atmosphere is considered as an input from lithosphere because nitrogenous salts enter plants through soil.

p N deposition from atmosphere: (2 MMT/yr, 1993 N-deposition flux) × (1.91 × 10⁶, farms in 1997) × (487 acres, average size of farm in 1997) / (1.93 × 10⁶, farms in 1993) / (491 acres, average size of farm in 1993) = 1.97 MMT/yr, 1997 N-deposition flux.

q N deposition from atmosphere: Mass flow rate (F) = 1.97 MMT/yr; exergy of $Mg(NO_3)_2$ (b) = 387J/g [17]; ICEC flow = $F \times b$ = 7.62×10^{14} J/yr.

r Return of detrital matter: (440 MMT/yr, 1993 flux) × (1.91 × 10⁶, farms in 1997) × (487 acres, average size of farm in 1997) / (1.93 × 10⁶, farms in 1993) (491acres, average size of farm in 1993) = 433.4 MMT/yr, 1997 flux; negative sign indicates flow from industry sector to lithosphere.

s Return of detrital matter: (433 MMT/yr of returned detritus residue) × (0.44g C/g of residue) × (11 Kcal/g C) × (4186 J/Kcal) = 8.77×10^{18} J/yr [4].

t (0.44g C/g of residue) × (11 Kcal/g C) × (4186 J/Kcal) × (11,068 sej/J, transformity of detritus production) = 2.24×10^8 sej/g residue.

Table 12.3 (continued)

Resource considered in this analysis	Industry sector receiving direct input and its NAICS code	Material or energy flow (F)	Data source for F	ICEC flow (J/yr)	Transformity (τ)	Data source for τ	ECEC flow $c = \tau \times F$ (sej/yr)
Biosphere							
Wood production	Logging (NAICS 113300)	520 MMT/yr of roundwood	[58]	8.27×10^{18u}	5.55×10^{8} sej/g^v	[4]	2.90×10^{23}
Pasture Grazing	Cattle Ranching and Farming (NAICS 112100)	200 MMT/yr of wet grass	[58]	1.67×10^{18w}	5.83×10^{19} sej/MMT of wet grass^x	[4]	1.17×10^{22}
Hydrosphere							
Water consumption	Water, sewage and other systems (NAICS 221300)	1.47×10^{14} gal/yr	[59]	2.73×10^{18y}	7.67×10^{8} sej/gal^z	[50]	1.13×10^{23}
Atmosphere							
CO_2 in 24-h net photosynthesis	Agricultural sectors (NAICS 1111A0–1119B0)	867 MMT/yr$^\alpha$	[58]	0^β	6.19×10^{7} sej/g $CO_2{}^\gamma$	[4]	5.37×10^{22}

[u] Wood production: 520 MMT/yr roundwood × 3.8 Kcal/g roundwood × 4186 J/Kcal = 8.27×10^{18} J/yr [4].

[v] (3.8 Kcal/g roundwood) × (4186 J/Kcal) × (34,900 sej/J) = 5.55 × 10^8 sej/g of roundwood.

[w] Pasture grazing: 440 MMT/yr of wet grass × 0.5 MMT of dry grass/MMT of wet grass × 10^{12} g/MMT × 1.86 × 10^{11} J/ha/yr of pasture evapotranspiration × 9 × 10^{-4} m^2/g × 10^{-4} ha/m^2 = 5.83 × 10^{19} sej/MMT of wet grass [4].

[x] (0.5 MMT of dry grass/MMT of wet grass) × (10^{12} g/MMT) × (1.86 × 10^{11} J/ha/yr of pasture evapotranspiration) × (6962 sej/J) × (9 × 10^{-4} m^2/g) × (10^{-4} ha/m^2) = 5.83 × 10^{19} sej/MMT of wet grass.

[y] Water consumption: 1.47 × 10^{14} gals/yr × 3785 cm^3/gal of water × 1 g of water/cm^3 of water × 4.94 J/g of water = 2.73 × 10^{18} J/yr [4].

[z] (3785 cm^3/gal of water) × (1 g of water/cm^3 of water) × (4.94J/g of water) × (4.1 × 10^4 sej/J) = 7.67 × 10^8 sej/gal of water.

[α] CO_2 in 24-h photosynthesis: (880 MMT/yr, 1993 CO_2 flux) × (1.91 × 10^6, farms in 1997) × (487 acres, average size of farm in 1997) / (1.93 × 10^6, farms in 1993) / (491 acres, average size of farm in 1993) = 866.8 MMT/yr, 1997 CO_2 flux.

[β] Atmospheric gases at reference state are ignored in ICEC analysis.

[γ] (12 g C/44 g CO_2) × (8 Kcal/g C) × (4186 J/Kcal) × (6780 sej/J) = 6.19 × 10^7 sej/g CO_2.

Note: MMT, million metric tons.

Table 12.4. *Pollutants, immediate destination of emission, and impact category*

Pollutant	Immediate destination of emission	Impact category considered	DALY/kg of emission[a]	ECEC/kg of emission (sej/kg)
SO_2	Air	Respiratory disorders	5.46×10^{-5}	1.86×10^{12b}
NO_2	Air	Respiratory disorders	8.87×10^{-5}	3.03×10^{12}
PM10	Air	Respiratory disorders	3.75×10^{-4}	1.28×10^{13}
CO_2	Air	Climate change[c]	2.1×10^{-7}	7.17×10^{9}
Methanol	Air	Respiratory disorders	2.81×10^{-7}	9.59×10^{9}
Ammonia	Air	Respiratory disorders	8.5×10^{-5}	2.90×10^{12}
Toluene	Air	Respiratory disorders	1.36×10^{-6}	4.64×10^{10}
1,1,1-TCE	Air	Ozone layer depletion	1.26×10^{-4}	4.30×10^{12}
Styrene	Air	Carcinogenic effect	2.44×10^{-8}	8.33×10^{8}
Styrene	Water	Carcinogenic effect	1.22×10^{-6}	4.16×10^{10}
Styrene	Soil	Carcinogenic effect	2.09×10^{-8}	7.13×10^{8}

[a] DALY values are based on hierarchist perspective.
[b] Human health impact of emission per kg of SO_2 emission = $(5.46 \times 10^{-5}$ DALY/kg of SO_2 emission) $\times$ (365 days/yr) $\times$ (9.35×10^{13} sej emergy associated with unskilled labor/workday) = 1.86×10^{12} sej/kg; emergy of unskilled labor is obtained from emergy literature [4], and is obtained by dividing total emergy budget of the United States $(7.85 \times 10^{24}$ sej/yr) by the total population of the United States (230×10^{6} people)
[c] Impacts are potential impacts in future [57].

Hence human resources incorporate natural capital flows between economy and human resources. Moreover, the per capita emergy budget of the United States can be used to represent unskilled labor as only half of the U.S. population was employed in 1997. The remaining half comprised minors, retirees, and unemployed people.

12.3.5 Impact of Emission on Human Health

Industrial emissions affect human health in myriad ways. The actual impact depends on the fate of a pollutant in the natural environment and its effect on human well-being. The fate itself depends on numerous physicochemical phenomena such as dispersion, diffusion, and atmospheric chemistry. There are several established procedures for calculating the impact of emissions on human health. The approach employed in this analysis represents the impact of several common pollutants on human health in terms of disability adjusted life years (DALY). This is an end-point impact-assessment methodology that considers several impact categories, including respiratory disorders, photochemical smog formation, ozone layer depletion, climate change, and carcinogenicity [61, 62]. Table 12.4 lists pollutants considered in this work, the impact categories they belong to, and corresponding DALY values per kg of emission. Emissions data were gathered from the U.S. Environmental Protection Agency's *Toxics Release Inventory* (TRI) [63, 64]. The approach for converting DALYs to ECEC was discussed in [64]. Work toward including more pollutants in this analysis is currently in progress. The analysis presented in this chapter does not consider the ecosystem impact of emissions, but the general approach could be applied for it if exergy loss that is due to ecosystem impact could be quantified.

12.3.6 Allocation Matrix for Interindustry Interactions

This analysis uses a monetary, interindustry transaction coefficient matrix to represent the U.S. economic system. In the United States, such a matrix is compiled periodically by the Department of Commerce's Bureau of Economic Analysis. More specifically, results presented in Section 12.4 are based on the 488-sector 1997 the U.S. interindustry benchmark model [65]. Similar results were published in the past for the 91-sector 1992 model, which is a more concise and aggregated representation of the U.S. economy [30]. An *allocation* matrix based on material or energy interactions between industry sectors would be more accurate than a monetary transaction matrix, but is not available at present. The "materials-count" initiative undertaken by the National Research Council [66] is an example of efforts that strive to compile a biophysical transaction matrix for the U.S. economy. If this initiative materializes, more accurate data could be used for interindustry allocation.

12.4 Cumulative Exergy Consumption in Industrial and Ecological System for Selected Natural Resources

The TIOA methodology can be applied to determine ECEC and ICEC requirements of individual industry sectors. This section primarily presents ECEC requirements of individual sectors for a few selected natural resources. These resources are selected for illustrative purposes only. Results for all the resources and emissions listed in Tables 12.2–12.4 are beyond the scope of this chapter, but can be found in Ukidwe [64] for the 488-sector 1997 model of U.S. economy and in Ukidwe and Bakshi [30] for the 91-sector 1992 model of the U.S. economy. An updated 1997 model is also available in Zhang et al. (2010). This work introduces the framework for *ecologically based LCA* (ECO-LCA) by extending the work in [64]. ICEC requirements can be plotted in a similar way. Figure 12.5, later in this section, shows ICEC requirements of industry sectors from the lithosphere as an illustrative example. ICEC results for all the resources are also beyond the scope of this chapter, but can be found in Ukidwe [64] and Ukidwe and Bakshi [30].

Sunlight. Figure 12.3 shows the contribution of sunlight. The agricultural sectors (NAICS 1111A0–1119B0) and the sector of forest nurseries, forest products, and timber tracts (NAICS 113A00) are the direct recipients of sunlight. Sectors of sawmills (NAICS 321113) and food services and drinking places (NAICS 722000) have prominent peaks in Fig. 12.3 on account of indirect consumption. In this chapter, solar inputs to the group of agricultural sectors and to the sector of forest nurseries, forest products, and timber tracts are determined by multiplying the average solar flux per unit area in the continental United States by the total land area of the two [52, 53]. To allocate solar inputs within the group of agricultural sectors, economic data were used. If data about land areas in individual agricultural sectors were available, it could have been used for allocation as well. Furthermore, the use of transformity values in this analysis ensures consideration of indirect routes of solar inputs to industry sectors. These indirect routes include biogeochemical cycles such as the hydrologic cycle and atmospheric circulation, which are driven by solar insolation. In that regard, the analysis presented in this chapter improves on Costanza [42],

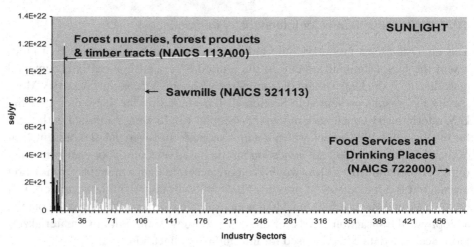

Figure 12.3. Contribution of sunlight, *y* axis, is annual flows of ECEC in solar equivalent joules (sej/yr), and the *x* axis is the sector serial number (the black part of each bar represents direct inputs, and the white part represents indirect inputs).

who considered only direct solar inputs to the U.S. economy to calculate energy intensities of industry sectors.

Table 12.2 underscores the shortcomings of MFA and ICEC analyses vis-à-vis ECEC analysis. Because MFA focuses only on material-interaction contributions from sunlight, hydropotential, geopotential, and wind energy are completely ignored in MFA. These energy flows are clearly important and must be considered in any joint analysis of economic and ecological systems. ICEC analysis considers both material and energy flows by considering the exergy of natural resources, but fails to appreciate the fact that some natural resources are readily available to the economic system whereas others are made available by global cycles. For instance, ICEC of the organic component of topsoil is almost 4 orders of magnitude smaller than that of sunlight. However, this part of soil is a much more concentrated form of natural resource that is produced by biogeochemical cycles as opposed to sunlight, which is freely available. This fact is captured by ECEC analysis through the use of transformity values. Consequently ECEC analysis does a better job in considering renewable and nonrenewable nature of natural resources than MFA or ICEC analyses.

Lithosphere. Figure 12.4 shows ECEC requirements of industry sectors from the lithosphere. Sectors of stone mining and quarrying (NAICS 212310), coal mining (NAICS 212100), sand, gravel, clay, and refractory mining (NAICS 212320), and oil and gas extraction (NAICS 211000) have prominent peaks because of direct inputs from the lithosphere. Sectors of power generation and supply (NAICS 221100), petroleum refineries (NAICS 324110), iron and steel mills (NAICS 332111), and automobile and light truck manufacturing (NAICS 336110) also have prominent peaks because of *indirect consumption* of lithospheric resources. Unlike mining sectors that extract resources from the lithosphere, the agricultural sectors (NAICS 1111A0–1119B0) add to the lithosphere on account of the return of detrital matter to agricultural soil. Consequently these sectors have *negative direct* ECEC requirements from the lithosphere. This is shown with the aid of the embedded graph in Fig. 12.4. The agricultural sectors, like other sectors in the economy, still have *positive indirect* ECEC requirements on account of consumption of fuels and

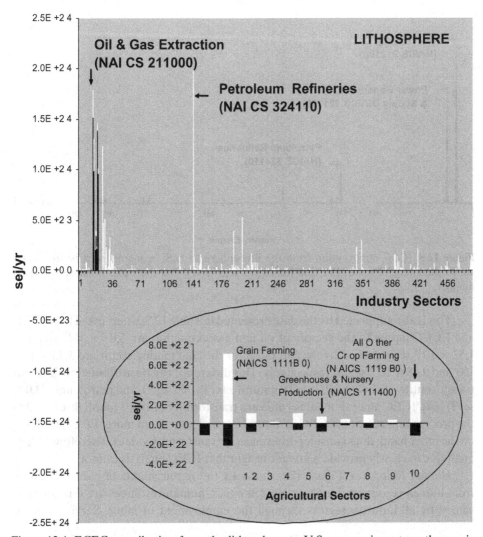

Figure 12.4. ECEC contribution from the lithosphere to U.S. economic sectors; the *y* axis shows the annual flows of ECEC in solar equivalent joules per year (sej/yr), and the *x* axis is the sector serial number (the black part of each bar represents direct inputs, and the white part represents indirect inputs).

electricity. Furthermore, the indirect requirements exceed the direct requirements, making the agricultural sectors net consumers of lithospheric resources. The sector of greenhouse and nursery production (NAICS 111400) is found to be the only exception in which direct requirements exceed indirect requirements, making it the net donor to the lithosphere.

Figure 12.5 shows ICEC requirements of industry sectors from the lithoshere. Coal mining (NAICS 212100), power generation and supply (NAICS 221100), petroleum refineries (NAICS 324191), and oil and gas extraction (NAICS 211000) have prominent peaks on account of direct inputs from the lithosphere. However, a comparison with Fig. 12.4 shows that coal mining and power-generation and supply sectors have replaced petroleum refineries and oil and gas extraction as the two most significant sectors in Fig. 12.5.

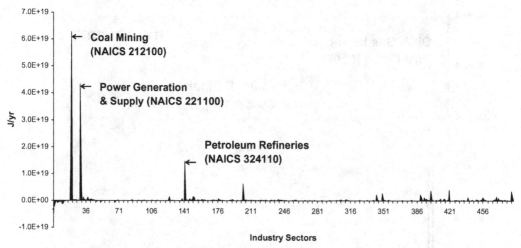

Figure 12.5. ICEC contribution from the lithosphere to U.S. economic sectors; the y axis shows the annual flows of ICEC in joules per year (J/yr), and the x axis is the sector serial number.

This is also supported by the data presented in Table 12.3, wherein the total ICEC and ECEC inputs to the sectors of oil and gas extraction are 3.05×10^{19} J/yr and 1.52×10^{24} sej/yr, respectively, and those for the coal mining sector are 5.73×10^{19} J/yr and 9.88×10^{23} sej/yr, respectively. This difference can be attributed to different transformities of coal (34,482 sej/J), natural gas (48,000 sej/J), and petroleum (53,000 sej/J) [64]. ICEC analysis assumes uniform transformity of 1sej/J for all three, and in the process suppresses the CEC of natural gas and petroleum more. ECEC analysis, on the other hand, does consider differences in transformity values. Therefore ECEC analysis can clearly provide a greater insight that ICEC analysis cannot.

Human Resources. Figure 12.6 shows ECEC requirements of industry sectors from human resources. Unlike other resources, human resources are directly consumed by all industry sectors through the employment of labor. Service sectors,

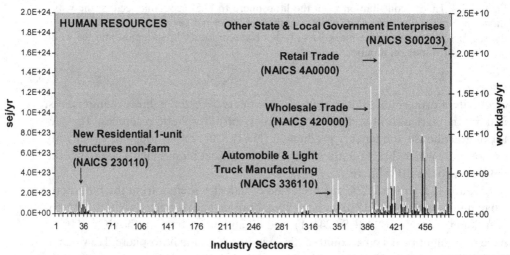

Figure 12.6. ECEC requirements from human resources; the y axes are annual ECEC flows in sej/yr and corresponding flows in workdays/yr; the x axis is the sector serial number (the black part of each bar represents direct inputs, and the white part represents indirect inputs).

in particular, have higher direct inputs than the rest of the economy. Sectors of other state and local government enterprises (NAICS S00203), retail trade (NAICS 4A0000), wholesale trade (NAICS 420000), and home health care services (NAICS 621600) have the highest consumption of human resources. These results also conform to those obtained for the 91-sector 1992 model of the U.S. economy. Other nonservice sectors with prominent peaks include automobile and light truck manufacturing (NAICS 336110), motor vehicle parts manufacturing (NAICS 336300), new residential one-unit structures, nonfarm (NAICS 230110), and commercial and institutional buildings (NAICS 230220). In this analysis, the contribution of human resources is determined from economic data that include the number of people employed and their average annual payrolls. A similar approach based on ICEC analysis can be used to evaluate industry-specific exergetic intensities of human labor in Sciubba's extended exergy accounting (EEA) [67]. Sciubba defines exergetic intensity of human labor as total exergetic resources into a portion of the society divided by the number of working hours sustained by it. This ratio can be calculated for the aggregate U.S. economy by dividing the emergy of unskilled labor, 9.35×10^{13} sej/workday [4] by the average ECEC/ICEC ratio for the 1997 U.S. economy, 2873 sej/J. This gives a value of 3.25×10^{10} J/workday that can be used in EEA.

Human Health Impact of SO_2. Figure 12.7(a) shows the impact associated with SO_2 in terms of ECEC and DALY. To convert the human health impact from *DALY*/yr to ECEC/yr, the former is multiplied by a factor of 3.42×10^{16} sej/yr [30, 64]. Power plants are the major emitters of SO_2. Consequently the sector of power generation and supply (NAICS 221100) has the most significant peak in Fig. 12.7(a). Other sectors with prominent peaks include petroleum refineries (NAICS 324110), real estate (NAICS 531000), and retail trade (NAICS 4A0000). The sector of petroleum refineries is one of the major suppliers to the sector of power generation and supply, whereas the sectors of real estate and retail trade are major consumers of electricity because of their large economic throughputs. Similar results can be obtained based on ICEC analysis by use of the exergetic intensity of human labor. In such a case, to convert human health impact from *DALY*/yr to ICEC/yr, the former needs to be multiplied by a factor of 1.19×10^{13} J/yr.

Human Health Impact of NH_3. Figure 12.7(b) shows the impact associated with the emission of NH_3. NH_3 is primarily emitted by the sectors of paper and paperboard mills (NAICS 3221A0), petroleum refineries (NAICS 324110), and iron and steel mills (NAICS 331111). As a result, these sectors also have the tallest peaks in Fig. 12.11(b) in Subsection 12.5.2.

Sectors of motor vehicle parts manufacturing (NAICS 336300) and automobile and light truck manufacturing (NAICS 336110) also have significant peaks on account of indirect effects. Similar results have been reported for other bulk pollutants such as CO_2, NO_2, and PM10 and nonbulk pollutants such as toluene, 1,1,1-trichloroethane, and styrene in Ukidwe and Bakshi [30] and Ukidwe [64].

12.5 Aggregate Metrics

One of the fortes of TIOA is its ability to provide separate results for each input and output category as obtained for the selected individual resources in Section 12.4 along with the *aggregate metrics*. Such aggregation is facilitated by the fact that all

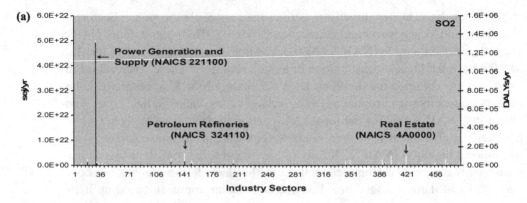

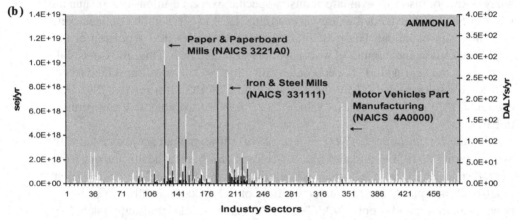

Figure 12.7. Human health impact of pollutants: (a) SO_2; (b) NH_3; the y axes show annual ECEC flows in sej/yr and the corresponding impact in DALYs/yr; the x axis is the sector serial number (the black part of each bar represents direct inputs, and the white part represents indirect inputs).

results for individual resources are expressed in a single consistent thermodynamic unit of solar equivalent joules while accounting for differences in their quality. In this regard, TIOA can be more useful than existing techniques such as EIO-LCA that report consumption and emission data in disparate units. It is then left to the user to distill these data into a smaller number of indices that are sufficiently representative and easy to use.

In the absence of a theoretically rigorous technique for combining disparate data, arbitrary valuation is often employed. TIOA is useful in this context as it presents the type of details of other methods as well as a systematic way of aggregating resource consumption and emission data. Detailed information is also available and hierarchical metrics with different levels of aggregation may be easily developed [68].

For combining results for individual natural resources, human resources, and human health impact of emissions, the algorithm of ECEC analysis is used. Such an algorithm avoids across-the-board addition that could lead to double counting. Rather, it divides the resources into two groups, additive and nonadditive, depending on whether they originate from dependent or independent sources. Renewable resources such as wind and rain originate as co-products from the same source, namely, sunlight, and hence cannot be added. On the other hand, nonrenewable resources such as coal and petroleum originate from independent stocks that have formed over geological time scales, and hence can be added without double counting. For a more in-depth discussion on allocation rules in ECEC analysis, the reader can refer to Hau and Bakshi [48] and Odum [4]. To calculate aggregate metrics in this chapter, inputs from atmosphere, hydrosphere, and ecosystem services are considered to be nonadditive whereas the rest are considered to be additive. This is so because, in the case of nonrenewable resources such as minerals and fossil fuels, allocation is possible and is typically done in proportion to their mass fraction in Earth's sedimentary cycle. In the case of renewable resources, however, such allocation is not possible as they are by-products of the same energy input to the Earth system. Because the choice of allocation rules is usually subjective, the sensitivity of the results to different allocation rules should be evaluated. It may also be possible to select system boundaries that avoid allocation altogether. The application of such techniques to the analysis presented in this chapter is a part of the ongoing work.

12.5.1 Total ECEC, Total ICEC, and ECEC/ICEC Ratios

The total ECEC of each industry sector is shown in Fig. 12.8, which is a semilog plot that shows relative contributions of renewable resources, nonrenewable resources, human resources, and human health impact of emissions to the total ECEC of each sector. The sector of stone mining and quarrying (NAICS 212310) is found to have the highest ECEC. Other sectors with high ECEC values are coal mining (NAICS 212100), power generation and supply (NAICS 221100), and sand, gravel, clay, and refractory mining (NAICS 212320). Sectors with the smallest ECEC are industrial pattern manufacturing (NAICS 332997), malt manufacturing (NAICS 311213), and tortilla manufacturing (NAICS 311830). Sectors with the smallest ECEC requirements are also among the sectors with the smallest economic activity.

The total ECEC requirement captures the CEC in all the links of the production network and, in principle, is equivalent to the concept of ecological cost. Unlike ecological cost that focuses on only industrial stages of the production network and nonrenewable resources, the total ECEC considers renewable resources along with nonrenewable resources and exergy consumed in the ecological links along with the industrial links of a production network. Figure 12.8(a) can be useful in determining an industry-specific proecological tax, as proposed by Szargut and others [69]. The ECEC by itself is of limited use for sustainable decision making. A normalized metric that compares ecosystem contribution with economic activity is more insightful and is discussed in Subsection 12.5.2.

Figure 12.9 shows the total ICEC of individual industry sectors. Agriculture and forestry sectors (NAICS 1111A0–113A00) and sectors relying on them for raw materials, such as the sector of sawmills (NAICS 321113), have some of the highest peaks in Fig. 12.9. The high ICEC requirement of agricultural and forestry sectors can be explained on account of inputs of sunlight. Furthermore, the sectors involved in extraction and processing of nonrenewable resources such as coal mining, oil and gas extraction, and petroleum refining do not appear prominently in Fig. 12.9. ICEC analysis considers a transformity of 1sej/J for all the resources and ignores the

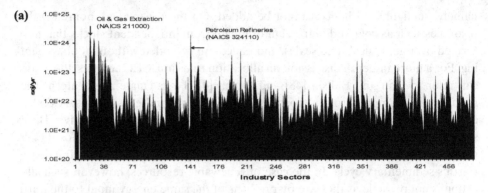

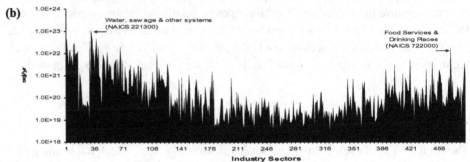

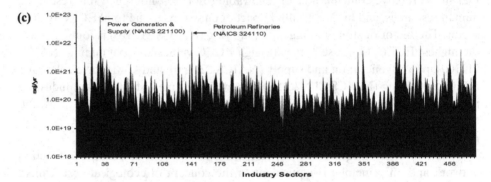

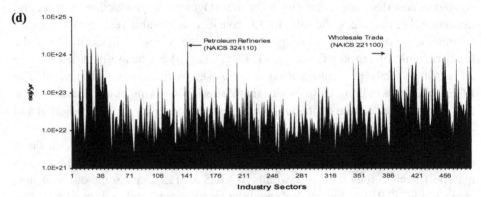

Figure 12.8. ECEC requirements of industry sectors; (a) nonrenewable resources, (b) renewable resources, (c) human health impact of emissions, (d) total; the *y* axes show the annual ECEC flows in sej/yr; the *x* axis is the sector serial number.

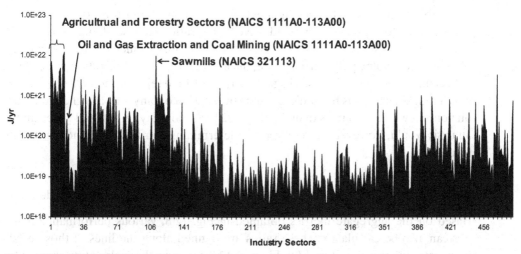

Figure 12.9. Total ICEC requirements of industry sectors.

fact that sunlight is practically free whereas a substantial amount of exergy needs to be expended by ecological processes to produce fossil fuels and to make them available to the economic system. Consequently, ICEC analysis tends to downplay the contribution from nonrenewable resources. Conversely, ECEC analysis does consider different transformity values of ecological resources.

Figure 12.10 depicts ECEC–ICEC ratios for industry sectors from the 1997 benchmark model of the U.S. economy. The ECEC–ICEC ratio indicates the extent to which ICEC analysis underestimates the contribution of ecological resources. As seen from Fig. 12.10, forest nurseries, forest products, and timber tracts (NAICS 113A00) have the lowest, whereas sand, gravel, clay, and refractory mining (NAICS 212320) have the highest ECEC–ICEC ratios among all sectors. In general, agricultural and forestry sectors have lower ECEC–ICEC ratios because of their reliance on

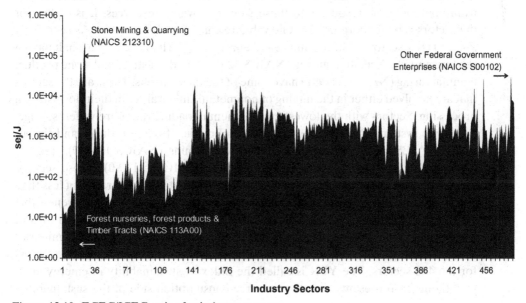

Figure 12.10. ECEC/ICEC ratios for industry sectors.

renewable resources such as sunlight, which have smaller transformity. Mining and extraction sectors, on the other hand, have higher ratios because of their reliance on nonrenewable resources. These results conform to those obtained for the 1992 model of the U.S. economy [30]. Furthermore, the average ECEC–ICEC ratio for the 1997 U.S. economy model is 2873 sej/J against a ratio of 1860 sej/J for the 1992 U.S. economy model. This plausibly indicates that the 1997 economy had a higher reliance on nonrenewable resources than did the 1992 U.S. economy. A detailed uncertainty analysis would, however, be required for determining confidence bounds on these results.

12.5.2 Yield Ratio, Environmental Loading Ratio, and Sustainability Index

Based on the aggregate metrics obtained in Fig. 12.8, various performance metrics can also be calculated. These ratios are defined along the lines of those used in emergy analysis literature [70]. Figure 12.11(a) shows the yield ratio for the 488 industry sectors. The yield ratio is defined as the ratio of total ECEC requirements to indirect ECEC requirements. Consequently, a peripheral sector that derives a large portion of its ECEC requirements directly from ecosystems or human resources has a higher yield ratio and vice versa. This is evident from Fig. 12.11(a), which shows high peaks for nonmetallic mineral mining sectors (NAICS 212310 and 212320) and water sewage and other systems sectors (NAICS 221300). Other federal government enterprises (NAICS S00102) and home health care services (NAICS 621600) also have prominent peaks as they rely heavily on human resources. Sectors with the lowest yield ratios include veterinary services (NAICS 541940), automotive repair and maintenance, except car washes (NAICS 8111A0), and religious organizations (NAICS 813100). These are service industries that are embedded in the economic network and have relatively lower direct reliance on ecological or human resources. Figure 12.11(b) shows the environmental loading ratio for the 488 industry sectors. The environmental loading ratio is defined as the ratio of total ECEC requirements from nonrenewable resources to those from renewable resources. It is higher for the sectors relying more on nonrenewable resources and vice versa. As seen from Fig. 12.11(b), sectors of stone mining and quarrying (NAICS 212310), asphalt paving mixture and block manufacturing (NAICS 324121), and cut stone and stone product manufacturing (NAICS 327991) have some of the tallest peaks. These are the sectors that are involved either in the mining of nonmetallic minerals or in their downstream processing. Sectors with the lowest environmental loading ratios are water, sewage, and other systems (NAICS 221300), forest nurseries, forest products, and timber tracts (NAICS 113A00), vegetable and melon farming (NAICS 111200), tree nut farming (NAICS 111335), and oilseed farming (NAICS 1111A0). These sectors, along with other agricultural sectors, have environmental loading ratios of less than unity, indicating that they rely on renewable resources more than on nonrenewable resources. All other sectors in the economy have environmental loading ratios of higher than unity because of a heavy reliance on metallic and nonmetallic minerals and fossil energy sources. Figure 12.11(c) shows the yield-to-loading ratios (YLRs) for the 488 sectors. The YLR is called the index of sustainability in emergy analysis, though it represents only the resource-consumption side of the sustainability

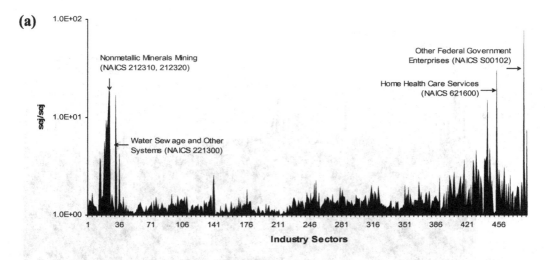

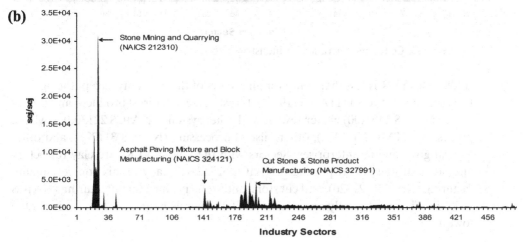

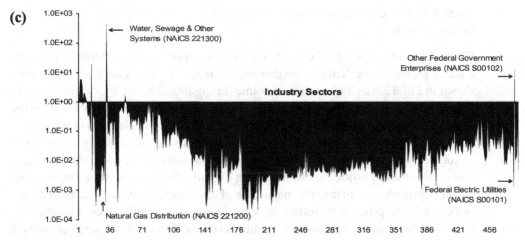

Figure 12.11. Performance metrics of industry sectors: (a) yield ratio, (b) environmental loading ratio, (c) sustainability index.

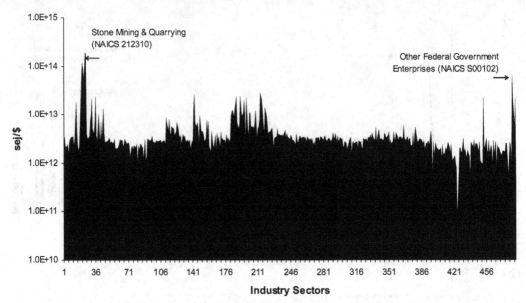

Figure 12.12. ECEC/money ratios for industry sectors.

riddle. The YLR is less than unity for all sectors of the economy except the agricultural sectors (NAICS 1111A0–1119A0), forest nurseries, forest products, and timber tracts (NAICS 113A00), water sewage and other systems (NAICS 221300), soybean processing (NAICS 311222), other oilseed processing (NAICS 311223), and other federal government enterprises. Sectors with the lowest YLRs are clay refractory and other structural clay products (NAICS 32712A), ready-to-mix concrete manufacturing (NAICS 327320), and cut stone and stone product manufacturing (NAICS 327991). Thus sectors relying on nonmetallic minerals in general have some of the lowest YLRs.

12.5.3 Ratios of ECEC–Money and ICEC–Money and Their Implications for Sustainability

Figure 12.12 shows the *ECEC–money ratio* of each of the 488 industry sectors on a semilog plot. It is calculated by dividing total ECEC throughput of each sector shown in Fig. 12.8(d) by its total economic throughput. The ECEC–money ratio is analogous to the emergy–money ratio used in emergy analysis and similar ratios suggested in exergy analysis [29, 69]. However, unlike the single ratio in emergy or exergy analysis for the entire economy, Fig. 12.12 provides a separate ratio for each sector. The variation in Fig. 12.12 confirms the heterogeneous nature of the economy. The ECEC–money ratio does not support or debunk any theory of value, but is rather meant to provide insight into the magnitude of discrepancy between thermodynamic work needed to produce a product or service and people's willingness to pay for it. ECEC–money ratios can be used to quantify ecological cumulative exergy contained in purchased inputs of industrial processes.

Such industry-specific ratios provide a more accurate alternative to the single emergy/$ ratio used in emergy analysis and similar ad hoc procedures used in thermoeconomics. Normalization with respect to money is possible because monetary

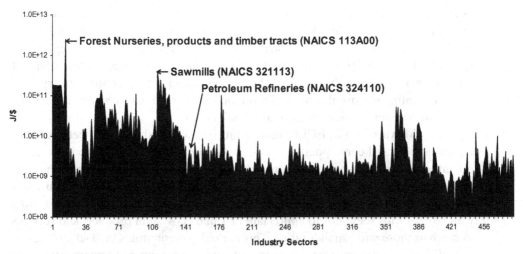

Figure 12.13. ICEC–money ratios for industry sectors.

outputs of industry sectors are well known. However, normalization with respect to exergy to determine transformity or cumulative degree of perfection (CDP) values of industry sectors is more difficult because of lack of information about exergetic outputs of industry sectors. The ECEC–money ratio is a measure of CEC in the production chain of an industry sector to generate $1 of economic activity.

Figure 12.13 shows the ICEC–money ratios of individual industry sectors. Agriculture and forestry sectors (NAICS 1111A0–113A00), in general, have high ICEC–money ratios. Sectors with direct inputs of sunlight such as forest nurseries, forest products, and timber tracts (NAICS 113A00), logging (NAICS 113300), and sectors that rely on forest products such as sawmills (NAICS 321113) and veneer and plywood manufacturing (NAICS 32121A) have some of the highest ICEC–money ratios. Service industries including employment services (NAICS 561300), insurance (NAICS 524100 and 524200), and monetary authorities (NAICS 52A000) have some of the lowest ICEC–money ratios. This observation can be explained based on higher economic capital generation vis-à-vis natural capital consumption of service industries. However, unlike results from Fig. 12.12, sectors involved in the extraction and processing of nonrenewable resources such as oil and gas extraction (NAICS 211000), ground or treated minerals and earth manufacturing (NAICS 327992), and natural gas distribution (NAICS 221200) also have low ICEC–money ratios. This again can be explained considering the ignorance of ecological processes in ICEC analysis and assumption of a uniform transformity of 1 sej/J for all resources.

As seen from Fig. 12.12, the mining sectors have the high ECEC–money ratios. Sectors of stone mining and quarrying (NAICS 212310), sand, gravel, clay and refractory mining (NAICS 212320), and iron ore mining (NAICS 212210) have some of the highest ECEC–money ratios. Sectors with the smallest ECEC–money ratios are lessors of nonfinancial intangible assets (NAICS 533000), owner-occupied dwellings (NAICS S00800), and all other miscellaneous professional and technical services (NAICS 5419A0). Sectors such as primary smelting and refining of copper (NAICS 331411) that rely on mining sectors also have high ECEC–money ratios. In general, more specialized sectors have lower ECEC–money ratios than the basic infrastructure sectors. For instance, the median ECEC–money ratio of finance, insurance, real estate, rental, and leasing sectors (NAICS 522A00–533000) is approximately 1/10 of that of mining and utilities (NAICS 211000–221300) sectors. Among sectors receiving direct inputs from ecosystems, agriculture, forestry, fishing, and hunting sectors (NAICS 1111A0–115000) have a median ECEC–money ratio that is 14% of that of the mining and utilities (NAICS 211000–221300) sectors. Agriculture, forestry, fishing, and hunting sectors also depend more on renewable resources than the mining and utilities sectors, which are primarily fossil based.

The wide variation in ECEC–money ratios indicates the discord between ecological activity and corresponding economic valuation. Because economic value is not inherent in objects but is a product of a variety of consumer judgments, the variation in this ratio may also reflect societal preferences because of a lack of consumer awareness about ecosystem contributions toward economic activity. Thus sectors with larger ratios seem not to appreciate or value ecosystem products and services as much as those with smaller ratios. This not only corroborates the lack of integration of the "ecoservices" sector with the rest of the economy but also quantifies the magnitude of this discrepancy [31, 71].

Furthermore, the ECEC–money ratio tends to decrease along supply chains of industrial processes. Basic infrastructure sectors that lie at the economy–ecosystem interface and the sectors that rely more heavily on nonrenewable resources have higher ECEC–money ratios. This suggests that sectors with high ECEC–money ratios consume natural capital in a manner that is disproportionate to their contribution to economic capital.

The resultant hierarchical structure of the economy resembles an ecological food chain wherein basic infrastructure industries constitute the base and are equivalent to photosynthetic tissue, whereas the value-added service industry constitutes the top and is equivalent to carnivores. These observations match other work on the relationship between environmental impact and economic value along supply chains of industrial processes [72].

Such a hierarchical structure for the 28 major subdivisions of the U.S. economy is depicted in Fig. 12.14. These subdivisions are defined by the Bureau of Economic Analysis [65], and have been used in EIO-LCA [47]. This aggregation scheme is preferred as it provides a more concise overview of the economy than the three-digit NAICS codes, and yet is more detailed than the two-digit NAICS codes. Figure 12.14 shows median ECEC–money ratios of each of the 28 subdivisions along with ratios of the constituent sectors in each subdivision. Basic extractive and infrastructure subdivisions such as mining and utilities, plastic, rubber and nonmetallic mineral products, and ferrous and nonferrous metal products constitute the base, whereas more specialized subdivisions such as finance, insurance, real estate, and professional and technical services constitute the top. Manufacturing sectors such as vehicles and other transportation equipment, textiles and leather products, and semiconductor manufacturing occupy the middle. This general trend is maintained even for other aggregation schemes.

These observations also provide a unique insight into sustainability of industrial supply chains from the standpoints of weak- and strong-sustainability paradigms, and other macroeconomic phenomena, including outsourcing, sustainable international trade, and corporate restructuring [73]. Because basic infrastructure industries are the underperformers of the economy in relation to the high value-added service and

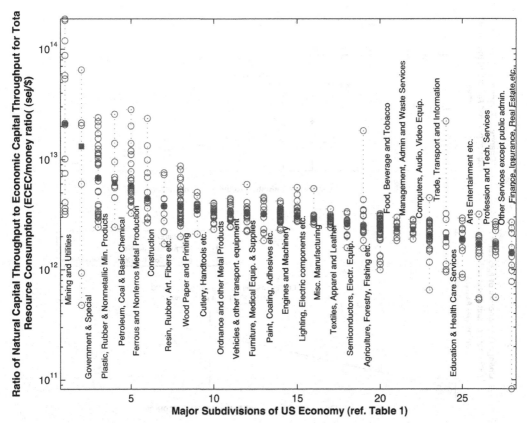

Figure 12.14. Subdivisions of the U.S. economy organized in descending order of median ECEC/$ ratios. Open circles represent ratio for individual sectors in each subdivision; filled squares represent the median for each subdivision.

advanced manufacturing industries, corporations have often sold off or outsourced such assets to gain a strategic advantage. Such actions may also allow them to move to trajectories of higher growth by switching to *emerging markets* and *new technologies* and positioning themselves favorably in market cycles of creative destruction [74–75]. For example, DuPont spun off Conoco and Monsanto divested its commodity chemicals business with this objective in mind [76]. The higher ECEC-to-money ratio for these basic industries means that getting rid of them will also improve commonly used sustainability and eco-efficiency metrics, at least for as long as natural capital remains undervalued. The approach used in this chapter provides a more holistic and multidimensional approach, which may be used for full cost accounting and corporate sustainability metrics by considering the broader implications of corporate decisions over multiple scales.

The replacement of less value-added industries with more value-added ones is also evident on a macroeconomic scale, wherein business enterprises in developed countries are increasingly outsourcing extractive and manufacturing-related activities abroad and are replacing them with service industries that are better at value addition, have higher growth prospects and returns on investment and lower risk perceptions and environmental costs. For instance, 50% of the manufactured goods bought by American people today are produced abroad, up from 31% in 1987 [77].

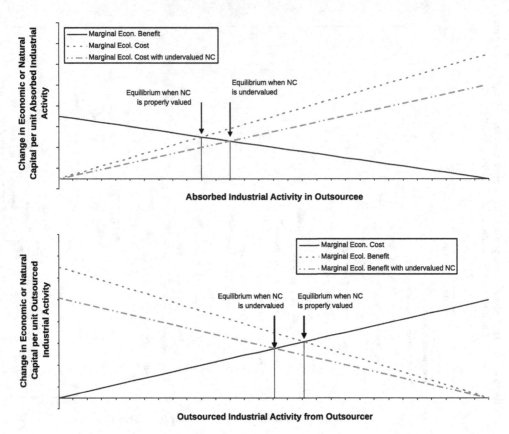

Figure 12.15. (a) Marginal changes in economic and natural capital as functions of absorbed industrial activity in outsourcees, (b) marginal changes in economic and natural capital as functions of outsourced industrial activity in outsourcers.

Even activities such as software writing and customer help desks that are being outsourced seem to have relatively less value addition in comparison with the activities that are higher up in their supply chains, namely, finance, health care services, banking, and insurance. As industrial activity in developed countries shifts toward the more value-added end of the spectrum, the result is reduced consumption of natural capital per unit of economic capital. The exact opposite situation occurs in developing countries, where absorption of the outsourced activity leads to creation of economic capital at the expense of natural capital. This is shown with the help of a hypothetical example in Fig. 12.15. In either case, a sustainability limit based on the weak-sustainability *paradigm* would follow the theory of comparative advantages and would coincide with the point where marginal changes in the net sum of economic, natural, and social capitals turn negative [78–80].

In Fig. 12.15 this coincides with the point where marginal benefit and marginal cost curves intersect. At this equilibrium point, the net capital base also reaches a maximum. Furthermore, the equilibriums for outsourcees and outsourcers may not coincide as valuation of economic and natural capitals may differ from region to region. Figures 12.15(a) and 12.15(b) also show equilibrium points for outsourcees and outsourcers, respectively, when natural capital is undervalued. In such a case

the marginal ecological cost curve for outsourcees and the marginal ecological benefit curve for outsourcers shift downward. Consequently the new equilibrium points
represent a higher sustainability limit for absorption of outsourced activity in outsourcees and a lower sustainability limit for outsourcing of industrial activity in
outsourcers.

From the viewpoint of strong sustainability, outsourcing may reduce sustainability of the outsourcees if their lost natural capital is irreplaceable or falls below
a critical limit. This loss of natural capital is likely to be even more than what is
indicated by the ratios calculated in our work if environmental regulations are weak
or not enforced. Identification and quantification of critical components of natural
capital that make a unique contribution to welfare and cannot be substituted by
other forms of capitals are important and active areas of research [80, 81]. Similarly,
criticality of natural capital also depends on various economic, ecological, political,
and social aspects that differ in space and time [82].

Most of the existing methods for quantifying industrial sustainability normal-
ize the environmental burden by monetary value added [83] and do not consider
effects on economic and natural capitals on larger scales. This can create an illusion
of sustainable development because such sustainability indicators can be improved
by simply becoming more profitable or moving up the economic food chain, while
actually eroding the net productive capital base they rely on for their future operations. Consideration of marginal changes in economic and natural capitals coupled
with identification and quantification of critical natural capital can be a more rigorous way of addressing sustainability issues. The data and approach used in this
analysis may be combined with process and life-cycle information to enable the
development of more holistic and hierarchical *sustainability metrics* [68, 84]. For
the outsourcees to enhance their sustainability, they must use the economic capital
available from outsourced activities to quickly move up the economic food chain
toward more economically value-added industries, without sending natural capital
below its critical limit. Similarly, compensating for any loss of economic capital and
jobs for outsourcers also requires them to move further up the economic food chain
via new innovations. In the short run, the sustainability of an existing economy may
be improved by small changes in the valuation of natural capital in sectors that
appear frequently in most supply chains. Such keystone sectors include real estate,
wholesale trade, and power generation and supply. However, from a global perspective, adjustment in market prices to reflect the contribution of natural capital is
ultimately necessary for sustainability. This will require combination of the type of
analysis presented in this section with economic principles and knowledge about the
crucial role of natural capital.

12.6 Hybrid Thermodynamic Life-Cycle Assessment

ICEC–money and ECEC–money ratios are particularly useful in hybrid thermodynamic life-cycle analyses of industrial systems. A *hybrid analysis* integrates process
models or product systems with economy-scale input–output models and, in the process, combines accurate, process-specific data with more uncertain economy-scale
data [85]. Consequently a hybrid analysis is more powerful as it combines the two

critical attributes of an environmental decision tool, specificity and a broad system boundary. ICEC–money and ECEC–money ratios can come in handy in this context as the interactions of a product system with the rest of the economy are routinely measured in monetary terms in normal accounting procedures.

This section applies the exergy, ICEC, and ECEC algorithms discussed briefly in Section 12.2.2 of this chapter as well as in the introductory chapters of this book to compare six alternative *electricity-generation systems*. These systems are *hydroelectric, wind, geothermal, natural gas, oil*, and *coal*. Further details about their purchased and direct environmental inputs and emissions can be found in [64]. Hybrid thermodynamic LCA was performed by combining detailed process-scale information with the ICEC–money and ECEC–money ratios discussed in Section 12.5.3 respectively. The resulting approach is similar to a tiered hybrid LCA [85]. Algorithms for combining process-scale information with ICEC–money and ECEC–money ratios are beyond the scope of this chapter, but can be found in Ukidwe [64].

The results are shown in Figs. 12.16 and 12.17. Based on the results shown in these figures, the following conclusions can be drawn.

- Power plants based on nonrenewable resources have higher exergetic efficiencies than those based on wind and geothermal heat. This is so because fossil fuels represent more concentrated forms of energy than wind or geothermal heat. A hydroelectric system is an exception, as it can be converted to shaft work more easily.
- As the system boundary is expanded to the scale of the economy, efficiency values of oil-, coal-, and ng-based power plants decrease to a larger extent than those of geothermal, hydro, and wind-based power plants. This can be explained considering the fact that oil-, ng-, and coal-based systems have relatively longer industrial supply chains than hydro, wind, and geothermal systems. Oil, coal, and ng are extracted and processed by other industry sectors before being consumed by utility companies. On the other hand, wind, geothermal heat, and hydropotential are directly obtained from the environment. A longer supply chain means more avenues for loss of exergy and a greater potential for a lower industrial (CDP). However, in spite of this decrease, oil-, ng-, and coal-based systems still remain more efficient than the wind-based power plant.
- As the system boundary is expanded further to the scale of the ecosystems, efficiency values of oil-, coal-, and ng-based power plants decrease to such an extent that they no longer remain more efficient than wind and geothermal systems. Lower ecological (ECDP) values for oil-, coal-, and ng-based systems can be attributed to higher transformities of these resources. In other words, ECDP values reflect not only the cumulative efficiencies of industrial supply chains that consume these natural resources but also those of the ecological networks that produce them. This is very significant as it allows ECDP values to acknowledge quality differences between natural resources – such as the degree of renewability – whereas exergetic efficiency and ICDP cannot.
- The yield ratio shown in Fig. 12.17(a) shows a clear distinction between power plants based on renewable and nonrenewable resources. Plants based on renewable resources have higher yield ratios than those based on nonrenewable

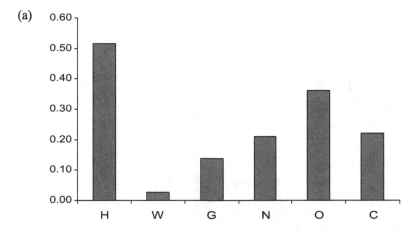

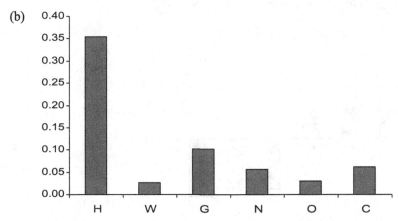

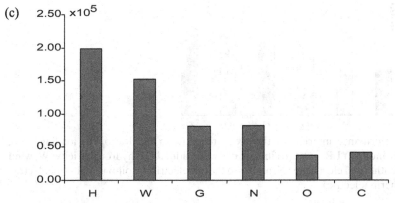

Figure 12.16. Efficiency values for the six systems: (a) exergetic efficiency, (b) ICDP; (c) ECDP (H, hydroelectricity; W, wind electricity; G, geothermal electricity; N, ng-based thermoelectric; O, oil-based thermoelectric; C, coal-based thermoelectric).

resources. Power plants based on renewable resources derive a major fraction of their energy requirements directly from the environment and rely less on economic activities. On the other hand, power plants based on nonrenewable resources derive a significant portion of their energy requirements from other

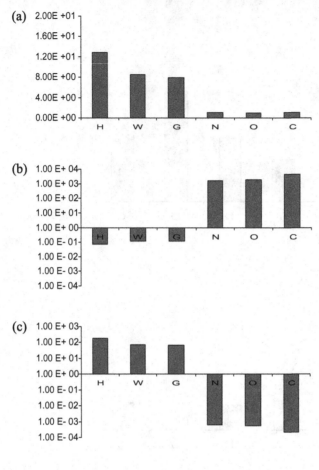

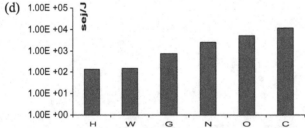

Figure 12.17. Performance metrics for the six systems: (a) Yield Ratio, (b) loading ratio, (c) sustainability index (YLR), and (d) impact per value added (H, hydroelectricity; W, wind electricity; G, geothermal electricity; N, ng-based thermoelectric; O, oil-based thermoelectric; C, coal-based thermoelectric).

industry sectors and have relatively longer industrial supply chains. These results also match those obtained by Ulgiati and Brown [86].

• The environmental loading ratio shown in Fig. 12.17(b) also shows a clear distinction between power plants based on renewable and nonrenewable resources. Power plants based on nonrenewable resources have a loading ratio of greater than 1 whereas those based on renewable resources have a loading ratio of less than 1. The environmental loading ratio is similar to the renewability indicator proposed by Berthiaume et al. [87]. If a contribution from nonrenewable resources (θ_{total}^{NR}) is assumed to be equal to the amount of work required to restore them to their original condition (W_R), and the difference between the yield (Y) and the restoration work (W_R) is assumed to be the useful work (W_P), an alternative definition of renewability indicator (I_r) can be derived as $I_r = (W_P - W_R)/W_P = (Y - 2 \times \theta_{total}^{NR})/(Y - \theta_{total}^{NR})$. Accordingly I_r values for hydro-, wind-, geothermal-, ng-, oil- and coal-based systems were evaluated as 0.94, 0.895, 0.889, -4.01, -7.07, and -19.23, respectively. An I_r value between 0 and 1 indicates a partially renewable process, whereas a negative I_r value indicates a process that needs more work for restoration than it produces. This trend in I_r corresponds well with that for the environmental loading ratios shown in Fig. 12.17(b).

- The YLR shown in Fig. 12.17(c) indicates that power plants based on renewable resources are more sustainable than those based on nonrenewable resources. Such a ratio is termed the index of sustainability in emergy literature, though in reality it reflects only the resource-consumption side of the sustainability riddle. Other attributes, including its economics and its impact on society, need to be considered before determining sustainability.
- Figure 12.17(d) shows that a coal-based power plant has the highest human health impact per unit electricity generation, whereas a hydroelectric power plant has the lowest. Power plants based on fossil energy sources emit CO_2, which affects human health through global warming and climate change. The combustion of coal, in particular, also emits SO_2 and NO_x that cause acidification and photochemical smog formation. Barring indirect emissions, power plants based on renewable resources do not have this problem and have a much lower impact on human health. At this point it is necessary to note that other impact categories such as an impact on ecosystem health and land use have not been considered in this analysis. Land use, in particular, is likely to be significant for the hydroelectric power plant. These effects must be considered to make a more comprehensive appraisal of the six electricity systems.

Figures 12.16(c) and 12.17(d) suggest a plausible correlation between ECDP and impact per value added that has been hypothesized but is yet unproven. Such a correlation, if statistically validated, is of great practical significance as it would enable decision makers to get an approximate idea about the impact of emissions based on material and energy inputs and without requiring detailed knowledge about process emissions and their toxicological aspects. This is particularly important in the early design stages, when knowledge about emissions and their impacts may not be readily available. This may imply following hypothesis.

Hypothesis: Among alternatives for making similar products, the one with a smaller thermodynamic efficiency has a larger life-cycle impact.

This hypothesis is based on the fact that all activities in ecological and industrial systems typically involve a reduction of entropy that is due to the creation of order in the desired products. As known from the second law of thermodynamics, such a reduction of entropy must cause a larger increase in the entropy of the surroundings. This increase may be interpreted as the creation of disorder or environmental impact. Because entropy increase corresponds to a decrease in the exergy, if the exergy loss during a life cycle can be quantified, it can provide a rough idea of the potential environmental impact of that life cycle. For such quantification to be meaningful, the inclusion of ecological systems in the analysis is essential. Otherwise processes that use a higher-quality natural resource, such as coal instead of sunlight for generating electricity, will be more thermodynamically efficient, and will often violate the hypothesis. Verification of this hypothesis requires many more studies that are part of the ongoing research.

12.7 Conclusion

This chapter presents data about inputs of ecosystem products and services, and human resources to and emissions of pollutants from the industry sectors defined according to their North American Industrial Classification System (NAICS) codes. Calculations for determining ICECs and ECECs associated with these flows were presented in detail along with the underlying assumptions at appropriate places in this analysis. The TIOA synthesizes concepts from systems ecology, engineering thermodynamics, and EIO analysis to study exergy flows in economic–ecological systems and to evaluate direct and indirect reliance of industry sectors on ecological resources and the impact of emissions from them. It treats the economic–ecological system as a single network of energy flows with exergy as the common currency. Although TIOA uses knowledge from systems ecology, it is free of all the controversial aspects associated with it, such as maximum empower principle, emergy theory of value, and reliance on prehistoric energy. Application of TIOA to the 1997 U.S. economic model yields unique results, with wide applications at microscales as well as macroscales. For instance, this study presents total ECEC requirements of industry sectors from non-renewable and renewable ecological resources, human resources, and human health impact of emissions. The total ECEC requirement is similar to the concept of ecological cost or the environmental footprint defined in exergetic terms. The analysis also presents industry-specific ECEC–money ratios. These ratios describe the discord between thermodynamic work required for an industrial operation and willingness of people to pay for it. Such discord is widely recognized to arise from lack of integration of ecosystems with the rest of the economy. The ECEC–money ratios presented in this chapter evaluate the magnitude of this discord for individual industry sectors and could prove useful in better internalization of ecological resources into economic policies and in formulating pre-ecological taxes. ECEC–money ratios are fundamentally identical to emergy/$ ratios used in systems ecology. However, unlike a single emergy/$ ratio for the entire economy, this analysis determines a separate ratio for each industry sector. Consequently ECEC–money ratios are not only readily applicable wherever emergy/$ ratios are used in emergy analysis, but also provide a more accurate and disaggregate alternative to the later. Industry-specific ICEC–money and ECEC–money ratios can also be used in hybrid thermodynamic LCA to compare alternative products and process. The unique insight provided by thermoLCA was illustrated by comparing six alternative electricity-generation systems. The results indicate a plausible correlation between cumulative degree of perfection at the scale of the ecosystems and the normalized human health impact of emission. Such a correlation, if valid, would be immensely useful in determining environmental implications of current and emerging technologies for which ecotoxicological data is not easily available. Such correlation would also enable construction of impact indicators based on material and energy inputs to a product or process and could be used for screening at an advance stage of design. Recent work [88] has strengthened these results by incorporating additional ecological resources. The resulting on-line software is available at http://resilience.osu.edu/ecolca.

12.8 Acknowledgments

The authors acknowledge financial support from National Science Foundation grant no. BES-9985554 and The Ohio State University Presidential Fellowship Program.

REFERENCES

[1] D. Tilman, K. G. Cassman, P. A. Matson, R. Naylor, and S. Polasky, "Agricultural sustainability and intensive production practices," *Nature (London)* **418**, 671–677 (2002).

[2] G. C. Daily, *Nature's Services: Societal Dependence on Natural Ecosystems, Ecosystem Services and Their Importance* (Island Press, Washington, D.C., 1997).

[3] R. Costanza, R. d'Agre, R. de Groot, S. Farber, M. Grasso, B. Hannon, K. Limburg, S. Naeem, R. V. O'Neil, J. Paruelo, R. G. Raskin, P. Sutton, and M. Van Den Belt, "The value of the world's ecosystem services and natural capital," *Nature (London)* **387**, 253–260 (1997).

[4] H. T. Odum, *Environmental Accounting: Emergy and Environmental Decision Making* (Wiley, New York, 1996).

[5] World Resource Institute, *World Resource 2000–2001: People and Ecosystems: The Fraying Web of Life* (World Resource Institute: Washington D.C., 2000), available at http://www.wri.org/wr2000 (accessed May 2004).

[6] WWF, *Living Planet Report 2000* (World Wildlife Foundation, 2000), available at http://panda.org/livingplanet/lpr00/ (accessed May 2004).

[7] UNEP, *Global Environmental Outlook 3: Past, Present and Future Perspectives* (Earthscan, Sterling, VA, 2002), available at http://www.unep.org/geo3 (accessed May 2003).

[8] Millenium Assessment, "Living beyond our means–natural assets and human well-being," Millennium Ecosystem Assessment, available at http://www.millenniumassessment.org/en/index.aspx (accessed April 2005).

[9] C. O. Holliday Jr., S. Schmidheiny, and P. Watts, *Walking the Talk: A Business Case for Sustainable Development* (Berrett-Koehler, San Francisco, CA, 2002).

[10] American Institute of Chemical Engineers, *Total Cost Assessment Methodology* (Center for Waste Reduction Technologies, 2004), available at http://www.aiche.org/cwrt/projects/cost.htm (accessed Sept. 2004).

[11] A. Balmford, A. Bruner, P. Cooper, R. Costanza, S. Farber, R. E. Green, M. Jenkins, P. Jefferiss, V. Jessamy, J. Madden, K. Munro, N. Myers, S. Naeem, J. Paavola, M. Rayment, S. Rosendo, J. Roughgarden, K. Trumper, and R. K. Turner, "Ecology – Economic reasons for conserving wild nature," *Science* **297**, 950–953 (2002).

[12] A. Adriaanse, S. Bringezu, A. Hammond, Y. Moriguchi, E. Rodenburg, D. Rogich, and H. Schutz, *Resource Flows: The Material Basis of Industrial Economies* (World Resource Institute, Washington, D.C., 1997).

[13] E. Matthews, C. Amann, S. Bringezu, M. Fischer-Kowalski, W. Huttler, R. Kleijn, Y. Moriguchi, C. Ottke, E. Rodenburg, D. Rogich, H. Schandl, H. Schutz, E. Van Der Voet, and H. Weisz, *The Weight of Nations: Material Outflows from Industrial Economies* (World Resource Institute, Washington, D.C., 2000).

[14] ConAccount, 2002, *Material Flow Accounting for Environmental Sustainability*, available at http://www.conaccount.net/ (accessed Dec. 2002).

[15] D. T. Spreng, *Net-Energy Analysis and the Energy Requirements of Energy Systems*, 1st ed. (Praeger, New York, 1988).

[16] B. Hannon, "Analysis of the energy cost of economic activities: 1963 to 2000," *Energy Syst. Policy J.* **6**, 249–278 (1982).

[17] J. Szargut, D. R. Morris, and F. R. Steward, *Exergy Analysis of Thermal, Chemical and Metallurgical Processes*, 1st ed. (Hemisphere, New York, 1988).

[18] R. L. Cornelissen and G. G. Hirs, "Exergetic optimization of a heat exchanger," *Energy Convers. Manage.* **38**, 1567–1576 (1997).

[19] M. Lenzen, "Errors in conventional and input–output-based life-cycle inventories," *J. Ind. Ecol.* **4**, 127–148 (2000).

[20] I. S. Ertesvag, "Society exergy analysis: A comparison of different societies," *Energy* **26**, 253–270 (2001).

[21] A. Hepbasli, "Modeling of sectoral energy and exergy utilization," *Energy Sources* **27**, 903–912 (2005).

[22] I. S. Ertesvag, "Energy, exergy, and extended-exergy analysis of the Norwegian society 2000," *Energy* **30**, 649–675 (2005).

[23] M. A. Rosen and I. Dincer, "On exergy and environmental impact," *Intl. J. Energy Res.* **21**, 643–654 (1997).

[24] T. P. Seager and T. L. Theis, "A uniform definition and quantitative basis for industrial ecology," *J. Cleaner Prod.* **10**, 225–235 (2002).

[25] J. DeWulf, H. Van Langenhove, and J. Dirckx, "Exergy analysis in the assessment of the sustainability of waste gas treatment systems," *Sci. Total Environ.* **273**, 41–52 (2001).

[26] M. T. Brown and R. A. Herendeen, "Embodied energy analysis and EMERGY analysis: A comparative view," *Ecol. Econ.* **19**, 219–235 (1996).

[27] J. L. Hau and B. R. Bakshi, "Promise and problems of emergy analysis," *Ecol. Model.* **178**, 215–225 (2004).

[28] R. Herendeen, "Energy analysis and EMERGY analysis – A comparison," *Ecol. Model.* **178**, 227–237 (2004).

[29] E. Sciubba, "Cost analysis of energy conversion systems via a novel resource-based quantifier," *Energy* **28**, 457–477 (2003).

[30] N. U. Ukidwe and B. R. Bakshi, "Thermodynamic accounting of ecosystem contribution to economic sectors with application to 1992 US economy," *Environ. Sci. Technol.* **38**, 4810–4827 (2004).

[31] R. U. Ayres, "The price-value paradox," *Ecol. Econ.* **25**, 17–19 (1998).

[32] K. Arrow, G. Daily, P. Dasgupta, S. Levin, K. G. Maler, E. Maskin, D. Starrett, T. Sterner, and T. Tietenberg, "Managing ecosystem resources," *Environ. Sci. Technol.* **34**, 1401–1406 (2000).

[33] W. W. Leontief, *Input–Output Economics* (Oxford University Press, New York, 1936).

[34] R. E. Miller and P. D. Blair, *Input–Output Analysis: Foundations and Extensions* (Prentice-Hall, Englewood Cliffs, NJ, 1985).

[35] J. H. Cumberland, "A regional inter-industry model for analysis of development objectives," *Reg. Sci. Ass. Papers* **17**, 65–94 (1966).

[36] S. B. Noble, "Material/energy accounting and forecasting models," in *Resources, Environment and Economics*, edited by R. U. Ayres (Wiley, New York, 1978).

[37] W. W. Leontief, J. Koo, S. Nasar, and I. Sohn, *The Production and Consumption of Non-Fuel Minerals to Year 2030 Analyzed Within an Input–Output Framework of the US and World Economy*, NSF/CPE-82002 (National Science Foundation, Washington, D.C., 1982).

[38] R. U. Ayres, "A material-process product model," in *Environmental Quality Analysis: Theory and Method in the Social Sciences*, edited by A. V. Kneese and B. T. Bower (Johns Hopkins University Press, Baltimore, MD, 1972).

[39] J. C. Saxton and R. U. Ayres, "The materials-process product model: Theory and applications," in *Mineral Materials Modeling: A State-of-the-Art Review*, edited by W. A. Vogley (Johns Hopkins University Press, Baltimore, MD, 1976).

[40] R. U. Ayres, *Resources, Environment and Economics: Applications of the Materials/Energy Balance Principle* (Wiley, New York, 1978).

[41] F. Duchin, "Input-output analysis and industrial ecology," in *The Greening of Industrial Ecosystems*, edited by B. R. Allenby and D. J. Richards (National Academy Press, Washington, D.C., 1994).

[42] R. Costanza, "Embodied energy and economic valuation," *Science* **210**, 1219–1224 (1980).

[43] R. Costanza and R. Herendeen, "Embodied energy and economic value in the United States economy – 1963, 1967 and 1972," *Resources Energy* **6**, 129–163 (1984).

[44] S. Casler and B. Hannon, "Readjustment potentials in industrial energy efficiency and structure," *J. Environ. Econ. Manage.* **17**, 93–108 (1989).

[45] C. W. Bullard and R. A. Herendeen, "The energy cost of goods and services," *Energy Policy* **3**, 268–278 (1975).

[46] L. B. Lave, E. Cobas-Flores, C. T. Hendrickson, and F. C. McMichael, "Using input–output-analysis to estimate economy-wide discharges," *Environ. Sci. Technol.* **29**, A420–A426 (1995).

[47] EIOLCA (Green Design Initiative, Carnegie Mellon University, Pittsburg, PA, 2004), available at http://www.eiolca.net (accessed Sept. 2004).

[48] J. L. Hau and B. R. Bakshi, "Expanding exergy analysis to account for ecosystem products and services," *Environ. Sci. Technol.* **38**, 3768–3777 (2004).

[49] J. Kay and H. Reiger, "Uncertainty, complexity and ecological integrity: Insights from an ecosystems approach," in *Implementing Ecological Integrity: Restoring Regional and Global Environmental and Human Health*, edited by P. Crabbe, A. Holland, L. Ryszkowski, and L. Westra, NATO Science Series, Environmental Security (Kluwer, Dordrecht, The Netherlands, 2000), pp. 121–156.

[50] M. T. Brown and E. Bardi, *Handbook of emergy evaluation: Folio 3; Emergy of Ecosystems* (Systems Ecology Center, University of Florida, Gainesville, FL, 2001), available at http://www.ees.ufl.edu/cep/downloads/Folio%203.pdf (accessed May 2004).

[51] S. L. Brandt-Williams, *Handbook of Emergy Evaluation: Folio 4; Emergy of Florida Agriculture* (Systems Ecology Center, University of Florida, Gainesville, FL, 2002), available at http://www.ees.ufl.edu/cep/downloads/Folio%204%20.pdf (accessed May 2004).

[52] USDOA, http://www.nass.usda.gov/census/census97/atlas97/summary1.htm (U.S. Department of Agriculture, 2004) (accessed May 2004).

[53] NASA, "NASA Surface Meteorology and Solar Energy," available at http://eosweb.larc.nasa.gov/sse, (accessed May 2004).

[54] USDOE, http://www.eia.doe.gov/emeu/mer/pdf/pages/sec7_5.pdf, (U.S. Department of Energy, 2004) (accessed October 2006).

[55] USDOE, http://www.eia.doe.gov/oil_gas/petroleum/info_glance/consumption.html (U.S. Department of Energy, 2004) (accessed May 2004).

[56] USDOE, http://tonto.eia.doe.gov/dnav/pet/pet_cons_top.asp (U.S. Department of Energy, 2004) (accessed Apr. 2006).

[57] USGS, http://minerals.usgs.gov/minerals/pubs/commodity/ (U.S. Geological Survey, 2004) (accessed May 2004).

[58] R. U. Ayres and L. W. Ayres, *Accounting for Resources I: Economy Wide Applications of Mass Balance Principles to Materials and Waste* (Elgar, Northampton, MA, 1998).

[59] USGS, http://water.usgs.gov/watuse/ (U.S. Geological Survey, 2004) (accessed May 2004).

[60] BLS, http://stats.bls.gov/oes/2000/oesi2_90.htm (U.S. Department of Labor, Bureau of Labor Statistics, 2004) (accessed May 2004).

[61] P. Hofstetter, *Perspectives in Life Cycle Impact Assessment: A Structured Approach to Combine Models of the Technosphere, Ecosphere, and Valuesphere* (Kluwer Academic, Boston, MA, 1998).

[62] M. Goedkoop and R. Spriensma, *The Eco-Indicator 99: A Damage Oriented Method For Life Cycle Impact Assessment, Methodology Report* (PRé Consultants, B. V., Plotterweg 12, 3821 BB, Amersfoort, The Netherlands, 1999).

[63] USEPA, *1999 Toxics Release Inventory: Public Data Release* (U.S. Environmental Protection Agency, 1999) available at http://www.epa.gov/tri/tri99/pdr/1999 pdr.pdf (accessed May 2004).

[64] N. U. Ukidwe, "Thermodynamic Input-Output Analysis Of Economic And Ecological Systems For Sustainable Engineering," Ph.D. dissertation (Ohio State University, Columbus, OH, 2005), available at http://www.ohiolink.edu/etd/view.cgi?osu1117555725 (accessed Apr. 2006).

[65] BEA, http://www.bea.doc.gov/bea/pn/ndn0180.exe (U.S. Department of Commerce, Bureau of Economic Analysis 2004) (accessed May 2004).

[66] National Research Council, *Materials Count: The Case for Material Flow Analysis* (National Research Council, National Academies Press, Washington, D.C., 2004).

[67] E. Sciubba, "Beyond thermoeconomics? The concept of extended exergy accounting and its application to the analysis and design of thermal systems," *Exergy Intl. J.* **1**, 68–84 (2001).

[68] H.-S. Yi, J. L. Hau, N. U. Ukidwe, and B. R. Bakshi, "Hierarchical thermodynamic metrics for evaluating the sustainability of industrial processes," *Environ. Progr.* **23**, 302–314 (2004).

[69] J. Szargut, "Application of exergy for the determination of the pro-ecological tax replacing the actual personal taxes," *Energy* **27**, 379–389 (2002).

[70] S. Ulgiati, M. T. Brown, S. Bastianoni, and N. Marchettini, "Emergy-based indices and ratios to evaluate the sustainable use of resources," *Ecol. Eng.* **5**, 519–531 (1995).

[71] R. U. Ayres, "Eco-thermodynamics: Economics and the second law," *Ecol. Econ.* **26**, 189–209 (1998).

[72] R. Clift and L. Wright, "Relationships between environmental impacts and added value along the supply chain," *Tech. Forecast. Soc. Change* **65**, 281–295 (2000).

[73] N. U. Ukidwe and B. R. Bakshi, "Flow of natural versus economic capital in industrial supply networks and its implications to sustainability," *Environ. Sci. Technol.* **39**, 9759–9769 (2005).

[74] S. Hart and M. B. Milstein, "Global sustainability and creative destruction of industries," *MIT Sloan Manage. Rev.* **41**, 1, 23–33 (1999).

[75] S. Hart and C. M. Christensen, "The great leap: Driving innovation from the base of the pyramid," *MIT Sloan Manage. Rev.* **44**, 1, 51–56 (2002).

[76] G. S. Hedstrom, J. B. Shopley, and C. M. Deluc, Realizing the Sustainable Development Premium Prism, Co **1**, 5–19 (2000).

[78] R. K. Turner, "Sustainability: Principles and practice," in *Sustainable Environmental Economics and Management: Principles and Practice*, edited by R. K. Turner (Belhaven, New York, 1993).

[79] P. Bartelmus, "Dematerialization and capital maintenance: two sides of the sustainability coin," *Ecol. Econ.* **46**, 61–81 (2003).

[79] L. Uchitelle, "Factories move abroad, as does US power," *New York Times*, 2003.

[80] P. Ekins, S. Simon, L. Deutsch, C. Folke, and R. de Groot, "A framework for the practical application of the concepts of critical natural capital and strong sustainability," *Ecol. Econ.* **44**, 165–185 (2003).

[81] R. De Groot, J. Van der Perk, A. Chiesura, and A. van Vliet, "Importance and threat as determining factors for criticality of natural capital," *Ecol. Econ.* **44**, 187–204 (2003).

[82] D. V. MacDonald, N. Hanley, and I. Moffatt, "Applying the concept of natural capital criticality to regional resource mangaement," *Ecol. Econ.* **29**, 73–87 (1999).

[83] G. Biswas, R. Clift, G. Davis, J. Ehrenfeld, R. Forster, O. Jolliet, I. Knoepfel, U. Luterbacher, D. Russell, and D. Hunkeler, "Econometrics: Identification, categorization and life cycle validation," *Intl. J. Life Cycle Anal.* **3**, 184–190 (1998).

[84] T. E. Graedel and B. R. Allenby, "Hierarchical metrics for sustainability," *Environ. Qual. Manage.* **12**, 21–30 (2002).

[85] S. Suh, M. Lenzen, G. J. Treloar, H. Hondo, A. Horvath, G. Huppes, O. Jolliet, U. Klann, W. Krewitt, Y. Moriguchi, J. Munksgaard, and G. Norris, "System boundary selection in life-cycle inventories using hybrid approaches," *Environ. Sci. Technol.* **38**, 657–664 (2004).

[86] S. Ulgiati and M. T. Brown, "Quantifying the environmental support for dilution and abatement of process emissions – The case of electricity production," *J. Cleaner Prod.* **10**, 335–348 (2002).

[87] R. Berthiaume, C. Bouchard, and M. A. Rosen, "Exergetic evaluation of the renewability of a biofuel," *Exergy Intl. J.* **1**, 256–268 (2001).

[88] Y. Zhang, A. Baral, and B. R. Bakshi, "Accounting for ecosystem services in life cycle assessment," *Environ. Sci. Technol.* **44**, 2624–2631 (2010).

13 Synthesis of Material Flow Analysis and Input–Output Analysis

Shinichiro Nakamura

13.1 Introduction

For a model of a production process to be valid in the real world, it has to take proper account of the principles of thermodynamics. Production models in economics are no exception. The conservation of mass requires that proper attention be paid to the mass balance between inputs and outputs entering and leaving a given production process. An increase of entropy implies the generation of process waste in the production phase and the reduction in the purity of materials in the use and end-of-life (EoL) phases. Because process waste is generated in the production phase, it should be classified as an output if the mass balance between inputs and outputs is to be established. The reduction in the purity of materials in the use phase is relevant for materials made of polymers, such as paper, textile, and plastics, whose chemical bindings loosen over time.

On the other hand, for metals such as iron, copper, or aluminum, such a decline in quality in the use phase will not occur [except for possible corrosions (oxidization)] because these metals are elements. In fact, it is not the use phase but the EoL phase in which a serious reduction in the quality of metal materials can occur because of the mixing-up of diverse metal elements or the "contamination" of pure elements with other elements in minor quantities (tramp elements). A typical example of metal contamination by tramp elements is the mixing of iron scrap with copper, which is known to reduce the quality of iron scrap [1, 2]. Because copper is chemically nobler than iron, once mixed in an electric arc furnace (EAF), copper cannot be separated from iron (in an economically reasonable way, at least). Figure 13.1 is an illustration of the decline in the quality of materials (or the destruction of resources) through the mixing and contamination in the EoL phase, in which a closed-loop material recycling process is hampered by the imperfect separation of materials.

Another example of the destruction of resources is the mixing of EoL steel products with different alloying elements, such as chromium, which have to be diluted by pig iron to make them acceptable as a source of iron for products of lower technical sophistication, such as steel bars used in construction, whose production does not require these alloying elements [3]. Scarce and costly alloying elements will then end up in the construction of roads or buildings without the possibility of ever being recovered.

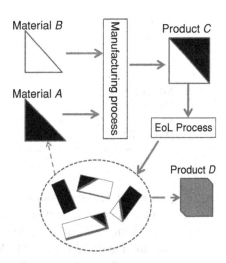

Figure 13.1. Decrease in the quality of materials over a product's life cycle: Only a part of material B can be recovered for closed-loop recycling. The rest of the materials are downcycled to produce product D, whose production requires materials of lower quality only.

The importance of the mixing and contamination of materials in the EoL phase of a product calls for the need of considering not a single material or substance at a time but the whole set of materials or substances constituting the product simultaneously. In this regard, one should note that most material flow analysis (MFA) and substance flow analysis (SFA) studies on individual materials or substances to date consider one material at a time [4, 5].

There is another stream of MFA, termed bulk MFA [6], in which broad ranges of materials, such as metals, fossil fuels, and industrial minerals are simultaneously considered at rather aggregated levels of resolution [7–9]. Because of the simultaneous consideration of the flows of many materials, these studies are conceptually close to the input–output (IO) analysis (IOA) in economics, and their flow tables resemble the IO tables (IOTs) [6, 10].

The currently available IOTs are mostly in monetary terms because of the ease of using a common unit to measure the flow of a large number of heterogeneous goods and services circulating in the whole economy. An IOT in monetary terms is called a monetary IOT (MIOT) and is distinguished from a physical IOT (PIOT) in which all the flows are measured in physical units. It is known that an MIOT can be converted to a PIOT under certain conditions [11]; this implies that the use of a monetary unit itself does not interfere with its usefulness as a physical model of production. However, from the point of view of thermodynamics, previously mentioned, the neglect (or no explicit consideration) of waste flows (including by-products) and waste treatment processes can be pointed out as a conceptual weakness in MIOT.

The currently available PIOTs, such as the German PIOT, consider waste flows. However, they tend to be highly aggregated and are hence not well suited to addressing the previously mentioned issues related to the mixing or contamination of materials, which require a level of resolution corresponding to standard MFA and SFA studies. Furthermore, these PIOTs are very (even prohibitively) costly to develop because they are mostly developed independently of MIOTs [12].

Insufficient development of analytical models can be mentioned as a common feature in MFA and SFA studies [13]. On the other hand, the currently available PIOTs tend to be mostly descriptive in nature (they were primarily developed as

Table 13.1. *A comparison of MFA methodologies*

Method	Material combinations	Resolution	Cost	Analytical model
MFA or SFA	No	High	Low–middle	Mostly descriptive
Bulk MFA	Yes	Low	?	Mostly descriptive
PIOT	Yes	Low	High	Mostly descriptive
WIO-MFA	Yes	High	Low–middle	Yes

part of accounting systems encompassing both the economy and the environment) and are hence not readily applicable for analytical purposes [11].

On the basis of these backgrounds, Nakamura et al. [14–16] developed a new IO-based methodology of MFA, termed waste IO MFA (WIO-MFA), which should address all these issues (Table 13.1). This chapter discusses the methodology of WIO-MFA and its application to the flow of metals and plastics in the Japanese economy.

13.2 The WIO-MFA Methodology

13.2.1 The Leontief Environmental IO Model With Waste Flows

In the previous section, the absence of waste flows was pointed out as a conceptual weakness in the standard MIOT. The attempts to account for waste flows in the IOA can be traced back to the seminal work of Leontief [17] on environmental IO (EIO), in which the flow of pollutants and the abatement acclivities thereof were introduced into an extended IO framework. Table 13.2 gives a prototype of the Leontief EIO with waste flows, in which X and f refer to the flow of goods and services, W and w

Table 13.2. *A prototype of Leontief EIO with waste flow*

From–To	Producing sectors 1 2 ... n	Waste-treatment sectors 1 2 ... k	Final demand
Production sector 1, Production sector 2, ... Production sector n	X_{I}	X_{II}	f_{I}
Absorption			
Waste 1, Waste 2, ... Waste k	$W_{\mathrm{I}}^{\mathrm{in}}$	$W_{\mathrm{II}}^{\mathrm{in}}$	w_f^{in}
Generation			
Waste 1, Waste 2, ... Waste k	$W_{\mathrm{I}}^{\mathrm{out}}$	$W_{\mathrm{II}}^{\mathrm{out}}$	w_f^{out}

refer to the waste flows, and subscripts I and II refer to producing sectors and waste-treatment sectors or waste, respectively. The superscripts in and out attached to W refer to the use (absorption) and generation of waste, respectively. It is important to note that $W_{\mathrm{II}}^{\mathrm{in}}$ refers to the use of waste in waste-treatment sectors, but not to the inflow of waste for treatment. It is also important to note that there is a one-to-one correspondence between wastes and waste treatments; for instance, waste k is treated only in waste-treatment sector k and vice versa (waste-treatment sector k treats no other waste but waste k).

Subtracting the waste-absorption matrices from the corresponding waste-generation matrices gives the net waste-generation matrices:

$$W_i = W_i^{\mathrm{out}} - W_i^{\mathrm{in}}, \quad i = \mathrm{I, II}, f. \tag{13.1}$$

Note that, because of the assumption of a one-to-one correspondence between wastes and waste-treatment processes, the matrix W_{II} is square, and hence the whole matrix of endogenous (production and waste-treatment) sectors is square as well.

Let us denote a $k \times 1$ vector of the activity level of waste-treatment sectors by x_{II}, with $(x_{\mathrm{II}})_i$ as its ith element. Denoting an $n \times 1$ vector of the activity levels of conventional producing sectors by x_{I}, with $(x_{\mathrm{I}})_j$ as its jth element, we can obtain the following balancing equations:

$$\sum_{j=1}^{n} (X_{\mathrm{I}})_{ij} + \sum_{j=1}^{k} (X_{\mathrm{II}})_{ij} + (f_{\mathrm{I}})_i = (x_{\mathrm{I}})_i, \quad i = 1, n, \tag{13.2}$$

$$\sum_{j=1}^{n} (W_{\mathrm{I}})_{ij} + \sum_{j=1}^{k} (W_{\mathrm{II}})_{ij} + (w_f)_i = (x_{\mathrm{II}})_i, \quad i = 1, k, \tag{13.3}$$

where $(Z)_{ij}$ refers to the ith row and the jth column element of Z. Note that (13.3) holds because of the one-to-one correspondence between wastes and waste treatments: Each waste is exclusively processed by a particular treatment sector that treats no other waste.

By use of input and waste-generation coefficient matrices, A_i and G_i, respectively, $i = \mathrm{I, II}$, the system of balance equations for goods and services and waste can be rewritten as

$$\sum_{j=1}^{n} (A_{\mathrm{I}})_{ij} (x_{\mathrm{I}})_j + \sum_{j=1}^{k} (A_{\mathrm{II}})_{ij}(x_{\mathrm{II}})_j + (f_{\mathrm{I}})_i = (x_{\mathrm{I}})_i, \quad i = 1, n, \tag{13.4}$$

$$\sum_{j=1}^{n} (G_{\mathrm{I}})_{ij} (x_{\mathrm{I}})_j + \sum_{j=1}^{k} (A_{\mathrm{II}})_{ij}(x_{\mathrm{II}})_j + (w_f)_i = (x_{\mathrm{II}})_i, \quad i = 1, m, \tag{13.5}$$

which can be represented by use of matrix notation as

$$\begin{bmatrix} A_{\mathrm{I}} & A_{\mathrm{II}} \\ G_{\mathrm{I}} & G_{\mathrm{II}} \end{bmatrix} \begin{pmatrix} x_{\mathrm{I}} \\ x_{\mathrm{II}} \end{pmatrix} + \begin{pmatrix} f_{\mathrm{I}} \\ w_f \end{pmatrix} = \begin{pmatrix} x_{\mathrm{I}} \\ x_{\mathrm{II}} \end{pmatrix}. \tag{13.6}$$

Solving for x_{I} and x_{II} gives

$$\begin{pmatrix} x_{\mathrm{I}} \\ x_{\mathrm{II}} \end{pmatrix} = \begin{bmatrix} I - A_{\mathrm{I}} & -A_{\mathrm{II}} \\ -G_{\mathrm{I}} & I - G_{\mathrm{II}} \end{bmatrix}^{-1} \begin{pmatrix} f_{\mathrm{I}} \\ w_f \end{pmatrix}. \tag{13.7}$$

Table 13.3. *A prototype of WIO*

From–To	Producing sectors 1, 2,..., n	Waste-treatment sectors 1, 2,..., k	Final demand
Production sector 1, Production sector 2, ... Production sector n	X_{I}	X_{II}	f_{I}
Net generation = Generation − Absorption			
Waste 1 ... Waste k ... Waste m	W_{I}	W_{II}	w_f

The one-to-one correspondence between wastes and waste-treatment processes is a prerequisite for solution (13.7). If this condition is not met, (13.3) does not hold, and hence (13.6) cannot be derived. In reality, however, wastes and waste-treatment processes will have neither a one-to-one correspondence nor will the correspondence be exclusive [18] for the following reasons:

1. A given treatment process can be applied to different types of wastes: Any solid waste can be landfilled.
2. A given waste can be subjected to different types of treatment: Garbage can be, among others, processes, incinerated, gasified, composted, or landfilled.
3. The number of types of waste exceeds the number of types of treatment processes.

The final point, although relevant in practice, is conceptually inessential: The satisfaction of the equality condition alone does not imply a one-to-one correspondence between wastes and waste treatments. A mere equality of the number of waste types and that of treatment types may make the matrix W_{II} square. However, solution (13.7) has no economic relevance because, in this case, (13.6) does not hold and x_{II} is not defined.

13.2.2 The WIOT and WIO Models

In a generalized version of the EIO table (EIOT), which does not assume the condition of a one-to-one correspondence between wastes and waste treatments, Nakamura [19] introduced the The WIO table (WIOT). Table 13.3 gives a prototype of this table for k waste-treatment sectors and m waste types, where $m > k$. Note that the matrix W_{II} referring to the waste generation from waste-treatment sectors is not square.

Table 13.4 then shows an aggregated version of a WIOT for Japan for the year 1995 with 13 producing sectors, 3 waste-treatment sectors, and 13 waste types. The equality between the number of wastes and the number of producing sectors is a mere coincidence and has no further meaning. With 3 waste-treatment sectors and

	Producing sectors													Waste-treatment sectors				
	AGR	MIN	FOD	WOD	CHE	CEM	MTL	MEP	MCN	CNS	UTL	SRV	TRN	INC	LND	SHR	FLD	Total
AGR	1715	1	7660	709	142	1	0	0	154	136	0	1097	2	0	0	0	4199	15815
MIN	0	3	0	8	132	593	47	2	2	800	42	1	0	0	0	0	23	1654
FOD	1081	7	7267	123	276	27	1	41	286	202	7	6787	108	0	0	0	33773	49985
WOD	184	7	1305	7277	705	246	6	246	1300	4045	130	9808	725	0	0	0	3883	29867
CHE	920	31	1827	1342	13108	322	337	661	5419	2345	934	8133	3554	46	12	3	11333	50329
CEM	20	0	176	98	206	959	121	127	998	5609	17	401	3	0	0	0	965	9700
MTL	0	0	0	0	33	13	1363	4673	215	2	0	3	0	0	0	0	176	6477
MEP	23	28	910	444	448	198	4	8790	9522	10368	29	867	105	0	0	0	3865	35600
MCN	84	12	175	86	67	36	0	125	40288	1837	27	7318	591	96	13	8	77856	128618
CNS	50	11	138	129	270	136	55	287	370	225	1141	4647	629	0	148	0	80030	88265
UTL	71	45	787	573	1412	357	323	875	1352	565	2303	5921	907	−9	9	31	7760	23283
SRV	1500	207	6906	4028	7714	1422	807	4054	19789	14469	3557	67413	15075	0	0	7	282052	429001
TRN	728	408	1827	1305	1476	810	386	1144	2755	5107	529	17158	6460	104	66	0	24533	64796
Net generation of waste																		
grb	0	0	0	0	0	0	0	0	0	0	0	6000	0	0	0	0	10000	16000
ppr	0	0	66	−11460	0	0	0	0	0	0	0	0	0	0	0	101	17829	6536
pls	249	1	314	553	−117	62	141	524	720	975	14	1236	144	0	0	229	3341	8386
mtl	2	1	24	730	−234	7	−35549	5013	9175	8424	11	306	21	0	0	3358	10679	1968
gls	0	0	−4426	18	126	−456	96	317	157	1758	56	4951	12	0	0	56	4561	7225
wds	19848	0	1454	2626	90	0	0	0	0	2818	0	0	0	0	0	1028	1023	28887
ash	0	4714	222	984	1627	−11726	2089	15079	1786	−9001	4757	39	5	2836	0	0	0	13412
sld	−5710	4333	2333	7003	4006	−894	320	1313	654	11750	3814	366	32	0	0	0	0	29319
oil	−6	3	1376	155	850	90	368	1594	1831	87	20	988	160	0	0	0	0	7528
cns	0	139	23	33	204	341	211	655	463	14192	154	675	382	0	0	0	0	17472
blk	0	0	0	0	0	0	0	0	0	0	0	0	0	0	0	0	3775	3775
vhc	0	0	0	0	0	0	0	0	0	0	0	0	0	0	0	0	5020	5020
dst	0	0	0	0	0	0	0	0	0	0	0	0	0	0	0	1377	0	1377

Source: [18]. Units: billion yen for goods and services, 10^6 kg for waste. AGR: agriculture, forestry, and fishery; MIN: mining; FOD: food, feed, and beverage; WOD: wood, pulp, and paper; CHE: chemical industry; CEM: cement, earth and glass; MTL: basic metals; MEP: metal products; MCN: machinery; CNS: construction and civil engineering; UTL: electricity, gas, and water; SRV: service; TRN: transport and communication; INC: incineration; LND: landfilling; SHR: shredding; FLD: final demand. Waste types: grb: garbage; ppr: waste paper and textile; pls: waste plastics; mtl: metal scraps; gls: waste glass and ceramics; wds: plant and animal waste; ash: ash, dust, and slag; sld: sludge; oil: waste oil, acid, and alkali; cns: construction debris; blk: bulky waste; vhc: discarded automobiles; dst: shredder dust.

13 waste types, the WIOT in Table 13.4 is fundamentally different from the square Leontief EIOT in Table 13.2. The waste generation from waste-treatment sectors refers to the results of waste transformation in these sectors: The incineration sector transforms waste into ash, fly ash, slag, and emissions, whereas shredding transforms waste into various types of scrap and shredding residues. More detailed and updated versions of WIOT can be found on the Internet [20].

Although Table 13.4 provides a comprehensive picture of the interindustry flow of goods and services and waste, its nonsquare nature (the nonsquareness of W_{II}) makes a modeling such as (13.7) not readily applicable. Nakamura [19] and Nakamura and Kondo [18] solved this problem by introducing an allocation matrix S that establishes a correspondence between wastes and waste treatments, with $s_{ij} = (S)_{ij}$ referring to the share of waste j that is treated by treatment sector i when the waste is to be subjected to waste treatment. In the Leontief EIO, S becomes an identity matrix. In the present case, S is a matrix of the order of 3×13. Because waste for treatment has to be subjected to at least one treatment process, it follows that

$$\sum_{i=1}^{3} s_{ij} = 1, \; j = 1, \ldots, 13. \tag{13.8}$$

Note that, for Table 13.4, the balance equation for waste (13.5) is given by

$$\sum_{j=1}^{13} (W_{\mathrm{I}})_{ij} + \sum_{k=1}^{3} (W_{\mathrm{II}})_{ik} + (w_f)_i = (w)_i, \quad i = 1, 2, 3, \tag{13.9}$$

where w_i refers to the amount of waste i for treatment. Using the matrices of waste-generation coefficients G_{I} (of the order of 3×13) and G_{II} (of the order of 3×3), we can write this in matrix form as

$$G_{\mathrm{I}} x_{\mathrm{I}} + G_{\mathrm{II}} x_{\mathrm{II}} + w_f = w. \tag{13.10}$$

The multiplication of S from the left of both sides gives

$$S G_{\mathrm{I}} x_{\mathrm{I}} + S G_{\mathrm{II}} x_{\mathrm{II}} + S w_f = S w = x_{\mathrm{II}}, \tag{13.11}$$

where the final equality follows from the definition of S. The system of balance equations for both goods and services and waste is then given by

$$\begin{bmatrix} A_{\mathrm{I}} & A_{\mathrm{II}} \\ S G_{\mathrm{I}} & S G_{\mathrm{II}} \end{bmatrix} \begin{pmatrix} x_{\mathrm{I}} \\ x_{\mathrm{II}} \end{pmatrix} + \begin{pmatrix} f_{\mathrm{I}} \\ S w_f \end{pmatrix} = \begin{pmatrix} x_{\mathrm{I}} \\ x_{\mathrm{II}} \end{pmatrix}. \tag{13.12}$$

Solving for x_{I} and x_{II} gives

$$\begin{pmatrix} x_{\mathrm{I}} \\ x_{\mathrm{II}} \end{pmatrix} = \begin{bmatrix} I - A_{\mathrm{I}} & -A_{\mathrm{II}} \\ -S G_{\mathrm{I}} & I - S G_{\mathrm{II}} \end{bmatrix}^{-1} \begin{pmatrix} f_{\mathrm{I}} \\ S w_f \end{pmatrix}. \tag{13.13}$$

This is the WIO (quantity) model.

13.2.3 From WIO to WIO-MFA

The WIOT in Table 13.4 consists of two parts: The upper part refers to the flow of goods and services, and the lower part refers to the flow of waste. The lower part is in physical units. The upper part, however, is in monetary units because it was taken from a MIOT (except for waste-treatment processes, which were obtained on the basis of engineering information).

Process waste refers to the discarded physical inputs that do not enter the products. The flow of process waste in the lower panel of a WIOT is thus closely connected to the flow of goods in the upper panel. However, because the upper panel is in monetary units, a relationship in physical terms cannot be established between them. Establishing this relationship calls for representing the flow of goods in terms of physical units. In particular, the physical flow of goods needs to be decomposed into the flow of individual materials or substances that constitute the goods. A mere conversion of the flow of goods in physical units, for example, kilograms, does not suffice.

The mass of a product consists of the mass of its material components. The decomposition of a product into its material components necessitates the estimation of a product's composition. This is the main task of WIO-MFA.

13.2.4 Classifying Inputs According to Whether They Become Physical Components of a Product

Not all inputs can become components of a physical product. By definition, non-physical inputs, such as trade, transportation, finance, and insurance, do not become physical components of a product. Being a physical input, however, does not mean that it can become a component of a product. Depending on whether they can become components of a product, physical inputs can be classified as primary or ancillary (or both) inputs. An example of an ancillary input, is limestone (CaO to be precise) used in blast furnaces for the production of pig iron, which leaves the process as slag without being embodied in pig iron. An example of a primary input is a metal, for example, copper, used in the production of a metal alloy, for example, brass. A given physical input can become both a primary and an ancillary input, depending on its use. For instance, acid is a primary input when used in the production of chemicals but is an ancillary input when used in the cleaning process of steel products.

Although a physical product consists of primary inputs, not all the primary inputs become components of the product. Some of them usually end up as process waste without entering the product. In other words, the yield or transfer ratio of a primary input, denoted by γ, is generally less than one. An ancillary input can be represented as a physical input, whose yield is zero. Of the input of primary input i per unit of product j, $a_{ij} = (A)_{ij}$, only $\gamma_i a_{ij}$ can become a component of the product, and $(1 - \gamma_i) a_{ij}$ is discarded as process waste. In sum, a product is composed of the primary inputs adjusted for the yield ratios (Fig. 13.2).

Let ϕ_i be an index that takes unity when input i is physical (has mass) and zero otherwise, and let $\gamma_{ij} \in [0, 1]$ be the yield ratio of input i used in the production of

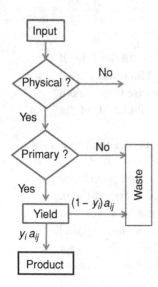

Figure 13.2. Inputs and the composition of a product.

product j. For a given input coefficient a_{ij}, the portion that becomes a component of product j, denoted by $\tilde{a}_{ij}$, is then given by

$$\tilde{a}_{ij} = \gamma_{ij}\,\phi_i\,a_{ij}. \tag{13.14}$$

Let an $n \times n$ diagonal matrix be denoted by Φ, the ith diagonal element of which is ϕ_i, and let it be termed the mass filter [15]. Furthermore, let the $n \times n$ matrix $\Gamma = (\gamma_{ij})$ be termed the yield matrix. The portions of the input coefficients matrix A that enter products are then given by

$$\tilde{A} = \Gamma \odot (\Phi\,A), \tag{13.15}$$

where $\odot$ refers to the Hadamard product (the element-wise product of two matrices).

13.2.5 Resources, Materials, and Products: A Formal Definition Based on Degrees of Fabrication

While simultaneously considering the flow of many materials, one must take care to avoid double counting. Double counting occurs when items of different degrees of fabrication (for example, metals and metal alloys) are counted as "materials" (the problem of double counting is addressed in 5.21 and 5.23 of [12]). This implies the need for a formal definition of "materials," which is not provided in the literature on MFA. Closely related to a formal definition of "materials" are formal definitions of "resources" and "products," because "materials" are made of "resources" and "products" are made of "materials."

Let us partition the physical inputs into the mutually exclusive and exhaustive sets of resources (R), materials (M), and products (P), and rewrite the matrix $\tilde{A}_I$ as follows:

$$\tilde{A}_I = \begin{bmatrix} \tilde{A}_{PP} & \tilde{A}_{PM} & \tilde{A}_{PR} \\ \tilde{A}_{MP} & \tilde{A}_{MM} & \tilde{A}_{MR} \\ \tilde{A}_{RP} & \tilde{A}_{RM} & \tilde{A}_{RR} \end{bmatrix} \tag{13.16}$$

This partitioning gives the definition of materials, resources, and products. In Naka-mura and Nakajima [15], this partition is defined as follows:

1. $[R]$ resources are not produced but provided from outside the system under consideration,
2. $[M]$ materials are made of resources,
3. $[P]$ products are made of products and materials.

This definition is based on the degree of fabrication, with resources and products at the lowest and highest level of fabrication, respectively. Note that, in this definition, resources have to be first transformed into materials before they enter the products: No resources can enter the products directly. Moreover, no materials are made out of materials; this is necessary to avoid the double counting of materials (see [16] for a formal proof on this point).

With this definition imposed, (13.16) becomes

$$\tilde{A}_I = \begin{bmatrix} \tilde{A}_{PP} & 0 & 0 \\ \tilde{A}_{MP} & 0 & 0 \\ 0 & \tilde{A}_{RM} & 0 \end{bmatrix}. \tag{13.17}$$

Note that the imposition of the preceding definition transforms the matrix of input coefficients into a triangular matrix because goods of lower degrees of fabrication are not used in producing goods of higher degrees of fabrication. In other words, for a given A_I, any partition of inputs that gives rise to the triangular structure (13.17) will give a valid definition of resources, materials, and products in the sense of the preceding definition.

The preceding definition of resources is a relative one in the sense that it depends on inputs whose levels of fabrication are chosen as materials, that is, as the objects, whose flow is to be accounted for by MFA. Accordingly, the term resources in this definition is a general one that refers to inputs that enter the fabrication stage, which immediately precedes that of materials and can hence considerably diverge from its conventional meanings. For example, if metal alloys are to be considered as materials, metals, not metal ores, should be counted as resources (see [16] for further details on this point). This feature merits that the preceding definition be classified as formal. Henceforth, the terms resources, materials, and products will be used in the sense of the preceding definition.

13.2.6 The Materials-Composition Matrix

When the rules of the inverse of partitioned matrices [21] are applied, the following expression for the Leontief inverse matrix of (13.17) is obtained:

$$(I - \tilde{A}_I)^{-1} = \begin{bmatrix} (I - \tilde{A}_{PP})^{-1} & 0 & 0 \\ \tilde{A}_{MP}(I - \tilde{A}_{PP})^{-1} & I & 0 \\ \tilde{A}_{RM}\tilde{A}_{MP}(I - \tilde{A}_{PP})^{-1} & \tilde{A}_{RM} & I \end{bmatrix} \tag{13.18}$$

The matrix $\tilde{A}_{MP}(I - \tilde{A}_{PP})^{-1}$ gives the material composition of products, with its ith row and jth column elements representing the amount of material i that is contained in a unit of product j [15]. When all the products are made of materials alone, that is, $\tilde{A}_{PP} = 0$, this reduces to $\tilde{A}_{MP}$. The term $(I - \tilde{A}_{PP})^{-1}$ is necessary when production requires products of lower degrees of fabrication, which is generally the case. In a similar vein, the matrix $\tilde{A}_{RM}\tilde{A}_{MP}(I - \tilde{A}_{PP})^{-1}$ at the southwest corner of the inverse matrix provides the resource composition of products. Because, we are concerned with the flow of materials in this chapter, this matrix is not discussed any further.

Some waste materials can also become primary inputs; a good example is iron scrap used in an EAF for the production of steel products. With $\tilde{G}_P^{in}$ for the matrix of waste input coefficients adjusted for yield ratios, the composition of waste materials in products is given by (see [16])

$$\tilde{G}_P^{in}(I - \tilde{A}_{PP})^{-1}. \tag{13.19}$$

The matrix of the overall material composition incorporating waste materials C then becomes

$$C = \begin{pmatrix} \tilde{A}_{MP} \\ \tilde{G}_P^{in} \end{pmatrix} (I - \tilde{A}_{PP})^{-1}. \tag{13.20}$$

Let us denote the sum of the elements of the jth column of C as c_j. When all the materials are measured in kilograms, c_j gives the weight of product j in kilograms per unit. If product j is also measured in kilograms, c_j will be unity.

The preceding result implies that by use of C (or its column sum c), MIOT can easily be converted to a PIOT. When the materials are measured in kilograms, the flow table of products in monetary units, X_{PP}^*, can be converted to a flow table in physical units, X_{PP}, by

$$X_{PP} = \hat{c}\, X_{PP}^*, \tag{13.21}$$

where $\hat{c}$ refers to a diagonal matrix with c_j as its jth diagonal element. Furthermore, if one is interested in the physical flow of a particular material, for example, material k, that is associated with the flow X_{PP}^*, it can be obtained by

$$\widehat{C_{k.}}\, X_{PP}^* \tag{13.22}$$

where $\widehat{C_{k.}}$ is a diagonal matrix with the kth row and jth column element of C, that is, C_{kj}, as its jth diagonal element. In other words, (13.22) enables one to retrieve from an MIOT the physical flow of any material under consideration: By use of the material composition matrix C, a MIOT can easily be converted to a physical flow table of materials or substances.

13.3 Application to the Flow of Metals and Plastics

Nakamura and Nakajima [15] applied the previously methodology to the flow of base metals in the Japanese economy. Eleven types of metals, comprising pig iron, ferroalloys, copper, lead, zinc, aluminum, and their scraps (except for ferroalloys), were considered as materials. Nakajima et al. [22] extended it further to include nine types of plastics [thermosetting resins, polyethylene (low density), polyethylene (high density), polystyrene, polypropylene, vinyl-chloride resins, high-performance resins, other resins, and waste plastics] as materials. The Japanese IO table for 2000 was used as a major source of data after it was extended or modified by use of detailed and mostly physical information on the production and supply of these materials. The resulting IO table consisted of 430 inputs, which included the 20 types of metals and plastics measured in 1000 kg.

13.3.1 The Composition of Metals and Plastics in Selected Products

Table 13.5 shows a sample of the estimated values of C for several products (the results for plastics are omitted owing to the lack of space). The 10 products in the upper panel are measured in 1000 kg, whereas the 10 products in the lower panel are measured in monetary units (1 million Japanese yen). For a product measured in kilograms, the row sum ("Total") is equal to unity, whereas for a product measured in monetary units, the row sum gives the weight (10^3 kg) per 1 million Japanese yen (or U.S $10,000 at the exchange rate of of 100 yen per dollar) of the product. Note that "Total" corresponds to the weight of the product, c.

A comparison of the compositions of iron and steel products shows substantial differences in the use of iron scrap among them. Although iron scrap constitutes 90% of the material composition of ordinary steel bars, it is less than 10% for coated steel. Special steel contains larger amounts of ferroalloys than ordinary steel does. With regard to the metal components, electric wires and cables consist almost exclusively of copper and aluminum, with plastics accounting for approximately 20% of its total material composition. Of the products listed in Table 13.5, electric wires and cables are the only products that have a noticeable share of polyvinyl chloride (PVC).

Of the products listed in the lower panel of Table 13.5, metal products for construction weigh the heaviest per unit of money (3160 kg per 1 million Japanese yen), whereas boilers, turbines, and refrigerators weigh the lightest (520–560 kg per 1 million Japanese yen). Compared with metal products for construction, boilers, refrigerators, and turbines are characterized by higher shares of copper and lower shares of iron scrap. In fact, no significant amount of copper is found in metal products for construction. Refrigerators are characterized by high shares of aluminum and plastics.

Metal products for architecture weigh substantially less than metal products for construction per unit of money. The substantially higher share of aluminum can be pointed out as a distinguishing feature of metal products for architecture. Machinist's precision tools are characterized by the highest component of ferroalloys, which reflect the use of special metals, such as tungsten, in these tools. As expected, car batteries are found to be the only products that are made almost exclusively out of lead.

13.3.2 Testing the Model

For a product for which reliable data are available on the weight, price, and material composition, the predictive power of the model can be tested by comparing the estimated weight and composition with its real counterparts. A good example for this purpose is a passenger car. Figure 13.3 compares the estimated weight and

Table 13.5. *The metal and plastics composition of products*

	Pig iron	Ferroalloys	Copper	Lead	Zinc	Aluminum	Plastics[a]	PVC	Iron[b]	Copper[b]	Aluminum[b]	Total
Products measured in 1000 kg												
Ordinary steel strip	0.90	0.01	0.00	0.00	0.00	0.00	0.00	0.00	0.09	0.00	0.00	1.00
Ordinary steel bar	0.06	0.01	0.00	0.00	0.00	0.00	0.00	0.00	0.93	0.00	0.00	1.00
Other hot-rolled ordinary steel	0.77	0.01	0.00	0.00	0.00	0.00	0.00	0.00	0.22	0.00	0.00	1.00
Hot-rolled special steel	0.58	0.06	0.00	0.00	0.00	0.00	0.00	0.00	0.35	0.00	0.00	1.00
Steel pipes and tubes	0.76	0.02	0.00	0.00	0.00	0.00	0.00	0.00	0.22	0.00	0.00	1.00
Cold-finished steel	0.87	0.01	0.00	0.00	0.00	0.00	0.00	0.00	0.11	0.00	0.00	1.00
Coated steel	0.87	0.01	0.00	0.00	0.02	0.00	0.00	0.00	0.10	0.00	0.00	1.00
Electric wires, cables	0.00	0.00	0.59	0.00	0.00	0.05	0.09	0.11	0.00	0.15	0.00	0.99
Rolled or drawn copper, alloys	0.00	0.00	0.37	0.00	0.08	0.00	0.00	0.00	0.00	0.55	0.00	1.00
Rolled or drawn aluminum	0.00	0.00	0.00	0.00	0.00	0.64	0.00	0.00	0.00	0.00	0.36	1.00
Products measured in monetary units												
Metal products for construction	2.00	0.06	0.00	0.00	0.00	0.01	0.01	0.00	1.07	0.00	0.00	3.16
Metal products for architecture	0.92	0.01	0.00	0.00	0.01	0.19	0.01	0.00	0.14	0.00	0.11	1.40
Bolts, nuts, rivets and springs	1.06	0.04	0.01	0.00	0.01	0.00	0.00	0.00	0.38	0.02	0.00	1.53
Boilers	0.36	0.01	0.01	0.00	0.00	0.00	0.00	0.00	0.13	0.01	0.00	0.53
Turbines	0.37	0.01	0.01	0.00	0.00	0.00	0.00	0.00	0.16	0.01	0.00	0.56
Engines	0.49	0.01	0.00	0.00	0.00	0.01	0.01	0.00	0.30	0.00	0.00	0.83
Pump and compressors	0.47	0.02	0.00	0.00	0.00	0.02	0.01	0.00	0.34	0.00	0.01	0.88
Machinists' precision tools	0.40	0.03	0.00	0.00	0.00	0.01	0.01	0.00	0.20	0.00	0.00	0.66
Car batteries	0.06	0.00	0.00	1.96	0.01	0.01	0.13	0.00	0.01	0.00	0.00	2.20
Refrigerators and air conditioners	0.26	0.01	0.02	0.00	0.00	0.02	0.18	0.02	0.00	0.00	0.01	0.52

[a] Exclude PVC; [b] scrap. Source [22].

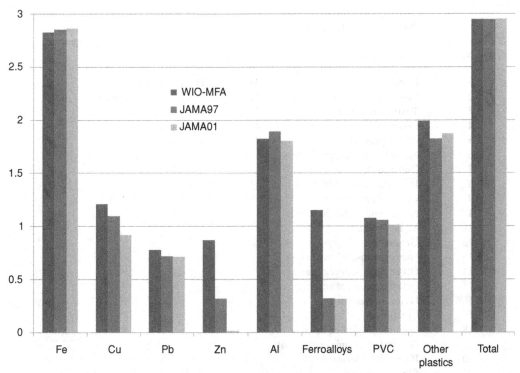

Figure 13.3. Testing the WIO-MFA model: the weight and composition of a passenger car in terms of metals and plastics. Units, kg; sources, [15, 22].

material compositions with the real values for a representative passenger car taken from Japan Automobile Manufacturers Association (JAMA) [23]. The estimated values compare fairly well with the real values, except for zinc and ferroalloys, whose shares are markedly smaller than those of other materials.

Although this result certainly indicates the need for improving the accuracy of our database regarding the materials having small shares, two observations should be made. First, the JAMA data show large fluctuations in the share of zinc over time, which possibly indicates large variations in the use of zinc among different car types. Second, there is a conceptual difference in the coverage of metal categories between the JAMA data and our model. In the JAMA data, the amount of zinc contained in coated steel is counted as iron, whereas in our computation it is counted as zinc. In fact, our estimate of the zinc component includes zinc that is found as additives in tires, pigments, and inorganic chemicals, in addition to that contained in coated steel. The same observations apply to ferroalloys as well.

13.3.3 Converting a MIOT to Material Flow Accounts

We now consider the application of (13.21) to convert a MIOT into a PIOT. For exposition purposes, the original MIOT was consolidated into a table with 38 industry sectors and 5 final-demand sectors (Table 13.6).

First, Fig. 13.4 shows the PIOT in terms of 11 metals and 9 plastics, in which five symbols are used to distinguish the relative size (weight) of each cell. A comparison

Table 13.6. *The sectoral classification of the Japanese IOT with 38 producing sectors and 5 final-demand sectors*

Number	Producing sector	Number	Final-demand sector
1	Primary sectors	39	Private consumption
2	Food and beverage	40	Government consumption
3	Textiles and paper	41	Capital accumulation
4	Chemicals	42	Export
5	Plastic products	43	Import
6	Cement and glass products		
7	Iron and steel		
8	Iron and steel products		
9	Cables and wires		
10	Nonferrous metal products		
11	Metal products		
12	General machinery		
13	Special industrial machinery		
14	Other general appliances		
15	Office equipment		
16	Home appliances		
17	Computers		
18	Communication equipment		
19	Applied electronic equipment		
20	IC, semiconductor		
21	Electronic parts		
22	Electrical equipment		
23	Other electrical devices and parts		
24	Automobiles		
25	Other transport machinery		
26	Precision machinery		
27	Miscellaneous products		
28	Construction		
29	Civil engineering		
30	Utilities		
31	Trade and banks		
32	House rent		
33	Transport		
34	Communication		
35	Services		
36	Education		
37	Research institute		
38	Health care		

with the original MIOT (Fig. 13.5) indicates that the physical flows are mostly concentrated in metals, machinery, and plastic products. It is indicated that the majority of metals and plastics enter household consumption in the form of food and beverage (containers), home appliances, automobiles, and physical services (such as repair). A look at the column of home appliances shows that the metals and plastics mostly enter home appliances in the form of iron and steel products, metal products, electronic parts, and plastic products. On the other hand, the service flows that are quite dominant in the monetary table are no longer significant.

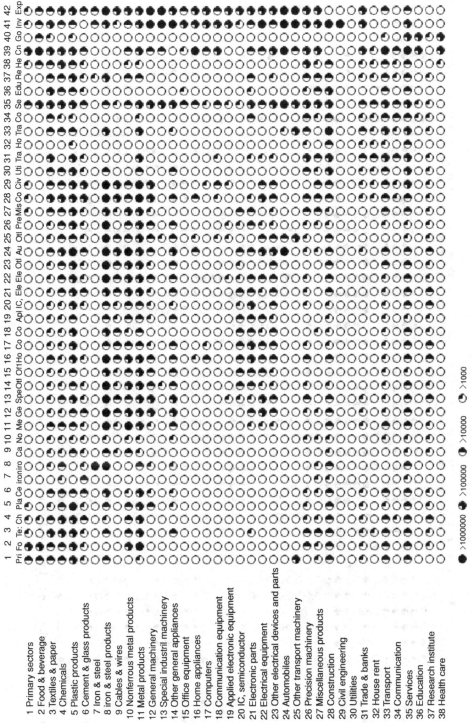

1 Primary sectors
2 Food & beverage
3 Textiles & paper
4 Chemicals
5 Plastic products
6 Cement & glass products
7 Iron & steel
8 iron & steel products
9 Cables & wires
10 Nonferrous metal products
11 Metal products
12 General machinery
13 Special industrʲl machinery
14 Other general appliances
15 Office equipment
16 Home appliances
17 Computers
18 Communication equipment
19 Applied electronic equipment
20 IC, semiconductor
21 Electronic parts
22 Electrical equipment
23 Other electrical devices and parts
24 Automobiles
25 Other transport machinery
26 Precision machinery
27 Miscellaneous products
28 Construction
29 Civil engineering
30 Utilities
31 Trade & banks
32 House rent
33 Transport
34 Communication
35 Services
36 Education
37 Research institute
38 Health care

● >1000000   ◑ >100000   ◔ >10000   ○ >1000
◑ >1000000   ● >10000   ● >10000   ◔ >1000

Figure 13.4. The physical IO flow: metals and plastics. Units, 1000 kg.

349

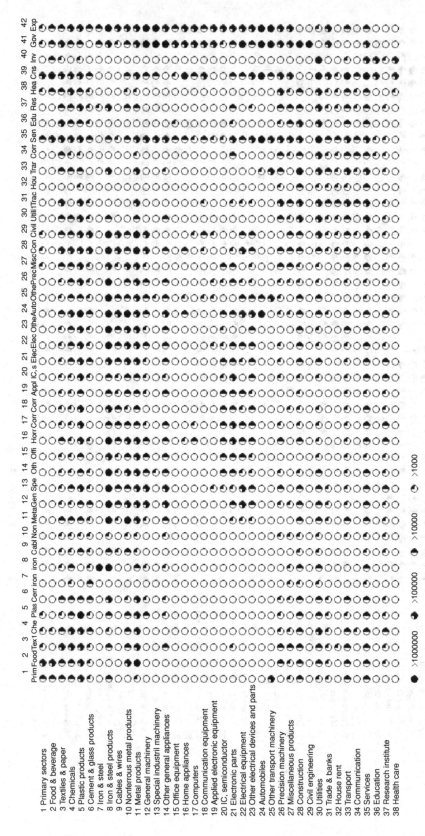

Figure 13.5. The monetary IO flow. Units, 1 million Japanese yen.

350

13.3.4 Metal Flow Analysis of Individual Metals

Next, we turn to the IO-based MFA of individual materials based on (13.22). First, Fig. 13.6 shows the flow of iron. A comparison with the flow shown in Fig. 13.4 indicates that the flow of metals and plastics is dominated by the flow of iron. The flow of iron mostly takes the form of iron and steel products, metal products, general machinery, automobiles, and construction.

The flow of copper (Fig. 13.7) gives a markedly different picture from the flow of iron. First, the flow spreads less widely and is concentrated in cables and wires, nonferrous metal products, electronics, and automobiles. The column of automobiles indicates that copper enters automobiles mostly in the form of cables and wires, nonferrous metal products, and electronics.

The flow of aluminum (Fig. 13.8) shows that it is used for more purposes than copper and somehow resembles the flow of iron. The similarity in the use patterns will indicates high possibilities of substitution between these metals.

The flow of lead in Fig. 13.9 is characterized by the smallest spread and is mostly concentrated in batteries and chemicals. It enters automobiles in the form of electrical devices, that is, batteries.

The flow of zinc (Fig. 13.10) resembles that of lead in that it is concentrated in a limited area, but differs in the areas where it is concentrated. It is noteworthy that a large portion of zinc flow occurs in the form of steel products, for which zinc is used as a coating element to protect steel from oxidization. Another major flow of zinc occurs as chemicals, for which zinc is mostly used not as a metal but as an oxide.

13.3.5 Tracing the Input Origin of Materials

The element C_{ij} gives the amount of material i that is contained in product j. When j consists of a large number of parts and combined materials such as an appliance or a car, C_{ij} consists of the sum of i that is contained in all these parts and combined materials. For instance, if j is a car and i is aluminum, C_{ij} refers to the sum of aluminum that is contained in, among other things, in an engine, body, electric parts, motors, and electric cables. C_{ij} is thus an aggregate of the composition of each of these components weighted by the respective weights.

The application of C allows one to trace the input origin of materials backward through the fabrication stage. Nakamura et al. [16] showed that the material composition of products that are used as direct inputs to j, that is, as inputs in the last fabrication stage of the product, is given by

$$C \operatorname{diag}(\tilde{A}_{PP_j}), \tag{13.23}$$

where $\tilde{A}_{PP_j}$ refers to the jth column of $\tilde{A}_{PP}$. For instance, let product j be a car, material k be aluminum, and product i be an automobile engine. The kth row and ith column element of (13.23) then refer to the weight of aluminum that enters a car in the form of an engine.

When a similar procedure is used, the material composition of a product can be further decomposed into inputs that are used in lower stages of fabrication. For

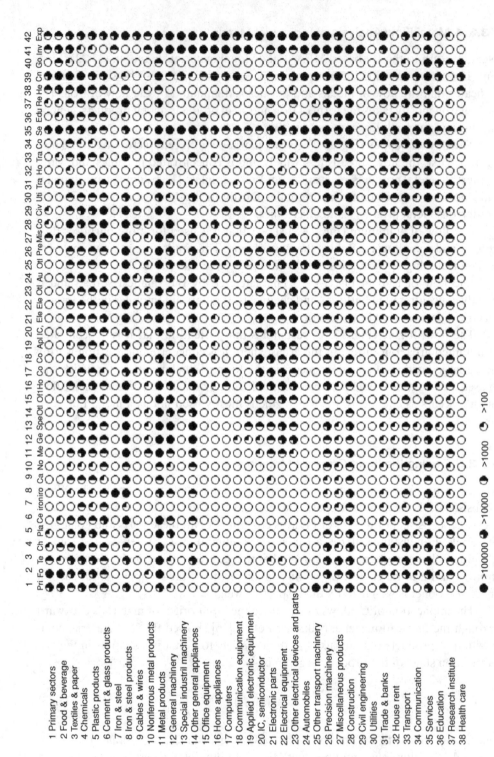

Figure 13.6. The physical IO flow: iron. Units, 1000 kg.

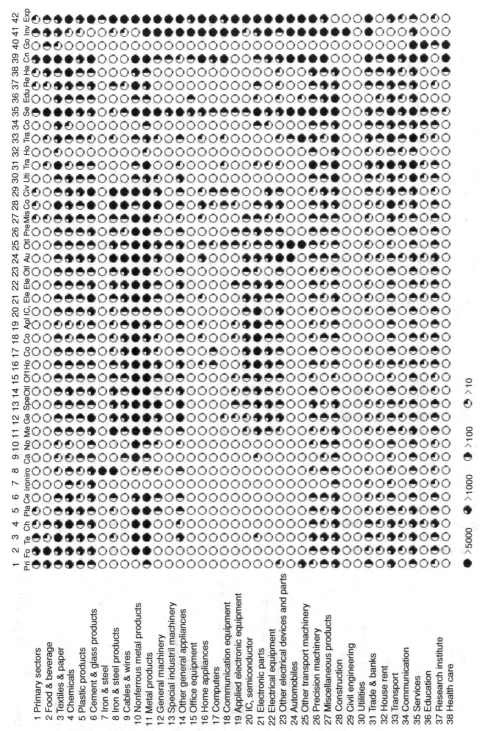

Figure 13.7. The physical IO flow: copper. Units, 1000 kg.

1 Primary sectors
2 Food & beverage
3 Textiles & paper
4 Chemicals
5 Plastic products
6 Cement & glass products
7 Iron & steel
8 Iron & steel products
9 Cables & wires
10 Nonferrous metal products
11 Metal products
12 General machinery
13 Special industrl machinery
14 Other general appliances
15 Office equipment
16 Home appliances
17 Computers
18 Communication equipment
19 Applied electronic equipment
20 IC, semiconductor
21 Electronic parts
22 Electrical equipment
23 Other electrical devices and parts
24 Automobiles
25 Other transport machinery
26 Precision machinery
27 Miscellaneous products
28 Construction
29 Civil engineering
30 Utilities
31 Trade & banks
32 House rent
33 Transport
34 Communication
35 Services
36 Education
37 Research institute
38 Health care

● >5000   ● >1000   ◐ >100   ◑ >10   ○ >10

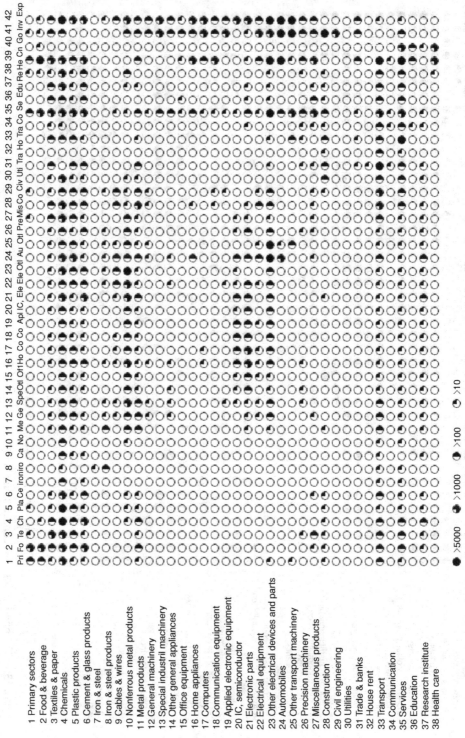

1 Primary sectors
2 Food & beverage
3 Textiles & paper
4 Chemicals
5 Plastic products
6 Cement & glass products
7 Iron & steel
8 Iron & steel products
9 Cables & wires
10 Nonferrous metal products
11 Metal products
12 General machinery
13 Special industril machinery
14 Other general appliances
15 Office equipment
16 Home appliances
17 Computers
18 Communication equipment
19 Applied electronic equipment
20 IC, semiconductor
21 Electronic parts
22 Electrical equipment
23 Other electrical devices and parts
24 Automobiles
25 Other transport machinery
26 Precision machinery
27 Miscellaneous products
28 Construction
29 Civil engineering
30 Utilities
31 Trade & banks
32 House rent
33 Transport
34 Communication
35 Services
36 Education
37 Research institute
38 Health care

● >5000   ◕ >1000   ◑ >100   ◔ >10

Figure 13.8. The physical IO flow: aluminum. units, 1000 kg.

354

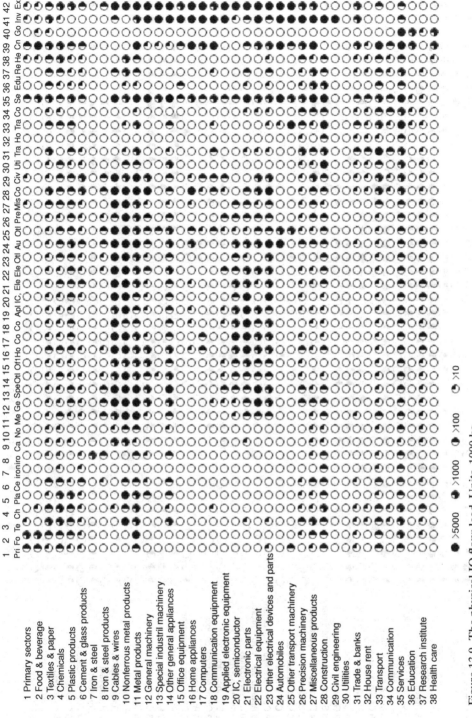

1 Primary sectors
2 Food & beverage
3 Textiles & paper
4 Chemicals
5 Plastic products
6 Cement & glass products
7 Iron & steel
8 Iron & steel products
9 Cables & wires
10 Nonferrous metal products
11 Metal products
12 General machinery
13 Special industril machinery
14 Other general appliances
15 Office equipment
16 Home appliances
17 Computers
18 Communication equipment
19 Applied electronic equipment
20 IC, semiconductor
21 Electronic parts
22 Electrical equipment
23 Other electrical devices and parts
24 Automobiles
25 Other transport machinery
26 Precision machinery
27 Miscellaneous products
28 Construction
29 Civil engineering
30 Utilities
31 Trade & banks
32 House rent
33 Transport
34 Communication
35 Services
36 Education
37 Research institute
38 Health care

Figure 13.9. The physical IO flow: lead. Units, 1000 kg.

● >5000   ● >1000   ● >100   ◑ >10

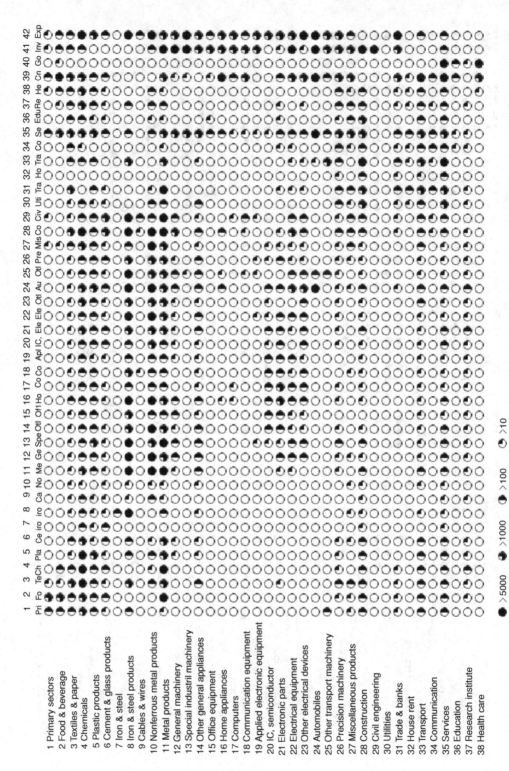

1 Primary sectors
2 Food & beverage
3 Textiles & paper
4 Chemicals
5 Plastic products
6 Cement & glass products
7 Iron & steel
8 Iron & steel products
9 Cables & wires
10 Nonferrous metal products
11 Metal products
12 General machinery
13 Special industril machinery
14 Other general appliances
15 Office equipment
16 Home appliances
17 Computers
18 Communication equipment
19 Applied electronic equipment
20 IC, semiconductor
21 Electronic parts
22 Electrical equipment
23 Other electrical devices
24 Automobiles
25 Other transport machinery
26 Precision machinery
27 Miscellaneous products
28 Construction
29 Civil engineering
30 Utilities
31 Trade & banks
32 House rent
33 Transport
34 Communication
35 Services
36 Education
37 Research institute
38 Health care

● >5000   ● >1000   ● >100   ◑ >10

Figure 13.10. The physical IO flow: zinc. Units, 1000 kg.

356

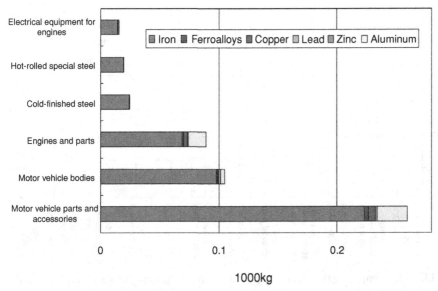

Figure 13.11. Decomposing the metal composition of a passenger car into its direct inputs.

instance, the material composition of the ith input that occurs in $\widetilde{A}_{PP_j}$, $\widetilde{a}_{PP_{ij}}$, is obtained from

$$C \operatorname{diag}(\widetilde{A}_{PP_i} \widetilde{a}_{PP_{ij}}). \qquad (13.24)$$

In the preceding example, let $\widetilde{a}_{PP_{ij}}$ be the amount of engine that is used per unit of a car. The direct inputs in the final stage of the production of an engine (or the elements of $\widetilde{A}_{PP_i}$) consist of, among others, cast and forged metals (denoted by r), special steels, electric parts, bearings, bolts, and electric cables. The kth row and rth column element of (13.24) then refer to the weight of aluminum that enters a car in the form of cast and forged metals embodied in the engine.

In an analogous manner, one can further decompose each column element of (13.24) into its product components at lower levels of fabrication, until the level is achieved at which all the inputs consist of only materials, and hence no further decomposition into products is possible.

In the following, some results of applying the preceding formulas to the metal composition of a passenger car are shown; the metal composition data were obtained for the original IO data. First, Fig. 13.11 gives the results of the decomposition into the direct inputs based on (13.23). The largest (heaviest) input item at the last stage of fabrication of a passenger car is parts and accessories, followed by vehicle bodies, and engines and parts. Of these input items, parts and accessories and engines and parts are characterized by significant shares of nonferrous metals, particularly aluminum. The composition of nonferrous metals in a passenger car is thus mostly attributable to parts and accessories and engines and parts.

Figures 13.12 and 13.13 then show the results of further decomposition by (13.24) of parts and accessories and engines and parts into input components. The large share of nonferrous metals in parts and accessories is attributable to nonferrous metal

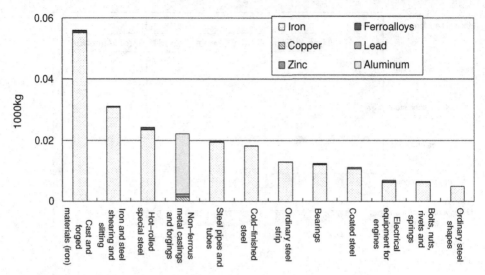

Figure 13.12. Decomposing the metal composition of passenger car parts and accessories into their direct inputs.

castings and forgings (Fig. 13.12), whereas the large share of nonferrous metals in engines and parts is attributable to nonferrous metal castings and forgings, electrical equipment for engines, rolled and drawn aluminum, and electric wires and cables (Fig. 13.13).

To ensure the recovery of high-quality iron scrap from EoL vehicles, it is necessary to avoid the mixing of copper. The preceding results indicate the importance of removing electrical equipment and electric wires (wire harness) in the disassembling process of EoL vehicles to avoid such mixing.

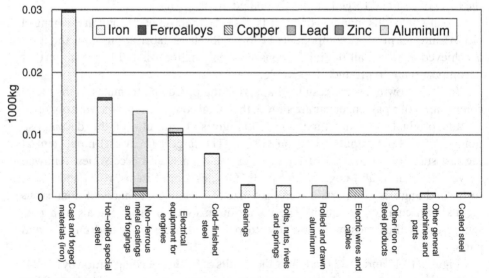

Figure 13.13. Decomposing the metal composition of a passenger car engine into its direct inputs.

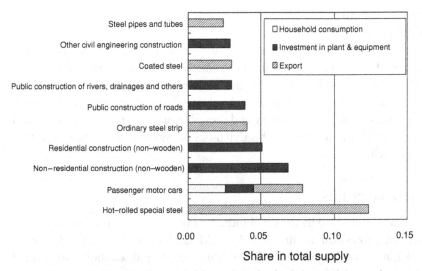

Figure 13.14. The final destination of ferroalloys by final-demand categories.

13.3.6 The Final Destination of Materials

An important task of MFA is to identify the location of crucial materials (from the point of view of scarcity, environmental concerns, and so forth) in the economy. All the final products are eventually directed to final-demand categories such as household consumption, investment in plants and equipment, and export, unless they are discarded in the intermediate stages of fabrication. Hence identifying the pattern by which materials are distributed among different categories of the final demand for products, that is, the final destination or user of materials by products, constitutes an important task of MFA. This identification is facilitated by converting the final demand for products by categories, F_P, into its material components:

$$\text{diag}(C_i.)F_P. \tag{13.25}$$

Figure 13.14 shows the destination of ferroalloys by products, which constitute approximately 62 % of the total supply (see [15] for other metals). It is found that the largest portion of ferroalloys is exported in the form of hot-rolled special steel. In sum, approximately a quarter of the total domestic supply is exported. On the other hand, the major portion of domestic accumulation of ferroalloys occurs in passenger cars, construction, and civil engineering.

13.4 Discussion

The PIOT provides a consistent accounting framework for material flow accounts, whereas the IOA provides a promising analytical foundation for MFA. The compilation of a PIOT is usually done independently of existing MIOTs, that is, almost from scratch without resorting to available MIOT resources (see Stahmer et al. [24], for instance). As previously shown, WIO-MFA makes it possible to derive a PIOT from readily available MIOT resources, once physical information on the use of materials as is used in life-cycle assessment is available. This can significantly economize on the otherwise prohibitive cost that is associated with the independent compilation of a PIOT ([12] p. 58).

The analysis in this chapter was a static one, because no attention has been paid to the dynamic flow of materials over time, which results from the life cycles of products. Because the products of today are the waste materials of tomorrow, the flows of materials at different time periods are related to one another by the generation of EoL products and the recovery and recycling of waste materials from them (Yokoyama et al. [25]). Addressing this dynamics provides an important direction for future research.

13.5 Acknowledgments

This research was supported by the Steel Industry Foundation for the Advancement of Environment Protection Technology. I would like to thank Tetsuya Nagasaka, Kenichi Nakajima, and Yasushi Kondo for many inspiring discussion. However, I remain solely responsible for any errors in this article.

REFERENCES

[1] I. Daigo, D. Fujimaki, Y. Matsuno, and A. Adachi, "Development of a dynamic model for assessing environmental impact associated with cyclic use of steel," presented at The Sixth International Conference on Eco Balance, Tsukuba, Japan, October, 2004.

[2] Y. Igarashi, I. Daigo, Y. Matsuno, and Y. Adachi, "Estimation of quality change in domestic steel production affected by steel scrap exports," *ISIJ Intl.* **47**, 753–757 (2007).

[3] E. Yamasue, K. Matsubae-Yokoyama, Y. Kondo, S. Nakamura, and K. Ishihara, "Hybrid (WIO) LCA of the introduction of active disassembling fasteners using hydrogen storing materials into electric and electronic equipments," presented at the Electronic Going Green Conference, Berlin, September, 2008.

[4] T. E. Graedel, D. van Beers, M. Bertram, K. Fuse, R. B. Gordon, A. Gritsinin, A. Kapur, R. Klee, R. Lifset, L. Memon, H. Rechberger, S. Spatari, and D. Vexler, "The multilevel cycle of anthropogenic copper," *Environ. Sci. Technol.* **38**, 1253–1261 (2004).

[5] T. Lanzano, M. Bertram, M. De Palo, C. Wagner, K. Zyla, and T. E. Graedel, "The contemporary European silver cycle," *Resources, Conserv. Recycl.* **46**, 27–43 (2006).

[6] S. Bringezu and Y. Moriguchi, "Material flow analysis," in *A Handbook of Industrial Ecology*, edited by R. Ayres and L. Ayres (Elgar, Cheltenham, UK, 2001).

[7] E. Matthews, C. Amann, S. Bringezu, M. Fischer-Kowalski, W. Huttler, R. Kleijn, Y. Moriguchi, C. Ottke, E. Rodenburg, D. Rogich, H. Schandl, H. Schutz, E. van der Voet, and H. Weisz, *The Weight of Nations* (World Resource Institute, Washington, D.C., 2000).

[8] H. Schandl and N. Schulz, "Changes in the United Kingdom's natural relations in terms of society's metabolism and land-use from 1850 to the present day," *Ecol. Econ.* **41**, 203–221 (2002).

[9] H. Weisz, F. Krausmann, C. Amann, N. Eisenmenger, K. Erb, K. Hubacek, and M. Fischer-Kowalski, "The physical economy of the European Union: Cross-country comparison and determinants of material consumption," *Ecol. Econ.* **58**, 676–698 (2006).

[10] Y. Moriguchi, "Material flow analysis and industrial ecology studies in Japan," in *A Handbook of Industrial Ecology*, edited by R. Ayres and L. Ayres (Elgar, Cheltenham, UK, 2001), pp. 301–310.

[11] H. Weisz and F. Duchin, "Physical and monetary input–output analysis: What makes the difference," *Ecol. Econ.* **57**, 534–541 (2006).

[12] Eurostat, *Economy-Wide Material Flow Accounts and Derived Indicators: A Methodological Guide* (Office for Official Publications of the European Communities, Luxembourg, 2001).

[13] P. Daniels, "Approaches for quantifying the metabolism of physical economies: A comparative survey, Part II: Review of individual approaches," *J. Ind. Ecol.* **6**, 65–88 (2002).

[14] S. Nakamura, S. Murakami, K. Nakajima, and T. Nagasaka, "A hybrid input–output approach to metal production and its application to the introduction of lead-free solders," *Environ. Sci. Technol.* **42**, 3843–3848 (2008).

[15] S. Nakamura and K. Nakajima, "Waste input–output material flow analysis of metals in the Japanese economy," *Mater. Trans.* **46**, 2550–2553 (2005).

[16] S. Nakamura, K. Nakajima, Y. Kondo, and T. Nagasaka, "The waste input–output approach to materials flow analysis concepts and application to base metals," *J. Ind. Ecol.* **11**, 50–63 (2006).

[17] W. Leontief, "Environmental repercussions and the economic structure: An input–output approach," *Rev. Econ. Statistics* **52**, 262–271 (1970).

[18] S. Nakamura and Y. Kondo, "Input–output analysis of waste management," *J. Ind. Ecol.* **6**, 39–63 (2002).

[19] S. Nakamura, "Input–output analysis of waste cycles," in *Proceedings of the First International Symposium on Environmentally Conscious Design and Inverse Manufacturing* (IEEE Computer Society, Los Alamitos, CA, 1999), pp. 475–480.

[20] S. Nakamura, "Waste input–output table," available at http://www.f.waseda.jp/nakashin/WIO.html.

[21] S. Searle, *Matrix Algebra Useful for Statistics* (Wiley, New York, 1982).

[22] K. Nakajima, Y. Yoshizawa, K. Yokoyama, S. Nakamura, and T. Nagasaka, "Material flow analysis based on WIO-MFA MODEL: Case study of PVC flow in Japan," paper presented at ConAccount 2008, Charles University Environment Center, Prague, September, 2008.

[23] Japan Automobile Manufacturers Association, *The Motor Industry of Japan* (JAMA, 2003).

[24] C. Stahmer, M. Kuhn, and N. Braun, "Physical input-output tables for Germany, 1990. Eurostat WP 2/1998/B/1," (European Commission, Luxembourg, 1998).

[25] K. Yokoyama, T. Onda, S. Kashiwakura, and T. Nagasaka, "Waste input-output analysis on landfill mining activity," *Mater. Trans.* **47**, 2582–2587 (2006).

ECONOMIC SYSTEMS, SOCIAL SYSTEMS, INDUSTRIAL SYSTEMS, AND ECOSYSTEMS

14 Early Development of Input–Output Analysis of Energy and Ecologic Systems

Bruce Hannon

14.1 Introduction

In the 1930s, Wassily Leontief was putting the finishing touches on his development of the input-output (IO) matrices of the U.S. economy, just in time for its strategic use in converting our industry to a war footing. His process allowed the direct and indirect demands of industry to be estimated for a given gross national product (GNP). The government stated its concepts of the needs for the items of war in terms of the numbers of airplanes, tanks, guns, explosives, and so forth, for each of the four or five years they expected the war to last. Leontief was able to determine, for this final bill of goods, the flows of steel, aluminum, energy, and such needed from each industry, directly and indirectly [1]. Then these flows were compared with the capital stocks needed in these industries to meet the wartime demands. What they found was that the output of war material and energy plus those of personal consumption was not possible given then-current capacities in any of the major sectors. Two major endeavors were soon undertaken: massive new construction programs in steel production and shipbuilding, including conversion of many industries to the production of military items, for example, the auto companies converting to the production of military vehicles, and the substantial reduction of personal consumption of cars, gasoline, tires, and certain kinds of food. How did he do it?

Imagine a matrix of exchanges between each entity involved in the production economy. The industries may be very materials oriented, such as the steel or the auto industries, or service oriented, such as the finance or transportation industries. The industries are listed down the rows of the matrix and in the same order across the columns of the matrix. At each intersection in the matrix the annual exchange between the row industry and the column industry was found from industry records and posted in dollars. The net outputs of each of these industries, their contribution to final demand, were also estimated for the year. The total of all of these net outputs was the gross national product (GNP), as it came to be called then [2]. This GNP, more recently named the gross domestic product or GDP, consisted of vectors of personal consumption items, government consumption items, new capital formation, net exports, and changes in inventories. Both Simon Kuznets, who designed this accounting framework, and Leontief would later get Nobel prizes in economics for this work.

The list or vector of net inputs to the economy was also estimated and arranged across the bottom of the exchange matrix. The net inputs were composed of separate vectors of the labor inputs to each industry, their profits, depreciation, and taxes paid. The dollar value of the total input of each industry (the column sum) was equal to the total output of each industry (the row sum). With this special set of data, Leotief showed how a total-requirements matrix could be calculated. Each element in this derived matrix was the total direct and indirect requirement from the row industry by the column industry per dollar of that industry's total output. The total required output of each of the industries was the product of a special form of the inverted exchange matrix times the GNP. By comparing these total outputs with the expanding demands (scenario versions of the GNP), a feasible transition to a wartime economy could be estimated. Private entrepreneurs, such as Henry J. Kaiser, were essentially given blank checks to build steel and ship construction enterprises at the estimated rates. At the same time, rationing of consumer goods and price controls were put into place. The overall success of this effort is legend.

After the war, these capacities for production remained, driving the ensuing economy in a variety of new and unique ways. No formal transition plan of reduction was forthcoming, even though the policy-generating process to quickly return to a peacetime economy was at hand. For example, the production of explosives required the production of nitrogen. After the war, these munitions companies desperately sought new uses for their nitrogen. One was found: the production of ammonia as a fertilizer of annual crops, particularly corn. This flood of cheap ammonia contributed to a glut in corn production, leading to the current billions in federal revenues spent annually on price support programs.

14.2 The Classic Input–Output Theory

The mathematics behind the Leontief idea is rather simple. Let this exchange matrix be $\mathbf{X}$ and the total output of each industry be vector $\mathbf{x}$. Divide the columns in matrix $\mathbf{X}$ by their respective total outputs to form matrix $\mathbf{A}$, the (normalized) direct-requirements matrix, so

$$\mathbf{X}^* \hat{\mathbf{x}}^{-1} = \mathbf{A} \tag{14.1}$$

where $\hat{\mathbf{x}}^{-1}$ is the inverted matrix of the diagnalized vector $\mathbf{x}$. Then

$$\mathbf{A}\mathbf{x} + \mathbf{y} = \mathbf{x}, \tag{14.2}$$

where vector $\mathbf{y}$ is the vector of net outputs (its sum is the GDP). Then

$$\mathbf{y} = [\mathbf{I} - \mathbf{A}]\mathbf{x}, \tag{14.3}$$

where $\mathbf{I}$ is the unity matrix.

Or, to find the vector of required industry outputs for a desired net output $\mathbf{y}$,

$$\mathbf{x} = [\mathbf{I} - \mathbf{A}]^{-1}\mathbf{y}, \tag{14.4}$$

requiring the inversion of matrix $[\mathbf{I} - \mathbf{A}]$.

The **X** matrix and **y** vector are routinely assembled by the Bureau of Economic Analysis (BEA) in the U.S. Department of Commerce and published about every 5 years.

14.3 Input–Output Energy Analysis

In 1969, a group of my students analyzed the direct and indirect energy use in making and delivering soft drinks and beer in refillable and throwaway containers. Our interest was to see if the system of refillable containers was more or less energy intensive than the system of throwaway containers. By that time, the glass-refillable containers were being rapidly displaced from the market by glass and aluminum throwaway containers. We had to trace the production of the glass container, for example, all the way back to the sand from which it was made. We labored long and hard to find as many inputs to this process chain as we could. We even traced back the chain of processing of the major inputs to the bottlemaking chain such as plastic and paper packaging. Along the way, we realized the enormous difficulty of the infinite regress we were involved in. There were several basic forms of energy, and the process had to be tracked for each of them. Nevertheless, we labored on, eventually producing the Center for Advanced Computation (CAC) document 23 and revising it until about 1972 [3]. The little pamphlet became extremely popular and the basis for many state legislative attempts to ban or tax these containers and many court challenges to the idea of requiring monetary deposits on them. We found that the total throwaway energy cost was about four times that of the 18-trip (typical) refillable container per unit of beverage. The direct and indirect labor costs of throwaways and refillables were calculated by Hugh Folk, an economics professor at the University of Illinois at Urbana-Champaign. He found that the refillable containers required more and different jobs than the throwaways [4].

Several very interesting new issues arose. First was the idea that energy conservation could be achieved by changing one's product choices, the idea of embodied energy in consumer products. Although we were only observers of the complex political process we had started, we noted the extreme reactions to the idea of a return to refillables from the industries that made the containers, those that made the material for the containers, the paper and plastic packaging of the containers, and the coalition of labor unions. All of them pushed for the throwaway containers. The labor union coalition reaction was especially interesting. There was then even a refillable bottle washers union in the coalition. Folk found that, because the total payments to labor under either system, refillable or throwaway, were roughly the same, the average wage within these systems would decline with a return to a refillable system. This was due to the increase in relatively low-wage jobs at the retail and wholesale levels and a decline in the relatively high-wage jobs in the glass, aluminum, and container industries. Thus, whereas the overall labor coalition effect would be an increase in jobs, the high union fees paid by the high-waged industries caused the coalition to lobby for throwaways.

We also found that, although the beer industry introduced throwaway containers in the 1930s, particularly for the outdoorsman, the major thrust came in WWII when beer was shipped to the troops. Because of the scarce shipping capacity, the return of the refillables was not feasible. After the war, to continue the movement toward throwaways, the container makers, along with the metals industries, primarily Reynolds, Alcoa, and Inland Steel, would form special contracts with bottlers such as Coke and Busch. The 5-year contracts yielded free throwaway containers in the first year, with increasingly small discounts as the years progressed. This incentive allowed the bottlers enough income to convert their bottling lines to throwaways and use one-way delivery trucks to wholesalers and retailers. The advent of the interstate highway system allowed overnight trucking from the new bottling plants, minimizing bottler inventory costs. Bottlers were then able to successfully invade the territories of smaller bottlers who used refillable containers exclusively. The big bottler's throwaways were of no use to the local bottlers, who soon went out of business. The prodigious use of the throwaway led to significant waste and littering issues. A strong public reaction to this waste was countered by an advertising campaign that spoke of wasteful consumer littering, turning the issue on its head, and of the ecology of the recycled aluminum container. The last holdout, the individual grocery shopper buying soft drinks in refillable bottles, succumbed to throwaways wherever local aluminum recycling was instituted by waste-conscious communities.

Our last effort in this area showed that the modern recycled aluminum soft drink or beer container required about twice the total direct and indirect system energy as its comparable refillable glass container. Today in the United States, the refillable container is a museum item.

The difficulty of finding the complete energy and labor embodied in these products and the fact that we found a significant energy–labor trade-off in our study caused me to go deeper into the issue.

By 1970, I had finished my doctorate and knew my various environmental concerns were far more interesting than engineering pursuits. Luckily I was able to join the University of Illinois' Center for Advanced Computation (CAC) and attract two other newly minted Ph.D.s, Clark Bullard (engineering) and Robert Herendeen (physics) to the study of embodied energy. Our underlying concern was the growing dependence of the economy on finite energy resources. We wanted to see if knowledge of the energy embodied in various products and services would change purchasing decisions. We needed a way to uniquely assign all of the energy of the various forms that moved from the finite resource base to all of the consumer products and services. The IO process seemed to hold promise, and fortunately there were appropriate economists at the CAC. We formed the Energy Research Group (with the acronym ERG, which is Greek for energy) that I directed in the Center. Bullard and I managed to get a significant National Science Foundation (NSF) grant to formally start us off in 1972.

Examination of the U.S. Bureau of Labor Statistics data revealed a particularly extreme variation (more than a factor of 10) in the apparent prices paid for energy across the economic sectors. It was abundantly clear that we would have to use physical units in the transactions matrix (X) rather than the given dollar values of the exchanges. This proved to be an enormous task but well worth it in the end. The current process used by those at Carnegie-Mellon to produce up-to-date energy intensities is significantly flawed for this reason. Our final work on the 1977 report on the U.S. economy is the most definitive [5]. We compared the change in the energy intensities between 1972 and 1977, a period in which energy prices increased suddenly and the economy made significant adjustments.

Figure 14.1. The energy balance for a specific type of energy for the jth sector of the economy for a given year. The energy intensities are the ε_i. E_j is the direct energy used in sector j. The x_{ij} are the elements of the economic transactions matrix except that the energy rows are in physical units (BTUs).

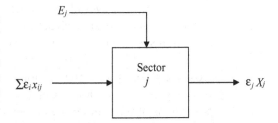

In the early 1970s, Bullard and Herendeen published the details of how these calculations were made [6, 7]. The result was a set of energy intensities, similar to economic prices. These intensities were the direct and indirect energy amounts (by type) that moved from the finite resource base per unit of output of each of 400 sectors of the economy during a given year. In the early 1970s, we had to take over the entire university computation facilities, usually at 3A.M., to allow inversion of the appropriate matrix!

The intensities were derived from the energy balance in Fig. 14.1 for each energy type.

In Fig. 14.1, the vector $\mathbf{E}$ is the list of the physical amount of energy of a specific type (coal, crude oil, refined petroleum, natural gas, electricity) used by each sector of the industry and commerce, and $\boldsymbol{\varepsilon}$ is the vector of energy intensities of this type (physical units, e.g., BTUs, per dollar) embodied in each of the (nonenergy) inputs x_{ij} to sector j, summed for the total direct and indirect embodied energy inputs to this sector. To calculate a total annual physical measure of the energy moving from the natural-resource base, we combined the coal intensity, the crude petroleum intensity, and a portion of the electricity intensity representing the contribution of nuclear and hydropower. To calculate the energy embodied in imports (e.g., Japanese autos), we assumed they had the same energy inputs as their domestic counterpart.

From Fig. 14.1, we get, in vector and matrix form, the energy balance,

$$\boldsymbol{\varepsilon}\mathbf{X} + \mathbf{E} = \boldsymbol{\varepsilon}\hat{\mathbf{x}}. \tag{14.5}$$

Solving for $\mathbf{E}$ gives, with Eq. (14.1),

$$\mathbf{E} = \boldsymbol{\varepsilon}\hat{\mathbf{x}} - \boldsymbol{\varepsilon}\mathbf{A}\hat{\mathbf{x}}, \tag{14.6}$$

leading to the vector of energy intensities,

$$\boldsymbol{\varepsilon} = \mathbf{E}\hat{\mathbf{x}}^{-1}(1-\mathbf{A})^{-1} \tag{14.7}$$

or

$$\boldsymbol{\varepsilon} = \mathbf{e}(\mathbf{I}-\mathbf{A})^{-1}, \tag{14.7}$$

where $\mathbf{e} = \mathbf{E}\hat{\mathbf{x}}^{-1}$, the vector of the physical energy of a specific type used by each sector per unit of total output of that sector.

We calculated these intensities with Eq. (14.7) by using the $\mathbf{A}$ modified to contain the physical units of energy in the energy sectors for the years in which the BEA had constructed the original $\mathbf{x}$ matrices that were usable during the years that our research existed (1963, 1972, 1977). The major effort on our part became the modification of the $\mathbf{A}$ matrix to substitute the physical units of energy for the dollar value and maintain the proper overall energy balance. This took at least two full time equivalent (FTE) years of effort for each data year. We had to make this substitution as a simple calculation showed extreme variation in the apparent energy prices being paid on average by each sector of the economy. The total energy intensities varied greatly across the 400 sectors of the economy.

The bulk of work done during the life of the ERG (1971–1980) was the various applications of these intensities to answer questions about the energy conservation potential of the individual sectors of the economy and of individual consumers. The federal support for our energy conservation work essentially disappeared in the early 1980s. In our 15 years of energy research, we noted a slightly lagged correlation between availability of federal funding of energy conservation research and energy price rises.

We were joined by economist Michael Reber in the mid-1970s, and the group of four professionals with graduate students reached about 30 FTEs at our peak. The ERG published over 200 papers during its lifetime, most of which found their way into appropriate professional journals. Over the 1970s, our group published 15 articles in *Science* alone. Unfortunately, almost all of this work was done before the normal historic reach of the current electronic library search programs and general knowledge of our work diminished steadily from the early 1980s. We did manage to establish a kind of record though, perhaps a permanent one, as the only research group to initiate and supply supporting documentation for *two* (unsuccessful) congressional attempts to establish an energy tax. Our group found that such a tax, placed on the physical energy as it moves from the natural-resource base, would produce widespread, effective, and equitable energy conservation. These results are subsequently discussed.

Economist Folk and I produced a paper on the trade-off between energy and labor for the 400 sectors of the U.S. economy [8]. We produced the labor intensities in a manner similar to the calculation of the energy intensities. Folk's student, Roger Bezdek, and I published a paper in *Science* in 1974 on the energy and labor differences between a unit of government spending on interstate highway construction and the alternatives of health care, criminal justice, and sewage treatment plant construction [9]. All these alternatives required less energy and created more jobs than highway construction. Analysis of a particular county in southern Illinois, where Interstate I-57 was being constructed while a large U.S. Army Corps of Engineers reservoir was also being constructed, revealed no correlation with unemployment there [10]. The major workforce for such construction comes from large cities at great distances just for the weekday work in the county. This result contrasted with the political justification for the highway and reservoir as a cure for high county unemployment.

Our first results predated the 1973 OPEC oil embargo and oil price tripling, and, needless to say, research funding ceased to be a problem for the rest of the decade. Our problem was the opposite. How should we avoid taking on so much research money that we became more projects managers than researchers and publishers? We restricted ourselves to funding from the chain of temporary federal energy offices, this evolution eventually leading to the cabinet-level Department of Energy.

To reproduce our results here would be of little value now because the actual numbers are so old. However, I summarize the kinds of work that we did and reveal our ulterior motive for this kind of work.

The original three professionals were from the hard sciences and had interests in the environment. It was easy for us to formulate a connection between energy use and environmental damage. We foresaw the problem faced today by those who advocate use of the environmental footprint to curb environmental impact. Many connections can be made between individual actions and their ultimate environmental impacts on land use, water use, energy use, the labor force, air pollution, water pollution, and so forth. We could even see the confusion this panoply of concerns would cause to caring consumers. We concluded that we needed a single, best indicator of the environmental impact of consumption choices. Being physical scientists, we naturally chose energy. We could see the air, water, climate, and resource base impacts as primary impacts. If a consumer could make alternative choices that lowered the direct and indirect energy use, chances were good that the environmental impacts could be lowered, providing the public was informed and acted on this information, a major additional problem. Our problem was gaining knowledge of that total energy use for each of the choices. We solved that problem by use of the modified IO analysis. Its drawbacks are manifold. It is a very dated picture of economic exchanges for a single year in a linear model of assumed equilibrium. However, the overall value of the IO approach was manifold. It yields the only way to assign the direct and indirect energy consumed per unit of each consumption item at a significant level of detail, completely, and without double counting [7].

We managed to address many of these drawbacks. We made the IO process dynamic, included the depreciation of capital in the resulting energy intensities, and made energy intensity comparisons between data years. This last work was especially useful in the 1972 versus 1977 comparison as these years bracket the first major jump in oil prices [11].

One of our most successful efforts was the comparison of the energy (combined types) intensities with the labor (total FTE jobs) intensities for a given year. We were able to show conclusively that energy and labor trade off or are substitutes . . . just as we had shown in the container study. We showed that, historically, when the price of electricity, for example, rose relative to the average (nonsalaried) wage, energy use decreased and employment increased per dollar of GDP. This would mean that labor productivity would decline and eventually so would real wages, all else staying the same [12].

We calculated the energy embodied in imports and exports, showing that the net energy dependency of the economy was greater than just the quantity of oil imports [13].

We did eventually manage to derive marginal energy intensities for the various sectors of the economy through a comparison of two consecutive IO data years [14]. The use of these intensities on the personal consumption part of the GDP revealed a slightly higher energy cost.

We showed how direct and indirect energy use varied with household incomes [15]. From this and further studies, we concluded that a tax on all energy as it moves from the resource base was the only equitable way to tax energy [16]. Our results showed that direct household energy use, primarily gasoline and heating fuels, saturated with rising income while the total direct and indirect energy use was linear with income and rose proportionately with income. These household energy studies also showed that, even though those living in New York City apartments, forswearing car use and claiming relatively low direct energy use, actually, because of their high relative incomes, used more than the average amounts of energy when the direct and indirect assignments were made to their consumption. Gasoline is a direct household energy use whereas airplane fuel is an indirect use.

In a 1975 *Science* paper, I laid out three dilemmas for the typical person wishing to unilaterally change their total energy footprint [17]. First, energy conservation can be achieved only by driving up the price of energy, especially electricity relative to wages, and this results in more employment. The increase in employment comes typically at the low-wage end of the spectrum whereas decreases in employment would occur at the high-wage end (as in the drink container example cited earlier.) The dilemma comes from the resistance of the high-wage, highly organized employees who can afford to pay high union dues. At the same time, no support for the change comes from the relatively disorganized low-wage employees. In addition, the jobs lost are those that exist in the electricity-intensive sectors whereas the jobs that would be gained do not yet exist. The second dilemma comes from the linear relationship of total direct and indirect energy use and household income previously noted: The only way to actually save energy is to reduce income. Even saving money results in energy use in the investment market. The third dilemma of the person determined to save energy comes from the fact that any voluntary energy-saving effort, such as riding a bus to work rather than driving a car, also saves money. This money savings is then spent on other goods or services, and those require energy (and jobs) to provide. Sometimes the energy demands of this alternative spending can completely offset the original savings, as when, for example, the dollar savings from a shift from meat to vegetable consumption is spent on gasoline.

The results of this paper only underscored the need for a rebated tax on all energy forms to steer consumers in the energy conservation direction. This was our main contribution. Today, because of concerns over global warning, the emphasis is on a carbon rather than an energy tax. But such a tax does not tax all energy whereas the energy tax covers most of the carbon releases to the atmosphere.

We also calculated the total energy and labor costs of many individual consumer choices, such as the use of disposable paper diapers versus washable cloth diapers, the use of mass transit versus the personal auto, of the consumption of vegetable versus animal protein, and so forth. All of these individual calculations required a combination of the intensities from the IO process with what we called process analysis. The latter consists of determining the total quantities of inputs to the last manufacturing step of a consumer good and then using the total energy and labor intensities on these inputs (summarized in [18]). This allowed us to be arbitrarily specific on the consumer good in question while still being able to calculate the total resource demand for it. We even calculated the amount of direct and indirect energy used to maintain overweight conditions in the U.S. population [19].

Wall Street used the energy intensities to show the energy price rise vulnerability of any corporation. We rebutted many industry public relations releases. Some corporations would, for example, report an energy conservation achievement by closing a small aluminum smelter on their property, only to buy that same quantity of aluminum from a large producer. The net energy saved in the country remained unchanged, but the press release did not mention the energy embodied in the purchased aluminum. The association of steel producers tallied the energy used (within the corporate boundary) to produce a ton of finished steel products. Because the very energy-intensive oxygen used in their blast furnaces was made by another company and there are hundreds of small (but collectively important) inputs, our estimate of their total energy demand was double their reported value.

One of our post-docs, Peter Kakela, studied the iron ore mining processes in his home state of Minnesota [20]. Before the end of WWII, the high-concentration ore was lightly processed and sent to the steel mills across the Great Lakes. Because of the mining disturbances, the state had taxed the mine companies. The war essentially drained these high-concentration ores, and the mining companies began to produce taconite iron ore from low natural ore concentrations. Because of the extensive cost of concentrating the iron into taconite pellets, they asked for and got the tax removed. Kakela extended his energy studies to the iron-making process and found that the taconite pellets greatly sped up the conversion of the ore into iron. The pellets (mixed with charcoal pellets) provided a porous mix in the blast furnace, allowing easy flow of the burning oil to filter uniformly up through the mix in the furnace. The original ore was a fine powder, and in the blast furnace, the burning oil channeled up through the mix with a much slower melting rate, resulting in a substantially higher cost of iron. When Kakela reported his results, the state of Minnesota reinstituted the mining tax. This study showed that, if the industry had used the lower grade ore in the first place, steel costs during WWII would have been significantly lower. Technical change can sometimes reverse the trend toward scarcity.

And sometimes, technical change can mask the growing resource scarcity, as Richard Norgaard of Berkeley showed in his studies of the production of gasoline in the late 1970s. The efficiency improvements in U.S. oil refineries concealed the rising price of crude such that the price of gasoline declined at the same time. Eventually, technical improvements at the refineries were exhausted, and gasoline prices began to reflect the increasing cost of crude oil.

While doing his doctoral thesis with our group, Robert Costanza, now Lund Professor at the University of Vermont, showed strong linear correlation between the embodied energy and the dollar value of each consumer good [21].

With Richard Stein, a New York architect, we studied the total energy cost of building construction, showing that, in 1967, the entire construction industry used 9% of all U.S. energy, with 80% of this amount being direct, onsite use [22]. Eric Hirst, then of Oak Ridge National Laboratory, and I evaluated the energy used and the conservation potential in U.S. commercial and residential buildings [23].

James Brodrick and I analyzed the steel industry for its optimum embodied energy intensity with use of more recycled metal [24]. We found that in 1967 the industry could increase its recycling by a factor of 6 and lower its energy intensity by 8.5%. We also found that the direct energy use reported by the industry was far less than the total energy embodied in a physical unit of steel. Some of these firms reported an energy savings by purchasing instead of making one of their smaller inputs, e.g., aluminum. The oxygen needed for steel making, made nearby but by another company, was considered by the industry to be available with no energy cost. The total energy analysis captures such indirect embodied energy inputs. Likewise, with more recycling, the total labor intensity of steel increased.

Reber studied the economics of western coal reserves in the United States [25]. The industry claimed that there were prodigious amounts of low sulfur coal in these reserves. Reber showed that, because this coal had a significantly lower energy content than the eastern coals, it would take much more western coal to make a unit of electrical energy. This step reduced the estimates of the western coal reserves by more than 25%. He also studied the cost of moving this coal to eastern and southern electric power plants [26]. The Bechtel Corporation favored a water-slurry coal pipeline from Wyoming to Texas. Reber showed that the existing railroads could handle the coal shipments at a significantly lower cost. Because of this paper, he subsequently lost his NSF grant (which the university funded to completion) and Congress decided the issue was worthy of a congressional hearing. Today, that coal moves by rail.

Although the papers of the ERG are dated and the working papers are difficult to find, they are available through the Engineering Grainger Engineering Library, University of Illinois at Urbana-Champaign. (The search page is at http://susanowo .grainger.uiuc.edu/engdoc/new/default.asp.)

Although the study of energy and the economy was fun for everyone in the group, a companion system, in many ways similar to the economy, caught the attention of Herendeen and me. In 1973, I published "The Structure of Ecosystems," a title clearly claiming more than I should have [27]. I developed an IO framework for the ecosystem in which the net output (the GDP of the ecosystem) was its net exports, inventory change, respiration, and new biocapital formation. Respiration was analogous with household and government consumption of the economic definition. The idea was applied to a specific ecosystem to show the direct and indirect energy dependence of the top carnivore on the solar input to the system. Robert Ulanowicz of the University of Maryland, Costanza, and I showed how such a matrix approach to ecosystem studies provided an excellent framework to collect the data from many different ecological research projects [28]. Such a framework could be used to show the direct and indirect connections of the various biological elements of the system. For example, if fishing was done on a system, the framework could show where and how much was needed to compensate and stabilize the system [29]. I could show that the energy intensities played the role of prices when one took such a view of the ecosystem [30]. Extending the economic analogy further into ecology, I showed how an ecological discount rate could be calculated; in a way, this was nature's time preference rate [31].

The linear IO system, when converted into a dynamic form, is unstable when run forward in time. However, Joseph Bentsman and I showed that oscillations of certain exchanges in the ecosystem could stabilize the system, provided these oscillations occurred within certain amplitude and frequency ranges [32].

The most recent work in the IO analysis arena let me demonstrate how the proper definition of the exchanges, the net inputs and outputs, for a combined economic and ecological IO matrix, allows the calculation of a system efficiency [33]. That thought was derived from another paper addressing the question of whether the addition of human activity to an ecosystem made the whole more or less efficient [34]. Surely, if the result is a decline in overall combined system efficiency, our position in the ecosystem is not a stable one.

These ecosystem studies turned out to provide the most enjoyment for me. Costanza and Herman Daly of the University of Maryland, along with many others (myself included in secondary roles), founded the Society of Ecological Economics out of this combination of ideas of the two economic and ecological systems. That society is now flourishing and publishes a prominent journal.

REFERENCES

[1] W. Leontief, *Input–output Economics* (Oxford University Press, New York, 1986).
[2] S. Kuznets, *National Income and Its Composition, 1919–1938* (National Bureau of Economic Research, New York, 1941).
[3] B. Hannon, "Bottles, cans, energy," *Environment* **14**(2), 23–31 (1972).
[4] H. Folk, "Two papers on the effects of mandatory deposits on beverage containers," PB 227–884 (National Technical Information Service, University of Maryland, College Park, MD, 1973).
[5] T. Blazeck, S. Casler, and B. Hannon, "Energy intensities for the U.S. economy–1977," Report to the Department of Energy (Office of Energy Conservation, Washington, D.C., 1985).
[6] C. Bullard, "Energy costs, benefits, and net energy," *Energy Syst. Policy* **1**, 367–382 (1976).
[7] C. Bullard and R. Herendeen, "Energy costs of goods and services," *Energy Policy* **3**, 263–278 (1975).
[8] H. Folk and B. Hannon, "An energy, pollution and employment policy model," in *Energy: Demand, Conservation, and Institutional Problems*, edited by M. Macrakis (MIT Press, Cambridge, MA, 1974).
[9] R. Bezdek and B. Hannon, "Energy, manpower and the Highway Trust Fund," *Science* **185**, 669–675 (1974).
[10] B. Hannon and R. Bezdek, "The job impact of alternatives to Corps of Engineers projects," *Eng. Issues, Am. Soc. Civil Eng.* **99**, 521–531 (1973).
[11] B. Hannon and S. Casler, "Readjustment potentials in industrial energy efficiency and structure," *J. Environ. Econ. Manage.* **17**, 93–108 (1989).
[12] B. Hannon, T. Blazeck, D. Kennedy, and R. Illyes, "A comparison of energy intensities: 1963, 1967 and 1972," *Resources Energy* **5**, 83–102 (1983).
[13] B. Hannon, "Analysis of the energy cost of economic activities: 1963–2000," *Energy Syst. and Policy* **6**, 249–278 (1982).
[14] B. Hannon, and T. Blazeck, "The marginal energy cost of goods & services," *Energy Syst. Policy* **8**(2), 85–112 (1984).
[15] R. Herendeen, C. Ford, and B. Hannon, "Energy cost of living, 1972–73," *Energy* **6**, 1433–1450 (1981).
[16] B. Hannon, R. Herendeen, and P. Penner, "An energy conservation tax: Impacts and policy implications," *Energy Syst. Policy* **5**, 141–166 (1981).
[17] B. Hannon, "Energy conservation and the consumer," *Science* **189**, 95–102 (1975).
[18] B. Hannon, "Energy and labor demand in a conserver society," *Technol. Rev.* **79**(5), 1–6 (1977).
[19] B. Hannon and T. Lohman, "The energy cost of overweight in the United States," *J. Public Health* **68**, 765–767 (1978).
[20] P. Kakela, "Iron ore: Energy, labor and capital changes with technology," *Science* **202**, 1151–1157 (1978).
[21] R. Costanza, "Embodied energy and economic valuation," *Science* **210**, 1219–1224 (1980).
[22] B. Hannon, R. Stein, B. Segal, and D. Serber, "Energy and labor use for building construction," *Science* **202**, 837–847 (1978).

[23] E. Hirst and B. Hannon, "Direct and indirect effects of energy conservation programs in residential and commercial buildings," *Science* **205**, 656–661 (1979).

[24] B. Hannon, and J. Brodrick, "Steel recycling and energy conservation," *Science.* **216**, 485–491 (1982).

[25] M. Reber, "Low sulfur coal: A revision of reserve and supply estimates," PB 248–062/AS (National Technical Information Service, University of Maryland, College Park, MD, 1973).

[26] M. Reber, "Coal transportation: Unit trains – slurry pipelines," PB 248–652/AS (National Technical Information Service, University of Maryland, College Park, MD, 1975).

[27] B. Hannon, "The structure of ecosystems," *J. Theor. Biol.* **41**, 535–546 (1973).

[28] B. Hannon, R. Costanza, and R. Ulanowicz, "A general accounting framework for ecological systems: A functional taxonomy for connectivist ecology," *Theor. Popul. Biol.* **40**, 78–104 (1991).

[29] B. Harmon, "Ecosystem control theory," *J. Theo. Biol.* **121**, 417–437 (1986).

[30] B. Hannon and C. Joiris, "A seasonal analysis of the Southern North Sea ecosystem," *Ecology* **70**, 1916–1934 (1989).

[31] B. Hannon, "Biological time value," *Math. Biol. Sci.* **100**, 115–140 (1990).

[32] J. Bentsman and B. Hannon, "Cyclic control in ecosystems," *Math. Biol. Sci.* **87**, 47–62 (1987).

[33] B. Hannon, "Ecological pricing and economic efficiency," *Ecol. Econ.* **36**, 19–30 (2001).

[34] B. Hannon, "How might nature value man?," *Ecol. Econ.* **25**, 265–280 (1998).

15 Exergoeconomics and Exergoenvironmental Analysis

George Tsatsaronis

15.1 Introduction

The objective evaluation and the improvement of an energy-conversion system from the viewpoints of thermodynamics, economics, and environmental impact require a deep understanding of

1. the real thermodynamic inefficiencies and the processes that caused them,
2. the costs associated with equipment and thermodynamic inefficiencies as well as the connection between these two important factors, and
3. possible measures that would improve the efficiency and the cost effectiveness and would reduce the environmental impact of the system being studied.

Exergoeconomics and exergoenvironmental evaluation provide methods for obtaining this information. Because an exergoenvironmental analysis and evaluation are conducted in complete analogy to the exergoeconomic ones, in the following sections more emphasis is placed on exergoeconomics, which has been significantly developed. Exergoeconomics consists of an exergy analysis, an economic analysis, and an exergoeconomic evaluation.

The term *exergoeconomics* was coined by the author in 1984 [1] to clearly characterize a combination of exergy analysis with economic analysis, when in this combination the *exergy-costing principle* (Subsection 15.4.1) is used. In this way, a distinction can be made between exergoeconomic methods and applications on one side and other numerous applications on the other side, in which results from a thermodynamic analysis (sometimes including an exergy analysis) and an economic analysis are presented (under the term thermoeconomic analysis) but without applying the exergy-costing principle. The case in which the exergy-costing principle is not used is in general of less interest to scientists and engineers and belongs to the broader field of thermoeconomics, which is defined as any possible combination or coexistence of a thermodynamic analysis with an economic one, and thus also includes exergoeconomics. It should be mentioned that, before 1984, all authors were using the term thermoeconomics to indicate what we call today exergoeconomics. Even after 1984, and still today, some authors continue to use the term thermoeconomics (even when the exergy-costing principle is used) instead of the more precise term, exergoeconomics. This practice, however, contributes to some confusion in the field.

The term *exergoenvironmental analysis* was coined in 2006 by L. Meyer and the author to characterize the new approach and to avoid confusion with other already existing terms and approaches.

15.2 Exergetic Analysis

An energy-based analysis identifies only the energy transfers to the environment as thermodynamic inefficiencies, fails to identify any inefficiency in an adiabatic process, and misleads the analyst by considering as an inefficiency the heat rejection to the environment dictated by the second law of thermodynamics. The additional concept that corrects these misconceptions is the exergy concept.

Exergy is the maximum theoretical useful work (shaft work or electrical work) obtainable from an energy-conversion system as this is brought into thermodynamic equilibrium with the thermodynamic environment while interacting only with this environment [2, 3]. Alternatively, exergy is the minimum theoretical work (shaft work or electrical work) required for forming a quantity of matter from substances present in the thermodynamic environment and for bringing the matter to a specified state. Thus exergy is a measure of the deviation of the state of the system from the state of this environment.

The *thermodynamic environment* in exergy analysis (also called *reference environment*) is a large thermodynamic system in equilibrium, in which the state variables (T_0, p_0) and the chemical potentials of the chemical components contained in it remain constant when, in a thermodynamic process, heat and materials are exchanged between another system and this environment. It is important to note that no chemical reactions can take place between chemical components contained in this environment because the latter is not in equilibrium. The thermodynamic environment is free of irreversibilities and its exergy is equal to zero. The thermodynamic environment, for which we need to use a model, should be as close as possible to, but is not identical with, the physical environment. The thermodynamic environment is part of the surroundings of any energy-conversion system. In the following discussion, the term environment refers to the thermodynamic environment when exergetic considerations are made, and to the physical environment when environmental impact is studied.

In the absence of nuclear, magnetic, electrical, and surface-tension effects, the *total exergy* of a system E_{sys} consists of four components: *physical exergy* E_{sys}^{PH}, *chemical exergy* E^{CH}, *kinetic exergy* E^{KN}, and *potential exergy* E^{PT} [3]:

$$E_{sys} = E_{sys}^{PH} + E^{CH} + E^{KN} + E^{PT} \tag{15.1a}$$

or

$$E_{sys} = m\left(e_{sys}^{PH} + e^{CH} + e^{KN} + e^{PT}\right). \tag{15.1b}$$

In the preceding equation, m is the mass of the system and e refers to the mass specific exergy.

The rate of physical exergy $\dot{E}_j^{PH}$ associated with the jth material stream is

$$\dot{E}_j^{PH} = \dot{m}e_j^{PH} = \dot{m}[(h_j - h_0) - T_0(s_j - s_0)]. \tag{15.2}$$

Here $\dot{m}$ is the mass flow rate and e, h, and s denote the specific exergy, enthalpy, and entropy, respectively, of the material stream. The subscript 0 refers to the property values of the same mass flow rate at temperature T_0 and pressure p_0 of the environment.

The physical exergy of a working fluid can be further split into its thermal (e^T) and mechanical (e^M) exergy components. This splitting may improve the accuracy of calculations and facilitates an exergoeconomic optimization:

$$e_j^{PH} = \underbrace{[(h_j - h_{j,X}) - T_0(s_j - s_{j,X})]_{p=\text{const}}}_{e^T} + \underbrace{[(h_{j,X} - h_{j,0}) - T_0(s_{j,X} - s_{j,0})]_{T_0=\text{const}}}_{e^M}.$$

$$(15.3)$$

In the preceding equation, the point X is defined at the given pressure p and the temperature T_0 of the environment.

The *chemical exergy* is defined as the maximum useful work obtainable as the system, being at temperature T_0 and pressure p_0, is brought into chemical equilibrium with the environment. Thus, for calculating the chemical exergy, not only the temperature and pressure but also the chemical composition of the environment has to be specified. Because our natural environment is not in equilibrium, there is a need to model an exergy-reference environment [3–5]. The use of tabulated *standard chemical exergy values* for substances contained in the environment at standard conditions ($T_{\text{ref}} = 298.15$ K, $p_{\text{ref}} = 1.013$ bar) facilitates the calculation of exergy values. The effect of small variations in the values of T_0 and p_0 on the chemical exergy of reference substances might be neglected in practical applications.

The chemical exergy of an ideal mixture of N ideal gases is

$$\bar{e}_{\substack{\text{mixture} \\ \text{ideal gases}}}^{CH} = \sum_{l=1}^{N} x_l \bar{e}_l^{CH} + \bar{R}T_0 \sum_{l=1}^{N} x_l \ln(x_l). \qquad (15.4)$$

Here $\bar{e}_l^{CH}$ is the standard molar chemical exergy of the lth substance, and x_l is the mole fraction of the lth substance in the system at T_0.

For solutions of liquids, the chemical exergy can be obtained with the aid of the activity coefficients γ_l:

$$\bar{e}_{\text{solution}}^{CH} = \sum_{l=1}^{N} x_l \bar{e}_l^{CH} + \bar{R}T_0 \sum_{l=1}^{N} x_l \ln(\gamma_l x_l). \qquad (15.5)$$

The standard chemical exergy of a substance not present in the environment can be calculated by considering a reversible reaction of the substance with other substances for which the values of standard chemical exergy are known [3].

The change in total exergy of a closed system undergoing a change from state 1 to state 2 ($E_2 - E_1$) is caused through transfers of energy in the form of work and heat between the system and its surroundings and is given by

$$E_2 - E_1 = E_{Q_{1-2}} + E_{W_{1-2}} - E_D. \qquad (15.6)$$

The exergy transfer $E_{Q_{1-2}}$ is associated with heat transfer Q_{1-2}, and the exergy transfer $E_{W_{1-2}}$ is associated with the transfer of energy by work W_{1-2}.

A part of the exergy supplied to a real energy-conversion system is destroyed because of irreversibilities within the system. Contrary to mass and energy, the exergy is not conserved in real systems. The exergy-destruction rate is equal to

$$E_{D,k} = T_0 S_{gen,k} = T_0 m_k s_{gen,k}. \tag{15.7}$$

Hence exergy destruction can be calculated either from the entropy generation [Eq. (15.7)] by use of an entropy balance or directly from an exergy balance [Eq. (15.6)]. The term $E_{D,k}$ is equal to zero only in ideal processes.

Thermodynamic processes are governed by the laws of conservation of mass and energy. These conservation laws state that the total mass and total energy can be neither created nor destroyed in a process. However, exergy is not generally conserved but is destroyed by irreversibilities within a system. Furthermore, exergy is lost when the energy associated with a material or energy stream is rejected to the environment.

An exergy balance for the kth component at steady-state conditions can be written as

$$\dot{E}_{F,k} = \dot{E}_{P,k} + \dot{E}_{D,k}, \tag{15.8}$$

where $\dot{E}_{P,k}$ is the exergy of product (the desired result, expressed in exergy terms, achieved by the kth component), and $\dot{E}_{F,k}$ is the exergy of fuel (the exergetic resources expended in the kth component to generate the exergy of product).

Here it is assumed that the system boundaries used for all exergy balances are at the temperature T_0 of the environment and therefore there are no exergy losses associated with a component [6]. Exergy losses ($\dot{E}_L$) appear only at the level of the overall system (subscript tot), for which the exergy balance becomes

$$\dot{E}_{F,tot} = \dot{E}_{P,tot} + \sum_{k=1}^{n} \dot{E}_{D,k} + \dot{E}_{L,tot}. \tag{15.9}$$

The following two dimensionless variables are used for the conventional exergetic evaluation of the kth component of a system [3, 6]:

• Exergetic efficiency,

$$\varepsilon_k = \frac{\dot{E}_{P,k}}{\dot{E}_{F,k}} = 1 - \frac{\dot{E}_{D,k}}{\dot{E}_{F,k}}; \tag{15.10}$$

• exergy-destruction ratio,

$$y_k = \frac{\dot{E}_{D,k}}{\dot{E}_{F,tot}}. \tag{15.11}$$

The exergetic efficiency of the overall system is

$$\varepsilon_{tot} = \frac{\dot{E}_{P,tot}}{\dot{E}_{F,tot}} = 1 - \sum_{k=1}^{n} y_k - \frac{\dot{E}_{L,tot}}{\dot{E}_{F,tot}}. \tag{15.12}$$

For the distinction between productive components, for which an exergetic efficiency is calculated, and dissipative components, for which no meaningful efficiency can be defined, see [7].

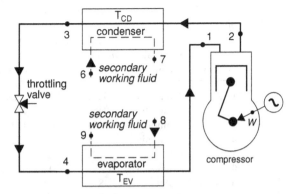

Figure 15.1. Schematic of a compression refrigeration machine.

The exergy concept complements and enhances an energetic analysis by calculating

1. the true thermodynamic value of an energy carrier,
2. the real thermodynamic inefficiencies in a system, and
3. variables that unambiguously characterize the performance of a system (kth component or overall system) from the thermodynamic viewpoint.

The real thermodynamic inefficiencies in an energy-conversion system are related to exergy destruction and exergy loss. All real processes are irreversible because of effects such as chemical reaction, heat transfer through a finite temperature difference, mixing of matter at different compositions, temperatures and pressure, unrestrained expansion, and friction. An exergy analysis identifies the system components with the highest thermodynamic inefficiencies and the processes that cause them.

15.3 A Compression Refrigeration Machine as an Example

The simple vapor-compression refrigeration machine shown in Fig. 15.1 is used here as an example for demonstrating the application of the methods discussed in this chapter. This machine consists of a compressor and motor, (CM), a condenser (CD), a throttling valve (TV), and an evaporator (EV). Ammonia is the primary working fluid for the refrigeration machine, whereas water is used as the secondary working fluid in the condenser, and air is the secondary working fluid in the evaporator. The product from the overall system is the cold rate $\dot{Q}_{cold} = 50$ kW, the exergy rate of which is kept constant in the analysis: $\dot{E}_{P,tot} = \dot{E}_9 - \dot{E}_8 = $ const. The isentropic efficiency of the compressor is assumed to be $\eta_{CM} = 0.85$. For simplicity, pressure drops are neglected in all heat exchangers.

Table 15.1 shows the material, mass flow rate, temperature, pressure, specific enthalpy, specific entropy, and specific physical exergy of all streams of matter shown in Fig. 15.1. The exergy destruction within each component of the refrigeration machine is calculated with Eq. (15.8):

$$\dot{E}_{D,CM} = \dot{W}_{CM} - (\dot{E}_2 - \dot{E}_1), \ \dot{E}_{D,CD} = (\dot{E}_2 - \dot{E}_3) - (\dot{E}_7 - \dot{E}_6)$$
$$\dot{E}_{D,EV} = (\dot{E}_4 - \dot{E}_1) - (\dot{E}_9 - \dot{E}_8),$$
$$\dot{E}_{D,TV} = (\dot{E}_3^M - \dot{E}_4^M) - (\dot{E}_4^T - \dot{E}_3^T) = \dot{E}_3 - \dot{E}_4.$$

Table 15.1. *Thermodynamic data for the vapor-compression refrigeration machine under real operating conditions*

Stream	Material stream	$\dot{m}$ (kg/s)	T (°C)	p (bars)	h (kJ/kg)	s (kJ/kg K)	e^{PH} (kJ/kg)	c^{PH} (€/GJ)
1	Ammonia	0.0454	−15	2.36	1444	5.827	126.3	66.69
2	Ammonia	0.0454	115	11.67	1716	5.934	366.8	66.43
3	Ammonia	0.0454	30	11.67	341.6	1.488	296.1	66.30
4	Ammonia	0.0454	−15	2.36	341.6	1.557	275.8	71.35
W								27.78
0^a	Ammonia		20	1	1536	6.572	0	
$6 = 0^a$	Water	2.98	20	1	83.93	0.296	0	0
7	Water	2.98	25	1	104.8	0.367	0.176	580.6
8	Air	9.94	0	1	273.3	6.776	0.719	0
9	Air	9.94	−5	1	268.3	6.757	1.138	84.73
0^a	Air		20	1	293.4	6.847	0	

[a] 0 is the reference point for calculating the exergy value of each material stream.

Table 15.2 shows the exergy rates associated with fuel, product, and exergy-destruction as well as the exergetic efficiency and the exergy-destruction ratio for each component and for the overall refrigeration machine. The results in Table 15.2 indicate that the condenser and the evaporator have the highest exergy-destruction ratios and the lowest exergetic efficiencies. According to these results, the efforts to improve the thermodynamic efficiency of the refrigeration machine should focus on these two components.

15.4 Economic Analysis

The *cost of the final products* is one of the most important factors affecting the selection of an option for the design or operation of an energy-conversion system. The *cost of a product* is the amount of money paid to acquire or produce it. The *market price of a product* is, in general, affected not only by the production cost of the product and the desired profit but also by other factors, such as demand, supply, competition, regulation, and subsidies. Exergoeconomics deals with costs.

The annual total revenue requirement (*total product cost*) for a system is the revenue that must be collected in a given year through the sale of all products generated by this system to compensate the system operating company for all expenditures

Table 15.2. *Conventional exergetic analysis for the vapor-compression refrigeration machine*

Component	$\dot{E}_{F,k}$ (kW)	$\dot{E}_{P,k}$ (kW)	$\dot{E}_{D,k}$ (kW)	ε_k (%)	y_k (%)
CM	12.340	10.912	1.428	88.43	11.57
CD	3.206	0.525	2.681	16.39	21.73
TV	7.953	7.028	0.925	88.37	7.50
EV	6.778	4.158	2.620	61.35	21.24
Overall system	12.340	4.158	7.653	33.70	62.0

Table 15.3. *Values of selected exergoeconomic variables for the vapor-compression refrigeration machine*

Component	$\dot{Z}_k$ (€/h)	$\dot{C}_{D,k}$ (€/h)	$\dot{Z}_k + \dot{C}_{D,k}$ (€/h)	$c_{F,k}$ (€/GJ)	$c_{P,k}$ (€/GJ)	r_k (−)	f_k (%)
CM	1.37	0.12	1.49	27.78	66.3	1.387	90.6
CD	0.32	0.65	0.97	67.00	580.6	7.666	33.5
TV	0.01	0.22	0.23	66.30	75.3	0.136	3.1
EV	1.61	0.71	2.32	75.30	230.4	2.060	69.4
Overall system	3.31	0.77	4.08	27.78	230.4	7.294	

Note: All values represent levelized costs over the assumed economic life (15 years) of this machine.

incurred in the same year and to ensure sound economic system operation. The total revenue requirement for an energy-conversion system consists of fuel cost, operating and maintenance (O&M) expenses, and the so-called carrying charges. The latter represent the costs associated with capital investment (CI) and include depreciation, returns on debt and equity, taxes, and insurances.

When evaluating the cost effectiveness and considering design modifications of an energy-conversion system, it is necessary to compare the annual values of carrying charges, fuel costs, and O&M expenses. These cost components may vary significantly within the economic life of the system. In general, carrying charges decrease while fuel and O&M costs increase with increasing years of system operation. Levelized annual values for all cost components should therefore be used when an energy-conversion system is analyzed.

The first column of Table 15.3 shows the contribution of investment costs (annual levelized carrying charges) associated with each component of the refrigeration machine presented in Section 15.3. The O&M expenses were neglected in this example. The investment costs were based on information provided by a company [8]. For the calculation of the levelized costs, an economic life of 15 years was used for the refrigeration machine as well as a general average annual inflation rate of 2.5% and an average cost of money of 10% were applied.

An economic analysis generally involves more uncertainties than a thermodynamic analysis. This fact, however, should not deter analysts from conducting detailed economic analyses using the best available data. Sensitivity studies are recommended to investigate the effect of major assumptions (e.g., cost of money, inflation rate, and real escalation rate of fuels) on the results of an economic analysis. Examples of detailed economic analyses conducted for energy conversion-systems are given in [3, 9–11].

15.5 Exergoeconomic Evaluation

Exergoeconomics is an exergy-based method that identifies and calculates the location, magnitude, causes, and costs of thermodynamic inefficiencies in an energy-conversion system. The real inefficiencies in such a system are the exergy destruction and the exergy loss. An exergoeconomic analysis is conducted at the component level of a system and identifies the relative cost importance of each component. If more than one product is generated by the overall installation being considered, then an exergoeconomic analysis also provides the cost allocation to the different products.

Exergoeconomics is based on two important principles that represent the fundamental connections between thermodynamics and economics. The first principle is common to all exergoeconomic approaches and applications, whereas the second principle refers only to applications in which new investment expenditures are required.

15.5.1 Exergy Costing

The exergy-costing principle states that exergy is the only rational basis for assigning monetary values to the transport of energy and to the inefficiencies within a system. In energy-conversion processes and installations, exergy represents the commodity of real thermodynamic value. Monetary values and environmental impacts (see Section 15.8) should be assigned only to real commodities of value. Mass, energy, or entropy should not be used for assigning monetary values because their exclusive use results in misleading conclusions [3, 12].

According to the exergy-costing principle, the cost stream ($\dot{C}_j$) associated with an exergy stream ($\dot{E}_j$) is given by

$$\dot{C}_j = c_j \dot{E}_j, \tag{15.13}$$

where c_j represents the average cost at which each unit of exergy is supplied to the jth stream in the system being studied. In exergoeconomics, the cost rates are calculated as costs per unit of time of system operation at the given capacity. Eq. (15.13) is applied to the exergy associated with material streams entering or exiting a system as well as to the exergy transfers associated with the transfer of work and heat. For the cost (C_k) associated with the exergy (E_k) contained within the kth component of a system we write

$$\dot{C}_k = c_k \dot{E}_k. \tag{15.14}$$

Here c_k is the average cost per unit of exergy contained within the kth component.

There are very few exceptions to the exergy-costing principle when the cost carried by a material stream is not associated with the exergy of the stream (for example, treated liquid water at ambient conditions before being used in a steam power plant or an inorganic chemical supplied to a chemical reactor). The treatment of these cases is discussed in [3, 7].

15.5.2 Exergy Destruction Affects the Investment Cost

In thermodynamics, the exergy destruction represents a major inefficiency and a quantity to be minimized when the overall system efficiency should be maximized. In the design of a new energy-conversion system, however, the exergy destruction within a component represents not only a thermodynamic inefficiency but also an opportunity to reduce the investment cost associated with the component being considered and thus with the overall system.

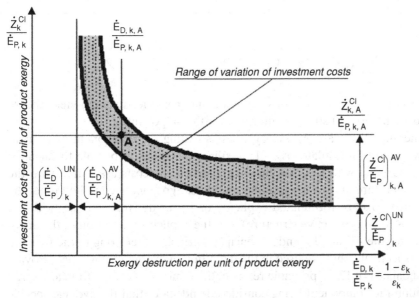

Figure 15.2. Expected relationship between investment cost and exergy destruction (or exergetic efficiency) for the kth component of an energy-conversion system.

Figure 15.2 refers to a component of the overall system and shows that the cost rate $\dot{Z}_k^{CI}$ associated with capital investment decreases with increasing exergy-destruction rate $(\dot{E}_{D,k})$ within the same component. Instead of a single curve, a shaded area is presented to denote that the investment cost could vary within a range for each given value of the exergy destruction. The effect of component size is taken into consideration in Fig. 15.2 by relating both $\dot{Z}_k^{CI}$ and $\dot{E}_{D,k}$ to the exergy rate of the product generated in this component $(\dot{E}_{P,k})$.

The vast majority of components in energy-conversion systems qualitatively exhibit the behavior between $\dot{Z}_k^{CI}$ and $\dot{E}_{D,k}$ shown in Fig. 15.2. Should the investment cost of a component increase or remain constant with increasing exergy destruction, then this component does not need to be considered in the optimization because in this case we would always select for this component the design point that has the lowest investment cost and, at the same time, the lowest thermodynamic inefficiencies (i.e., the highest exergetic efficiency).

For practical applications, the area in which significant trade-offs between investment cost and exergy destruction can be realized is of particular importance.

15.5.3 Cost Balances and Auxiliary Equations

A cost balance shows that the sum of cost rates associated with all exiting exergy streams equals the sum of cost rates of all entering exergy streams plus the appropriate charges (cost rates) that are due to capital investment as well as to operating and maintenance expenses. The sum of the last two terms is denoted by $\dot{Z}_k$ and is calculated with the aid of a detailed economic analysis. In exergoeconomics a cost balance is formulated separately for each system component [1, 3, 13]. Thus, for the kth component receiving a heat transfer and generating power, for example, we write

$$\sum_{e}^{N_e} (c_e \dot{E}_e)_k + c_{w,k} \dot{W}_k = c_{q,k} \dot{E}_{q,k} + \sum_{i}^{N_i} (c_i \dot{E}_i)_k + \dot{Z}_k. \tag{15.15}$$

Here, N_e is equal to the number of streams exiting from the kth component and N_i is equal to the number of streams entering the kth component.

In general, if there are N_e exergy streams exiting the component being considered, we need to formulate $(N_e - 1)$ auxiliary equations to be able to calculate the costs associated with the existing streams when the costs associated with the entering streams are known. Depending on the exergoeconomic method used, the auxiliary equations are formulated explicitly or implicitly (for example, with the aid of the so-called productive structure). For the explicit formulation of the auxiliary equations, the so-called F and P principles (principles referring to the fuel and product, respectively, that are used in the definition of exergetic efficiency) provide general guidance [7]: The F principle refers to the removal of exergy from an exergy stream within the component being considered and states that the average specific cost (cost per exergy unit) associated with this removal of exergy (which is part of the exergy of fuel) must be equal to the average specific cost at which the removed exergy has been supplied to the same stream in upstream components. The P principle refers to the supply of exergy to a stream or to the generation of an exergy stream within the component being considered and states that each exergy unit is supplied to any stream associated with the exergetic product of the component at the same average cost (c_P). Examples of auxiliary equations are found in every publication dealing with exergoeconomics. It should be noted that the formulation of auxiliary equations should be meaningful from both the exergetic and the economic viewpoints.

With the aid of cost balances and auxiliary equations, the cost rate and the cost per unit of exergy are calculated for each exergy stream (i.e., for each material and energy stream) in the overall system. In this way we associate with each stream not only mass, energy, entropy, and exergy but also cost. This is the first step in understanding the cost-formation process and the real cost sources and thus in making a well-informed decision for improving the cost effectiveness of an energy conversion system.

The cost balance for the kth system component can also be written as

$$\dot{C}_{P,k} = \dot{C}_{F,k} + \dot{Z}_k \tag{15.16}$$

or

$$c_{P,k} \dot{E}_{P,k} = c_{F,k} \dot{E}_{F,k} + \dot{Z}_k. \tag{15.17}$$

Here $\dot{C}_{P,k}$ and $\dot{C}_{F,k}$ are the cost rates associated with product and fuel, respectively, $c_{P,k}$ and $c_{F,k}$ represent the costs per unit of exergy associated with product and fuel, respectively, and $\dot{Z}_k$ is the sum of capital-investment-related cost and O&M expenses for the kth component:

$$\dot{Z}_k = \dot{Z}_k^{CI} + \dot{Z}_k^{OM}. \tag{15.18}$$

The cost associated with exergy destruction is hidden in a cost balance and is charged to the product. The last column in Table 15.1 shows the specific costs for the streams of the refrigeration unit. These costs are obtained by solving a system of equations consisting of cost balances and auxiliary equations [3, 14].

15.5.4 Exergoeconomic Variables and Iterative Improvement

The exergoeconomic evaluation is conducted at the component level with the aid of a series of exergoeconomic variables. The following variables can be used for evaluating the cost effectiveness of the kth component in an iterative conventional exergoeconomic optimization:

- Cost rate associated with exergy destruction within the component being considered,

$$\dot{C}_{D,k} = c_{F,k}\dot{E}_{D,k}, \tag{15.19}$$

- relative cost difference,

$$r = \frac{c_{P,k} - c_{F,k}}{c_{F,k}} = \frac{1 - \varepsilon_k}{\varepsilon_k} + \frac{\dot{Z}_k}{\dot{C}_{D,k}}; \tag{15.20}$$

- exergoeconomic factor

$$f_k = \frac{\dot{Z}_k}{\dot{Z}_k + \dot{C}_{D,k}}. \tag{15.21}$$

In what follows, we discuss an iterative improvement technique that can be applied to design optimization, which is the most general optimization case. This technique consists of the following steps:

1. A workable design (i.e., a design that fulfils mass and energy balances) is developed first. Several guidelines presented in the literature may assist in developing a workable design that is relatively close to the optimal one. For example, the values of $\dot{m}$, T, and p given in Table 15.1 represent a workable thermal design for the refrigeration machine shown in Fig. 15.1.
2. A detailed exergoeconomic analysis and an evaluation are conducted for the design developed in the previous step. The results from the exergoeconomic analysis are used to determine design changes (changes in the structure and in the process variables) that are expected to improve the design currently being considered. In this step, we consider only changes that affect both the exergetic efficiency and the investment cost: For the most important system components, we ask whether it is cost effective to reduce the capital investment at the expense of component efficiency or whether the component efficiency (and the investment cost) should be increased. Any subprocesses that contribute to the exergy destruction or exergy loss from the overall system should be eliminated if they do not contribute to the reduction of either capital investment or fuel costs for other components.

   Table 15.3 summarizes the results from the exergoeconomic analysis of the refrigeration machine shown in Fig. 15.1. The sum $\dot{Z}_k + \dot{C}_{D,k}$ indicates that the evaporator is by far the most important component from the viewpoint of costs.

The relatively high values of the exergoeconomic factor f_k for the compressor and evaporator suggest that the cost effectiveness of the refrigeration machine might be increased if we would decrease the investment costs associated with these two components at the expense of their efficiency. The relatively low value of f_k for the condenser indicates that we should decrease the minimum temperature difference in this component. It must be noted that no design changes are possible for a throttling valve.

3. Based on the results from the previous step, a new design is developed and the value of the objective function is calculated for the new design. If in comparison with the previous design this value has been improved, we may decide to proceed with another iteration that involves step 2. If, however, the value of the objective function has not improved in the new design compared with that of the previous one, we may either revise some changes conducted in the new design and repeat step 2 or proceed with step 4.

4. In this final step we optimize the decision variables that affect the cost but not the exergetic efficiency. After obtaining the cost optimal design, it is always advisable to conduct a parametric study to investigate the effect on the optimization results of some parameters used and of the assumptions made in the optimization procedure.

When applying this methodology, the designer should recognize that the values of all exergoeconomic variables depend on the component type: heat exchanger, compressor, turbine, pump, chemical reactor, etc. Accordingly, whether a particular value is judged to be high or low can be determined only with reference to a particular class of components. In that respect, combining knowledge-based and fuzzy approaches with an iterative exergoeconomic optimization technique may be useful for the system designer [15].

15.6 Advanced Exergetic Analysis

Exergy analysis is without a doubt a powerful tool for developing, evaluating, and improving an energy-conversion system. The lack of a formal procedure in using the results obtained by an exergy analysis is, however, one of the reasons why exergy analysis is not very popular among energy practitioners. Such a formal procedure cannot be developed as long as the interactions among components of the overall system are not being taken properly into account [16].

A conventional exergetic or a conventional exergoeconomic analysis, as discussed in Section 15.2, cannot evaluate the mutual interdependencies among the system components. This becomes possible in an advanced exergetic analysis, in which the exergy destruction in each component is split into endogenous and exogenous parts. An additional splitting of the exergy destruction into avoidable and unavoidable exergy destruction enables a realistic assessment of the potential for improvement. Finally, a combination of these two splitting approaches provides the designer or operator of an energy-conversion system with unambiguous and valuable detailed information with respect to options for improving the overall efficiency.

15.6.1 Endogenous and Exogenous Exergy Destruction

The exergy destruction within the kth component depends not only on the irreversibilities occurring within the same component but also on the exergy destructions within the remaining system components. When $\dot{E}_{P,\text{tot}}$ remains constant, these exergy destructions lead to an increase in the mass flow rate through the kth component and thus in the value of $\dot{E}_{D,k}$. The effect of the remaining components on $\dot{E}_{D,k}$ can be quantified by calculating the so-called exogenous exergy destruction that occurs within the kth component.

The total exergy destruction within the kth component can be split into *endogenous* and *exogenous* parts $\dot{E}_{D,k} = \dot{E}_{D,k}^{\text{EN}} + \dot{E}_{D,k}^{\text{EX}}$ [6]. Here $\dot{E}_{D,k}^{\text{EN}}$ is the *endogenous* part of exergy destruction. This is defined as the exergy destruction occurring within the kth component when all other components operate in an ideal way and the kth component operates with its current efficiency. To calculate the endogenous exergy destruction in a component of an exergy-conversion system we must first define an ideal (or a so-called theoretical) process for the overall system [14, 17, 18]. $\dot{E}_{D,k}^{\text{EX}}$, the *exogenous* part of exergy destruction within the kth component, is the result of the irreversibilities that occur in the remaining components and is calculated as the difference between $\dot{E}_{D,k}$ and $\dot{E}_{D,k}^{\text{EN}}$.

This splitting [17–19] enables engineers working in system optimization to estimate the exergy destruction within a component caused by the component itself on the one hand and by the remaining components on the other hand. This information can be used to decide whether engineers should focus on the component being considered or on the remaining system components in order to effectively improve the overall performance.

The first two columns (I and II) in Table 15.4 contain the endogenous and exogenous parts of the exergy destruction within each component of the refrigeration machine shown in Fig. 15.1. The endogenous part lies between 42% for the throttling valve and 100% for the evaporator. These results show that the evaporator, which has now the highest endogenous exergy destruction, is more important than the conventional exergetic analysis (Section 15.3 and Table 15.2) indicated.

15.6.2 Unavoidable and Avoidable Exergy Destruction

Only a part of the exergy destruction within a component can be avoided. The exergy destruction that cannot be reduced due to technological limitations such as availability and cost of materials and manufacturing methods is the *unavoidable* $(\dot{E}_{D,k}^{\text{UN}})$ part of the exergy destruction. The remaining part represents the *avoidable* $(\dot{E}_{D,k}^{\text{AV}})$ part of the exergy destruction. Thus splitting the exergy destruction into unavoidable and avoidable parts in the kth component $\dot{E}_{D,k} = \dot{E}_{D,k}^{\text{UN}} + \dot{E}_{D,k}^{\text{AV}}$ provides a realistic measure of the potential for improving the thermodynamic efficiency of a component [20, 21].

The curves and the shaded area shown in Fig. 15.1 are usually not known. However, even then we can estimate the two asymptotic lines that determine the unavoidable exergy destruction per unit of product exergy $(\frac{\dot{E}_{D,k}}{\dot{E}_{P,k}})^{\text{UN}}$ and the unavoidable investment cost per unit of product exergy $(\frac{\dot{Z}^{\text{CI}}}{\dot{E}_P})_k^{\text{UN}}$.

Table 15.4. *Advanced exergetic analysis for the vapor-compression refrigeration machine (Fig. 15.1)*

Component	$\dot{E}_{D,k}^{EN}$ (kW)	$\dot{E}_{D,k}^{EX}$ (kW)		$\dot{E}_{D,k}^{UN}$ (kW)	$\dot{E}_{D,k}^{AV}$ (kW)	Splitting $\dot{E}_{D,k}^{real}$ (kW) $\dot{E}_{D,k}^{UN}$ (kW) $\dot{E}_{D,k}^{UN,EN}$ (kW)	$\dot{E}_{D,k}^{UN,EX}$ (kW)		$\dot{E}_{D,k}^{AV}$ (kW) $\dot{E}_{D,k}^{AV,EN}$ (kW)		$\dot{E}_{D,k}^{AV,EX}$ (kW)
	I	II	III	IV	V	VI	VII	VIII	IX	X	XI
CM	0.928 (65%)	0.5 (35%)	**CD** 0.150 **TV** 0.007 **EV** 0.331 **mexo** 0.012	0.288 (20%)	1.14 (80%)	0.280	0.008	**CD** 0.001 **TV** 0.002 **EV** 0.005 **mexo** 0	0.648	0.492	**CD** 0.149 **TV** 0.005 **EV** 0.326 **mexo** 0.012
CD	1.814 (68%)	0.867 (32%)	**CM** 0.226 **TV** 0.018 **EV** 0.423 **mexo** 0.2	0.791 (30%)	1.89 (70%)	0.727	0.064	**CM** 0.043 **TV** 0.005 **EV** 0.013 **mexo** 0.003	1.087	0.803	**CM** 0.183 **TV** 0.013 **EV** 0.410 **mexo** 0.197
TV	0.388 (42%)	0.537 (58%)	**CD** 0.161 **CM** 0 **EV** 0.317 **mexo** 0.059	0.407 (44%)	0.518 (56%)	0.388	0.019	**CD** 0.006 **CM** 0 **EV** 0.013 **mexo** 0	0	0.518	**CD** 0.155 **CM** 0 **EV** 0.304 **mexo** 0.059
EV	2.62 (100%)		0	0.604 (23%)	2.016 (77%)	0.604		0	2.016		0
Overall system	5.75 (75%)	1.903 (25%)		2.090 (27%)	5.563 (73%)	1.999		0.091	3.751		1.813

All design improvement efforts should focus only on the avoidable parts of exergy destruction and investment costs. These parts are calculated by subtracting the unavoidable exergy destruction from the total exergy destruction within the component being studied.

Columns IV and V in Table 15.4 show the unavoidable and avoidable parts of the exergy destruction within each component of the refrigeration machine (Fig. 15.1). Here the unavoidable exergy destruction is calculated by defining a cycle in which only unavoidable exergy destruction occurs in each component. The results show that at least 70% of the exergy destruction in the most important components and in the overall system could be avoided.

15.6.3 Combination of the Two Splittings

Using an appropriate technique, we can calculate the unavoidable exergy destruction within the kth component when all the remaining components operate without irreversibilities. This calculation allows us then to obtain the unavoidable endogenous exergy destruction and subsequently the avoidable endogenous, the unavoidable exogenous, and the avoidable exogenous parts of exergy destruction within the kth component [14].

The *avoidable endogenous* part of the exergy destruction can be reduced by improving the efficiency of the kth component. The *avoidable exogenous* part of the exergy destruction can be reduced by a structural improvement of the overall system or by improving the efficiency of the remaining components and of course by improving the efficiency in the kth component.

The *unavoidable endogenous* part of the exergy destruction cannot be reduced because of technical limitations for the kth component. The *unavoidable exogenous* part of the exergy destruction cannot be reduced because of technical limitations in the other components of the overall system for the given structure.

It is apparent that the sum of all four parts equals the total exergy destruction within the kth component:

$$\dot{E}_{D,k}^{UN,EN} + \dot{E}_{D,k}^{UN,EX} + \dot{E}_{D,k}^{AV,EN} + \dot{E}_{D,k}^{AV,EX} = \dot{E}_{D,k}. \qquad (15.22)$$

This information is extremely useful for the iterative exergoeconomic optimization of energy-conversion systems (Subsection 15.4.4). The analyst is guided

- to focus only on the endogenous avoidable and exogenous avoidable exergy destructions, and
- to consider the appropriate measures (referring to the component being considered, to the efficiency of the remaining components, or to the structure of the overall system) that have the potential for reducing the exergy destruction within the kth component and the remaining ones.

The results for the refrigeration machine (Fig. 15.1) are presented in columns VI, VII, IX, and X of Table 15.4. The avoidable endogenous exergy destruction in the evaporator is significantly higher that the corresponding values for the other components. Thus efforts to improve the thermodynamic efficiency of the refrigeration machine should focus on the evaporator.

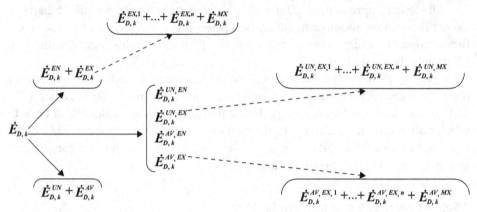

Figure 15.3. Options for splitting the exergy destruction within the kth component in an advanced exergoeconomic analysis.

To better understand the interactions among components, we further split the exogenous exergy destruction and particularly the avoidable exogenous exergy destruction within the kth component. We calculate (a) the part ($\dot{E}_{D,k}^{EX,r}$) of the exogenous exergy destruction within the kth component that is caused by the irreversibilities occurring within the rth component and (b) the part ($\dot{E}_{D,r}^{EX,k}$) of the exogenous exergy destruction within the rth component that is caused by the exergy destruction taking place within the kth component. The sum of all $\dot{E}_{D,k}^{EX,r}$ terms is lower than the exogenous exergy destruction within the kth component. The difference is caused by the simultaneous interactions of all the remaining $(n-1)$ components. We call this difference *mexogenous exergy destruction* within the kth component ($\dot{E}_{D,k}^{MX}$) and calculate it through

$$\dot{E}_{D,k}^{MX} = \dot{E}_{D,k}^{EX} - \sum_{\substack{r=1 \\ r \neq k}}^{n-1} \dot{E}_{D,k}^{EX,r}. \qquad (15.23)$$

The approach for splitting the exogenous exergy destruction within the kth component may also be applied to the avoidable exogenous exergy destructions separately. An advanced exergoeconomic evaluation should consider in the first place the values obtained from a splitting of the avoidable exogenous exergy destructions. Figure 15.3 summarizes all options for splitting the exergy destruction within the kth component.

The detailed splitting for the refrigeration machine (Fig. 15.1) is shown in columns III for the exogenous and XI for the avoidable exogenous exergy destruction in Table 15.4. These values clearly show that among all components the evaporator has always the highest contribution to the exogenous (or avoidable exogenous) exergy destruction within all other components. This emphasizes the importance of the evaporator for improving the refrigeration machine.

15.7 Advanced Exergoeconomic Analysis

In addition to an exergy analysis, the concepts of avoidable or unavoidable as well as endogenous or exogenous exergy destruction can be applied to the investment

Table 15.5. *Splitting the capital investment cost for components of the vapor-compression refrigeration machine*

| | | | | | Splitting $\dot{Z}_k^{\text{real}}$ (€/h) | | | |
| | | | | | $\dot{Z}_k^{\text{UN}}$ (€/h) | | $\dot{Z}_k^{\text{AV}}$ (€/h) | |
Component	$\dot{Z}_k^{\text{EN}}$ (€/h)	$\dot{Z}_k^{\text{EX}}$ (€/h)	$\dot{Z}_k^{\text{UN}}$ (€/h)	$\dot{Z}_k^{\text{AV}}$ (€/h)	$\dot{Z}_k^{\text{UN,EN}}$	$\dot{Z}_k^{\text{UN,EX}}$	$\dot{Z}_k^{\text{AV,EN}}$	$\dot{Z}_k^{\text{AV,EX}}$
CM	1.05	0.32	1.09	0.28	0.97	0.12	0.08	0.20
CD	0.28	0.04	0.29	0.03	0.27	0.02	0.01	0.02
TV	0.01	0	0.01	0	0.01	0	0	0
EV	1.61	0	0.65	0.96	0.65	0	0.96	0
Overall system	2.95	0.36	2.04	1.27	1.90	0.14	1.05	0.22

and fuel costs in exergoeconomics. Thus the exergoeconomic evaluation can be based on

- avoidable endogenous costs and
- avoidable exogenous costs.

In the iterative exergoeconomic improvement (Subsection 15.5.4) when advanced exergoeconomics are applied, the decisions about changes in a system are based on the values of

- exergy destruction ($\dot{E}_{D,k}^{\text{AV,EN}}$ and $\dot{E}_{D,k}^{\text{AV,EX}}$) and cost of exergy destruction ($\dot{C}_{D,k}^{\text{AV,EN}}$ and $\dot{C}_{D,k}^{\text{AV,EX}}$) associated with only avoidable endogenous exergy destruction and avoidable exogenous exergy destruction, and
- the investment cost ($\dot{Z}_{D,k}^{\text{AV,EN}}$) that is associated only with the endogenous avoidable exergy destruction.

Here, the value of $\dot{Z}_k$ must be split in a way similar to splitting the value of $\dot{E}_{D,k}$ [14]. For the refrigeration machine, the results of splitting the capital investment costs associated with all components are presented in Table 15.5. Here the unavoidable endogenous costs dominate.

In advanced exergoeconomics, some modified exergoeconomic variables based on avoidable endogenous variables can be used to better characterize the performance of the component being considered:

- modified exergetic efficiency,

$$\varepsilon_k^{\text{AV,EN}} = \frac{\dot{E}_{P,k}}{\dot{E}_{F,k} - \dot{E}_{D,k}^{\text{UN}} - \dot{E}_{D,k}^{\text{AV,EX}}} = 1 - \frac{\dot{E}_{D,k}^{\text{AV,EN}}}{\dot{E}_{F,k} - \dot{E}_{D,k}^{\text{UN}} - \dot{E}_{D,k}^{\text{AV,EX}}}; \quad (15.24)$$

- modified exergoeconomic factor,

$$f_k^{\text{AV,EN}} = \frac{\dot{Z}_k^{\text{AV,EN}}}{\dot{Z}_k^{\text{AV,EN}} + \dot{C}_{D,k}^{\text{AV,EN}}} = \frac{\dot{Z}_k^{\text{AV,EN}}}{\dot{Z}_k^{\text{AV,EN}} + c_{F,k}\dot{E}_{D,k}^{\text{AV,EN}}}. \quad (15.25)$$

The variables $\varepsilon_k^{\text{AV,EN}}$ and $f_k^{\text{AV,EN}}$ provide more reliable information with respect to whether the investment cost should be reduced or the exergetic efficiency should be improved for the kth component. Thus the value of $f_{\text{CM}}^{\text{AV,EN}}$ presented in Table 15.6

Table 15.6. *Values of selected exergoeconomic variables for the compressor and evaporator obtained from an advanced exergoeconomic analysis*

Component	$\varepsilon_k^{AV,EN}$ (%)	$\dot{Z}_k^{AV,EN}$ (€/h)	$\dot{C}_{D,k}^{AV,EN}$ (€/h)	$\dot{Z}_k^{AV,EN} + \dot{C}_{D,k}^{AV,EN}$ (€/h)	$f_k^{AV,EN}$ (%)
CM	94.4	0.08	0.06	0.14	57.1
EV	67.3	0.96	0.55	1.51	66.1

(57.1%) shows that the conclusion obtained from the high value of f_{CM} in Table 15.3 (90.6%) might not be correct, and therefore the investment cost and the efficiency of the compressor do not need to be reduced.

Finally, some information obtained from the two processes with (a) only unavoidable exergy destructions and (b) minimum investment costs: The unavoidable specific cost of cold for this process is $c_{cold}^{UN} = 181.1$ €/GJ. The specific cost of cold for the same process under real operating conditions is $c_{cold}^{real} = 230.4$ €/GJ. Thus the potential for reducing the cost of cold is less than 21%.

15.8 Exergoenvironmental Analysis

15.8.1 General Concept

To improve the performance of an energy-conversion system from the environmental (ecological) point of view, it is very helpful to understand the formation of environmental impact at the component level. The term environment here refers to the physical environment, which is different from the thermodynamic environment used in exergy analysis. A so-called exergoenvironmental analysis rests on the notion that exergy, being the commodity of value, is the only rational basis for assigning not only monetary values but also environmental impact values to the transport of energy and to the inefficiencies within a system [22]. We refer to this approach as *exergoenvironmental costing*.

The *exergoenvironmental analysis* consists of three steps: The first step is an exergetic analysis (Chaps. 2 and 6) of the overall energy-conversion system. In the second step, a life-cycle assessment (LCA) of (a) each relevant system component, and (b) all relevant input streams to the overall system is carried out.

In the last step, the environmental impact obtained from the LCA is assigned to the exergy streams in the system: *exergoenvironmental variables* are calculated and an *exergoenvironmental evaluation* is conducted. With the aid of an exergoenvironmental analysis, the most important components with the highest environmental impact and the sources of this impact are identified. In the following subsection we briefly discuss the last two steps of an exergoenvironmental analysis.

15.8.2 Life Cycle Assessment

A LCA is conducted for the input streams, supplied to the overall system, especially the fuel stream(s), and for the full life cycle of components. Inventories of elementary flows (i.e., consumption of natural resources and energy as well as emissions) are calculated following the guidelines of standard approaches. The environmental impact assessment can be performed with an indicator. In the refrigeration machine (Fig. 15.1) the Eco-Indicator 99 [23] was applied, as an example. It is apparent that other environmental impact metrics could be used instead of the Eco-Indicator 99. The latter is based on the definition of three damage categories: human health, ecosystem quality, and natural resources. The numbers obtained from the three categories are aggregated and the result is expressed as Eco-Indicator points (pts), where a higher damage corresponds to a higher Eco-Indicator value.

15.8.3 Exergoenvironmental Evaluation

In analogy to the assignment of costs to exergy streams in exergoeconomics (Section 15.5), an environmental impact rate $\dot{B}_j$ and an environmental impact per unit of exergy b_j are assigned to exergy streams in an exergoenvironmental evaluation. The *environmental impact rate* $\dot{B}_j$ is the environmental impact expressed, for example, in Eco-Indicator points per unit of time of system operation (Pts/s or mPts/s). The *specific (exergy-based) environmental impact* b_j (also called the specific *environmental cost*) is the average environmental impact associated with the production of the jth stream per unit of exergy of the same stream [Pts/(kJ exergy) or mPts/(MJ exergy)]. The *environmental impact rate* $\dot{B}_j$ of the exergy stream j is the product of its exergy rate $\dot{E}_j$ and the specific environmental impact b_j:

$$\dot{B}_j = \dot{E}_j b_j. \tag{15.26}$$

The environmental impact balance for the kth component states that the sum of environmental impacts associated with all input streams plus the component-related environmental impact is equal to the sum of the environmental impacts associated with all output streams:

$$\sum_e^{N_e} \dot{B}_{e,k} + \dot{Y}_k = \sum_i^{N_i} \dot{B}_{i,k} \tag{15.27}$$

or

$$\sum_e^{N_e} (b_e \dot{E}_e)_k + \dot{Y}_k = \sum_i^{N_i} (b_i \dot{E}_i)_k. \tag{15.28}$$

The F and P equations are formulated in analogy to the same equations for an exergoeconomic evaluation: The specific environmental impact of the exergy streams associated with fuel remains constant between inlet and outlet; each exergy unit is supplied to all exergy streams associated with the product at the same average specific environmental impact $b_{P,k}$. In this way the environmental impacts associated with the component being considered and with the exergy destruction within it are charged to the exergy streams associated with the product and finally to the products generated by the overall system.

With the aid of exergoenvironmental variables, the environmental performance of system components can be evaluated. These variables are defined for every system component in analogy to the definition of exergoeconomic variables in exergoeconomics.

The average specific (exergy-based) environmental impacts of product and fuel for the kth component are given respectively by

$$b_{P,k} = \frac{\dot{B}_{P,k}}{\dot{E}_{P,k}}, \tag{15.29}$$

$$b_{F,k} = \frac{\dot{B}_{F,k}}{\dot{E}_{F,k}}. \tag{15.30}$$

The relative position of the kth component and its interconnections with other components affects the values of $b_{P,k}$ and $b_{F,k}$. In general, these values are lower for components closer to the fuel of the overall system and higher for components closer to the product stream(s) for the overall system. This is due to decreasing exergy rates and increasing environmental impact rates as we move from the fuel of the overall system to the product of the system.

The environmental impact rate $\dot{B}_{D,k}$ associated with the exergy destruction $\dot{E}_{D,k}$ within the kth component can be calculated by

$$\dot{B}_{D,k} = b_{F,k} \dot{E}_{D,k}. \tag{15.31}$$

The exergoenvironmental approach assesses the total environmental impact associated with the kth component by calculating the environmental impact of exergy destruction $\dot{B}_{D,k}$ and the component-related environmental impact $\dot{Y}_k$. The sum of these quantities $(\dot{Y}_k + \dot{B}_{D,k})$ identifies the relevance from the environmental point of view of the kth component within the system being studied.

The relative difference $r_{b,k}$ defined by

$$r_{b,k} = \frac{b_{P,k} - b_{F,k}}{b_{F,k}} \tag{15.32}$$

is an indicator of the potential for reducing the environmental impact associated with a component, particularly when it is calculated with only avoidable costs.

The sources for the formation of environmental impact in a component are compared using the exergoenvironmental factor $f_{b,k}$, which expresses the relative contribution of the component-related environmental impact $\dot{Y}_k$ to the sum of environmental impacts associated with the kth component:

$$f_{b,k} = \frac{\dot{Y}_k}{\dot{Y}_k + \dot{B}_{D,k}}. \tag{15.33}$$

The formation of environmental impact at the system component level of energy conversion systems can be studied with the aid of an exergoenvironmental analysis. The objective is to generate information that serves as a basis for the development of improved options so that the environmental impact of the overall system can be reduced. A systematic evaluation can be conducted as follows [22].

First the environmentally relevant system components are identified using the sum of environmental impacts $(\dot{Y}_k + \dot{B}_{D,k})$. Among these components, we select the ones that have the highest improvement potential, as this is indicated by the relative difference of the specific environmental impacts $r_{b,k}$. The exergoenvironmental factor $f_{b,k}$ reveals the main source of the environmental impact associated with these components. Finally, suggestions for reducing the overall environmental

Table 15.7. *Values of selected exergoenvironmental variables for the vapor-compression refrigeration machine (Fig. 15.1)*

Components	$\dot{Y}_k$ (mPts/h)	$\dot{B}_{D,k}$ (mPts/h)	$\dot{Y}_k + \dot{B}_{D,k}$ (mPts/h)	$b_{F,k}$ (mPts/MJ)	$b_{P,k}$ (mPts/MJ)	$r_{b,k}$ (−)	$f_{b,k}$ (%)
CM	10.850	38.556	49.406	7.500	8.758	0.168	22.0
CD	1.146	85.392	86.538	8.848	54.580	5.168	1.3
TV	0.115	29.164	29.279	8.758	9.915	0.132	39.3
EV	56.664	93.528	150.192	9.915	19.950	1.012	37.7
Overall system	68.775	264.640	315.415	7.500	19.950	0.624	26.0

impact can be developed based on the results of the LCA if the component-related impact dominates the overall impact, or with the aid of the exergy analysis if the thermodynamic inefficiencies are the dominant source of environmental impact for the component being considered. The results from the exergoenvironmental analysis of the refrigeration machine are summarized in Table 15.7. The sum $\dot{Y}_k + \dot{B}_{D,k}$ shows that the highest environmental impact occurs in the evaporator. The variables $f_{b,k}$ and $f_{b,\text{tot}}$ indicate that a reduction of the overall environmental impact associated with the components and the overall refrigeration machine can be obtained mainly by reducing the exergy destruction.

It is apparent also that an advanced exergoenvironmental analysis can be conducted. This analysis employs avoidable and unavoidable as well as endogenous and exogenous environmental impacts and costs and is conducted in analogy to the advanced exergoeconomic analysis discussed in Section 15.7.

15.9 Closure

The cost improvement of energy-conversion systems requires, among others, appropriate trade-offs between fuel cost and investment expenditures or between the cost of thermodynamic inefficiencies and investment costs. An exergy analysis complements and enhances an energy analysis and identifies the location, magnitude, and causes of thermodynamic inefficiencies in each system component. A detailed cost analysis provides the investment cost and the operating and maintenance expenses associated with each system component. The appropriate trade-offs between fuel cost and investment expenditures can be realized with the aid of an exergoeconomic approach that calculates the costs of thermodynamic inefficiencies and compares them with the required investment cost at the system component level. Thus exergoeconomics identifies and properly evaluates the real cost sources.

An exergoeconomic approach applies the principle of exergy costing to enable the objective costing of energy carriers and to provide effective assistance in the decision-making process and the optimization or improvement of energy-conversion systems. In addition, exergoeconomics demonstrates that the exergy destruction plays a positive role in the design of energy-conversion systems by helping engineers to keep the investment costs associated with the system components at an acceptable level.

From the thermodynamic viewpoint, the value of each unit of exergy destruction is the same. Exergoeconomics demonstrates that the average cost per unit of exergy destruction is different for each component and depends on the relative position of the component within the system: Components closer to the supply of exergy to the overall system have, in general, a lower cost per unit of exergy destruction than components closer to the point of supply of the product streams from the overall system.

By revealing the true connections between thermodynamics and economics in the design of an energy-conversion system, exergoeconomics enhances the knowledge, experience, intuition, and creativity of engineers and provides students with the appropriate tools to understand the cost interactions involved in system improvement.

An exergoenvironmental analysis is conducted in a way very similar to an exergoeconomic analysis and reveals the environmental impact associated with each system component and the real sources of the impact. Thus the approaches reported here allow an integrated detailed evaluation of an energy-conversion system from the viewpoints of thermodynamics, economics, and protection of the environment.

The example of a vapor-compression refrigeration machine used throughout this chapter has demonstrated, among all components, the importance of the evaporator from all these viewpoints. These findings are also important for the allocation of research funds among single components, particularly those of new energy-conversion concepts.

Advanced exergetic, exergoeconomic, and exergoenvironmental analyses are based on a splitting of exergy destruction, cost, and environmental impact into avoidable endogenous, avoidable exogenous (with further splitting to include the effects of the remaining components), unavoidable endogenous, and unavoidable exogenous values. Through these techniques, our understanding of energy-conversion processes, of the interactions among system components, and of the interactions among thermodynamics, economics and environmental impact is greatly improved.

15.10 Nomenclature

B environmental impact (points)
b specific environmental impact (points/J)
c cost per unit of exergy (€/J)
C cost associated with an exergy stream (€)
E exergy (J)
e specific exergy (J/kg)
h specific enthalpy (J/kg)
f exergoeconomic factor (−)
m mass (kg)
N total number of system components
p pressure (Pa)
r relative cost difference (−)
s specific entropy (kJ/kg K)
T temperature (K)

x mole fraction $(-)$
Y component-related environmental impact
y exergy destruction ratio $(-)$
Z cost rate associated with investment expenditures (€)

Greek Symbols

γ activity coefficient $(-)$
ε exergetic efficiency $(-)$
η isentropic efficiency $(-)$

Subscripts

b refers to environmental impact
D destruction
F fuel
e exiting stream
gen   entropy generation
i entering stream
j jth stream
k kth component
n number of system components
P product
r rth component
ref   refers to the state used to calculate standard chemical exergy values
tot   refers to the total system

Superscripts

$\cdot$ time rate
AV   avoidable
CH   chemical exergy
CI   capital investment
EN   endogenous
EX   exogenous
KN   kinetic exergy
M mechanical exergy
MX   mexogenous
OM   operating and maintenance expenses
P potential exergy
T thermal exergy
UN   unavoidable

15.11 Acknowledgment

The author would like to thank Tatiana Morosuk for her help in preparing the manuscript and for conducting the calculations for the example of the refrigeration machine presented here.

REFERENCES

[1] G. Tsatsaronis, "Combination of exergetic and economic analysis in energy-conversion processes," in *Energy Economics and Management in Industry*, Proceedings of the European Congress, Algarve, Portugal, Apr. 2–5, 1984 (Pergamon, Oxford, England, 1984), Vol. 1, pp. 151–157.

[2] G. Tsatsaronis, "Definitions and nomenclature in exergy analysis and exergoeconomics," *Energy Intl. J.* **32**, 249–253 (2007).

[3] A. Bejan, G. Tsatsaronis, and M. Moran, *Thermal Design and Optimization* (Wiley, New York, 1996).

[4] J. Szargut, D.R. Morris, and F. R. Steward, *Exergy Analysis of Thermal, Chemical, and Metallurgical Processes* (Springer-Verlag, Berlin, 1988).

[5] J. Ahrendts, "Reference states," *Energy Intl. J.* **5**, 667–677 (1980).

[6] G. Tsatsaronis, "Design optimization using exergoeconomics," in *Thermodynamic Optimization of Complex Energy Systems*, edited by A. Bejan and E. Mamut (Kluwer Academic, Dordrecht, The Netherlands, 1999), pp. 101–115.

[7] A. Lazzaretto and G. Tsatsaronis, "SPECO: A systematic and general methodology for calculating efficiencies and costs in thermal systems," *Energy Intl. J.* **31**, 1257–1289 (2006).

[8] www.bitzer.de. Bitzer Kühlmashchinenbau GmbH, Sindelfingen, Germany, 2008.

[9] G. Tsatsaronis and M. Winhold, "Thermoeconomic analysis of power plants," EPRI AP-3651, RP 2029–8, Final Rep. (Electric Power Research Institute, Palo Alto, CA, 1984).

[10] G. Tsatsaronis, M. Winhold, and C.G. Stojanoff, "Thermoeconomic analysis of a gasification-combined-cycle power plant," EPRI AP-4734, RP 2029–8, Final Rep. (Electric Power Research Institute, Palo Alto, CA, 1986).

[11] G. Tsatsaronis, L. Lin, J. Pisa, and T. Tawfik, "Thermoeconomic design optimization of a KRW-based IGCC power plant," Final Rep. submitted to Southern Company Services and the U.S. Department of Energy, DE-FC21 –89MC26019 (Center for Electric Power, Tennessee Technological University, 1991).

[12] R. A. Gaggioli and W. J. Wepfer, "Exergy economics," *Energy Intl. J.* **5**, 823–838 (1980).

[13] G. Tsatsaronis and M. Winhold, "Exergoeconomic analyses and evaluation of energy conversation systems," *Energy Intl. J.* **10**, 69–94 (1985).

[14] T. Morosuk and G. Tsatsaronis, "Exergoeconomic evaluation of refrigeration machines based on avoidable endogenous and exogenous costs," in *Proceedings of the 20th International Conference on Efficiency, Cost, Optimization, Simulation and Environmental Impact of Energy Systems*, June 25–28, 2007, Padova, Italy, edited by A. Mirandola, O. Arnas, and A. Lazzaretto, Vol. 2, pp. 1459–1467.

[15] F. Cziesla and G. Tsatsaronis, "Iterative exergoeconomic evaluation and improvement of thermal power systems using fuzzy inference systems," *Energy Convers. Manage.* **43**, 1537–1548 (2002).

[16] G. Tsatsaronis, "Strengths and limitations of exergy analysis," in *Thermodynamic Optimization of Complex Energy Systems*, edited by A. Bejan and E. Mamut (Kluwer Academic, Dordrecht, The Netherlands, 1999), pp. 93–100.

[17] T. Morosuk and G. Tsatsaronis, "Splitting the exergy destruction into endogenous and exogenous parts – Application to refrigeration machines," in *Proceedings of the 19th International Conference on Efficiency, Cost, Optimization, Simulation and Environmental Impact of Energy Systems*, July 12–14, 2006, Aghia Pelagia, Crete, edited by C. Frangopoulos, C. Rakopoulos and G. Tsatsaronis, Vol. 1, pp. 165–172.

[18] T. Morosuk and G. Tsatsaronis, "The 'Cycle Method' used in the exergy analysis of refrigeration machines: From education to research," in *Proceedings of the 19th International Conference on Efficiency, Cost, Optimization, Simulation and Environmental Impact of Energy Systems*, July 12–14, 2006, Aghia Pelagia, Crete, edited by C. Frangopoulos, C. Rakopoulos and G. Tsatsaronis, Vol. 1, pp. 157–163.

[19] G. Tsatsaronis, S. Kelly, and T. Morosuk, "Endogenous and exogenous exergy destruction in thermal systems," in *Proceedings of the ASME International Mechanical Engineering Congress and Exposition*, November, 5–10, 2006, Chicago (American Society of Mechanical Engineers, New York, 2006), CD-ROM, file 2006–13675.

[20] G. Tsatsaronis and M.-H. Park, "On avoidable and unavoidable exergy destructions and investment costs in thermal systems," *Energy Convers. Manage.* **43**, 1259–1270 (2002).

[21] F. Cziesla, G. Tsatsaronis, and Z. Gao, "Avoidable thermodynamic inefficiencies and costs in an externally fired combined cycle power system," *Energy Intl. J.* **31**, 1472–1489 (2006).

[22] L. Meyer, "Exergiebasierte Untersuchung der Entstehung von Umweltbelastungen in Energieumwandlungsprozessen auf Komponentenebene: Exergoökologische Analyse," Ph.D. thesis (Universität Darmstadt, Darmstadt, Germany, 2006).

[23] M. Goedkoop and R. Spriensma, "The Eco-Indicator 99: A damage oriented method for life cycle impact assessment," *Method. Rep.*, Amersfoort, The Netherlands (2000) (available at http:\\www.pre.nl]).

16 Entropy, Economics, and Policy

Matthias Ruth

> If Thought is capable of being classed with Electricity, or Will with chemical affinity, as a mode of motion, it seems necessary to fall at once under the second law of thermodynamics as one of the energies which most easily degrades itself, and, if not carefully guarded, returns bodily to the cheaper form called Heat. Of all possible theories, this is likely to prove the most fatal to Professors of History.
>
> Henry Brooks Adams (1838–1918), U.S. historian. *The Degradation of the Democratic Dogma*, p. 195.

16.1 Introduction

Modern, mainstream economics is as much a product of social and political history as it is of the scientific method. The decline of centralized power in the late middle ages, held previously by kings, feudal lords, bishops, or priests, increasingly provided opportunities for individuals and their communities to determine their own fate. As the roles and responsibilities of individuals were redefined and as decision making became decentralized, concerns were raised whether a society, driven by individuals and their self-interests, could and would be able to reach some stable outcome [1, 2]. As some of the societies in Western Europe experienced rapid change, new technologies broadened the resource base – domestic and foreign – with opportunities for seemingly unbounded expansion of the human enterprise. Most notably, the technological innovations of the agricultural and industrial revolutions fueled, and were driven by, economic growth and social change.

The interrelated rapid changes in the physical, engineering, and health sciences influenced economists' attempts to understand the increasing complexity of society [3]. By the end of the 19th century, highly abstract, mathematical descriptions of economic activity were available with natural-law-like character to be applied anywhere and anytime to gain insights into the interactions among producers and consumers. Objective functions for households and firms were postulated, and, on the basis of those postulates, conditions were derived under which the decisions of all actors in the economy came to a general equilibrium – a state at which profits of firms are maximized, consumers reached a maximum level of satisfaction, and all markets cleared, given the resource endowments and technologies available to society [4, 5]. Where conditions for such optimality are not met, interventions are sought to get closer to, or actually achieve, economic equilibrium. The main instruments for interventions are changes in relative prices of goods and services, for example, by imposing taxes, or changes in the forms of markets themselves, such as by deregulating monopolies or providing incentives for increased competition [6].

One of the most frequently mentioned definitions of economics, reaching back to the 19th century, is that of a science that seeks to identify the optimal allocation of scarce resources to meet the needs of humans. Initially labor and capital were considered scarce, but subsequent expansions began to include natural resources as well as the environment's waste assimilation and absorption capacities. The notion of economic equilibrium and stability was contrasted with that of a "steady state" in which the economy develops within the constraints given by a finite resource endowment and a near-constant influx of energy into and out of the earth system [7, 8].

The broadening of boundaries around the economy to explicitly include resource and waste streams also prompted closer attention to ecological processes. Instead of separate from its environment, the economy increasingly is considered a subsystem of the ecosystem, with each, and the relation between them, always changing [9].

The early fascination of economists with state-of-the-art mathematics and physics has imparted undeniable influence over the development of economic theory, often prompting a reference to economics as the physics of social sciences (e.g., [10] and [11]). With the stature that comes from a high degree of scientific formalism, the role of economists in providing decision support has increased. Some have gone so far as to refer to economists as the new high priests of society [12], offering advice, making predictions, and consorting with the secular world much as priests in antiquity may have done.

As mathematics and physics advanced, and as the roles of natural resources and the environment for economic activity surfaced on the radar of economic analysis, new concepts and tools were imported from physics to advance economic model development. One such example is Jevons' attempt to draw parallels between classical mechanics and his economic model of the exchange of goods among households and firms [13], which "does not differ in general character from those which are really treated in many branches of physical science." He proceeds to compare the equality of the ratios of marginal utility of two goods and their inverted trading ratio to the law of the lever, where, in equilibrium, the point masses at each end are inversely proportional to the ratio of their respective distances from the fulcrum.

Edgeworth ([14], p. 9) goes a step further by stating that "pleasure is the concomitant of Energy. *Energy* may be regarded as the central idea of Mathematical Psychics (economics); "maximum energy" the object of the principal investigation in that science."

With the general acceptance of physical concepts in economic theory came an application of mathematical tools developed in classical mechanics for the analysis of economic processes. Mirowski [15] suggests that it was the analogy of energy and utility that provided the inspiration behind the neoclassical revolution, and Christensen ([16], p. 77) points out that

neoclassicals simply substituted utility for energy in the equations of analytical mechanics. Treating utility like energy provided economics with a powerful metaphor for individual action, a rigorous set of mathematical techniques (the calculus of variations), a theory of economizing (in the principle of least effort), and a theory of optimality.

Such application of mathematical techniques developed originally for the solution of physical problems is based on the "recognition that certain aspects of production and exchange are amenable to mathematical representations in terms of previously explored functional forms" ([17], p. 156). Processes in the economic system are treated analogously to processes in physical and engineering systems, assuming that the functioning of an economic system and its interactions with the surroundings follows the same principles as the physical science and engineering counterparts.

As the laws of thermodynamics became well established in physical and engineering sciences, these laws slowly proceeded to find their way into economic theory. Much as with classical mechanics, insights from thermodynamics were borrowed to explain economic processes on the basis of analogies without altering fundamentally the theoretical concepts of economic theory to account for the newly found laws (for illustrations see, e.g., [18–21]).

Going beyond analogies, some uses of thermodynamic concepts for the explanation of economic processes were spurred by the recognition of the role of the entropy law in the determination of upper bounds on efficiencies of material and energy transformations. Many of the studies that import the laws of thermodynamics into economic modeling condemn economic theory for neglecting upper limits on resource availability [22] or for disregarding the entropy law in the representation of economic processes [7, 23–25].

In response to the neglect of thermodynamic laws in economic models, some efforts were made to represent economic production and consumption processes consistently with the laws of thermodynamics [24, 26–28]. However, none of these studies provides a comprehensive representation of economy–environment interactions or generates evidence for the relevance of thermodynamic laws for the analysis of economic processes. Although it is readily apparent that all processes occurring in nature must obey physical laws, it has not yet been shown convincingly whether the laws of thermodynamics impose constraints that are significant enough to be considered explicitly in economic analysis. Rather, the sometimes inaccurate adoption of thermodynamic concepts in economic theory led, time and again, to considerable confusion among economists, who continue to ask whether the entropy law has any relevance to economics at all (see the debate among Young [29], Daly [30], and Townsend [31] as a prominent example).

First among the efforts to move from analogy in model development to import of physical constraints into economic analysis are efforts by Victor [32], who expanded traditional input–output models – in essence, representations in matrix format of sales of goods and services among all sectors of the economy – by the law of conservation of matter. Similarly, Bullard and Herendeen [33] converted input–output tables of the U.S. economy to capture energy flows. Both cases, and the large and growing literature that ensued, were motivated by needs to address environmental constraints on economic growth and development through incorporation of material and energy constraints on the exchanges among firms, households, and the physical environment. They are sebsequently described in more detail.

More recently, as the economy has become redefined as part of the ecosystem and as ecosystem change has been intricately linked to economic change, notions of evolution and co-evolution have slowly come to the fore [34]. But yet again, the work has remained largely conceptual and based on analogies, rather than led to a reformulation of the systems of equations used to describe economic and environmental changes in their interrelationships. Although there is, in principle, nothing wrong with starting from analogies to gain new insights, the appeal of analogy-based reasoning quickly declines as economists seek to quantify functional relationships in their models and make numeric predictions about possible economic outcomes under alternative investment and policy interventions.

None of the recent expansions of economic theorizing and modeling have fundamentally altered mainstream economics, but have rather taken place at the fringes of the discipline, largely ignored or disregarded by the majority [35]. Economic models continue to focus almost exclusively on the description of monetary exchanges among households and firms. Because the unit of measurement is monetary, it is easy to miss or disregard the underlying physical constraints on the material and energy flows that accompany money flows. For example, a set of equations describing extraction of a mineral deposit may concentrate on payments for labor and capital because these two have to be acquired in the marketplace, and may consider the removal of mineral deposits as a loss of assets under the ownership of the firms. It would typically not capture the environmental harm done by unintended releases of by-products from the mining operation. If there is no economic value attached to these by-products, they, for all intents and purposes, do not exist on the economist's radar. Conversely, if harm done by those releases feeds back to affect the economy, the damage done, as judged in economic and policy decision making, can be used to adjust prices in ways that may reduce release. In contrast, starting from a thermodynamics-based analysis of the mining operation, material and energy balances would help quantify and trace releases of all materials and energy from sinks to sources, provide a basis for guiding economic decision making before unintended consequences are felt, and thus help decision makers guide investments and policies toward an allocation of resources that is optimal from a broader systems perspective rather than the narrow confines of monetary-based analysis.

From a macroeconomic perspective, the extraction of materials and energy is only a small and shrinking fraction of overall economic activity compared with the (monetarily valued) contributions of labor and capital. The rather erroneous conclusion has then often been drawn that labor and capital can substitute for materials and energy [36].

Despite the challenges associated with changing economic theory formation, economists who have argued for a revision of mainstream economics with an eye toward the biophysical realities that govern material and energy use are gaining in numbers and influence. Their critique of the mainstream is founded, in part, on the notion that disregard of recent advances in physics and biology renders answers from mainstream economics to the challenges of modern society irrelevant or misleading. Proponents of this "new" economics suggest, as a correlate, that investment and policy decision-making processes need to be updated with the insights from modern physics and biology, rather than based on the economic models crafted after 19th- and early 20th-century natural sciences (see, e.g., [37–39]).

This chapter, although not intending to provide a history of economic thought, traces some of the conceptual changes in economic modeling to developments in physics, particularly thermodynamics. Because thermodynamics has increasingly influenced biology and ecology, and insights from the latter, in turn, are shaping modern analysis of complex human–environment interactions, additional attention is given to the relation of economics to biology (see also [40]). While tracing some of the relations among economics, ecology, and thermodynamics, the chapter lays out important ways in which the mindsets behind economic analysis are shaping policy making and new ways in which policy may develop, given recent advances at the interface of these three disciplines. Here, notions of stakeholder involvement, adaptive and anticipatory management, and planning in the light of uncertainty and surprise surface in contrast to expert-driven economic advice grounded on partial or general equilibrium models.

The structure of this chapter closely follows this line of argument. In the next section, I briefly review the influences of equilibrium thermodynamics on economics and biology. Section 16.3 outlines the basic tenets of nonequilibrium analysis and Section 16.4 addresses insights for investment and policy making. Section 16.5 closes with a brief summary and conclusions.

16.2 Mass and Energy Flows Through the Economy

16.2.1 The Material Connection of the Economy to the Environment

16.2.1.1 Mass and Energy Balances in Economic Modeling

Traditional economic analysis concentrates on the exchange of wealth among the members of an economy, focusing on the role of consumer preferences, technologies, and capital endowments for the existence and stability of market equilibria – situations in which the interplay of producers and consumers lead to prices that clear markets and in which firm profits and consumer utility are maximized. The underlying worldview is one of circular, monetary flows among the members of the economy (Fig. 16.1): purchases of goods and services result in profits for firms, which can then be used to purchase labor and other inputs to the production process. Savings of households, for example, provide means for investment in capacity and productivity of firms, which in turn requires inputs and leads to outputs available to consumers. Households receive interest payments for their savings, and firms pay interest for the capital they borrow [41]. Resource flows into the economy and waste flows from it are considered to the extent that they have monetary values associated with them, for example, for their purchase as inputs into production by firms or for the compensation to households for loss of welfare from harm being done to them. Where those monetary flows do not exist, yet the physical flows have an impact on the economy, internalization of externalities may be achieved, a posteriori, by establishing markets for those flows, as has been attempted, for example, for water allocation or sulfur emissions [42].

Among the earliest and most influential efforts to begin extending the monetary-flow model of economics to account for material and energy flow was Georgescu-Roegen's [24, 43–45] work on the biophysical basis of the economic process. Georgescu-Roegen emphasized the need for qualitative, physical changes in the

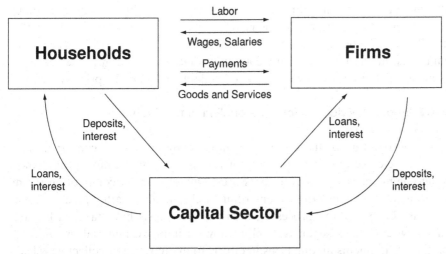

Figure 16.1. Basic circular flow model of the economy, with capital accumulation (modified from [41]).

material and energy bases of an economy to inform the prices of materials and energy as conversions in production and consumption take place. His efforts occurred independently of, and at the same time as, Boulding's [23] celebrated demonstration of the environmental implications of the mass-balance principle, Odum's [25] energy flow analysis that places economic processes within the broader ecosystem, and Ayres and Kneese's [46], Converse's [47], Victor's [32], and d'Arge and Kogiku's [48] materials-balance approaches that place physical constraints on materials use and release within economic input–output models. Instead of simply focusing on the monetary flows among economic sectors, these expanded input–output models capture – consistent with mass-balance descriptions – material flows between the environment and the economy and show not only the direct impacts and ripple effects in monetary terms but also mass flows that occur in the economy with changes in technology, resource availability or investment, and policy interventions. Similarly, Bullard and Hannon's [33] and Hannon's [49, 50] modified input–output models trace interactions among economic sectors and between ecological and economic systems in terms of energy units. The insights generated by their early models helped quantify impacts of energy constraints experienced during the oil price shocks of the 1970s and 1980s in ways that were not possible with simply monetary-based input–output analyses as the latter are limited in their abilities to capture the many subtle, yet pervasive, flows that connect economic sectors with each other and their environment.

Much of the attention surrounding the law of conservation of mass and energy is paid to limits imposed by the law on the growth of economic systems. In contrast, little attention has been paid by these early studies to the fact that the generation of waste products by the economy and their release into the environment leads to environmental change that in turn demands change in production processes over time. Turning to that issue, Perrings [51] developed a model of an economy that is constrained by the law of conservation of mass and exhibits the evolution of production processes in response to changes in the environment. His model contrasts with the model by Ayres and Kneese [46] and its successors that attempt to examine the implications of the conservation of mass for general economic equilibrium within a static allocative framework. Perrings's model stresses the necessity for an economic system to respond to nonequilibrium conditions that are caused by processes in the environment that are not reflected in, or controllable through, the price system.

16.2.1.2 Empirical Evidence for a Decoupling of the Economy from its Material Base

Increasing recognition of the constraints imposed by the law of conservation of mass has directed attention to the role of resource-saving technological change. Although engineers have – under the pressures of increased price competition among products and increasingly stringent environmental standards – attempted to reduce material use in production processes, industrial ecologists have taken a broader systems view and have begun to explore how to improve material use through symbiotic developments among firms in efforts to minimize their collective release of wastes into the environment [52–54]. Economists have, at the economy-wide level, then asked whether development has indeed led to dematerialization. Here, dematerialization is taken to mean a reduction in the quantity of materials used or the quantity of waste generated in the production of a unit of economic output. There has been a steady stream of research that suggests that the United States and other industrialized economies have dematerialized [55–59]. Well-documented examples include metal use in the beverage container industry [60], materials use in automobile manufacture [57], and communications [61]. Technical changes that improve material efficiency include advances in not only engineering and materials science, but also in the organization and management of production itself, such as computer-aided production processes and just-in-time production [62, 63]. More general economy-wide examples include the substitution of coal, oil, and natural gas for wood as a fuel source [64], and the substitution of iron and steel, aluminum, cement, and plastic for wood as a construction material.

Many attribute dematerialization to a "natural" or "evolutionary" process driven by the maturation of economies or rising incomes [63, 65, 66]. The apparent dematerialization has led some to hypothesize that the human economy can decouple itself from energy and material inputs by a factor of 10 [67].

Other analysts have a less sanguine interpretation of the historical record for a number of reasons. First, aggregate material use often is measured in terms of weight, which, when compared with the gross national product (GNP) in an index of the "efficiency" of material use, probably has little economic meaning because weight is only one of many attributes that users consider when choosing materials. Second, many analyses of dematerialization do not explicitly represent demand, technological change, or structural change, e.g., the effects of a shift from manufacturing to services on materials use, and they do not use methodologies that can test for the presence and relative strength of these forces [68]. Third, the techniques used to test for dematerialization "trends" in time series and cross-sectional data often lack statistical rigor, with a few notable exceptions. Fourth, a reduction in the quantity (weight) of material use per unit output is not necessarily better from an environmental perspective because every change in the pattern of material use has

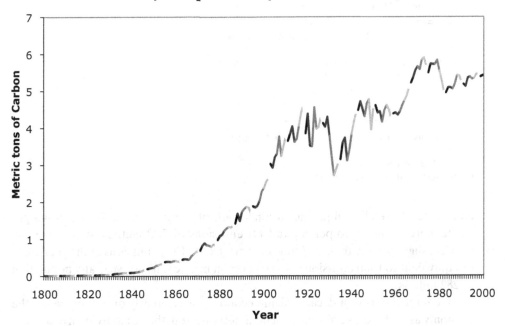

Per capita CO₂ emissions (metric tons of carbon)

Figure 16.2. Historical per-capita CO_2 emissions in the United States.

a unique impact on the quantity and quality of waste generation [69]. Fifth, little attention has been paid to the "rebound effect," the potential for improvements in efficiency to actually increase material use. Finally, although the efficiency of use of individual materials could rise, overall aggregate economic growth could increase total material consumption. Thus the picture of the pattern and implications for broader classes of materials and for aggregate material use is much less complete and much more ambiguous than has been suggested by the bulk of the research on dematerialization [70]. What has become clear, though, is the fact that a "straight-up" use of physical concepts – be it naive measurements of the weight of materials or more sophisticated measures of exergy flows – is inadequate to simultaneously capture the physical reality as well as the role that material flows have for the struc-ture and function of economic and ecological systems. What is needed is a matrix of indicators that capture a diverse set of performance criteria, such as price, weight, energy intensity, toxicity, and the like, because no one criterion – economic, biolog-ical or physical – can adequately capture the complexities surrounding material and energy use decisions.

More recent incarnations of the dematerialization debate – under the heading of "decarbonization" – focus on changes in the fuel mix and increased energy effi-ciency of economies. Figure 16.2 illustrates the decarbonization claim with times series data of per-capita CO_2 emissions in the United States. Table 16.1 shows the estimated parameters of a simple regression equation with per-capita CO_2 emissions as the dependent variables and the gross domestic product (GDP) per capital and its squared value as the independent variables. The resulting equation indeed shows the long-run relationship between per-capita CO_2 emissions and per-capita GDP as

Table 16.1. *Per-capita CO_2 emissions as functions of per-capita GDP, United States, 1800–2000*[*]

	Parameter estimates	t statistic
Per capita GDP	0.0004927	7.28
(per-capita GDP)^2	-8.63×10^{-9}	-5.39
R^2	0.39	
n	200	

[*] Model uses heteroskedasticity-corrected standard errors.
Durbin–Watson d statistic = 2.482.
Cochrane–Orcutt AR (1) regression.

an inverted-U relationship. The turning point of the inverted U is at $28,546 per capita, corresponding to per-capita CO_2 emissions of 7.03 metric tons per person, which is slightly more than 1.37 times the per-capita CO_2 emissions in the year 2000, or equivalent to total emissions of 1983 million metric tons per year at a population of 282 million.

General claims of a decarbonizing economy are similarly questionable for the economy as a whole, as are studies of dematerialization, though individual countries, sectors, or periods of development may show trends toward lower carbon intensities [71]. For example, decarbonization by one country may be at the expense of higher energy use and carbon emissions in other countries if decarbonization is achieved through the export of energy-intensive industries and subsequent import of their products.

Other empirically based inquiries into aggregate environmental behavior of individual economies have further stirred, rather than settled, the debate on whether there have been economy-wide improvements in resource use and environmental impact. Among these inquiries are efforts to detect "environmental kuznets curves" (EKCs), which characterize the relationship between pollution levels and income: The working hypothesis is that pollution levels will increase with income, but some threshold of income will eventually be reached, beyond which pollution levels will decrease (see Fig. 16.2 and Table 16.1). The link between the original Kuznets curve, which posited a similar relationship between income and inequality, and its pollution-concerned offspring lies primarily with the shape of both curves (an upside-down U) and the central role played by income change [72, 73]. The underlying logic behind the EKC presumes environmental quality to be a normal good, i.e., one where demand increases as income increases. Economies of scale, resource-saving technological change in the extractive and manufacturing sectors, trade liberalization leading to "out-migration" of dirty processes, and development of regulatory mechanisms and institutions are all seen to contribute to a country's improved environmental quality as economic development takes place [74–77].

The cross-sectional empirical work, on which much of the earlier EKC literature rests, was dismissed by some for doing injustice to the unique development paths of individual countries or regions of a country [78–80]. For example, while one group of countries or regions may undergo increases in pollution with increased development, others may exhibit the reverse behavior (Fig. 16.3). Grouping them together in a

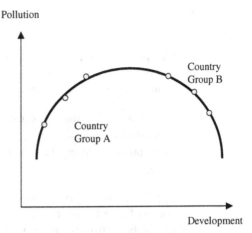

Figure 16.3. The Kuznets curve in cross-sectional analysis.

cross-sectional analysis will consequently lead to an inverted-U pattern that does not adequately describe the behavior of any of the countries. Additional problems reside with differences in the quality of data used in cross-sectional studies, the high degree of sensitivity of statistical results to variable choice and model specification [81], the possibility of zero or negative pollution when the inverted-U is extended, the fact that income is not a measure of development, or neglect of potential rebound effects, leading in the long run to an N-shaped pollution-development relationship [82]. A return to thermodynamic principles and theory may perhaps help shed light on the role of materials for long-run economic development. As a first step, special attention would need to be given to the establishment of careful mass and energy balances that link the economy with its environment and minimum material and energy requirements for specific processes.

16.2.1.3 Georgescu-Roegen's Fourth Law of Thermodynamics
The oil price shocks of the 1970s and 1980s, as well as those more recently, heightened the awareness of social and economic dependence on adequate energy supplies, and the unfolding of climate change begins to show limits on environmental waste absorption at global scales. In the debate about economic challenges and possible solutions, energy has moved front and center. Yet much of the energy is used to change the thermodynamic state of materials, and lack of attention to the material base of economic growth and development may unnecessarily confine both problem definition and solution space.

The most visible defender in economics of the argument that "matter matters too" may be Georgescu-Roegen [83, 84]. He argued that the principle of entropy applied to materials as well as to energy – any material flow carries an entropy flow, though not all energy flows are associated with entropy flows (e.g., heat is, but work is not). He further claimed existence of a "dual" of the first law of thermodynamics, namely that no mechanical work can be performed without the use of some matter, and that it is because of "imperfections in matter" (i.e., because there are no frictionless materials, no perfect insulators, no perfect conductors, no perfectly elastic materials, etc.) that there is no perfect conversion of energy into mechanical work without generation of waste heat. This suggested to Georgescu-Roegen that a full understanding of material and energy transformations requires explicit attention to matter, and led him to the postulation of a fourth law of thermodynamics – or law of matter entropy:

> [A] system that can exchange only energy with its outside and performs work indefinitely at a constant rate [...] is another thermodynamic impossibility. [...S]ooner or later, some elements will become totally dissipated ([85], pp. 53–54).

The bottom line for Georgescu is that, because of material dissipation and the generally declining quality of resource utilization, materials in the end may become more crucial than energy. He illustrates this by the now infamous example of a pearl necklace:

> [O]ne may argue that we can certainly reassemble the pearls of a broken necklace scattered over the floor. Is not recycling such a type of operation? To see the error in extrapolating from the molar to the molecular level, let us suppose that the same pearls are first dissolved in some acid and the solution is spread over the oceans – an experiment which depicts what actually happens to one material substance after the other. Even if we had as much energy as we pleased, it will take us a fantastically long, practically infinite time, to reassemble the pearls ([44], p. 269).

What is neglected here, though, is the fact that pearls are originally assembled from diffuse materials in the ocean. Carbon dioxide from the atmosphere is used to make the thread and ores from the Earth's crust are mined and refined to make the ties that close the string. All this, of course, did take place in less than "infinite time." Other counterexamples to the postulate of "imperfect materials" include phenomena of superconductivity and superfluidity. That said, a law of matter entropy may perhaps be invoked for some isolated systems, but real-world economies clearly are not isolated at all, allowing, in principle, for the collection of dissipated materials, given enough energy, information, and time, and given that not all of the materials are simultaneously in "active service" [86, 87].

16.2.2 Economy and the Second Law

The concept of the economy as a physical system subject to the laws of physics can be illustrated by defining the system boundary to encompass producers and consumers of goods and services and the intermediate inputs and final output. The environment, in turn, can be defined as the system containing all natural resources and the sinks that receive all wastes (Fig. 16.4). Thus the combined system consisting of the economy and the environment is defined as a closed system with no materials crossing the system boundary. Energy stocks are depleted for the purpose of changing the thermodynamic state of materials from their natural state to one that is more highly valued by humans. Because materials are conserved rather than "consumed," the analysis traditionally focused on the use of energy to affect the desired changes in the state of material resources and the associated buildup of capital within the economy [37, 89].

As capital is built up, as the availability of (skilled) labor increases, and as the use of both of them in production is improved, opportunities exist for economic growth, although the influx of solar energy and the limited stocks of materials place bounds on the overall size of the economy in the long run [36]. Understanding what contributes to economic growth and modeling the growth process in efforts to

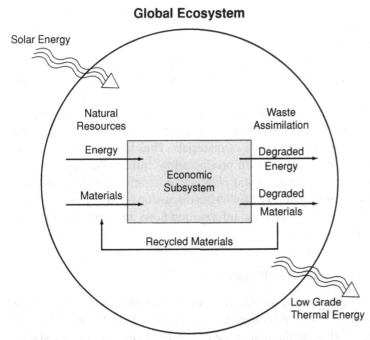

Figure 16.4. The economy is an open subsystem of the larger closed environmental system. The economic process is sustained by the irreversible, unidirectional flow of low-entropy energy and materials from the environment, through the economic system, and back to the environment in the form of high-entropy, unavailable energy and materials (modified from [88]).

forecast future economic conditions has long been of interest to economists. With their fixation on capital and labor as the prime inputs into production, early growth models attempted to establish relationships of the form laid out in Eq. (16.1), without any regard to physical constraints:

$$Y(t) = A(t) [K(t)]^{\beta} [L(t)]^{1-\beta}, \qquad (16.1)$$

where $Y(t)$ represents total production of the economy (typically measured as GDP) in year t, $K(t)$ and $L(t)$ are, capital and labor available in the economy, and $A(t)$ captures changes in technology (see, for example, [90, 91]). The differentiation of Eq. (16.1) with respect to time yields insights into the various contributors to growth in output Y. Using-time series data for K and L can then help to empirically estimate the parameter β and A.

Modifications of Eq. (16.1) are used to differentiate among the relative roles that capital and labor may have in their contribution to technical change, in which that part of growth that is not explained by growth in capital and labor is termed the residual. Because that residual has turned out to be quite sizable, considerable effort has been put in place to explain it.

One set of efforts focused on amending Eq. (16.1) to account for contributions of energy or materials (or both) in the production process, using various means of quantifying them on economic (monetary) and physical (mass and energy units, exergy, etc.) bases. An example here is expansions of traditional economic models of optimal mineral resource extraction and processing that quantify minimum material and energy requirements and thus constrain the optimality space [92]. Another set of efforts attempted to better portray the growth process itself as a function of changes in inputs as well as in the structure of the economy. Both sets of efforts strongly suggest that the contribution of materials and energy to the growth process can help explain a considerable part of technical change (i.e., reduce the residual), particularly when measured by the contributions of useful work (see, for example, [93, 94]).

In addition to the buildup of material endowments with desirable thermodynamic properties, humans accumulate knowledge about the processes (technology) for affecting desired changes in the state of materials. Thus high-quality energy is degraded to produce work, which in turn produces low-entropy (highly ordered) configurations of molecules (goods, capital plant and equipment) and information (services, technological know-how). The following discussion turns to potential connections between economic processes and information flows and storage, as well as associated changes in knowledge.

16.2.3 Economy, Information and Knowledge

The relation between energy and information has served as a means to confirm the validity of the second law of thermodynamics. Shannon [95], Wiener [96], and Shannon and Weaver [97], were the first in the late 1940s, to explore rigorously the relationship between information and order in the context of communication theory. Information was defined as a measure of uncertainty that caused an adjustment in probabilities assigned to a set of answers to a given question. In quantifying the uncertainty introduced by random background noise in a communications signal, and calling it entropy, they set the stage for Brillouin [98] to identify negative entropy with knowledge. Evans [99] and Tribus and McIrvine [100] formalized the connection between Shannon's work and thermodynamic information, in which entropy differences distinguish a system from its reference environment. Stressing the nonequilibrium character of systems that are distinguishable from their reference environment, Berg [101] uses information as a measure of the order of a product.

Going a step further, Spreng [102] ranks economic activities by the relative importance of their output, measured by information, and compares over time various production processes by their efficiency. The choice among alternative technologies used to "speed up the pace of life" or conserve valuable resources must be made by society, informed by the physical and ecological processes that are associated with economic activity.

The thermodynamic state is in fact what distinguishes one product from others and from its surroundings. Yet, from a physical perspective, a product has no intrinsic value, although it is possible to quantify the amount of energy required for changing the thermodynamic state of the input materials from their initial to a final state. The value of "goods" (materials in highly valued thermodynamic states) is determined by humans from sensory inputs received by the brain. For example, warm air molecules in a heated room produce valuable sensory inputs, and humans minimize the cost of those inputs by selecting optimal combinations of furnaces, fuels, insulating materials, and clothing. Similarly, other goods and services can be modeled physically as materials in particular thermodynamic states that produce audio and visual sensations or smells and tastes that human "consumers" value.

Although this physically based model does not encompass the richness of psychologically driven or socially constructed valuation of goods and services, it captures one set of aspects relevant for their valuation. By focusing on the fact that the net output of the economic system is information, it illustrates the potential for materials conservation (e.g., artificial sweetener technology, solid-state electronics). Because the theoretical minimum energy required for producing a bit of information is so small [103], the potential for energy-conserving technological change is correspondingly large. For example, a comparison of the energy cost of information-handling processes using character-recording technology such as an electric typewriter with digital-recording technologies such as computer output reveals differences of an order of magnitude [100], with both being significantly larger than the theoretical minimum thermodynamic entropy change associated with 1 bit of information (4.11 $\times$ 10^{-21} J at ambient temperature) (calculated from [103]).

An information-based approach to resource use was chosen in a theoretical analysis by Ayres and Miller [104] and later applied to the U.S. energy sector by Ayres [105]. Natural resources, labor, physical capital, and knowledge are all treated as forms of information and, within limits, mutually substitutable. Accumulation of knowledge and its embodiment in physical capital and labor skills lead to changes in the processing efficiency of the economic system and thus to decreases in the release of waste materials and heat into the environment. Yet little attention was given to the fate of waste products in the environment and the connection between waste generation and information as a measure describing products, technologies, and technical change.

With the theoretical background provided by studies on thermodynamic information and applications to energy and material use in economic processes, Ruth [40] formulated a model of a simple economic system in which all production functions and consumption processes were explicitly constrained by the laws of conservation of mass and energy. Ruth analyzed quantitatively the use of high-quality energy inputs for increasing the order of materials inside the economic system boundary. In addition, production functions changed as learning occurred, asymptotically approaching the theoretical maximum levels of materials and energy efficiency. The mass of the system remained constant. The net effect of energy input to the system was to increase the order of the materials (goods) and to change the state of knowledge (technology).

The same physically based analysis may be applied to the entire ecosystem of which economic systems are a subset. Incident solar energy is captured and concentrated by plants and animals and used to perform the work required for maintaining materials in low-entropy forms that provide the infrastructure for survival. Materials are conserved in the system and tend, in the absence of energy channeled through living organisms, toward a less-ordered (high-entropy) state.

16.2.4 A Brief Conceptual Assessment of Equilibrium Thermodynamics in Economic Theory Building

Significant headway has been made in the last few decades to provide a biopysical foundation for economics and to interpret economic processes from the perspective of thermodynamics. The studies previously cited and similar efforts may be grouped

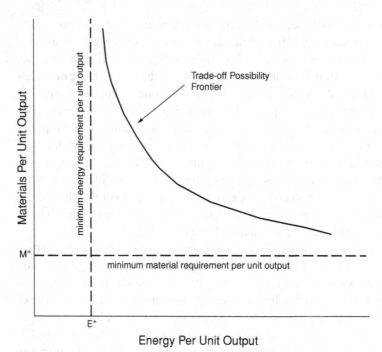

Figure 16.5. Trade-off possibility frontier for energy and material inputs per unit of output (modified from [40]).

into two broad categories – the ones that attempt to establish formal mathematical models of the economy in analogy to thermodynamics, and those directly applying thermodynamic laws to economic processes, such as the use of materials and energy for the concentration of materials [106, 107]. Comparable observations can be made for modern ecology, with formal model development in analogy to thermodynamics (e.g., [108]) and straightforward applications of thermodynamic laws to individual ecosystem processes (e.g., [109]).

Motivations for the development of economic theory in analogy to thermodynamics lies in the power of establishing isomorphic theories that enable transfer of insights from one area of scientific inquiry to another. However, isomorphism of formal structures between economic theory and thermodynamics does not imply that economic theory complies with thermodynamic laws. For example, as Sousa [110] beautifully pointed out, "[t]he formal equilibrium considered for the consumer problem is not the thermodynamic equilibrium of the consumer. The thermodynamic equilibrium of the consumer would be a dead consumer."

Although analogy-based theory building may help economize on theory building itself, the studies directly applying thermodynamics to economic processes gain power over traditional economic approaches because they explicitly identify and account for thermodynamic limits on the transformation of materials and energy. As an example, the production function represented in Fig. 16.5 illustrates these constraints. The minimum material and energy inputs required for producing a desired output are defined by M^* and E^*, respectively. The function, which describes possible substitutions between M and E, is bounded by these lower limits. This approach has

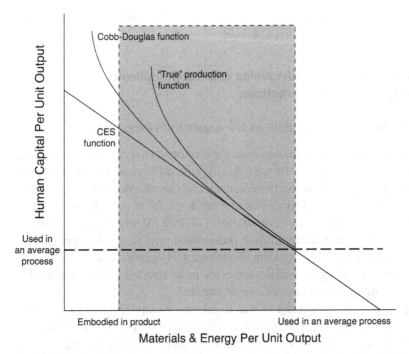

Figure 16.6. Possibilities for substitution between human and natural capital implied by different production functions. Because of the complementarity between human and natural capital, the substitutions are limited to the shaded area (modified from [26]).

been used in empirical analyses of material and energy use in individual processes such as copper extraction [111, 112], and copper and aluminum processing [92].

But what about the substitution possibilities at the level of an economy? The thermodynamic analysis described in Fig. 16.5 is based on mass, energy, and exergy balances of individual production processes. It requires detailed information on material and energy flows and as a result is difficult to apply accurately at the level of an entire economy. Many different forms of energy and materials are used as raw materials. Materials in particular are extremely heterogeneous at a macrolevel because humans use all 92 naturally occurring elements. Capital itself exists in myriad forms at larger scales, making it difficult to generalize about the material and energy requirements of producing and maintaining capital. By the same token, the abilities of workers to perform processes are inherently linked to their social, cultural, and educational backgrounds. It was this heterogeneity of funds and flows (machines, labor, materials, energy, knowledge) that prompted their aggregation in monetary terms and further masked the physical realities of the production process.

Despite these problems, the complementarity between human-made and natural capital enables us to distinguish production functions that are consistent with physical reality from those that are not (Fig. 16.6). In certain applications or interpretations, widely used models such as the Cobb–Douglas or constant elasticity of substitution (CES) production functions [113] embody the physically impossible assumption that a given output can be maintained as energy or material inputs vanishes as long as human-made capital can be increased sufficiently. In reality, the substitution possibilities are limited to the quantity of energy and materials embodied in final goods and in capital itself [26, 40, 51, 114–116].

16.3 Nonequilibrium Thermodynamics and the Complexity of Economy–Environment Interactions

16.3.1 Resource Use and Pollution as Nonequilibrium Processes

Most of the studies that use thermodynamics for economic analysis have been based on equilibrium thermodynamics, though clearly nonequilibrium thermodynamics would be more appropriate because the economy and the ecosystem within which it is embedded are thermodynamically open systems kept out of equilibrium by mass and energy flows across their system boundaries [117, 118]. As with equilibrium thermodynamics, the import of nonequilibrium thermodynamics strives for applications to specific processes or occurs in the form of analogies. However, actual applications are limited in number and scope, and analogies are more structural than formal. This section of the chapter addresses both kinds of attempts.

Assuming an open system is not in, but is close to, its equilibrium position, the entropy generated inside the system by naturally occurring processes per unit of time, $d_i S/dt$, is the sum of the products of the rates J_k and the corresponding n forces X_k $(k = 1{:}n)$ such that:

$$P = \frac{d_i S}{dt} = \sum_{k=1}^{n} J_k X_k > 0. \tag{16.2}$$

The rates J_k may characterize, for example, heat flow across finite temperature differences, diffusion or inelastic deformation, accompanied by generalized forces X_k such as affinities and gradients of temperatures or chemical potentials [119]. In the steady state, the entropy production inside the open system must be accompanied by an outflow of entropy into the system's surroundings. Systems approaching the steady state are characterized by a decrease in entropy production, i.e., $dP/dt < 0$.

Equation (16.2) holds close only to a local equilibrium. Far from equilibrium, it is typically not appropriate to assume such linearity. There is considerable debate in how far the "close-to-equilibrium" assumption can be maintained for the analysis of production and consumption processes in living systems, engineering systems, or economies. If the corresponding systems are far from equilibrium, using Eq. (16.2) to assess system change may be circumspect at best.

In the nonlinear realm, systems can be characterized by a generalized potential, the excess entropy production. Excess entropy production is defined as

$$P_E = \sum_{k}^{m} \delta J_k \delta X_k, \tag{16.3}$$

with δJ_k and δX_k as deviations from the values J_k and X_k at the steady state. Unlike for systems in, or close to, equilibrium, the sign of P_E is generally not well defined. However, close to equilibrium the sign of P_E is equal to that of P [120].

Calculations of excess entropy production show decreases in P_E in open systems moving toward steady state. For example, an increase in temperature gradients in a far-from-equilibrium system may trigger increasingly complex structures, as Bénard's experiments clearly demonstrate [119]. The evolution toward these structures is typically not smooth but accompanied by discontinuities and instabilities. In the critical transition point between stability and instability, a more complex structure emerges and $P_E = 0$ [120]. Excess entropy production can therefore serve as a measure of changes in the structure and stability of a system. However, its applicability to real, living systems, and especially to economic systems, is severely limited by the lack of data sufficient for meaningful calculations in nonlaboratory systems, such as ecosystems or economic systems [116].

Much movement, however, can be observed for the use of general insights that nonequilibrium thermodynamics provides – by analogy – for the study of complex systems. The concepts of nonlinearity, complexity, chaos, catastrophe, criticality, resilience, and adaptation all deeply resonate with scientists and practitioners interested in reevaluating and reshaping human–environment interactions. The following subsection briefly addresses these concepts before I explore their value as guiding principles in investment and policy decision support.

16.3.2 Complex Economy–Environment Interactions

From a systems perspective, both the economy and the ecosystem, within which the economy is embedded, can be understood in terms of nonlinear, dynamic, often time-lagged processes that connect individual system components [121]. Some of these processes are self-reinforcing, for example when environmental impacts on the economy reduce its ability to cope with such impacts. Others are counteracting, for example when increases in stocks of renewable resources prompt increased harvest, which in turn diminishes the stocks. The presence of both positive and negative feedback processes, and the fact that their strengths change through time, add to the challenge of trying to understand overall system behavior in space and time.

The connections among system components often take place across different system hierarchies. For example, employees in firms, firms in industries, and industries in economies or individual organisms within populations and populations within ecosystems interact with each other through exchanges of materials, energy, and information. As a consequence, analyses often need to take into account specifically the processes at system hierarchies above and below those of actual interest, lest important influences and impacts be neglected.

Because the economy interacts with its environment, nonlinear time-lagged feedback processes occur not only within and among the hierarchies of each of the systems, but connect to the outside as well. The discovery and use of fossil fuels stimulated economic activity and triggered the large-scale deployment not only of combustion technology around the world, but also of changes in society at large. The extractive processes themselves changed local ecosystems, as has become painfully apparent for oil and gas extraction in Louisiana or coal extraction in China, for example. Emissions of greenhouse gases from combustion, in turn, have an impact on ecosystems from local to global scales. Recognizing not only the adverse impacts from fossil fuel use but also the opportunities that may come from the developments of alternatives, far-reaching technological, behavioral, and institutional changes are sought, which may redefine modern society.

Taking a long view and drawing on the insights generated in systems theory, one may interpret such developments from an entropic perspective. For example, Hannon et al. [122] and others [123] hypothesized that ecosystems evolve toward climax states that are the most massive and highly ordered structures that can be maintained on a limited energy budget. As these systems evolve, knowledge accumulates in the genetic material that serves as blueprints for material- and energy-transforming processes. Similarly, the increasingly rapid entropy-generating activities of fossil fuel use through evermore sophisticated pathways developed in modern societies may be perceived as increasingly effective means to degrade existing environmental gradients. The self-organization of the globalizing industrial system shows similarities to the buildup of structure and organization in physical and biological systems and, like those, may experience critical thresholds that will lead to reorganization [124, 125]. Given the complexity of the system, it is not knowable where those thresholds are, when exactly they may be encountered, or what subsequent development paths may look like. Yet, if past system performance is a guide, a new buildup of structures is likely to occur.

Self-organized criticality is wrought with fundamental uncertainties. A challenge to decision makers in industry and policy, for example, will be to identify actions and promote the development of mechanisms that foster resilience of individual systems with which they deal, and the overall sustainability of the larger system within which those are embedded. Adaptive and anticipatory management have been promoted as responses to uncertain, ever-changing relations between society and its environment.

16.4 Adaptive and Anticipatory Management

Efforts to draw on concepts from thermodynamics to better understand and manage natural resource use by and environmental impacts of social and economic systems have been motivated by the fact that purely economic criteria are insufficient in capturing the full extent to which human action imparts change. Noting the role of feedbacks between economic and environmental changes across space and time, broadly drawing on insights from nonequilibrium thermodynamics has become a hook on which to hang concepts and models of human–environment relationships. Yet these efforts, much like the ones they attempt to replace, tend to fall short in guiding action toward more sustainable practices *because* they too limit the dimensions of analysis.

The complexity of economy–environment interactions is not simply a result of the connections of a myriad of system components, but is fundamentally and inherently related to our ability to comprehend and explain them. One disciplinary perspective, be it economic, engineering, biological, or other, will not be sufficient to encompass the relevant system features.

To illustrate the need for multiple perspectives, consider the role of nuclear power in energy supply. A physical description, though highly relevant, provides only basic insights into the processes of nuclear fission or fusion. Engineering perspectives offer technical descriptions of resource extraction, power generation, or waste disposal. Economics helps quantify opportunity costs of alternatives. Biology helps trace environmental impacts, and medicine describes human health implications of alternatives. But without, for example, an understanding of psychology, sociology, history, ethics, and law, any assessment of the opportunities for and constraints on nuclear power as an energy source are incomplete at best. The complexity of the issue comes not only from the many possible interactions at the physical and technological levels, but also from the many pathways through which their ramifications permeate environmental, economic, and social systems. The possibility for bifurcations extend in multiple dimensions when moving away from simple physical descriptions. Likewise, more degrees of freedom exist to promote self-organization and resilience outside the simply physical world.

Major challenges to modern society and to the scientific enterprise facing complex economy–environment interactions persist. Institutions and mechanisms have developed to reward discipline-specific advancement. Universities and science foundations still struggle with assessments and quality control of interdisciplinary research and education. A parallel culture of investment and policy advisors feeds off discipline-specific knowledge based on the reduction of complexities. No comparable social processes are in place to help embrace complexity. Deliberative democracy remains more of an ideal than a lived reality. Among the few pragmatic approaches are adaptive and anticipatory management, which have found widespread recognition in the field of ecosystem management and are slowly permeating investment and policy making.

The notion of adaptive management has been spawned by the recognition that, because boundary constraints for management continuously change, no one action will ever necessarily achieve desired long-term results [87, 126, 127]. Impacts of decision making need to be observed and fed back to inform and influence subsequent rounds of assessment and decision making, *ad infinitum*. The larger the time lags between system intervention and system response, the more important it will be to base interventions on anticipated future, rather than observed past or current, system change. Notable examples are investments in infrastructure, which are often lumpy and irreversibly alter the economic, social, and biophysical environment. The time between investment, observation of impacts, and revision or refinement is often too large to allow for meaningful adaptation, making anticipation and identification of robust strategies – strategies that are desirable under a wide range of alternative futures – all the more important.

A goal of adaptive and anticipatory management is not only a continuously fine-tuned approach to problem solving but also, as a prerequisite, a continuously improved ability to look forward into the future and to develop understanding and appreciation of the system dynamics, as the system itself evolves. One way to improve that knowledge is through the use of formal, transparent computer models that embed as much of the relevant system attributes as possible, establish material and energy balances and economic descriptions of transformation processes that are consistent with physical laws, and then use those models in a structured and iterative discourse with stakeholders.

Stakeholders from academia, policy, industry, nonprofit organizations, the public, and many other walks of life may make valuable contributions to the modeling process, and their inclusion in that process can set the stage for constructive dialog about alternative investment and policy decisions and actions. That dialog can be an essential ingredient for anticipatory management of socioeconomic systems in light of many complex settings in which decisions need to be made, in which the stakes are high, and uncertainties abound.

Although efforts are increasing the diversity of perspectives – from different academic disciplines to a wide range of stakeholder inputs – in formal assessments of economy–environment relations, the resulting models often are fairly basic with respect to the material and energy flows they capture, the interlinkages among system components, and the richness of dynamic processes overall. Although true in spirit to the insights from thermodynamics and complexity, very little substantive import of physical laws can be found in these models. Much room exists to offer a substantive basis.

16.5 Summary and Conclusions

Over the past five decades, two main strands of research have developed to provide a physical basis and interpretation of socioeconomic development in its relation to environmental change. One of these strands applies physical concepts and laws to quantify material and energy use from a thermodynamic perspective. Given the difficulties of doing so, analyses typically limited themselves to individual processes or discrete process chains. Examples include efforts to quantify exergetic requirements for the extraction and refinement of ores, or the entropic nature of human consumptive processes. The second strand of research has been guided by analogies and systems-theoretic insights and has tended to be largely conceptual and qualitative, though not necessarily less appealing as guides to decision making than their quantitative, process-specific counterparts. Examples include interpretations of economic and social change from the perspective of entropy theory, chaos theory, catastrophe theory, or self-organized criticality.

Together these two strands of research have contributed to the ongoing sustainability debate by juxtaposing hitherto standard static, linear, equilibrium-focused analyses common in traditional economics, with the notion of non-linear dynamics in a world where irreversibilities dominate, and adaptation and anticipation are key to success. Rather than dealing with isolated systems for which experts provide probabilistic forecasts as inputs into planning and management, the postmodern approach recognizes openness and hierarchies as essential system features that contribute to complexity, cause surprise, and require scenario-based exploration of the impacts of alternative system interventions. The institutional implications of these developments are clear – the basis for decision making needs to be shifted from limited expert advice to diverse, consensus-based approaches that promote responsibility and stewardship. As the sciences provide quantified and structural insights into changing human–environment relations, institutional innovation will need to explore alternative means by which to translate those insights into action. Here may also lie the frontier of 21st-century environmentalism.

REFERENCES

[1] A. Smith, *The Wealth of Nations* (1776; reprinted by Random House, New York, 1937).
[2] H. Brems, *Pioneering Economic Theory, 1630–1980: A Mathematical Restatement* (Johns Hopkins University Press, Baltimore, London, 1986).

[3] J. Martinez-Alier and K. Schlupmann, *Ecological Economics: Energy, Environment and Society* (Blackwell, New York, 1991).

[4] E. Malinvaud, *Lectures on Microeconomic Theory* (North-Holland Amsterdam, and Elsevier, New York, 1972).

[5] A. Mas-Colell, *The Theory of General Economic Equilibrium: A Differentiable Approach* (Cambridge University Press, Cambridge, 1985).

[6] M. L. Katz and H. S. Rosen, *Microeconomics*, 3rd Edition (Irwin McGraw-Hill, Boston, 1998).

[7] H. E. Daly, *Toward a Steady-State Economics* (Freeman, San Francisco, 1973).

[8] H. E. Daly, *Steady-State Economics* (Island Press, Washington, D.C., 1991).

[9] R. Costanza and L. Wainger, *Ecological Economics: The Science and Management of Sustainability* (Columbia University Press, New York, 1991).

[10] C. Freeman and C. Perez, "Structural crisis and adjustments, business cycles and investment behaviour," in *Technical Change and Economic Theory*, edited by G. Dosi and C. Freeman (Pinter Publishing, London, and New York, 1988).

[11] J. Sarkar, "Technological diffusion: Alternative theories and historical evidence," *J. Econ. Surv.* **12**, 131–176 (1998).

[12] R. H. Nelson, *Economics as Religion – From Samuelson to Chicago and Beyond* (Pennsylvania State University Press, University Park, PA, 2001).

[13] W. S. Jevons, *The Theory of Political Economy* (London and New York, Macmillan and Co., 1871).

[14] F. Y. Edgeworth, *Mathematical Psychics* (Kegan, London, 1881).

[15] P. Mirowski, *More Heat than Light: Economics as Social Physics* (Cambridge University Press, Cambridge, 1989).

[16] P. Christensen, "Driving forces, increasing returns and ecological sustainability," in *Ecological Economics: The Sci, & Mngmt. of Sustainability*, edited by R. Constanza, Columbia University Press, New York, 75–87 (1991).

[17] J. L. R. Proops, "Thermodynamics and economics: From analogy to physical functioning," in *Energy and Time in Economics and Physical Sciences*, edited by W. van Gool and J. Bruggink (Elsevier, North Holland, Amsterdam, 1985), pp. 155–174.

[18] H. T. Davis, *The Theory of Econometrics* (Principia, Bloomington, IN, 1941).

[19] H. C. Lisman, "Econometrics and thermodynamics: A remark on Davis' theory of budgets." *Econometrica* **12**, 59–62 (1949).

[20] A. G. Pikler, "Optimum allocation in econometrics and physics," *Weltwirtschaftliches Archiv*, No. 66, 97–132 (1951).

[21] J. Bryant, "A thermodynamic approach to economics," *Energy Econ.* **4**, 36–50 (1982).

[22] D. A. Underwood and P. G. King, "The ideological foundations of environmental policy," *Ecol. Econ.* **1**, 315–334 (1989).

[23] K. Boulding, "The economics of the coming spaceship earth," in *Environmental Quality in a Growing Economy*, edited by H. Jarrett (Johns Hopkins University Press, Baltimore, 1966), pp. 3–14.

[24] N. Georgescu-Roegen, *The Entropy Law and the Economic Process* (Harvard University Press, Cambridge, MA, 1971).

[25] H. T. Odum, *Environment, Power and Society* (Wiley-Interscience, New York, 1971).

[26] R. U. Ayres and I. Nair, "Thermodynamics and economics," *Phys. Today* **37**, 62–71 (1984).

[27] M. Faber, "A biophysical approach to the economy: Entropy, environment and resources," in *Energy and Time in Economics and Physical Sciences*, edited by W. van Gool and J. Bruggink (Elsevier Science, North Holland, Amsterdam, 1985).

[28] N. Georgescu-Roegen, "Process analysis and the neoclassical theory of production," *Am. J. Agricul. Econ.* **54**, 279–294 (1972).

[29] T. Young, "Is the entropy law relevant to the economics of natural resource scarcity?" *J. Environ. Econ. Mngmt.* **21**, 169–179 (1991).

[30] H. E. Daly, "Is the entropy law relevant to the economics of natural resource scarcity? – Yes, of course it is " *J. Environ. Econ. Mngmt.* **23**, 91–95 (1992).

[31] K. N. Townsend, "Is the entropy law relevant to the economics of natural resource scarcity? Comment," *J. Environ. Econ. Mngmt.* **23**, 96–100 (1992).

[32] P. A. Victor, *Pollution: Economy and Environment* (Allen and Unwin, London, 1972).

[33] C. W. Bullard and R. Herendeen, "Energy impact of consumption decisions," *Proc. IEEE*, **63**, 484–493 (1975).

[34] R. B. Norgaard, *Development Betrayed: The End of Progress and a Coevolutionary Revisioning of the Future* (Routledge, New York, 1994).

[35] C. J. Cleveland and M. Ruth, "When, where and by how much does thermodynamics constrain economic processes? A survey of Nicholas Georgescu-Roegen's contribution to ecological economics," *Ecol. Econ.* **22**, 203–223 (1997).

[36] H. E. Daly and J. C. Farley, *Ecological Economics: Principles and Applications* (Island Press, Washington, D.C., 2004).

[37] M. Faber, H. Niemes, and G. Stephan, *Entropy, Environment, and Resources: An Essay in Physio-Economics* (Springer-Verlag, New York, 1987).

[38] M. Faber, R. Manstetten, and J. L. R. Proops, *Ecological Economics: Concepts and Methods* (Elgar, Cheltenham, UK, 1996).

[39] S. Funtowicz, J. Ravetz, and M. O'Connor, "Challenges in the use of science for sustainable development," *Intl. J. Sustainable Develop.* **1**, 99–107 (1998).

[40] M. Ruth, *Integrating Economics, Ecology and Thermodynamics* (Kluwer Academic, Dortrecht, The Netherlands, 1993).

[41] R. L. Heilbroner and L. C. Thurow, *Economics Explained* (Prentice-Hall, Englewood Cliffs, NJ, 1982).

[42] W. J. Baumol, W. E. Oates, V. S Bawa, and D. Bradford, *The Theory of Environmental Policy* (Cambridge University Press, Cambridge, 1994).

[43] N. Georgescu-Roegen, *Energy and Economic Myths* (Pergamon, New York, 1976).

[44] N. Georgescu-Roegen, "The steady-state and ecological salvation," *Bioscience* **27**, 266–270 (1977).

[45] N. Georgescu-Roegen, "The entropy law and the economic process in retrospect," *East. Econ. J.* **12**, 3–23 (1986).

[46] R. Ayres and A. Kneese, "Production, consumption, and externalities," *Am. Econ. Rev.* **59**, 282–297 (1969).

[47] A. O. Converse, "On the extension of input–output analysis to account for enviromental externalities," *Am. Econ. Rev.* **61**, 197–198 (1971).

[48] R. C. d'Arge and K. C. Kogiku, "Economic growth and the environment," *Rev. Econ. Stud.* **59**, 61–77 (1973).

[49] B. Hannon, "The structure of ecosystems," *J. Theor. Biol.* **41**, 535–546 (1973).

[50] B. Hannon, "Energy conservation and the consumer," *Science* **89**, 95–100 (1975).

[51] C. Perrings, *Economy and Environment: A Theoretical Essay on the Interdependence of Economic and Environmental Systems* (Cambridge University Press, Cambridge, New York, 1987).

[52] J. H. Ausubel and H. E. Sladovich, editors, *Technology and Environment* (National Academy Press, Washington, D.C., 1989).

[53] T. E. Graedel and B. R. Allenby, *Industrial Ecology* (Prentice-Hall, Englewood Cliffs, NJ, 1995).

[54] M. R. Chertow, "Industrial symbiosis: Literature and taxonomy," *Annu. Rev. Energy Environ.* **25**, 313–337 (2000).

[55] A. P. Carter, "The economics of technological change," *Sci. Am.* **214**, 25–31 (1966).

[56] W. Malenbaum, *World Demand for Raw Materials in 1985 and 2000* (McGraw-Hill, New York, 1978).

[57] E. D. Larson, M. H. Ross, and R. H. Williams, "Beyond the era of materials," *Sci. Am.* **254**, 34–41, (1986).

[58] M. Jänicke, H. Monch, T. Ranneberg, and U. E. Simonis, "Structural change and environmental impact. Empirical evidence on thirty-one countries in east and west," *Environ. Monitor. Assess.* **12**, 99–114 (1989).

[59] F. Hinterberger and E. Seifert, "Reducing material throughput: A contribution to the measurement of dematerialisation and sustainable human development," in Environment, Technology and Economic Growth: The Challenge to Sustainable Development, edited by J. v. d. Straaten and A. Tylecote (Edward Elgar, Cheltennam, UK, 1997), pp. 75–91.

[60] C. Nappi, "The food and beverage container industries: Change and diversity," in *World Metal Demand, Trends and Prospects*, edited by J. E Tilton (Resources for the Future, Washington, D.C., 1990), pp. 217–254.

[61] P. L. Key and T. D. Schlabach, "Metals demand in telecommunications," *Mater. Soc.* **10**, 433–451 (1986).

[62] P. R. Devine, "The effects of economic growth, technology change, consumption pattern change and foreign trade on domestic demand for primary metals, 1963–82," PB89–204002 (Bureau of Mines, Washington, D.C., 1988).

[63] O. Bernardini and R. Galli, "Dematerialization: Long-term trends in the intensity of use of materials and energy," *Futures* **25**, 431–448 (1993).

[64] N. Nakicenovic, "Freeing energy from carbon," *Daedalus* **125**, 434–440 (1996).

[65] R. U. Ayres, "Industrial metabolism," in *Technology and Environment*, edited by J. H. Ausubel and H. E. Sladovich (National Academy Press, Washington, D.C., 1989), pp. 23–49.

[66] A. Grübler, "Industrialization as a historical phenomenon," in *Industrial Ecology and Global Change*, edited by R. H. Socolow, C. Andrews, F. Berkhout, and V. Thomas (Cambridge University Press, Cambridge, 1994), pp. 43–68.

[67] Factor 10 Club, Statement to Government and Business Leaders. Factor 10 Institute, Carnoules, 1997.

[68] R. Auty. "Materials intensity of GDP: Research issues on the measurement and explanation of change," *Resources Policy* **11**, 275–283 (1985).

[69] R. Herman, S. A. Arkekani, and J. H. Ausubel, "Dematerialization," in *Technology and Environment*, edited by J. H. Ausubel and H. E. Sladovich (National Academy Press, Washington, D.C., 1989), pp. 50–69.

[70] C. J. Cleveland and M. Ruth, "Indicators of dematerialization and the materials intensity of use," *J. Ind. Ecol.* **2**, 15–50 (1999).

[71] M. Ruth, "Energy use and CO_2 emissions in a dematerializing economy: Examples from five US metals sectors," *Resources Policy* **24**, 1–18 (1998).

[72] G. M. Grossman and A. B. Krueger, "Economic growth and the environment," *Q. J. Econ.* **110**, 353–377 (1995).

[73] G. M. Grossman, "Pollution and growth: What do we know?," in *The Economics of Sustainable Development*, edited by I. Goldin and L. A. Winters (Cambridge University Press, New York, 1995), pp. 19–45.

[74] T. Panayotou, "Empirical tests and policy analysis of environmental degradation at different stages of economic development," World Employment Programme Research, Working Paper 238 (International Labor Office, Geneva, 1993).

[75] M. Komen, S. Gerking, and H. Folmer. "Income and Environmental R&D: Empirical Evidence from OECD Countries," *Environ. Develop. Econ.* **2**, 505–515 (1997).

[76] V. Suri and D. Chapman, "Economic growth, trade and energy: Implications for the environmental Kuznets curve," *Ecol. Econ.* **25**, 195–208 (1998).

[77] J. Andreoni and A. Levinson, "The simple analytics of the environmental kuznets curve," *J. Public Econ.* **80**, pp. 269–286 (2001).

[78] D. I. Stern, "Progress on the environmental kuznets curve?," *Environ. Develop. Econ.* **3**, 173–196 (1998).

[79] J. A. List and C. A. Gallet, "The environmental Kuznets curve: Does one size fit all," *Ecol. Econ.* **31**, 409–423 (1999).

[80] G. C. Unruh and W. R. Moomaw, "An alternative analysis of apparent EKC-type transitions," *Ecol. Econ.* **25**, 221–229 (1998).

[81] W. T. Harbaugh, A. Levinson, and D. M. Wilson, "Reexamining the empirical evidence for an environmental kuznets curve," *Rev. Econ. Stat.* **84**, 541–551 (2002).

[82] S. Borghesi, *The Environmental Kuznets Curve: A Survey of the Literature*, (International Monetary Fund Washington, D.C., 1999).

[83] N. Georgescu-Roegen, "Comments on the papers by Daly and Stiglitz," in *Scarcity and Growth Reconsidered*, edited by V. K. Smith (Johns Hopkins University Press, Baltimore, 1979), pp. 95–105.

[84] N. Georgescu-Roegen, "Energetic dogma, energetic economics, and viable technology," in *Advances in the Economics of Energy and Resources*, edited by J. R. Moroney (JAI Press, Greenwood, CT, 1982).

[85] N. Georgescu-Roegen, "Energy, matter, and economic valuation: Where do we stand?," in *Energy Economics and the Environment: Conflicting Views of an Essential Interrelationship*, edited by H. D. Daly and A. F. Umana (American Association for the Advancement of Science, Washington, D.C., 1981), pp. 43–79.

[86] R. U. Ayres, "The second law, the fourth law, recycling and limits to growth," *Ecol. Econ.* **29**, pp. 474–483 (1999).

[87] M. Ruth, "The economics of sustainability and the sustainability of economics," *Ecol. Econ.* **56**, 332–342 (2006).

[88] C. Hall, C. Cleveland, and R. Kaufmann, *Energy and Resource Quality: The Ecology of the Economic Process* (Wiley-Interscience, New York, 1986).

[89] M. Faber, J. L. R. Proops, M. Ruth, and P. Michaelis, "Economy–environment interactions in the long-run: A neo-Austrian approach," *Ecol. Econ.* **2**, 27–55 (1990).

[90] R. M. Solow, "A contribution to the theory of economic growth," *Q. J. Econ.* **70**, 65–94 (1956).

[91] D. Romer, *Advanced Macroeconomics* (McGraw-Hill/Irwin, New York, 2000).

[92] M. Ruth, "Thermodynamic constraints on optimal depletion of copper and aluminum in the United States: A dynamic model of substitution and technical change," *Ecol. Econ.* **15**, 197–213 (1995).

[93] B. Hannon and J. Joyce, "Energy and technical progress," Energy Research Group Document, ERG 308 (Office of the Vice Chancellor for Research, University of Illinois, Urbana-Champaign, IL, 1980).

[94] R. U. Ayres and J. C. J. M. Van Den Bergh, "A theory of economic growth with material/energy resources and dematerialization: Interaction of three growth mechanisms," *Ecol. Econ.* **55**, 96–118 (2005).

[95] C. E. Shannon, "A mathematical theory of communications," *Bell Syst. Technol. J.* **27**, 379–623 (1948).

[96] N. Winer, *Cybernetics* (Wiley, New York, 1948).

[97] C. E. Shannon and W. Weaver, *The Mathematical Theory of Information* (University of Illinois Press, Urbana, IL, 1949).

[98] L. Brillouin, *Scientific Uncertainty and Information* (Academic Press, New York, London, 1964).

[99] R. B. Evans, "A proof that essergy is the only consistent measure of potential work," Ph.D. dessertation (Dartmouth College, Hanover, NH, 1969).

[100] M. Tribus and E. C. McIrvine, "Energy and information," *Sci. Am.* **225**, 179–188 (1971).

[101] C. A. Berg, "The use of energy: A flow of information, paper presented at the International Institute for Applied Systems Analysis," *Conference on Socioeconomic Impacts of Regional Integrated Energy Systems*, Prague, CSSR, Oct. 10–12, 1988.

[102] D. T. Spreng, *Net-Energy Analysis* (Praeger, New York, 1988).

[103] L. Szilard "Über die Entropieverminderung in einem thermodynamischen System bei eingriffe intelligenter Wesen," *Z. Phys.* **53**, 840–960 (1929).

[104] R. U. Ayres and S. M. Miller, "The role of technical change," *J. Environ. Econ. Manage.* **7**, 353–371 (1980).

[105] R. U. Ayres, "Optimal investment policies with exhaustible resources: An information-based model," *J. Environ. Econ. Manage.* **15**, 439–461 (1988).

[106] J. Martinez-Alier, "Some issues in agrarian and ecological economics," *Ecol. Econ.* **22**, 225–238 (1997).

[107] S. Baumgärtner, "Thermodynamic models," in *Modelling in Ecological Economics*, edited by J. Proops and P. Safonov (Elgar, Cheltenham, UK, 2004), pp. 102–129.

[108] S. E. Jørgansen and Y. M. Svirezhev, *Towards a Thermodynamic Theory for Ecological Systems* (Amsterdam; Boston: Elsevier, 2004).

[109] J. C. Luvall and H. R. Holbo, "Thermal remote sensing methods in landscape ecology," in *Quantitative Methods in Landscape Ecology*, edited by M. Turner and R. H. Gardner (Springer-Verlag, New York, 1991), pp. 127–152.

[110] T. A. Sousa, "Is neoclassical economics formally valid? An approach based on an analogy between equilibrium thermodynamics and neoclassical microeconomics," *Ecol. Econ.* **58**, 160–169 (2006).

[111] M. Ruth, "Thermodynamic implications for natural resource extraction and technical change in US copper mining," *Environ. Resource Econ.* **6**, 187–206 (1995).

[112] S. Gössling-Reisemann, "Entropy as a measure for resource consumption – Application to primary and secondary copper production," in *Sustainable Metals Management*, edited by A. von Gleich, R. Ayres, and S. Gössling-Reisemann (Springer, Amsterdam, 2006), 195–235.

[113] R. W. Shephard, *Cost and Production Functions* (Springer-Verlag, New York, 1981).

[114] P. S. Dasgupta and G. M. Heal, *Economic Theory and Exhaustible Resources* (Cambridge University Press, Cambridge, 1979).

[115] N. Meshkov and R. S. Berry, "Can thermodynamics say anything about the economics of production?," in *Changing Energy Use Futures*, edited by R. A. Fazzolare and C. B. Smith (Pergamon, New York, 1979), pp. 374–382.

[116] M. Ruth, "Information, order and knowledge in economic and ecological systems: Implications for material and energy use," *Ecol. Econ.* **13**, 99–114 (1995).

[117] J. A. Reiss, "Comparative thermodynamics in chemistry and economics," in *Economics and Thermodynamics: New Perspectives on Economic Analysis*, edited by P. Burley and J. Foster (Kluwer Academic, Dortrecht, The Netherlands, 1994), pp. 47–72.

[118] T. A. Sousa, "Equilibrium econophysics: A unified formalism for neoclassical economics and equilibrium thermodynamics," *Physica A* **371**, 492–512 (2006).

[119] I. Prigogine, *From Being to Becoming: Time and Complexity in the Physical Sciences* (W.H. Freeman and Company, New York, 1980).

[120] P. Glansdorf and I. Prigogine, *Thermodynamic Theory of Structure, Stability and Fluctuations*, Second Edition (John Wiley & Sons Ltd., New York, NY, 1971).

[121] J. J. Kay, "Self-organization in living systems," Ph.D. dessertation, (Systems Design Engineering, University of Waterloo, Waterloo, Ontario, Canada, 1984).

[122] B. Hannon, M. Ruth, and A. Delucia, "A physical view of sustainability," *Ecol. Econ.* **8**, 253–268 (1993).

[123] J. J. Kay, H. A. Regierb, M. Boylec, and G. Francisa, "An ecosystem approach for sustainability: Addressing the challenge of complexity," *Futures* **31**, 721–742 (1999).

[124] P. Bak, *How Nature Works: The Science of Self-Organized Criticality* (Copernicus, New York, 1996).

[125] H. J. Jensen, *Organized Criticality: Emergent Complex Behavior in Physical and Biological Systems* (Cambridge University Press, New York, 1998).

[126] C. S. Holling, editor, *Adaptive Environmental Assessment and Management* (Wiley, London, 1978).

[127] L. C. Gunderson, S. Holling, and S. Light, editors, *Barriers and Bridges to the Renewal of Ecosystems and Institutions* (Columbia University Press, New York, 1995).

17 Integration and Segregation in a Population – a Thermodynamicist's View

Ingo Müller

17.1 Introduction

When resources become exhausted – for whatever reasons – dearth and starvation occur. Impending crises of that sort are foreshadowed by sociological changes, and the oft-deplored phenomenon of segregation of sociological groups is one of them. However, that phenomenon may not be only a random concomitant of an economic crisis: Like a fever in an infected body, when a sick body runs a temperature, segregation may be a symptom that shows that the society is trying to survive and that it makes the best of a bad situation in expectation of better times.

Social behavior is largely dictated by the competition of sociological groups for a limited amount of resources, essentially and ultimately food. Such groups may represent social classes, or ethnic and racial groups, or religious sects, etc.

If resources are abundant and consequently prices are low, the competition is more or less friendly and relaxed, and there is room and occasion for social niceties and tolerant conduct between sociological groups. Granted that there is always competition, yet in times of abundance the competitive strategy is dictated by good will and a population finds it easy to *integrate* members of different groups.

When resources are scarce and therefore expensive, the competition becomes more serious, or even fierce. A new strategy – a more competitive one – may be employed by all sociological groups, and the mutual tolerance between groups is strained or altogether abandoned. Those are the conditions under which *segregation* occurs in a population. The population falls apart into different colonies so that members of a group have social contact primarily among themselves.

In some intuitive way we all know this and accept it, even though segregation is usually considered undesirable politically. Segregation complicates the life of politicians because it forces them to make conflicting promises in the different colonies belonging to their domain. Therefore they try to discourage segregation, advise against it, and lament over it. However, usually their disapproval achieves nothing, and in the present investigation I attempt to find out why that is so. It is *not* due to obstinacy on the part of the population.

Rather the reason is – in my opinion – that in economically difficult times, when resources are rare and expensive, segregation into colonies with different strategies is more conducive to individual gain for members of all groups than an integrated society with either strategy, the relaxed one or the fierce one. I present a sociological model to substantiate that claim.

To avoid the minefield of a discussion of complex human relations, I take recourse to a behavioral model from game theory. The model is one of two species, hawks and doves, both competing for the same resource or, to be specific, for food. It is unnecessary, perhaps, for me to say that the species are metaphorical, and their members are not real birds; in particular, the hawks do not eat the doves; they are more like the hawks and doves often attributed to the State Department, the foreign ministry of the United States.[1] Also my hawks and doves cannot fly, so their habitat is a plane.

My treatment of the hawk–dove population is analogous to a large extent to the thermodynamics of binary solutions in two phases. It is well known to the thermodynamicist – more specifically to the chemical engineer or the metallurgist, but also to ordinary cooks – that miscibility gaps do occur at low temperatures, like when specks of fat form on a watery soup. The segregation of fat and water lowers the free energy of the solution. At high temperatures the miscibility gaps are diminished or disappear altogether. Analogously in my model, a population may segregate into hawk-rich colonies and dove-rich ones when resources are scarce and expensive. Different contest strategies are then usually followed in the two types of colonies and that increases the gain for all. For abundant resources – low prices – the colonies disappear and integration prevails.

The conclusion is that, if the king of birds wishes to rule over an integrated population of hawks and doves, he should make resources cheap – if he can. Anyway, his efforts in that direction should be more effective than to lambaste the constituent groups for obstinacy.

17.2 Prologue

17.2.1 Model From Game Theory

We let ourselves be motivated by an often-discussed model of game theory for a mixed population of hawks and doves who compete for the same resource; see [1, 4]. The value of the resource, or its price, is denoted by τ.[2] Prices are out of control for the birds, but they do change, much like the weather or the stock market, and they influence their behavior. Indeed, in their competition the birds may assume different strategies, A or B, which we define as follows.

---

[1] My hawk-and-dove model is an adaptation of the sociobiological model invented by Maynard-Smith and Price [1] to show that a mixed-species population can evolve to a stable state with *both* species present in a certain proportion. See also [2]. Having said this, I stress that my model is different in both quality and intention. In particular, there is no evolution in my model. The attraction of my model – if it has any – does not come from evolution; rather it comes from the possibility of the birds to choose between different competitive strategies. The model was first described in my article [3].

[2] Let us speak of τ "points." Throughout this paper, except for the properly thermodynamic part in Subsection 17.3.3.1, I forgo units; all quantities are made dimensionless in some proper manner.

Strategy A

If two hawks meet over the resource, they fight until one is injured. The winner gains the value τ, whereas the loser, being injured, needs time for healing his wounds. Let that time be such that the hawk must buy resources, at price 2, to feed himself during convalescence. Two doves do not fight. They merely engage in a symbolic conflict, posturing and threatening, but not actually fighting. One of them will eventually win the resource – always with the value τ – but both lose time such that after every dove–dove encounter they need to catch up by buying part of a resource, worth 0.2. If a hawk meets a dove, the dove walks away, and the hawk wins the resource; there is no injury, nor is any time lost.

Assuming that winning and losing the fights or the posturing game is equally probable, we conclude that the elementary expectation values for the gain per encounter are given by the arithmetic-mean values of the gains in winning and losing, i.e.,

$$e_{HH}^A = 0.5\,(\tau - 2),$$

$$e_{HD}^A = \tau,$$

$$e_{DH}^A = 0,$$

$$e_{DD}^A = (0.5\tau - 0.2), \tag{17.1}$$

for the four possible encounters: HH, HD, DH, and DD.

Note that both the fighting of the hawks and the posturing of the doves detract from the gain. Thus both species would do better without these activities. From the point of view of gain, that aspect of the birds' behavior must be considered a luxury. Its rationale is based on ulterior motives like, perhaps, the desire to impress a possible sexual partner. However, when times are bad and food is scarce and expensive, the birds may feel that it is more important to eat than to be sexually attractive. So they may decide to cut down on fighting and posturing or abandon that behavior altogether. Also, the meekness of a dove confronted with a hawk may be regarded as overcautious in bad times.

Such observations lead to the formulation of strategy B.

Strategy B

The hawks adjust the severity of the fighting, and thus the gravity of the injury, to the prevailing price τ. If the price of the resource is higher than 1, they fight less, so that the time of convalescence in the case of a defeat is shorter and the value to be bought during convalescence is reduced from 2 to $2[1 - 0.3(\tau - 1)]$. Likewise, the doves adjust the duration of the posturing so that the payment for lost time is reduced from 0.2 to $0.2[1 - 0.3(\tau - 1)]$. But that is not all: To be sure, in strategy B the doves will still not fight when they find themselves competing with a hawk, but they will try to grab the resource and run. Let them be successful 4 out of 10 times. However, if unsuccessful, they risk injury from the enraged hawk and may need a period of convalescence at the cost $2[1 + 0.8(\tau - 1)]$.

Thus the elementary expectation values for gains under strategy B may be written as[3]

$$e_{HH}^B = 0.5\{\tau - 2[1 - 0.3(\tau - 1)]\} = 0.8\ \tau - 1.3,$$
$$e_{HD}^B = 0.6\tau,$$
$$e_{DH}^B = 0.4\tau - 0.6 \times 2[1 + 0.8(\tau - 1)] = -0.56\tau - 0.24,$$
$$e_{DD}^B = 0.5\tau - 0.2[1 - 0.3(\tau - 1)] = 0.56\tau - 0.26. \quad (17.2)$$

In the interest of reaching a definite result quickly, I forego motivation and discussion of the strategies at this point. Indeed, at this point, for motivation I rest content with reference to [1] and [5] for strategy A, and with the previous few words preceding strategy B. However, some discussion may be found in Subsection 17.2.4, at the end of this prologue, when the theory can be evaluated in light of its results.

17.2.2 Expectations of Gain

Now, let z_H and $z_D = 1 - z_H$ be the fractions of hawks and doves respectively, and let all hawks and doves employ either strategy A or B. Therefore the gain expectations e_H^i and e_D^i ($i =$ A, B) of a hawk and a dove per encounter with another bird may be written as

$$e_H^i = z_H e_{HH}^i + (1 - z_H)e_{HD}^i, \quad e_D^i = z_H e_{DH}^i + (1 - z_H)e_{DD}^i \quad (17.3)$$

in terms of the elementary expectation values. And the gain expectations e^i for strategy i per bird and per encounter read

$$e^i = z_H e_H^i + (1 - z_H)e_D^i, \quad (17.4)$$

or, explicitly,

$$e_H^i = z_H^2 \left(e_{HH}^i + e_{DD}^i - e_{HD}^i - e_{DH}^i\right) + z_H \left(e_{HD}^i + e_{DH}^i - 2e_{DD}^i\right) + e_{DD}^i,$$

or

$$e^i = e_{HH}^i z_H + e_{DD}^i(1 - z_H) - \left(e_{HH}^i + e_{DD}^i - e_{DH}^i\right)z_H(1 - z_H). \quad (17.5)$$

Specifically, in our case we have

$$e^A = (0.5\tau - 1)z_H + (0.5\tau - 0.2)(1 - z_H) + 1.2z_H(1 - z_H), \quad (17.6)$$
$$e^B = (0.8\tau - 1.3)z_H + (0.56\tau - 0.26)(1 - z_H) - 1.32(\tau - 1)z_H(1 - z_H). \quad (17.7)$$

The graphs of these functions of z_H are parabolas in a (e^i, z_H) diagram that, for some values of τ, are plotted in Figs. 17.1(a)–17.1(e).

The interpretation of those graphs is contingent on the reasonable assumption that the population chooses *the* strategy that provides the maximal gain expectation. Obviously for $\tau = 0.6$ and $\tau = 1$ that strategy is strategy A. At that price level the

---

[3] $\tau = 1$ is a reference price in which both strategies coincide, except for the grab-and-run feature of strategy B. Penalties for either fighting or posturing should never turn into rewards for whatever permissible value of τ. This condition imposes a constraint on the permissible values of τ: $0 < \tau < 4.33$. That condition is merely a technical point; it could be avoided by allowing the gain expectations to be nonlinear in τ.

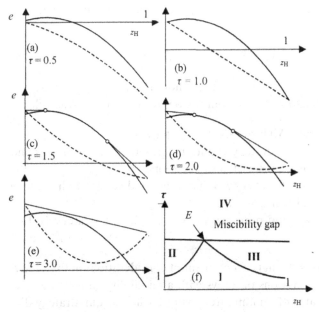

Figure 17.1. Expectation values e^A (solid) and e^B (dashed) as functions of fraction z_H for some values of the price τ. Concavification, straight lines; (f) strategy diagram.

hawks and the doves will therefore all choose strategy A, irrespective of the hawk fraction z_H in the population.

For higher price levels the situation is more subtle, because the graph max $[e^A, e^B]$ is not concave.[4] This provides the possibility of concavification, cf. Figs. 17.1(c) and 17.1(d): There are intervals of z_H where the concave envelope of max $[e^A, e^B]$ – represented by straight lines in the figure – lies higher than that graph itself. The population thus has the possibility of increasing the expected gain by unmixing; it *segregates* into homogeneous *colonies* with hawk fractions corresponding to the end points of the concavifying tangents, the straight lines in the figures. In Figs. 17.1(c) and 17.1(d), the adopted strategies are A and B in the colonies, and the species are mixed in the colony with strategy A, whereas the colony with strategy B is pure dove or pure hawk, depending on whether the extant overall hawk fraction lies below the left or right tangent, respectively. For $\tau > 2.91$ the concave envelope connects the end points of the parabola e^B so that hawks and doves are completely segregated in two colonies, both employing strategy B.

The result of segregation is that, on average, the birds have a higher gain expectation from an encounter than in the integrated population with whatever strategy, A or B. That fact is, of course, the underlying motivation – and a strong incentive – for segregation.

17.2.3 Strategy Diagram

It is convenient to summarize this discussion in a (τ, z_H) *strategy diagram*; see Fig. 17.1(f). The tangents are projected to the horizontal line in that diagram with

---

[4] max $[e_A, e_B]$ is the graph following the curves e_A or e_B, whichever is higher.

the appropriate τ, and the end points of the projections are connected. Thus we create two parabolic curves, viz.,

$$\tau = \left(20 z_H^2 + 1\right), \quad \tau = \left(4 z_H^2 - 8 z_H + 5\right), \tag{17.8}$$

which intersect in the point $(\tau, z_H) = (2.91, 0.31)$, denoted by E.

In this way the strategy diagram exhibits four regions to be interpreted as follows:

- **I.** Full integration of species. All birds employ strategy A.
- **II.** Colony of pure doves with strategy B and integrated colony of hawks and doves with strategy A. This means partial segregation.
- **III.** Colony of pure hawks with strategy B and integrated colony with strategy A. Again this means partial segregation.
- **IV.** Colonies of pure doves and pure hawks both employing strategy B. Full segregation.

Mutatis mutandis, all this is strongly reminiscent of considerations in thermodynamics of *phase* diagrams of solutions or alloys with a miscibility gap. To be sure, phase diagrams represent states of minimal free energy, whereas our strategy diagrams represent states of maximal gain. But that is a superficial difference. We call E the eutectic point in analogy to phase diagrams of alloys.

17.2.4 Discussion

Much of the following discussion was written in response to comments by reviewers of a previous draft of my manuscript. I appreciated those comments, and I thank the reviewers for the care with which they have read my paper; in most instances I changed the manuscript in accord with the reviewers' suggestions. Those suggestions, however, have also made me realize the need to explain the scientific method that I have employed, in particular in regard to the formulation of the rules of my game and the choice of its parameters. Let me explain this.

The invention of a strategy and the assignment of numbers to tactical measures within a strategy are always problems in game theory, because they are liable to contain some arbitrariness. I tried to minimize that feature by adapting, for strategy A, a popular characterization of animal behavior, which is discussed by some eminent biologists, namely Maynard-Smith and Dawkins. Trusting their biological expertise, I consider the fact sufficient motivation for adopting strategy A, and indeed, the only feature that I have added to that strategy myself is a variable price. Strategy B is my own, however, and although it is clearly a modification of strategy A, in a more or less straightforward manner, the aforementioned biologists must not be blamed, if it is found objectionable. Both strategies, including the numbers, represent what I think is a conceivable idea of the behavior of the species – and they lead to segregation! Let us consider some features of this.

A person might believe that the hawks are likely to fight *more* rather than *less* when times become harder and food is more expensive. This is not so in my model. Indeed, for $\tau > 1$, the *intra*species squabbles are reduced in strategy B with growing price. However, a new *inter*species aggressiveness in the hawks is introduced in strategy B as a response to the grab-and-run policy of the doves, and that aggressiveness grows with growing price, because the hawks will exert more violence against the impertinent doves when the stolen resource is more valuable.

It is obvious from the numbers that the grab-and-run policy is not a wise one for the doves, because, on average, they get punished for it. So why do they adopt that policy? We may explain that by assuming that doves are no wiser than people, who have often in history started a conflict with the expectation of a quick gain and then met disaster.

I have been accused by a reviewer of having *rigged* the strategies so as to obtain the desired results. This is true. I admit it, or rather, I state it proudly: I formulated a framework of strategies *that can be rigged* in order to describe the ubiquitous phenomenon of segregation in a society under stress, and that was my intention all along, and, to the best of my knowledge, it has not been done before.

The situation is much as in molecular physics. Thus it is impossible to *rig* the ideal-gas laws so as to provide a miscibility gab for a mixture of ideal gases. On the other hand, if we allow for an energetic interaction between the molecules of two constituents of a mixture, we may obtain miscibility gaps under the condition that we *rig* the interaction energies so as to provide a *malus* for unequal next neighbors. In other words, there has to be sufficient complexity in a model for it to be *riggable* so as to provide complex results. That is the case for fluids with intermolecular forces – and it is also the case for my behavioral strategies.

Some criticism of the rules of my game comes from people who know the works of Maynard-Smith and Dawkins and do not like my modification. They do have a point. Indeed, the biological game of those researchers concerns evolution toward a society with a hawk fraction of maximal gain. In my game there is no evolution, and the hawk fraction is fixed. In this model there is not enough time for evolution, which takes many generations to become effective. Here, however, although the birds do not evolve, they do change their strategy in reaction to the economic climate that may radically alter many times in a lifetime.[5]

Another point criticized by adherents of Maynard-Smith and Dawkins, particularly Dawkins, concerns selfishness. Selfishness – the selfishness of the genes – may be the rule of conduct in biology, in fact, I believe that it probably is. But it is forced into recession under the laws of a modern democratic society. Indeed, during my lifetime I have been obliged to pay millions in taxes that were then used to support inefficient industries or unemployable people *for the common good* – and for a thriving burocracy that administers the common good. So therefore I find nothing uncommon in the feature implied by Fig. 17.1(d), say, where, for $z_H \approx 0.1$, in a single population, the members of the colony with strategy A are clearly less well off than the members of the pure-dove colony employing strategy B. The "common good" is still well served in that case because, *on average*, the members of that partially segregated society are better off than in a homogeneous society with either strategy A or B.

When I started this work, I was under the illusion that economists and sociologists might be interested in the results. That was not the case. Indeed, economists and sociologists firmly shut their eyes and ears, and have nothing but disdain for the

---

[5] It is possible to consider evolution *and* change of strategy at the same time; see Section 4 for first results.

application of methods of physics or, more appropriately, thermodynamics, in their fields. I must not complain, however, because the disdain is entirely reciprocal.

Maybe social scientists cannot be blamed individually because, in the culture in which we live, the interpretation of phase diagrams does not belong to their curriculum; they have never seen a phase diagram in their lives.

17.3 Sociothermodynamics

17.3.1 The System and Its Functions of State

Although the preceding argument is reminiscent of arguments in thermodynamics of solutions, it cannot qualify as sociothermodynamics because it is too simple. With some good will, we may admit an analogy of the tendency toward maximal gain with the thermodynamic tendency toward a minimal free energy, but that should not be enough. There is so far no equivalent in the sociological system to working and heating, temperature and internal energy, and pressure and volume. And there is no first law and no second law. Those missing features in the population of hawks and doves are now described.

At this stage – until later in Subsection 17.3.3.4 – we assume that all birds exercise one strategy homogeneously, either A or B.

The population lives in a habitat of size A, and it consists of the constant number N of birds. They are generally not distributed homogeneously over the habitat of area A, and the distribution may depend on time. We have

$$N = \int_A n(x, t)\, dx, \tag{17.9}$$

where $n(x, t)$ is the number density of both species, i.e., hawks and doves, so that

$$N_\alpha = \int_A n_\alpha(x, t)\, dx = \int_A n(x, t) z_\alpha(x, t)\, dx \, (\alpha = \mathrm{H}, \mathrm{D}) \tag{17.10}$$

holds, where $z_\alpha(x, t)$ are the ratios of hawks and doves, locally and instantaneously.

Resources are distributed on the plane of which the habitat covers the area A. Thus the available total resource R is an integral over A, viz.,

$$R = \int_A r(x, t)\, dx, \tag{17.11}$$

where r is the resource density.

We stipulate that the value $\tau(x, t)$ depends monotonically on $r(x, t)$ such that the price decreases when r increases. That stipulation seems to correspond to the well-known everyday economic experience about supply, demand, and price. In our case the demand is constant, because the number of birds and their need does not change; therefore the supply determines the price.

The birds, walking randomly in A, fill that area whatever its size. Therefore, if the population is constrained by a boundary ∂A, it feels a pressure p_0 that the boundary exerts on it. The population reacts by exerting a counterpressure – the

Table 17.1. *Pressure coefficients a_α^i*

	H	D
A	$\dfrac{1}{5}$	$\dfrac{3}{25}$
B	$\dfrac{2}{5}$	$\dfrac{9}{25}$

population pressure p – on the boundary ∂A that depends on its number density and the price. We assume the pressure p_α^i of a species α in strategy i to be given by

$$p_\alpha^i = a_\alpha^i n_\alpha^i \tau, \tag{17.12}$$

because more birds of species α employing strategy i are pressing against the boundary when their partial population is dense, and because the random walk becomes more frantic for higher prices. We call this relation the *equation of state for the pressure*. The equation of state is much like the ideal-gas law of thermodynamics, except that the factor a is not universal. Indeed, there are four such relations corresponding to the two species ($\alpha = $ H, D) and to the two strategies ($i = $ A, B), and each one has its own coefficient a. Therefore we have four coefficients a_α^i, which we choose as shown in Table 17.1.

We adopt Dalton's law of mixtures for our mixed population of hawks and doves and set the total pressure equal to the sum of the partial pressures p_H^i and p_D^i, and we denote it by p:[6]

$$p = \sum_{\alpha=\mathrm{H}}^{\mathrm{D}} p_\alpha^i. \tag{17.13}$$

The pressure ratios are obviously

$$\frac{p_\alpha^i}{p} = \frac{a_\alpha^i z_\alpha}{\sum_{\beta=\mathrm{H}}^{\mathrm{D}} a_\beta^i z_\beta} \quad (\alpha = \mathrm{H, D}),\ (i = \mathrm{A, B}). \tag{17.14}$$

The population pressure can be felt on any wall that is obstructing the passage of the birds, even on hypothetical walls in the interior of A. Therefore p_H^i, p_D^i, and p are fields.

We assume a priori that the fields $\tau(x, t)$ and $p(x, t)$ become homogeneous when the population is left alone, so that, in slow processes, in which $\tau(x, t)$ and $p(x, t)$ change slowly, they pass through homogeneous values.

If there are two possible gains e^A and e^B appropriate to different strategies, as discussed in Section 17.2, and if the population chooses the strategy with the higher gain, there are also two *shortfalls*,

$$u^i = \tau - e^i \quad (i = \mathrm{A, B}), \tag{17.15}$$

and the population chooses the strategy with the smaller shortfall. The shortfall is so named because τ is the maximal gain, achieved by a bird in the unlikely case that

---

[6] Recall that at this stage the population exercises only *one* strategy.

he wins the resource in each encounter, and neither fights nor postures. Therefore u^i measures the extent to which the actual gain, per bird and encounter, falls short of the maximal one. u^i is introduced in place of e^i in order to bring the subsequent arguments closer to thermodynamics. Because, by Eqs. (17.6) and (17.7), e^i are linearly increasing functions of τ, so are u^i, and we write

$$u^A = \underbrace{(0.5\tau + 1)\, z_H}_{u_H^A} + \underbrace{(0.5\tau + 0.2)(1 - z_H)}_{u_D^A} \underbrace{-1.2 z_H(1 - z_H)}_{u_{Mix}^A} \tag{17.16}$$

$$u^B = \underbrace{(0.2\tau + 1.3)\, z_H}_{u_H^B} + \underbrace{(0.44\tau + 0.26)(1 - z_H)}_{u_D^B} + \underbrace{1.32(\tau - 1) z_H(1 - z_H)}_{u_{Mix}^B}. \tag{17.17}$$

u^A and u^B are fields, and the last term in each one may be called *shortfalls of mixing* because they are proportional to the probability $z_H(1 - z_H)$ of hawk–dove encounters in a mixed-species population. The parts denoted by $u_\alpha^i(\tau)$ in (17.16) and (17.17) are the shortfalls in the pure-α species exercising strategy i. We call Eqs. (17.16) and (17.17) *equations of state for the shortfalls*.

The total shortfalls are obtained by integration over the fields $u^i(\tau, z_H)$, viz.,[7]

$$U^i = \int_A u^i(\tau, z_H) n(x, t)\, dx. \tag{17.18}$$

17.3.2 First Law of Sociothermodynamics

The expected gains E^i or the shortfall U^i may change in time because of trading $\dot{T}$ and working $\dot{W}$. Therefore we write the *first law of sociothermodynamics* in the form

$$\frac{dU^i}{dt} = \dot{T} + \dot{W}. \tag{17.19}$$

Trading means the exchange of goods across the boundary, and the goods are paid for in resources. If $\dot{T} > 0$ holds, goods are imported and resources inside A diminish. Thus prices increase, and so do the expected gain e^i and the shortfall u^i. If $\dot{T} < 0$ holds, goods are exported; this brings more resources into the habitat, and therefore prices decrease and so do gain and shortfall.

Working occurs when the boundary is moved. We set

$$\dot{W} = -\int_{\partial A} p_0 w_q n_q\, ds, \tag{17.20}$$

where p_0, as before, is the pressure on the population exerted by the boundary ∂A, w_q is the velocity of the boundary element ds, and n_q is its outer normal. If p_0 is homogeneous on ∂A, the working may be written as

$$\dot{W} = -p_0 \frac{dA}{dt}, \tag{17.21}$$

---

[7] Recall footnote 6.

and in a slow process, where p is homogeneous in A and hence equal to p_0, we have

$$W = -p\frac{\mathrm{d}A}{\mathrm{d}t}, \quad \text{where } p = \sum_{\alpha=\mathrm{H}}^{\mathrm{D}} p_\alpha^i. \tag{17.22}$$

This is tantamount to saying that a given $\mathrm{d}A/\mathrm{d}t$ entails a faster change of shortfall, or price, for bigger values of p; such an assumption is eminently sensible, because the newly available area is obviously filled more rapidly by the randomly walking birds if the density is larger and the price is high.

In a slow process, if there is no trading, the first law may be integrated by use of the equations of state for the pressures and shortfalls. We obtain

$$\tau\frac{\partial u^i}{\partial \tau} A^{\frac{\sum_{\alpha=\mathrm{H}}^{\mathrm{D}} a_\alpha^i z_\alpha}} = \text{constant}, \tag{17.23}$$

so that the price decreases when the area of the habitat grows. We anticipated this result previously when we stipulated that a greater supply means lower prices, because the increase of A increases the supply.

On the other hand, again for a slow process, a possible price change that is due to trading may be offset by working so that the price does not change in the process.

The thermodynamicist recognizes (17.23) as the equivalent of the "adiabatic equation of state" and the process with constant τ as the equivalent of an isothermal process.

It is now possible for a population to gain power by a process of importing, expanding, exporting, and contracting, and come back to the same state (p, A) that it started from, in the manner of thermodynamic cycles. The work gained is W_o – the time integral over the working or power W – and it has to be compared with the expenditure T_{import} – the time integral over T for imports. Thus an efficiency may be defined as

$$e = \frac{|W_o|}{T_{\mathrm{import}}}, \tag{17.24}$$

in complete analogy to the efficiency of a cycle in thermodynamics.

In particular, if the cycle has four branches, two at constant price and two without trade, we come to the equivalent of a Carnot cycle and the efficiency is

$$e = 1 - \frac{\tau_{\mathrm{export}}}{\tau_{\mathrm{import}}}; \tag{17.25}$$

this follows easily by a calculation of W_o and T_{import}, when we use the equations of state.

It would be tempting to extend the model and to address the question of what the population might do with the power gained. The mostly likely answer is that it would use the power to impose on neighboring populations or on its trade partners. But here I abstain from making such considerations explicit.

17.3.3 Second Law of Sociothermodynamics

17.3.3.1 A Reminder of Thermodynamics Proper

In order that we appreciate fully the formal analogy of sociothermodynamics and thermodynamics proper, I insert a brief reminder of the thermodynamics of fluids. This should be particularly helpful in regard to the second law, which we have not yet discussed, but it may not be useless either to the reader for the proper appreciation of the first law.

In a fluid mixture, with N particles and two constituents $\alpha = 1, 2$, particle numbers N_α, which are homogeneously in one phase, either as a liquid L or as a vapor V, the rate of change of the internal energy U^i $(i = \text{L,V})$ is equal to heating $\dot{Q}$ and working $\dot{W}$:

$$\frac{dU^i}{dt} = \dot{Q} + \dot{W} \ (i = \text{L, V}).$$

If the fluid is subject to a homogeneous pressure p_0 on the boundary ∂V, the working is given by

$$\dot{W} = -p_0 \frac{dV}{dt},$$

where V is the volume of the fluid. In a slow process, characterized by homogeneous fields of temperature Θ and pressure p throughout V, we have

$$\dot{W} = -p \frac{dV}{dt} \quad \text{with } p = \sum_{\alpha=1}^{2} p_\alpha^i \ (i = \text{L, V}).$$

The partial pressures p_α^i and the specific internal energy u^i are functions of the temperature and of the particle number densities n_α^i. The relations are called the *thermal* and *caloric equations of state*.

All this is strictly analogous to the first law of sociothermodynamics, as described in Subsection 17.3.2. I emphasized the analogy in the notation by denoting the shortfalls by U^i and the partial pressures by p_α^i. Obviously the heating $\dot{Q}$ of the fluid and the trading $\dot{T}$ of the population correspond to each other. The constituents of the mixture are the species of the population, and the phases of the mixture correspond to the strategies employed by the population. The volume of the mixture V is replaced with the area A of the habitat.

There are several equivalent formulations of the second law of thermodynamics and they all concern the heating. Thus Rudolf E. Clausius, the discoverer of the second law, expressed it as an axiom in these words:

Heat cannot pass by itself from cold body to hot.

The statement incorporates the perceived irreversibility of thermodynamic processes in that it forbids heat to pass, or be transferred, from a cold body to a hot one, but permits the passage in the reverse direction. Applied to heat conduction, the axiom implies that eventually, given time, the temperature approaches a homogeneous value. Thus in slow-enough processes – with slowly applied heating and working – it is reasonable to assume that a body passes through states of homogeneous temperature and homogeneous pressure.[8] This assumption, in fact, is usually anticipated in most thermodynamic exercises that exploit the first law,[9] like the operation of a compressor or a steam engine.

In this chapter I assume that the reader is familiar with thermodynamics and, in particular, knows how Clausius managed to cast the second law into a mathematical form, an inequality.[10] The inequality introduces the entropy S^i ($i = L,V$) of a fluid and relates its rate of change to the heating $\dot{Q}$ through the boundary that is supposed to have a homogeneous temperature Θ_0:

$$\frac{dS^i}{dt} \geq \frac{\dot{Q}}{\Theta_0} \quad \text{(equality for slow processes).}$$

This inequality represents the mathematical expression of the second law.

For slow processes, when $\dot{Q}$ is eliminated between the first and second laws, the Gibbs equation results:

$$\frac{dS^i}{dt} = \frac{1}{\Theta}\left(\frac{dU^i}{dt} + p\frac{dV}{dt}\right) \quad (i = 1, 2).$$

Insertion of the thermal and caloric equations of state provides the equation of state $S^i = S^i(\Theta, V)$ for the entropies of the phases $i = L,V$. The equations are obtained by integration to within an additive constant.

Another obvious consequence of the second law for a *rapid*, adiabatic process is that S tends to a maximum; the maximum is attained in equilibrium, the stationary final state. Alternatives of this conclusion result from the elimination of $\dot{Q}$ between the first and second laws, viz.,

$$\frac{d(U^i - \Theta_0 S^i)}{dt} \leq -S^i\frac{d\Theta_0}{dt} - p_0\frac{dV}{dt},$$

$$\frac{d(U^i + p_0 V - \Theta_0 S^i)}{dt} \leq -S^i\frac{d\Theta_0}{dt} + V\frac{dp_0}{dt},$$

so that, for Θ_0 and V constant, $U^i - \Theta_0 S^i$ tends to a minimum, and, for Θ_0 and p_0 constant, $U^i + p_0 V - \Theta_0 S^i$ tends to a minimum.

When, in the approach to equilibrium, the fields Θ and p are already both homogeneous – and equal to the constant-boundary values Θ_0 and p_0 – the Gibbs free energy $G^i \equiv U^i + pV - \Theta S^i$ satisfies the last inequality trivially, as an equality, because U^i, S^i, and V all depend on Θ and p.

This is true if the fluid is homogeneously in *one* phase. However, when the fluid is composed of *two* phases, the decrease of the free energy toward equilibrium retains a nontrivial meaning, even when the temperature and the pressure are already homogeneous. Let us consider the following:

---

[8] Homogeneity of pressure means that mechanical equilibrium prevails during the slow process.
[9] Notable exceptions are rapid processes, e.g., when a gas is suddenly expanding into a formerly evacuated chamber or when a piston drops into a gas-filled cylinder.
[10] That derivation is replayed in most textbooks on thermodynamics, e.g., in the book [5] by the author and Wolf Weiss.

If there are two phases, liquid L and vapor V, say, each occupying the part V^i of V, the Gibbs free energy consists of two additive terms

$$G = \sum_{i=L}^{V} G^i,$$

$$G = \sum_{i=L}^{V} (U^i + pV^i - \Theta S^i),$$

or, more explicitly,

$$G = \sum_{i=L}^{V} Nx^i \left[u^i \left(\tau, p, z_1^i \right) + p\frac{1}{n^i \left(\tau, p, z_1^i \right)} - \Theta s^i \left(\tau, p, z_1^i \right) \right],$$

where x^i are the phase fractions, $n^i = \sum_{\alpha=1}^{2} n_\alpha^i$ are the particle number densities in the phases, u^i and s^i are the specific values of internal energy and entropy of the phases, and z_1^i is the fraction of particles of constituent 1 in phase i. In this case the G is no longer a function of τ and p only; there are three new variables, viz., x^L, say, and z_1^L and z_1^V. And their values adjust themselves so that G becomes minimal.

17.3.3.2 Second Law

We recall that, in sociothermodynamics, the value is measured by the price τ, and we formulate the second law in this form:

Value cannot pass by itself from cheap to dear.

For motivation and explanation, this consideration is offered: Imagine a weekend market in which two stalls are selling the same good at different prices. The cheap stall will sell quickly, so that the owner comes to realize that he may raise the price and still sell all his goods. The dear stall sells little, and the owner will therefore lower the price so as to sell more. Thus there is a transfer of value from dear to cheap, even without direct exchange of goods. The reverse will not happen, or so our second law maintains. If the goods hold out, eventually the price will be homogeneous.

The analogy with thermodynamics is obvious. If this axiom is accepted, we find ourselves in the same position as Clausius, and by using his arguments[11] we may give the second law the mathematical form of an inequality, viz.,

$$\frac{\mathrm{d}S^i}{\mathrm{d}t} \geq \frac{\dot{T}}{\tau_0} \quad \text{(equality for slow processes)}. \tag{17.26}$$

S^i is the entropy of a population employing strategy i and τ_0 is the price imposed on the boundary ∂A.

---

[11] Clausius's arguments on this occasion are intricate and not short. I skip them and refer the reader to books on thermodynamics, e.g., my own book [5] with W. Weiss: *Entropy and Energy – A Universal Competition.*

17.3.3.3 Gibbs Equation. Equations of State for the Entropies s_α^i. The Entropy of Mixing

Just as in thermodynamics, we derive the Gibbs equation for a slow process by elimination of $\hat{T}$ between second law (17.26) and first law (17.19) with (17.22). We obtain

$$\frac{dS^i}{dt} = \frac{1}{\tau}\left(\frac{dU^i}{dt} + p\frac{dA}{dt}\right). \tag{17.27}$$

For a pure species s_α^i, the specific entropy of the pure species α with strategy i, the Gibbs equation reads

$$\frac{ds_\alpha^i}{dt} = \frac{1}{\tau}\left(\frac{du_\alpha^i}{dt} + p_\alpha^i\frac{d\frac{1}{n_\alpha^i}}{dt}\right), \tag{17.28}$$

where n_α^i is the number density and u_α^i and p_α^i may be read off from (17.16), (17.17), and (17.12), respectively. Integration gives

$$s_\alpha^i\left(\tau, p_\alpha^i\right) = s_\alpha^i(\tau_R, p_R) + \left(\frac{du_\alpha^i}{d\tau} + a_\alpha^i\right)\ln\frac{\tau}{\tau_R} - a_\alpha^i\ln\frac{p_\alpha^i}{p_R}, \tag{17.29}$$

where $s_\alpha^i(\tau_R, p_R)$ is the entropy corresponding to some arbitrary reference state with τ_R, p_R. We add the entropy densities of the mixed population to obtain the specific entropy of the mixture as

$$s^i = \sum_{\alpha=H}^{D} z_\alpha\left[s_\alpha^i(\tau_R, p_R) + \left(\frac{du_\alpha^i}{d\tau} + a_\alpha^i\right)\ln\frac{\tau}{\tau_R} - a_\alpha^i\ln\frac{p_\alpha^i}{p_R}\right], \tag{17.30}$$

or by (17.14),

$$s^i = \sum_{\alpha=H}^{D} z_\alpha\left[s_\alpha^i(\tau_R, p_R) + \left(\frac{du_\alpha^i}{d\tau} + a_\alpha^i\right)\ln\frac{\tau}{\tau_R} - a_\alpha^i\ln\frac{p}{p_R} - a_\alpha^i\ln\frac{a_\alpha^i z_\alpha}{\sum_{\beta=H}^{D} a_\beta^i z_\beta}\right], \tag{17.31}$$

$$s^i = \sum_{\alpha=H}^{D} z_\alpha s_\alpha^i(\tau, p) - \underbrace{\sum_{\alpha=H}^{D} a_\alpha^i z_\alpha \ln\frac{a_\alpha^i z_\alpha}{\sum_{\beta=H}^{D} a_\beta^i z_\beta}}_{s_{\mathrm{Mix}}^i}. \tag{17.32}$$

The last term represents the entropy of mixing, as indicated.

17.3.3.4 Minimal Gibbs Free Energy in Sociothermodynamics

Again, just as in thermodynamics (see Subsection 17.3.3.2), second law (17.26) may be rewritten by use of first law (17.19), (17.21), (3.13) by elimination of $\hat{T}$. We obtain

$$\frac{d(U^i - \tau_0 S^i)}{dt} \leq -S^i\frac{d\tau_0}{dt} - p_0\frac{dA}{dt} \quad \text{or also} \quad \frac{d(U^i + p_0 A - \tau_0 S^i)}{dt}$$

$$\leq -S^i\frac{d\tau_0}{dt} + A\frac{dp_0}{dt}, \tag{17.33}$$

so that in particular the sociothermodynamic equivalent to the Gibbs free energy, namely, $U^i + p_0 A - \tau_0 S^i$, tends to a minimum if equilibrium is approached for constant homogeneous boundary values τ_0 and p_0. This is the case that we wish to consider.

If τ and p are the only variables, the inequality is obviously trivially satisfied by homogeneous constant fields τ and p. The situation is as described in Subsection 17.3.3.2 within thermodynamics proper, and in the same spirit we now allow for *colonies* of sizes A^i, in which the birds assume different strategies. In analogy to the thermodynamic formula for the Gibbs free energy, we write

$$G = \sum_{i=A}^{B} Nx^i \left[u^i \left(\tau, p, z_H^i\right) + p \frac{1}{n^i \left(\tau, p, z_H^i\right)} - \tau s^i \left(\tau, p, z_H^i\right) \right], \qquad (17.34)$$

where x^i is the fraction of birds employing strategy i. This is the point where we give up the assumption that all birds exercise one strategy homogeneously. G is a function of the three variables x^A, say, z_H^A and z_H^B, but there is an obvious constraint between the variables, namely,

$$z_H = \sum_{i=A}^{B} x^i z_H^i, \qquad (17.35)$$

because the overall hawk fraction is fixed. In equilibrium the variables adjust themselves so that the Gibbs free energy assumes a minimum. We proceed to determine the values.

17.3.4 Strategy Diagram

17.3.4.1 Minimizing G
For abbreviation, we introduce g^i

$$g^i \left(\tau, p, z_H^i\right) \equiv u^i \left(\tau, p, z_H^i\right) + p \frac{1}{n^i(\tau, p, z_H^i)} - \tau s^i \left(\tau, p, z_H^i\right), \qquad (17.36)$$

so that

$$\frac{G}{N} = \sum_{i=A}^{B} x^i g^i \left(\tau, p, z_H^i\right). \qquad (17.37)$$

We minimize

$$\frac{G}{N} = \sum_{i=A}^{B} x^i g^i \left(\tau, p, z_H^i\right) - \lambda \sum_{i=A}^{B} x^i z_H^i,$$

thus taking care of constraint (17.35) by use of the Lagrange multiplier λ.

The three equilibrium conditions come out as

$$\lambda = \left.\frac{\partial g^A}{\partial z_H^A}\right|_E = \left.\frac{\partial g^B}{\partial z_H^B}\right|_E = \frac{g^A|_E - g^B|_E}{z_H^A|_E - z_H^B|_E}, \qquad (17.38)$$

which, together with the constraint, provide four equations for the four unknowns x^A, z_H^A, z_H^B, and λ.

An analytic solution of the equations is not possible and, in fact, is not needed for our purposes, because there is a very instructive graphical solution. Indeed, conditions $(17.38)_{1,2}$ imply that equilibrium requires z_H^A and z_H^B to be abscissae of *the* points where $g^A(\tau, p, z_H^A)$ and $g^B(\tau, p, z_H^B)$ have tangents of the same slope. And $(17.38)_3$ means that the tangents represent the common tangent of $g^A(\tau, p, z_H^A)$ and $g^B(\tau, p, z_H^B)$. Once the common tangent has been constructed, and thus knowing $z_H^A|_E$ and $z_H^B|_E$, we may use (17.35) to calculate

$$x|_E = \frac{z_H - z_H^B|_E}{z_H^A|_E - z_H^B|_E},\qquad(17.39)$$

and hence the equilibrium value of the Gibbs free energy, viz.,

$$\left.\frac{G}{N}\right|_E = g^B|_E + \frac{z_H - z_H^B|_E}{z_H^A|_E - z_H^B|_E}\, g^A|_E.\qquad(17.40)$$

Thus the equilibrium value of G lies on the common tangent above a given value of z_H because (17.40) represents the straight line of the common tangent.

17.3.4.2 The Gibbs Free Energies $g^i(\tau, p, z_H^i)$

To make the graphical method for the determination of $z_H^A|_E$ and $z_H^B|_E$ practical, we need to have explicit functions $g^i(\tau, p, z_H^i)$. These are obtained by insertion of $u^i(\tau, p, z_H^i)$ from (17.16) and (17.17), of $s^i(\tau, p, z_H^i)$ from (17.31), and of p from (17.12) and (17.13) into (17.36). In a fairly explicit form, the functions read as follows

$g^A(\tau, p, z_H^A)$
$$= (0.5\tau + 1)z_H^A + (0.5\tau + 0.2)(1 - z_H^A) - 1.2z_H^A(1 - z_H^A) + a_H^A z_H^A \tau + a_D^A(1 - z_H^A)\tau$$
$$-\tau\left\{\left[(0.5 + a_H^A)\ln\frac{\tau}{\tau_R} - a_H^A \ln\frac{p}{p_R} - a_H^A \ln\frac{a_H^A z_H^A}{a_H^A z_H^A + a_D^A(1 - z_H^A)}\right]z_H^A\right.$$
$$\left.+ \left[(0.5 + a_D^A)\ln\frac{\tau}{\tau_R} - a_D^A \ln\frac{p}{p_R} - a_D^A \ln\frac{a_D^A(1 - z_H^A)}{a_H^A z_H^A + a_D^A(1 - z_H^A)}\right](1 - z_H^A)\right\}$$
$$-\tau\left[s_H^A(\tau_R, p_R)z_H^A + s_D^A(\tau_R, p_R)(1 - z_H^A)\right],\qquad(17.41)$$

$g^B(\tau, p, z_H^B)$
$$= (0.2\tau + 1.3)z_H^B + (0.44\tau + 0.26)(1 - z_H^B) + 1.32(\tau - 1)z_H^B(1 - z_H^B)$$
$$+ a_H^B z_H^B \tau + a_D^B(1 - z_H^B)\tau$$
$$-\tau\left\{\left[(0.2 + a_H^B)\ln\frac{\tau}{\tau_R} - a_H^B \ln\frac{p}{p_R} - a_H^B \ln\frac{a_H^B z_H^B}{a_H^B z_H^B + a_D^B(1 - z_H^B)}\right]z_H^B\right.$$
$$\left.+ \left[(0.44 + a_D^B)\ln\frac{\tau}{\tau_R} - a_D^B \ln\frac{p}{p_R} - a_D^B \ln\frac{a_D^B(1 - z_H^B)}{a_H^B z_H^B + a_D^B(1 - z_H^B)}\right](1 - z_H^B)\right\}$$
$$- \tau\left[s_H^B(\tau_R, p_R)z_H^B + s_D^B(\tau_R, p_R)(1 - z_H^B)\right].\qquad(17.42)$$

Given the values a_α^i, see Table 17.1, we see that these functions are entirely explicit, *except for the last lines* with the reference entropies $s_\alpha^i(\tau_R, p_R)$ of the pure species.

17.3.4.3 Reference Values of Entropy

We proceed to determine these constants or to restrict their values to the extent needed. First, to simplify our task without essential loss of generality, we take the reference pressure p_R equal to the prevailing pressure p. Also, and again without loss of generality, we take the reference price τ_R equal to 1.

Now we assume that the price $\tau_\alpha(p_R)$ of a strategy change in the pure populations of either hawks or doves is known to us from observation. That price must satisfy the relations

$$g^A[\tau_D(p_R), p_R, 0] = g^B[\tau_D(p_R), p_R, 0]$$

for a pure-dove population, and

$$g^A[\tau_H(p_R), p_R, 1] = g^B[\tau_H(p_R), p_R, 1]$$

for a pure-hawk population.

These equations provide the differences of the entropy constants for the species

$$s_D^A(1, p_R) - s_D^B(1, p_R) = 0.06 + \left(a_D^A - a_D^B\right)[1 - \ln \tau_D(p_R)] - 0.06\frac{1}{\tau_D(p_R)},$$

$$s_H^A(1, p_R) - s_H^B(1, p_R) = 0.3 + \left(a_H^A - a_H^B\right)[1 - \ln \tau_H(p_R)] - 0.3\frac{1}{\tau_H(p_R)} \quad (17.43)$$

in terms of known values.

To simplify the equations further we take $\tau_D(p_R) = \tau_H(p_R) = 1$. This assumption, in contrast to the previous special choices, restricts the generality of our argument. Indeed it is quite unlikely that a pure population of doves changes strategy from A to B at the same price as a pure population of hawks does, just as it is unlikely that in a mixture of fluids the pure constituents have the same temperature of evaporation. But that is possible as a special case, and we consider that special case.

With a little juggling of the Eqs. (17.43) it is possible to arrive at the final results:

$$g^A(\tau, p_R, z_H) = (0.5\tau + 1)z_H + (0.5\tau + 0.2)(1 - z_H) - 1.2z_H(1 - z_H)$$

$$- \tau \left\{ \left[(0.5 + a_H^A) \ln \tau - a_H^A \ln \frac{a_H^A z_H}{a_H^A z + a_D^A(1 - z_H)} \right] z_H \right.$$

$$+ \left[(0.5 + a_D^A) \ln \tau - a_D^A \ln \frac{a_D^A(1 - z_H)}{a_H^A z_H + a_D^A(1 - z_H)} \right] (1 - z_H) \right\}$$

$$- \tau \left[s_H^B(1, p_R)z_H + s_D^B(1, p_R)(1 - z_H) - a_H^B z_H - a_D^B(1 - z_H) \right],$$

$$(17.44)$$

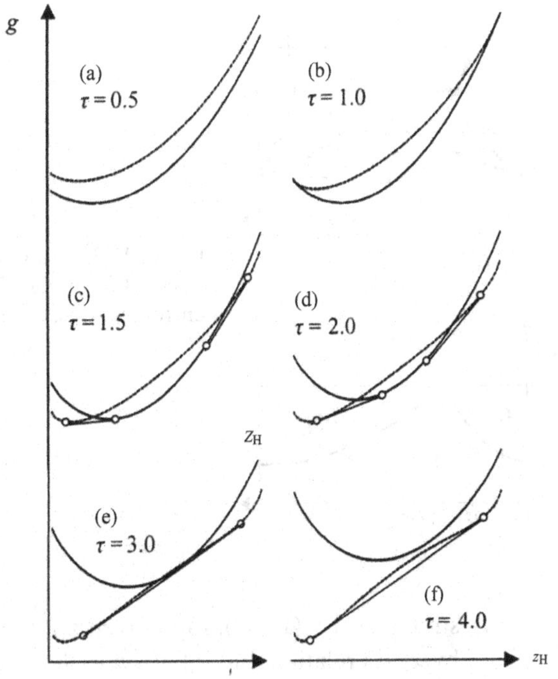

Figure 17.2. Gibbs free energies for a population employing strategy A (solid) and strategy B (dashed) as functions of z_H with $0 < z_H < 1$ at different temperatures. Convexification, straight lines.

$$g^B(\tau, p_R, z_H) = (0.2\tau + 1.3)z_H + (0.44\tau + 0.26)(1 - z_H) + 1.32(\tau - 1)z_H(1 - z_H)$$

$$- \tau \left\{ \left[(0.2 + a_H^B) \ln \tau - a_H^B \ln \frac{a_H^B z_H}{a_H^B z + a_D^B(1 - z_H)} \right] z_H \right.$$

$$+ \left[(0.44 + a_D^B) \ln \tau - a_D^B \ln \frac{a_D^B(1 - z_H)}{a_H^B z_H + a_D^B(1 - z_H)} \right] (1 - z_H) \right\}$$

$$- \tau \left[s_H^B(1, p_R)z_H + s_D^B(1, p_R)(1 - z_H) - a_H^B z_H - a_D^B(1 - z_H) \right].$$

$$(17.45)$$

The last line in both equations is the *same* unknown linear function of τ and z_H. Because we are interested in *differences* of the Gibbs free energies, we ignore that function in the further argument.

For different values of τ, the Gibbs free energies $g^A(\tau, p_R, z_H)$ and $g^B(\tau, p_R, z_H)$ are plotted in Fig. 17.2, which we proceed to discuss.

17.3.4.4 Convexification and Strategy Diagram

According to the analysis of Subsection 17.3.4.1, we have drawn common tangents to the graphs g^A and g^B, wherever possible, and have thus convexified the graph $\min[g^A, g^B]$. In Figs. 17.2(c) and 17.2(d), the common tangents are lower than either g^A or g^B and therefore the population occupies points on the tangents, i.e. it falls apart into separate colonies with hawk fractions corresponding to the points, where the tangents touch the graphs g^A and g^B. Accordingly the colonies employ different

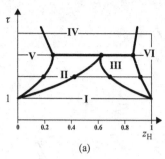

(a)

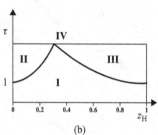

(b)

Figure 17.3. Strategy diagrams: (a) Full sociothermodynamics, constructed from Fig. 17.2; (b) sociothermodynamics with omission of entropic terms, constructed from Fig. 17.4.

strategies. By (17.39), for a given z_H, the fraction of birds with strategy A is given by $x|_E$, a relation that is known in the jargon of students of metallurgy as the *law of opposite lever arms*.

The tableau of graphs ranges from small values of τ, in which strategy A has a lower Gibbs free energy than strategy B (i.e., is more favorable), to high values of τ, in which g^B itself permits convexification. In the latter case the population segregates into hawk-rich and dove-rich colonies, both employing strategy B.

For a convenient summary of this tableau, it is appropriate to draw a (τ,z_H) strategy diagram. This is constructed by projection of the common tangents between the dots onto the corresponding horizontal line $\tau = $ const in that diagram and connection of the end points of those projections. Figure 17.3(b) shows the strategy diagram appropriate to the curves of Fig. 17.2.

It seems instructive to investigate the separate roles of shortfall and entropy in their effect on the strategy diagram. For that purpose I have drawn the graphs of Fig. 17.4, which ignore the entropic effects and represent shortfalls only. The corresponding strategy diagram is shown in Fig. 17.4(a); it is identical to the one in Fig. 17.1. A comparison of Figs. 17.4(a) and 17.4(b) shows by their differences the effect of entropy. The gross qualitative aspect is the same in both figures, with full integration at a low price and a the miscibility gap at a high price. But there is a qualitative difference in the fact that the narrow lateral regions of Fig. 17.4(b), marked V and VI, indicate that integration is possible at *all* prices, *provided that the hawk fraction is either very low or very high*. In other words: Populations with a large majority of one species are capable of integrating a minority of the other species. On close inspection of the preceding formulas we may see that that type of integration is entirely due to the entropy of mixing, which becomes a powerful force when threatened with complete segregation.

Before we leave the technical subject for a general final discussion, it seems appropriate to look back to the Prologue, i.e., Section 17.2, and the strategy diagram of Fig. 17.1 drawn there. The Prologue was intended to provide a first glance at the

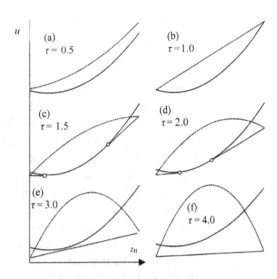

Figure 17.4. Shortfalls for a population employing strategy A (solid) or strategy B (dashed) as
functions of z_H. Convexification, straight lines.
(Same as Fig. 17.2, but without the entropic
contributions).

final aim of this chapter before any technicalities could obstruct the view onto the
essentials. I exploited the plausible ad hoc axiom that the birds adjust their strategy
to maximal gain. Therefore I came to two *concave* curves of gain that had to be
concavified. Now, in the context of a proper sociothermodynamics, within which
I introduced the shortfall and entropy – and first and second laws – we arrive at
two largely *convex* curves that have to be *convexified*. Surely the intelligent – and
interested – reader cannot be confused by the difference, I hope.

3.5 Conclusions

I finish this chapter with a summary of the phenomena implied by Figs. 17.2–17.4,
and compare them with the phenomena occurring in gas or vapor mixtures and
solutions:

- In good times, when resources are abundant and therefore prices are low, the
  species are fully mixed and the whole population employs strategy A, the strategy
  with a low level of competitiveness.
  - In thermodynamics of gas mixtures this corresponds to a homogeneous mix-
    ture of both constituents that, in that field, prevails at high temperature. Note
    that high temperature in thermodynamics corresponds to low price in socio-
    thermodynamics.
- When a shortage of resources occurs, prices increase. A population of *pure*
  hawks or *pure* doves reacts by changing its strategy of contest from A to B,
  where B is the more competitive strategy. This happens at a fixed price and –
  in the present model – at the same price for both pure populations, namely at
  $\tau = 1$.
  - In thermodynamics there is an analogous behavior in a pure vapor: When the
    temperature drops, the vapor becomes a liquid at a fixed temperature.
- The reaction of a mixed population of hawks and doves to shortages and high
  prices is different. The population segregates into two colonies with different
  hawks fractions and one colony, the one with the smaller minority of either

hawks or doves, employs strategy B, whereas the colony with the larger minority continues to employ strategy A, the low-price strategy. The reason for this segregation lies in the fact that in the segregated population the gain per bird and per competitive encounter is higher than it would be if the population remained to be homogeneously mixed and employed a single strategy, either A or B. In a manner of speaking we may thus say that segregation improves the provision of the population with resources. In that view segregation may be regarded as a strategy of surviving bad times.

- The thermodynamic equivalent is the decomposition of a mixture into a mixture of vapors and a liquid solution when the temperature is lowered. Both phases coexist and the mixture as a whole attains a lower Gibbs free energy than if it assumed a homogeneous phase, either gas, or rather vapor, or liquid.

- On a further increase of price, meaning still rarer resources, there comes the point where strategy A is eliminated. It is then better for the average gain of each bird if the population falls apart into two colonies, each with only a small minority, and both employing strategy B, the high-price strategy. Once again that situation is more profitable for the population than remaining homogeneous with strategy A or B.

- In thermodynamics a low-enough temperature eliminates the vapor phase altogether and forms two immiscible liquid phases, each with a different constituent in low concentration.

- A peculiarity occurs for either very low or very high hawk fractions, i.e., small minorities in the population. In that case a homogeneous population can exist at nearly all price levels, excepting only the immediate neighborhood of the price of strategy change in a pure population. Thus at nearly all price levels a population can accommodate small minorities. A comparison of Figs. 17.2 and 17.4 and of Figs. 17.3(a) and 17.3(b) shows that the root cause of this accommodation is an entropic effect, notably the entropy of mixing.

- In thermodynamics a similar phenomenon occurs for dilute solutions with a low concentration of either constituent. In that case a homogeneous solution can exist at nearly all temperatures. The reason in both cases – the mixed population or the dilute solution – is entropic: The entropy of mixing imposes an infinitely strong bias when total unmixing threatens the system.

All this is similar enough – in the sociology of our birds and in thermodynamics – that this conclusion cannot be avoided: *Segregation in a population and coexisting phases in a mixture are analogous phenomena, which serve the purpose of a better accommodation under adverse conditions.*

17.4. Extrapolation

17.4.1. Evolutionarily Stable Strategy

As I mentioned in the introduction, the rules of contest, particularly of strategy A, are adapted from a model of game theory for a biological evolution of mixed populations, see [1]. Let us consider this:

To present the argument as simply as possible, I restrict the attention to strategy A only and set $\tau = 1$. In that case Eqs. (17.1), (17.3), and (17.5) reduce to

$$e_{HH} = -0.5, \quad e_{HD} = 1, \quad e_{DH} = 0 \quad e_{DD} = 0.3,$$
$$e_H(z_H) = 1 - 1.5z_H, \quad e_D(z_H) = 0.3 - 0.3z_H,$$
$$e(z_H) = 0.3 + 0.4z_H - 1.2z_H^2. \tag{17.46}$$

We conclude that $e_H(z_H \approx 0) \approx 1$ and $e_D(z_H \approx 0) \approx 0.3$, so that few hawks in a predominantly dove population have an advantage over the doves. Likewise we have $e_H(z_H \approx 1) \approx -0.5$ and $e_D(z_H \approx 1) \approx 0$, so that few doves in a predominantly hawk population are better off than the hawks. Both populations are therefore evolutionarily unstable, because sociobiologists assume that the species with the higher gain will have more progeny; therefore the phase fractions will change away from $z_H \approx 0$ to higher values of z_H – or from $z_H \approx 1$ to smaller values of z_H in subsequent generations.

The strategy defined by (17.46) is evolutionarily stable for *the* value of z_H, where $e_H(z_H) = e_D(z_H)$, and that happens for

$$z_H^{ESS} = \frac{7}{12}. \tag{17.47}$$

The superscript ESS stands for evolutionarily stable strategy. In the ESS the gains of hawks and doves are given by

$$e_H\left(z_H^{ESS}\right) = e_D\left(z_H^{ESS}\right) = \frac{1}{8}. \tag{17.48}$$

We may also ask where the gain per encounter and per bird is maximal, i.e., for which z_H does the function $e(z_H)$ in (17.46) have a maximum? Obviously that happens for

$$z_H^{max} = \frac{1}{6}, \tag{17.49}$$

so that

$$e_H\left(z_H^{max}\right) = \frac{3}{4}, \quad e_D\left(z_H^{max}\right) = \frac{3}{20}, \quad e\left(z_H^{max}\right) = \frac{1}{3}. \tag{17.50}$$

Thus it happens that both hawks and doves and the "average bird" are better off for z_H^{max} than for z_H^{ESS}. But, of course the hawks are 6 times better off and the doves only 1.2 times. Therefore the well-fed hawks with their consequently larger progeny will undermine the population with the hawk fraction $z_H^{max} = \frac{1}{6}$, growing with respect to the doves, thereby hurting their own gain and that of the doves.

That type of behavior is selfish, but then the inventors of the game were also the promoters of the concept of the selfish gene,[12] which is the motor of evolution in their view.

Another way of discussing the numbers is by saying that cartels may be good for *all* dealers, but they are prone to collapse because of the selfish behavior of *some*.

---

[12] The concept was made popular by Richard Dawkins, the author of the book *The Selfish Gene* [4].

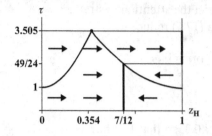

Figure 17.5. Directions of evolutionary change in the strategy diagram depend on hawk fractions and price levels.

17.4.2 Economic and Evolutionary Equilibria

In my case, presented in Sections 17.2 and 17.3, the hawk fraction could not change. One might say that the economic changes – the up and down of prices – were all happening in one generation, and that there was no time for producing the next generation with a possibly altered fraction of hawks.

Of course, the possible extrapolation is obvious now: Let us lift the constraint of constant hawk fraction, and let the z_H change depending on the relative gains of hawks and doves. Together with a former student I worked out a model of that type [6]. We assume that the population reacts quickly to price changes – too quick to suffer an appreciable change of z_H – and then it reacts slowly to different gains of hawks and doves by changing z_H *and*, possibly, the strategy. We formulate rate laws and determine the directions in the (τ, z_H) strategy diagram, in which a population moves toward evolutionarily stable states, see Fig. 17.5. It turns out that the target state depends on the economic conditions, namely the price of the resource: At low prices the population tends to approach the hawk fraction $z_H = \frac{7}{12}$, as in the simple case of sociobiology just described.[13] But at high prices, the doves lose out altogether, and the population eventually becomes an all-hawk population unless, of course, during the evolution the economic situation improves again. A theory like this may deserve more attention than it has received so far.

REFERENCES

[1] J. Maynard-Smith and G. R. Price, "The logic of animal conflict," *Nature (London)* **246**, 15–18 (1973).
[2] P. D. Straffin, "Game theory and strategy," *New Math. Library, Math Assn. Am.* **36** (1993).
[3] I. Müller, "Socio-thermodynamics – integration and segregation in a population," Cont. Mech. Thermodyn. **14**, 389–404 (2002).
[4] R. Dawkins, *The Selfish Gene* (Oxford University Press, New York, 1989).
[5] I. Müller and W. Weiss, *Energy and Entropy – A Universal Competition* (Springer-Verlag, Heidelberg, 2005).
[6] J. Kalisch and I. Müller, "Strategic and evolutionary equilibrium in a population of hawks and doves," Suppl. Rend. Circ. Mater. Palermo, Ser. II, Number **78**, 163–171 (2006).

---

[13] There was a slightly different assignment of elementary gain expectations in [6]. In the present case, $z_H = \frac{7}{12}$ holds only for $\tau = 1$ and z^{ESS} depends on τ weakly.

Exergy Use in Ecosystem Analysis:
Background and Challenges

Roberto Pastres and Brian D. Fath

18.1 Introduction to *Ecological Thermodynamics*

Ecology covers many spatial, temporal, and organizational scales, from individual to population, communities to ecosystems. At the broadest scale, *ecosystem ecology* is mostly concerned with how energy and matter enter, circulate, and are discharged through a biotic and abiotic open-system complex. This aggregate approach often overlooks and obscures the minute details of how individual organisms grow, reproduce, and respond within their environment. Therefore, there exists a divide between evolutionary ecology, dealing with the internal, reductionistic, genetic workings of individuals, and ecosystem ecology at the holistic, integrative scale. Attempts have been made to span this chasm by specifically relating the way large-scale ecosystem processes develop alongside evolutionary processes (e.g., see [1–4]).

As open and far-from-equilibrium systems, all ecosystems are dependent on a continual input of exergy, defined, as in Chap. 1, as energy available to perform work, in order to maintain their complex and adaptive functionality. At the same time, ecosystems need to discharge into the surrounding both matter and energy in degraded forms to keep the matter and energy balance. Even though the concept of exergy was introduced in 1953, this need was recognized much earlier and expressed in the framework of classical thermodynamics (CT) as the necessity of acquiring free energy. This is an evolutionary challenge according to Lotka [5], who stated that those organisms, and the systems in which they exist, which are better equipped to capture and utilize this energy, should have an evolutionary advantage over ones that are less capable of doing so. These concepts were recognized as early as 1905 by Ludwig von Boltzmann [6], who asserted that life is a struggle for free energy.

The original source for exergy on Earth is, of course, solar radiation. Although it is possible that the first life on Earth was heterotrophic, living off the store of available complex chemical compounds in the primordial soup, this was not a sustainable situation unless abiotic processes could produce the chemical molecules faster or equal to the rate of heterotrophic consumption. A more sustainable situation arose with the development of autotrophic processes. Therefore, the first design requirement in the autotrophic system's evolution was the ability to capture and utilize this exergy. These organisms not only employed the solar radiation for immediate tasks such as overcoming gradients or speeding chemical reactions, but also they developed mechanisms to store this energy in organic protoplasm for later use – photosystems arose. These energy battery packets could also perform much more specialized work such as active ion pumping, protein synthesis, reproduction, and, once incorporated into the primary producers, biomass became available for other heterotrophic or saprophagous organisms. The exergy that originated as solar radiation was repackaged into chemical energy in the many complex biochemical compounds in the living organisms. Estimates are that 1 out of every 500 photons hitting the Earth is captured by autotrophic photosystems [20].

Physicist Erwin Schrödinger, in his seminal essay "What is Life" [7], recapitulated and refined Boltzmann's ideas, stating, "the device by which an organism maintains itself stationary at a fairly high level of orderliness (= fairly low level of entropy) really consists of continually sucking orderliness from its environment." Thermodynamically, biological organisms are able to build complex structures within their boundaries by diverting low-entropy inputs (sunlight and food) and exporting high-entropy (heat and higher-entropy chemicals) outputs. The organisms that were more successful at capturing and processing this exergy would have an advantage over others that were less successful. In this manner life is able to create order from disorder, at least within the system boundaries.

Biologist Alfred Lotka [5] proposed the principle that, in the struggle for existence, the advantage goes to organisms whose energy-capturing devices are most efficient in directing available energy into channels favorable to the preservation of the species. This concept was furthered by Howard Odum [8, 9] and formed the basis for his *maximum power principle*, which states that, during self-organization, ecosystem designs develop and prevail that maximize power intake, energy transformation, and those uses that reinforce production and efficiency.

These theoretical contributions represent different attempts at finding a general principle, which ultimately governs the ecosystem development, in accordance with the laws of thermodynamics. If the existence of such a principle could be rigorously proved, then it would be possible to predict the future state of an ecosystem without following individual constituent dynamics. The evident analogy with the second law of thermodynamics, which allows one to make quite general predictions about the "final," or equilibrium, state of abiotic systems, has led to the postulation of the existence of the so-called fourth law of thermodynamics, which deals specifically with the issue of reconciling evolution and thermodynamics. Lokta's principle can be regarded as the first tentative formulation of this fourth law, followed by a series of others. For example, Jørgensen [10] has proposed a fourth law variation that deals specifically with the *eco-exergy* (EEx) storage of an ecosystem (for at least 15 variations of the *fourth law* we recommend you to see http://www.humanthermodynamics.com/4th-Law-Variations.html).

One criticism of an evolutionary fourth law is that directed change implies nature is teleological. But, even in the absence of a director, there is an a posteriori observation that ecosystems have tended to follow a dynamic trajectory over time for example, moving further from equilibrium. The question then is how to measure this deviation from equilibrium.

It is not our intention here to promote or validate these fourth law formulations, but, rather, to investigate the foundations on which the concept of EEx rests. This quantity, which was introduced in [11], has been recognized as one of the more useful "orientors," which could provide valuable insights into the evolution of ecosystems.

18.2 *Ecological Goal Functions*

The search for thermodynamically based organizing principles in ecology has produced a variety of *orientors* or *goal functions* [12–14]. For example,

> the central idea of the orientor approach ... refers to self-organizing processes, that are able to build up gradients and macroscopic structures from the microscopic disorder" of non-structured, homogeneous element distributions in open systems, without receiving directing regulations from the outside. In such dissipative structures the self-organizing process sequences in principle generate comparable series of constellations that can be observed by certain emergent or collective features. Thus, similar changes of certain attributes can be observed in different environments. Utilizing these attributes, the development of the systems seems to be oriented toward specific points or areas in the state space. The respective state variables which are used to elucidate these dynamics are termed *orientors*. Their technical counterparts in modeling are called *goal functions* [15, p. 15].

Many of these orientors follow from the seminal work of Odum [1], in which he hypothesized on the trends to be expected in ecosystem development. The exergy concept, in particular, has provided fertile ground for two of these approaches, namely maximizing *exergy* dissipation [16–19] and maximizing exergy storage [11, 20]. Other important ecological goal functions include entropy-production minimization [21, 22] or maximization [23], which has been refined by Aoki [24] as the min–max principle of entropy production; energy-cycling maximization [25]; power maximization [5, 9]; ascendency maximization [26, 27]; and residence-time maximization [28], to name a few.

In this chapter, those goal functions dealing with exergy are most relevant. In ecosystems, an increase in available exergy corresponds to an increase in the overall biomass and an increase in the species' internal organization – this informational component is currently measured as genetic complexity [29]. In the ecological context, exergy dissipation refers to the energy given off by breaking down the high-quality, low-entropy chemical compounds for both growth and maintenance of the system. As an ecosystem develops, it will import and capture more exergy, up to a point, and therefore be able to dissipate or degrade more exergy. In other words, exergy dissipation is dependent on the exergy capture or the ability of the ecosystem to divert a greater amount of low-entropy energy across its border.

This book deals with exergy applications in many fields, and the application as proposed by Jørgensen in ecosystem evolution has a unique development that diverges from CT, and herewith is called eco-exergy. In this chapter, we (1) present a transparent, even didactic, explanation of the development of EEx, (2) demonstrate some applications of EEx, and (3) offer some suggestions for improvement of this metric both on thermodynamic and ecological grounds.

18.3 From Classical Thermodynamics to Exergy

18.3.1 Thermodynamics Overview

The laws of thermodynamics (see Chap. 1) are foundational for understanding the physical world and are viewed as the most important constraints for biological development. Whereas the zeroeth and third laws are less important in providing useful direction for ecosystem development, the first law, which allows one to write energy-balance equations, and the second, requiring dissipative losses, have brought much to theoretical ecology and ecological modeling [19].

In the framework of generalized thermodynamics, the *second law* is stated in terms of the existence of a unique and stable equilibrium state for isolated systems, i.e., for systems that have given values of energy, amount of constituents, and volume [30]. This statement implies the existence of the entropy function as a "property" of any isolated system that cannot decrease as it approaches the state of equilibrium. During the path toward equilibrium, the way in which energy and matter are distributed within the system changes, being accompanied by an increase in entropy, which reaches its maximum value at equilibrium.

In this section, we provide an exergy definition that is based on the framework of CT: This setting may be more familiar to ecologists than to the more general approach to thermodynamics presented in Chap. 1. In CT, the application of the second law in the entropy formulation to isolated systems allows one to deduce that the values of all the intensive variables, T, P, and μ_i, in all parts of the system are the same at equilibrium. Furthermore, from the so-called "stability conditions," one can derive a set of inequalities that allow one to predict how a system responds to perturbations from the equilibrium state [31]. The extensive variables, U, V, and n_i, are the true independent variables in an isolated system: They are called the characteristic variables. The state function, which embodies all the thermodynamic "information" concerning a given system, is called the *"characteristic function"*: Entropy is the characteristic function for isolated systems.

There is an important point, which is sometimes overlooked in textbooks: Any *characteristic function* captures the complete picture of a given system at equilibrium only if it is expressed as a function of the set of its own characteristic variables. As a result, if the system is not isolated, then entropy is no longer the "characteristic function," and we cannot use it to make predictions about the equilibrium state. The preceding difficulty is overcome by the introduction of other characteristic functions, which can be derived from entropy or from internal energy, because it can be demonstrated that the second law implies that the internal energy of a system kept at constant entropy, volume, and mass reaches a minimum at equilibrium. The most frequently used characteristic functions, called thermodynamics potentials, have the same property of the internal energy, i.e., they decrease as long as spontaneous processes occur in the system and reach their minimum value at equilibrium. Thermodynamic potential can be expressed in finite form by applying the Euler theorem, thus obtaining the relationships listed in Table 18.1. It may be worth noting that the set of independent "characteristic variables" specify how the system exchanges energy and matter with a "reservoir," as defined in Chap. 1. Namely, in the case of Helmholtz free energy, the system is thought to be in contact with a heat reservoir at a temperature T; likewise, for enthalpy, system + volume work reservoir, and Gibbs free energy, system + heat and work reservoir.

The last equation in Table 18.1 [Eq. (18.4)] is important because it states that the *Gibbs free energy* is given by a "weighted sum" of the masses of the chemical components. The "weights" are chemical potentials that depend on the temperature and pressure of the reservoir.

Table 18.1. *Thermodynamic potentials with chemical potential terms*

Internal energy	$U(S, V, n_i) = TS - PV + \sum \mu_i n_i$	(18.1)
Helmholtz free energy	$A(T, V, n_i) = U - TS = -PV + \sum \mu_i n_i$	(18.2)
Enthalpy	$H(S, P, n_i) = U + PV = TS + \sum \mu_i n_i$	(18.3)
Gibbs free energy	$G(T, P, n_i) = U + PV - TS = H - TS = + \sum \mu_i n_i$	(18.4)

T = temperature, S = entropy, P = pressure, V = volume, μ_i = chemical potential of ith species, n_i = number of particles of ith species.

For *pure substances* at standard temperature ($T_0 = 298.15$ K) and pressure ($P_0 = 1$ bar), the chemical potential can be expressed as

$$\text{gas}: \mu_i = \mu_i^*(T_0, P_0) \equiv \mu_i^0 = \text{standard potential}.$$

Because $G(T, P, n_i) = \sum_i \mu_i n_i$, the standard potential of a pure substance represents the molar free energy of formation, i.e., the free energy that is "embodied" in one mole of a pure chemical: These values can be found in physical chemistry handbooks or on dedicated websites.

Chemical potentials depend, in general, on the temperature, pressure, and *composition* of each phase into which a system can be subdivided. The general expression of the chemical potential for each component of a gas mixture is

$$\mu_i = \mu_i^*(T) + RT \ln(f_i), \tag{18.5}$$

in which f_i is the fugacity.

The general expression of the chemical potential for the components of a liquid solution is

$$\mu_i = \mu_i^*(T) + RT \ln(a_i), \tag{18.6}$$

in which a_i is the activity.

Fugacity and activity *quantify the influence of the interactions among the components on the chemical reactivity of a given component*. When these interactions are strong, these quantities depend strongly on the actual composition of the system and must be empirically measured or deduced from statistical mechanics models.

Simplified expressions of activity can be theoretically deduced *for systems in which interactions are very weak*.

1. Mixtures of ideal gases:

$$\mu_i = \mu_i^*(T_0, P_0) + RT \ln(p_i/p_0), \tag{18.7}$$

where p_i is the partial pressure and P_0 is the standard-state pressure.

2. "Ideal" solutions:

$$\mu_i = \mu_i^*(T) + RT \ln(\chi_i) \text{ for the solvent}, \tag{18.8}$$

$$\mu_i = \mu_i^+(T) + RT \ln(\chi_i) \text{ or}, \tag{18.9a}$$

$$\mu_i = \mu_i^+(T) + RT \ln\left(c_i/c_i^0\right) \tag{18.9b}$$

for either the solvent and the solutes, where $\chi_i = n_i / \sum n_i$ is the molar fraction and c_i^0 is the reference concentration. The meaning of $\mu_i^*(T)$ in Eq. (18.8) is clear: It represents the chemical potential of the pure solvent, because the second term vanishes if $\chi_i = 1$. Regarding Eqs. (18.9), it is important to bear in mind that the standard potential $\mu_i^+(T)$ is constant at constant temperature but is numerically different in (18.9a) and (18.9b). The concentration c_i^0 is a reference concentration *that must always be specified, because the numerical values of $\mu_i^+(T)$ and $\mu_i^*(T)$ depend on the reference concentration and also the activity is defined as* $a_i \to c_i / c_i^0$ for $c_i \to 0$.

18.3.2 Definition of Exergy

The concept of *exergy* was introduced in 1953 by Zoran Rant [32] in order to quantify the maximum "useful" energy that could be extracted from a system in contact with a reservoir at constant temperature, pressure, and chemical potentials – *these intensive variables define the* reference state – of all the chemicals that could be exchanged. Exergy can also be interpreted as the minimum work that should be supplied to a system at equilibrium with the reference reservoir in order to change its thermodynamic "state" to any desired one.

It is assumed that the system and the reservoir are always internally at equilibrium, *but the system may not be in equilibrium with the reservoir*. In this case, it is characterized by the set of state variables: U, V, n_i, T, P, and μ_i. On the other hand, the intensive properties of the reservoir, T_R, P_R, and μ_{iR}, by definition, do not change when the system and the reservoir are brought into contact. Therefore, exergy must depend on the set of state variables that characterize the system and on the intensive variables of the reservoir. When a system that can exchange heat, work, and constituents is at equilibrium with the reservoir, its intensive variables are equal to T_R, P_R, and μ_{iR}: These values therefore determine the volume V_0 and the amounts of energy U_0 and mass n_{i0} when the composite system is at equilibrium. This state is also referred to as "*dead state*," because no more useful work can be extracted from such a system.

As shown by Evans [33, 34], exergy E can be expressed in terms of the system and reservoir state variables as follows:

$$E = U - \left(T_R S - P_R V + \sum_{i=1}^{m} \mu_{iR} n_i \right). \tag{18.10}$$

Equation (18.10) is consistent with the more general expression of the availability function given in Chap. 1 and can be rewritten in two ways, relating exergy:

1. to the differences between the system and reservoir intensive variables, and
2. to the differences between the actual values of the extensive variables and the equilibrium ones.

If in Eq. (18.10) we substitute U from Table 18.1, we obtain

$$E = \left(TS - PV + \sum_{i=1}^{m} \mu_i n_i \right) - \left(T_R S - P_R V + \sum_{i=1}^{m} \mu_{iR} n_i \right),$$

$$E = S(T - T_R +) - V(P - P_R) + \sum_{i=1}^{m} n_i (\mu_i - \mu_{iR}). \tag{18.11a}$$

If, starting again from Eq. (18.10), we add and subtract U_0 and then substitute U_0 by using Eq. (18.1) in Table 18.1, we get

$$E = U - U_0 + U_0 - \left(T_R S - P_R V + \sum_{i=1}^{m} \mu_{iR} n_i \right),$$

$$E = U - U_0 + \left(T_R S_0 - P_R V_0 + \sum_{i=1}^{m} \mu_{iR} n_i \right) - \left(T_R S - P_R V + \sum_{i=1}^{m} \mu_{iR} n_i \right),$$

$$E = U - U_0 - T_R (S - S_0) + P_R (V - V_0) - \sum_{i=1}^{m} \mu_{iR} (n_i - n_{i0}). \qquad (18.11b)$$

Both expressions (18.11a) and (18.11b) vanish when the system reaches equilibrium with the reservoir or reference environment. The more general expression of a system exergy derived in Chap. 1 takes into account other modes of interaction between the system and the reservoir. However, it is also important to point out that, based on Eq. (18.10), thermodynamic potentials can be thought of as exergy functions for particular system–reservoir composite systems. Therefore their physical meaning in terms of available energy is straightforward:

A (Helmholtz): represents the maximum amount of reversible work that can be obtained from the system + heat reservoir composite at a given temperature

H (Enthalpy): represents the heat absorbed or released by the system + volume work reservoir, at a given pressure

G (Gibbs): represents the maximum amount of reversible work that can be obtained from the system + heat and work reservoir composite system at given temperature and pressure.

18.4 Exergy in Ecology

The idea of using exergy as a way of comparing ecosystems and as an ecosystem orientor was introduced in 1979 by Jørgensen and Mejer [11]. To extend the concept of exergy to living matter, it was assumed that a given organism can be treated as a chemical, and therefore in principle the chemical potential is an intensive property of any biological species. After introducing this postulate, Jørgensen and Mejer considered the hypothetical equilibrium between a present state ecosystem and a system *at the same temperature and pressure* that contains the same amount of each element but in which all the chemicals are found in their most stable form: thermodynamic equilibrium. This second, hypothetical, system is labeled the "inorganic soup" and is thought to represent the pre-biotic system before life appeared on Earth. Alternatively, it can be regarded as the state the ecosystem would reach in the long term if the fluxes of matter and low-entropy energy, mainly solar radiation, would cease. Therefore the "composite system" considered here is made up of the ecosystem and of a reservoir that can exchange only heat and mechanical work: This means that the ecosystem is implicitly viewed as a closed system. In this setting, the equilibrium value of the chemical potentials for the composite system depends on the temperature and pressure of the reservoir and on the amount of each chemical and biological species present in the ecosystem. The spontaneous chemical reactions, which would occur in the ecosystem if the energy–matter fluxes that keep it far from equilibrium ceased, would lead to a change in their chemical form, but could not change the total mass of each chemical element.

In accordance with Eq. (1.11b) and Chap. 1, in this case, exergy is the difference in Gibbs free energy between the two states, as Jørgensen and Mejer also stated in their paper. However, *the authors skipped the problem of defining an equivalent of the "Gibbs energy of formation"* or standard potential of a certain biological species. In applying the exergy concept to a lake, they assumed that the *"chemical potentials"* of phytoplankton, μ_a, and that of inorganic phosphorus, μ_s, were given by

$$\mu_a = RT\ln\left(P_a/P_a^{eq}\right), \quad \mu_s = RT\ln\left(P_s/P_s^{eq}\right)$$
$$\mu_a = RT\ln\left(c_a/c_a^{eq}\right), \quad \mu_a = RT\ln\left(c_s/c_s^{eq}\right), \tag{18.12}$$

in which c_a and c_a^{eq} are the actual and hypothetical equilibrium concentrations of phosphorus in phytoplankton and c_s and c_s^{eq} are the actual and hypothetical equilibrium concentrations of inorganic phosphorus. A comparison of Eqs. (18.7) and (18.12) suggests that, from the thermodynamic point of view, Eqs. (18.12) are not entirely accurate, even if one accepts the approximation $a_i \approx c_i/c_i^0$, which means that the ecosystem is thought of as a mixture of noninteracting particles. In fact [see Eq. (18.9b)], the chemical potential for the inorganic component should be written as

$$\mu_s = \mu_s^+(T) + RT\ln\left(c_s/c_s^0\right),$$

in which c_s^0 is a *reference*, and not an equilibrium, concentration. With Eqs. (18.12) used as a starting point, the following expression of exergy per unit volume is found:

$$E/V = RT\sum_{i=1}^{m} c_i \ln(c_i/c_i^0), \tag{18.13}$$

in which V represents the volume of the system, c_i is the concentration of a given organism or chemical species in the actual ecosystem and c_i^0 is the equilibrium concentration.

The exergy concept was then revisited: the derivation of a second expression, currently called "*eco-exergy*," EEx in this chapter, is summarized in [36]. In this paper, the authors take as a starting point Eq. (18.11a), *which expresses the exergy of an open system* that has already reached thermal and mechanical equilibrium with a reservoir at $T = T_R$ and $P = P_R$:

$$EEx = \sum_{i=1}^{m} n_i \left(\mu_i - \mu_{iR}\right). \tag{18.14}$$

In the following derivation, however, the authors assumed that μ_{iR} was not the chemical potential of the species in an actual reservoir in contact with the system, but rather the virtual chemical potential of the n_i moles in the "inorganic soup," which is obtained after the complete degradation of the organic matter into its inorganic constituents. In fact, this means that Eq. (18.14) is used again for analyzing *a closed system*. For this reason, in the following discussion we denote as c_{i0} the concentration of the chemical in the end state of the composite system considered in [35] and with μ_{i0} as the corresponding chemical potential.

The exergy expression in Eq. (18.13) is derived from (18.14) by substituting the chemical potentials $\mu_{iR} \equiv \mu_{i0}$ with Eq. (18.8) and dividing by the volume V.

As pointed out in [35], (hypothetical) equilibrium concentrations of currently living organisms in the inorganic soup are very small.

The estimation of the hypothetical equilibrium concentration, c_{i0}, is based on the idea that the formation of an organism can be regarded as a "consecutive" chemical reaction:

inorganic soup $\rightarrow$ dead organic matter (detritus) $\rightarrow$ living organism.

Subsequently, the probabilities P_i of finding a given organism in the soup were introduced: These probabilities can be related with the hypothetical molar fraction of the organism in the soup:

$$P_i = \chi_i = \frac{c_{i0}}{\sum_{i=0}^{m} c_{i0}}. \tag{18.15}$$

Because in the inorganic soup, c_{00} – the concentration of the inorganic component – is much larger than all living ones, the probability P_i could be written as

$$P_i = \frac{c_{i0}}{c_{00}} \rightarrow c_{i_0} = P_i c_{00}, \quad i = 1, N. \tag{18.16}$$

Equation (18.16) can be used for substituting c_{i0} in Eq. (18.13), thus obtaining

$$EEx \approx RT \sum_{i=0}^{m} c_i \ln (c_i / P_i c_{00}). \tag{18.17}$$

Equation (18.17) is simplified by use of two main arguments: (1) the probability P_0 is close to one, and therefore its logarithm is close to 0, and (2) the probabilities P_i, $i \geq 1$, are all extremely small, and thus the absolute values of their logarithms are much higher than the logarithms of the ratio c_{00}/c_i. In fact, the probability P_i is taken as proportional to the product of the probability P_1 of finding detritus and the probability $P_{i,a}$ of obtaining by chance the right sequence of amino acids in the proteins of a given organism from the detritus:

$$P_i = P_1 P_{i,a}. \tag{18.18}$$

The probability $P_{i,a}$ of obtaining by chance the right genetic sequence in the proteins of each organism from detritus was estimated as a function of the permutation of the four nucleotides:

$$P_{1,a} = 4^{-a_i(1-g_i)},$$

where a_i is the number of structured nucleotides and g_i is the percentage of repeating genes. For example, it is estimated that bacteria DNA are composed of 6.7 million genes, of which only 16% are repeating ones [37]. In fact, therefore

$$RT \sum_{i=0}^{m} c_i \ln (c_i / P_i c_{00}) \approx RT \sum_{i=1}^{m} c_i \ln (c_i / P_i c_{00}),$$

$$RT \sum_{i=1}^{m} c_i \ln(c_i / P_i c_{00}) = RT \sum_{i=1}^{m} c_i [-\ln P_i - \ln(c_{00}/c_i)] \approx RT \sum_{i=1}^{m} c_i (-\ln P_i).$$

Based on the preceding assumptions, the following EEx is proposed:

$$EEx \approx RT \sum_{i=0}^{N} c_i \ln(1/P_i). \tag{18.19}$$

As explicitly stated in [38], a comparison of Eqs. (18.19) and (18.13) shows that $\ln P_i$ represents an estimate of $\ln c_{i0}$, because $c_{i0} \ll c_i$ and therefore $|\ln c_{i0}| \gg |\ln c_i|$:

$$EEx/V \approx RT \sum_{i=0}^{N} c_i \ln \left(c_i / c_{i0} \right) = RT \sum_{i=0}^{N} c_i \left(\ln c_i - \ln c_{i0} \right) \approx RT \sum_{i=0}^{N} c_i \left(- \ln c_{i0} \right).$$

Because the probability depends on the type of organism and *not on its concentration in any actual ecosystem*, $\ln(1/P_i)$ is a constant weight that multiplies the concentration of each species. Equation (18.19) can be rearranged in order to scale all the weights in respect to detritus:

$$EEx = RT \ln(1/P_1)c_1 + RT \sum_{i=2}^{N} c_i \ln(1/P_1 P_{i,a}). \tag{18.20}$$

The probability P_1 is related to the difference in Gibbs free energy per unit mass by means of [35]

$$P_1 = e^{\frac{[-(\mu_1 - \mu_{10})]}{RT}}, \tag{18.21}$$

in which μ_1 represent the chemical potential of the detritus in present state ecosystems, which, of course, varies across ecosystems, and μ_{10} is its chemical potential at the hypothetical equilibrium concentration of detritus in the inorganic soup. However, the difference between μ_1 and μ_{10} *is taken as a constant*, 18 kJ/g, which, in fact, is the free energy released per gram of organic matter when the latter is turned into inorganic components [35]. Introducing Eq. (18.21) into Eq. (18.20) yields:

$$EEx/RT = \frac{(\mu_1 - \mu_{10})}{RT} c_1 + \sum_{i=2}^{N} c_i \ln(1/P_1 P_{i,a}). \tag{18.22}$$

The weights are normalized on detritus by assuming a molecular weight of 100,000, the normalizing factor being 18,000 J g$^{-1} \times 10^5$ (g mol^{-1})/(8.4 J mol^{-1} K$^{-1} \times$ 300 K) = 7.14 $\times 10^5$;

$$EEx/RT = c_1 + \sum_{i=2}^{N} \beta_i c_i, \tag{18.23}$$

in which $\beta_i = - \ln(P_1 P_{i,a})/7.14 \times 10^5$ and exergy is expressed in grams per liter.

According to this treatment, the β values become species-specific weighting factors: see [37] for a recent review. Examples of these coefficients are given in Table 18.2.

Equation (18.23) has been used in a number of papers as an ecosystem orientor because, in accordance with Jørgensen and Nielsen [36], ecosystems tend to maximize exergy in the course of their evolution.

Table 18.2. *List of typical β values used in EEx calculations (from [37])*

Organism	
Detritus	1
Bacteria	8.5
Algae	20
Plant	437
Insect	167
Fish	499
Reptile	833
Bird	980
Mammalia	2127
Human	2173

18.4.1 Eco-Exergy Case Studies

As previously stated, the primary initiative for developing EEx was to provide a thermodynamic ecosystem metric that tracked the distance of the ecosystem from equilibrium. If one assumes that ecosystems through evolution and succession are moving farther away from thermodynamic equilibrium, as was postulated in various formations of the fourth law, then EEx could be used to measure this displacement. The one Jørgensen proposed for ecological applications states, "an ecosystem receiving solar radiation will attempt to maximize [eco-]exergy storage such that if more than one possibility is offered, then in the long-run the one which moves the system furthest from thermodynamic equilibrium will be selected" [11]. This orientation of ecosystem growth and development toward greater biomass and genetic complexity can be measured empirically, modeled, and tested [38]. There have been many applications of this metric over the past decade, and through continued use and testing of this tentative law further improvements continue to be made. Here we briefly discuss three recent examples.

18.4.1.1 Eco-Exergy Change in Ecological Networks

First, in an effort to understand the relation between EEx and system network changes, Jørgensen and Fath [39] conducted an examination of various network configurations. In other words, a fixed network model was used as baseline situation (Fig. 18.1) and a series of systematic changes were made to the network structure (i.e., trophic length, connectance pattern) and the new EEx values calculated for these conditions. The results show that all changes that increase or retain the energy in the system longer increase overall EEx, although certain changes had a greater overall effect on the EEx level (Table 18.3). In fact, other than doubling the input, which corresponds to increased importation and better utilization by the primary producers, the change with the greatest increase in EEx was adding a bacterial food chain. Although bacteria has a low overall EEx weighting factor, the presence of this new energy-feeding pathway retains the energy in the system longer, making it available for other species as well. The only change that lowered the EEx value was the elimination of the top carnivore species, which is explainable by the fact that those compartments typically have the highest overall EEx weighting factors.

Table 18.3. *Results of network changes on EEx*

Network configuration	EEx
Baseline model	54,463
Doubling of input exergy to 2	105,636
Removing top carnivores	52,818
Link plant A to herbivores B (specific rate 0.25)	57,424
Link herbivores A to carnivores B	55,567
Link carnivores B to top carnivores A	54,652
Transfer of exergy from A to B	54,463
Adding bacterial food chain	61,938
Detritus to carnivores B, not to herbivores B	54,762

18.4.1.2 Eco-Exergy Change During Ecological Restoration

A second example is taken from Patricio et al. [40], in which EEx – as well as several other holistic indicators – was measured in an empirical setting in which portions of an aquatic substrate were scrapped clean and others left for control. The ecological

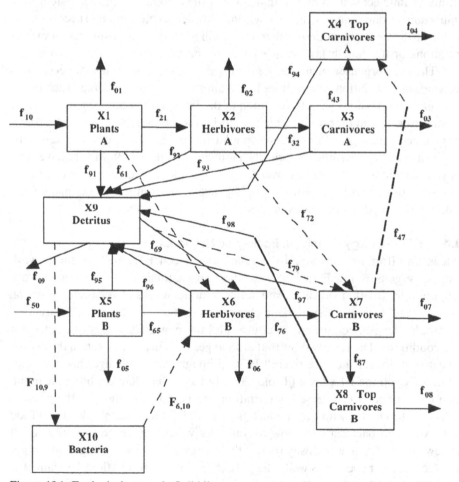

Figure 18.1. Ecological network. Solid lines represent baseline model, and dashed lines represent modifications to the model (reprinted with permission from Jørgensen and Fath [39]).

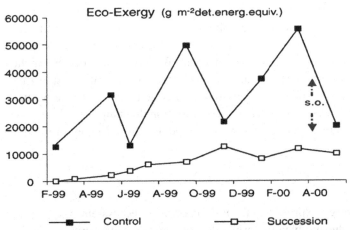

Figure 18.2. Recovery of EEx (open boxes) after disturbance, compared with undisturbed control (reprinted with permission from Patricio et al. [40]).

communities were collected and identified monthly for a period of 18 months. The results showed a steady increase in EEx during the recovery phase of the ecosystem (see Fig. 18.2).

18.4.1.3 Eco-Exergy in Structurally Dynamics Models: Darwin's Finches

In the last example, we demonstrated how EEx was used as an ecological orientor in structurally dynamic models [41] for parameter estimation [42]. Physical models traditionally have constant parameters, and early ecological models assumed the same approach, but there is good reason to believe that parameters in ecological systems vary over time. The question is one of determining when and how the parameters should be altered. In this manner, adopting an ecological orientor or specific goal function in the model assumes that the ecosystem development follows a dynamic trajectory determined by the orientor, be it maximizing throughflow, EEx, dissipation, or minimizing specific entropy, for example (see [13]). In keeping with the theme of this chapter, we consider an application in which EEx is used as the goal function in a structurally dynamic model.

When studying dynamic trajectories, evolutionary processes quickly come to mind, and one of the best-studied empirical examples of evolution is the work by Peter Grant and colleagues [43–47], working with Darwin's finches on the Galapagos Islands. In particular, one dataset dealing with a drought period showed how the average bird's beak depth changed rapidly in response to the available food source – drier conditions resulted in larger, harder seeds. Jørgensen and Fath [48] developed a simple model with three state variables: adult finches, juvenile finches, and seeds to simulate these changes.

The bird's diet utilized an empirical relation between beak size and seed hardness, such that birds with a "better" beak depth for the available resource were able to ingest more food. The model was initialized with data for a normal weather period and was run during the dry spell. During the simulation, the model employed a standard structural dynamic modeling algorithm, checking the sensitivity of a key variable – here determined to be beak depth – against the chosen goal function –

maximization of EEx. In other words, the model simulation was paused at set intervals in order to see if altering the beak depth by $\pm 10\%$ had an impact on the goal function. If the modification yielded greater EEx, then the change was accepted and the model continued using the new parameter value. With only two species (finches and seeds), there were only two beta values to consider in this example. Results show that the model accurately tracked the beak-depth enlargement during the drought period and did so with a conceptually and computationally simple model (to get similar results using a nonstructurally dynamics model required many more state variables).

18.5 Discussion

As was shown, the exergy expression currently used as an ecosystem orientor was derived starting from thermodynamic principles. However, several critical approximations and assumptions were introduced that may need further development or consideration. The main discussion points are as follows:

1. The "standard" chemical potentials and the standard chemical exergies for organisms have not been properly defined. This point is crucial for applying an exergy balance to any technical process occurring in, respectively, closed or *open systems*, which can exchange "biological consituents" with the surroundings. As stressed in Szargut et al. [49], the definition of standard chemical exergies requires the choice of a set of "common components" in the environment and the estimation of their concentrations and molar fractions.

2. It is not clear whether the EEx expressions, which were, in fact, derived for both a composite system made up of an ecosystem and a reservoir at the same temperature and pressure, should then be applied to analyze the evolution ecosystems, which are open systems, as has been done in a number of papers.

3. As an open system, an ecosystem readily exchanges energy – matter with neighboring ecosystems. This results in a transfer of both biomass and information such that response to perturbations is influenced mostly by these "reservoir" ecosystems.

4. The EEx weighting factor operates at an evolutionary time scale, yet we observe rapid changes in ecosystems, which are driven by changes in the energy – matter fluxes at their boundaries: Is the current expression of EEx a good orientor also at this short time scale?

18.5.1 Derivation of Standard "Chemical Potentials" for Organisms

As far as point 1) is concerned, one can notice that the problem of the proper definition of standard potential was skipped, both in [50] and in the later EEx derivation [35, 51]. In the last two papers, the estimation of P_1 represents, in our opinion, a critical step, because the preceding probability was related to the *constant* difference between the Gibbs free energy of a unit mass of detritus and that of the mixture of inorganic components, which are the products of detritus degradation.

In fact, in terms of CT, this process can be regarded as a nonequilibrium chemical reaction: In this case, the amount of free energy released per unit mass is given as the difference between the chemical potentials of the detritus and that of the pool of inorganic components resulting from detritus decomposition *in certain standard conditions:*

$$\text{organic detritus} \rightarrow \text{simple inorganic molecules.} \tag{18.24}$$

From the preceding assumption, probability P_1 should be expressed in terms of the "equilibrium constant" of reaction (18.24):

$$\ln K = -\frac{\Delta_r G^0}{RT} = \frac{-\left(\mu_0^0 - \mu_1^0\right)}{RT} = \ln\left(\frac{1}{P_1}\right), \tag{18.25}$$

in which μ_1^0 and μ_0^0 represent, respectively, the standard potentials of detritus and of the pool of inorganic compounds. The difference $(\mu_0^0 - \mu_1^0)$ is negative, and its absolute value is high because the Gibbs free energy of the inorganic component mixture is much lower than that for detritus.

This interpretation is consistent with the EEx derivation given in [49]. In this paper, Jørgensen and Fath proposed also a stoichiometric composition for the detritus and therefore explicitly introduced more than one inorganic species. On this basis, one is led to conclude that Eq. (18.21) is not correctly stated or, at least, is not consistent with the estimation of P_1 presented in [48] and, instead, Eq. (18.25) should be used.

One may overcome some of these difficulties by explicitly stating upfront which kind of system one wishes to study and then using the correct exergy function. In the following subsections, an alternative derivation of EEx for the composite system considered in [35], more consistent with thermodynamic principles, is given and an extension to open systems is outlined.

18.5.2 Eco-Exergy of a Composite System at Fixed Temperature and Pressure

It is interesting to note that Eqs. (18.13) and (18.23) were both derived for a composite system that is made up of an actual ecosystem and a reservoir at the same temperature and pressure. In fact, the inorganic soup represents the dead state of such a composite system that would be ultimately reached if all the organisms were degraded to the most stable inorganic compounds: This would be the fate of any ecosystem, unless some external exergy sources were available. Because for such a composite system exergy becomes equal to Gibbs free energy, one possibility is to use the difference in Gibbs free energy between the current and the end state as an estimate of EEx.

Let us assume that an aquatic ecosystem can be considered as an aqueous solution and that we are able to specify the chemical potential of any organism of interest. In this case, if the system is closed such that total mass is constant, then we can compute the difference between the Gibbs free energy of this system and the one that would be obtained if the fluxes of free energy that keeps it far from equilibrium would cease. The latter will contain only simple inorganic molecules.

Let us define

$p + m =$ number of components, where p are inorganic chemicals, and m are detritus plus organisms;

$j = 1, \ldots, p \rightarrow$ inorganic species, (water, dissolved nitrate, dissolved phosphorus, etc.);

$i = 1, \ldots, m \rightarrow$ detritus and organisms;

μ_i = chemical potential of the component i in the actual ecosystem;

$\mu_{i,\text{eq}}$ = chemical potential of the component i at equilibrium;

n_i = mass (mol number or grams) of the i component in the actual ecosystem;

$n_{i,\text{eq}}$ = mass (mol number or grams) of the i component at equilibrium;

$N = \sum\limits_{j=1}^{p} n_j + \sum\limits_{i=1}^{m} n_i$ = total mass;

G = eco-Gibbs free energy of the actual ecosystem;

G_{eq} = eco-Gibbs free energy of the system in the hypothetical equilibrium state;

V = volume of the system, assumed to be constant.

Equation (18.4) can be straightforwardly applied for computing the difference ΔG between the two states of the system. This difference is, by definition [52], equal to the system exergy:

$$Ex = \Delta G = G - G_{\text{eq}} = \sum_{i=1}^{p+m} \mu_i n_i - \sum_{i=1}^{p+m} \mu_{i,\text{eq}} n_{i,\text{eq}}. \qquad (18.26)$$

Because the decomposition of any organism into inorganic components is a spontaneous process, we can think of it as a nonequilibrium, or quantitative, chemical reaction. In this case, $n_{i,\text{eq}}$, $i = 1, m$ are *extremely small and can be neglected in* G_{eq}. On the other hand, at equilibrium, all elements will be present as inorganic species, such that $\sum_{i=1,p} n_{i,\text{eq}}$, $= N$. Therefore, we obtain:

$$\Delta G = G - G_{\text{eq}} = \sum_{j=1}^{p} \mu_j n_j + \sum_{i=1}^{m} \mu_i n_i - \sum_{j=1}^{p} \mu_{j,\text{eq}} n_{j,\text{eq}}. \qquad (18.27)$$

Let us now introduce the following quantity:

$$\bar{\mu} = \frac{\sum\limits_{j=1}^{p} \mu_j n_j}{\sum\limits_{j=1}^{p} n_j},$$

which represents the average chemical potential of the inorganic species. Equation (18.27) can then be rewritten as

$$\Delta G = \bar{\mu} \sum_{j=1}^{p} n_j + \sum_{i=1}^{m} \mu_i n_i - N \bar{\mu}_{\text{eq}}$$

because

$$\sum \mu_{i,\text{eq}} n_i = \bar{\mu}_{\text{eq}} \sum i n_i.$$

We can use the fact that the total mass is constant for rewriting this equation in terms of the differences between the chemical potentials of any biological component and that of the average chemical potential of the inorganic components, because

$$\sum_{i=1}^{p} n_i = N - \sum_{i=1}^{m} n_i,$$

$$\Delta G = \bar{\mu}\left(N - \sum_{i=1}^{m} n_i\right) + \sum_{i=1}^{m} \mu_i n_i - N\bar{\mu}_{eq} = \sum_{i=1}^{m} (\mu_i - \bar{\mu})\, n_i + (\bar{\mu} - \bar{\mu}_{eq})N.$$

$$(18.28)$$

In Eqs. (18.28), the first term represents the increase in Gibbs free energy that is due to the formation of all the species, i.e., the "products," from the inorganic component, i.e., the "reactants." This term embodies all the work done by evolution. The second term represents the decrease in Gibbs free energy that is due to the decrease in the concentration of the inorganic component, because the total mass is constant. It can be noted that these two terms, in principle, are computable, because we assumed as known the total mass of the system, its volume, and its actual composition. In particular, $\bar{\mu}$ and $\bar{\mu}_{eq}$ can be evaluated within the boundary of CT.

In the exergy expression currently employed, the contribution of inorganic components is not usually considered. To compare Eq. (18.28) and Eq. (18.23), we can then focus on the first term and expand the Gibbs free-energy difference per unity volume by introducing Eq. (18.5):

$$\Delta G/V \approx \sum_{i=1}^{m} \{\mu_i^* + RT \ln(a_i) - [\bar{\mu}_0^* + RT \ln(a_0)]\} n_i /V$$

$$\approx \sum_{i=1}^{m} (\mu_i^* - \bar{\mu}_0^*)\, c_i + RT \sum_{i=1}^{m} c_i \ln\left(\frac{a_i}{a_0}\right),$$

$$(18.29)$$

where $a_0 = \prod_{i=1}^{P} a_i$.

Equation (18.29) has an interesting structure: The first term is a weighted sum of the concentrations, which represents the work done by evolution for producing a "standard" ecosystem. As we shall see soon, this term corresponds to Eq. (18.23). The second term, instead, accounts for the interaction among the species in an actual ecosystem: In fact, it has the same structure as Eq. (18.13), if we assume $a_i \approx c_i$.

In aqueous solutions, standard chemical potentials are usually referred to a solution of a given concentration, for example 1 mol of solutes per liter. To be consistent with the previous exergy formulation, let us assume 1 g/L to be the standard concentration. Subsequently we can think of the formation of a given component as a consecutive reaction:

inorganic component → organic detritus → biotic component.

The inverse reactions are spontaneous and quantitative: Therefore their equilibrium constants are related with the differences between the standard chemical potential of the products and the reactants:

$$\Delta G/V \approx (\mu_1^* - \bar{\mu}_0^*)\, c_1 + \sum_{i=2}^{m} [(\mu_i^* - \bar{\mu}_1^*) + (\mu_1^* - \bar{\mu}_0^*)]\, c_i + RT \sum_{i=1}^{m} \ln\left(\frac{a_i}{a_0}\right) c_i.$$

$$(18.30)$$

The first term is known, because $(\mu_1^* - \bar{\mu}_0^*) = 18{,}000$ J g^{-1} [35]. The difference $(\mu_i^* - \mu_1^*)$, which is due to the formation of a biotic component from the detritus, is related to the probability $P_{i,a}$ of obtaining the biotic component from a mixture of amino acids. If we treat the formation of an organism from detritus as a chemical equilibrium, we get

$$-\frac{(\mu_i^* - \mu_1^*)}{RT} = \ln K_{eq},$$

$$K_{eq} = \frac{a_{i,eq}}{a_{1,eq}}.$$

And, using the approximation $a_i \approx c_i$ and the fact that the volume is constant, we obtain

$$K_{eq} = \frac{c_{i,eq}}{c_{1,eq}} = \frac{n_{i,eq}}{n_{1,eq}} = P_{i,a},$$

$$(\mu_i^* - \mu_1^*) = -RT \ln P_{i,a}. \qquad (18.31)$$

Substituting Eqs. (18.31) into Eq. (18.30), we eventually obtain

$$\frac{\Delta G}{V} \approx (\mu_1^* - \mu_0^*) c_1 + \sum_{1=2}^{m} [(\mu_i^* - \mu_0^*) - RT \ln P_{i,a}] c_i + RT \sum_{i=1}^{m} c_i \ln \left(\frac{c_i}{c_0} \right). \qquad (18.32)$$

This expression can be renormalized by dividing by the difference between the standard Gibbs free energy of detritus and of the inorganic compounds:

$$Ex = \frac{\Delta G}{V (\mu_1^* - \bar{\mu}_0^*)} \approx c_1 + \sum_{i=2}^{m} \beta_i c_i + (1/\beta_1) \sum_{i=1}^{m} c_i \ln \left(\frac{c_i}{c_0} \right), \qquad (18.23)$$

where $\beta_1 = [(\mu_1^* - \bar{\mu}_0^*)/RT] = 18{,}000$ J g$^{-1} \times 10^5$ (g mol^{-1})/(8.4 J mol^{-1} K$^{-1} \times 300$ K) $= 7.14 \times 10^5$ and $\beta_i = 1 - (\ln P_{i,a})/\beta_1$.

Therefore the weights β_i can be computed with the same information used for computing Eq. (18.33). The treatment previously given underlines the fact that these coefficients can be interpreted as "standard" Gibbs free energies of formation or standard chemical exergies. For example, the coefficient β_1 is, in fact, an estimate of the Gibbs free energy of formation of a mixture of amino acids from inorganic molecules in an aqueous solution. However, the concentrations of the reactants and products in their standard state should be rigorously defined.

The matter is more complicated for species: What is the "standard state" of an organism? Because $(1/\beta_1) \approx 10^{-5}$, the contribution of the second term to the Gibbs free energy is very small: This is the reason why it is usually not taken into account. This means that the interaction among species in a given ecosystem bring a small contribution to the ecosystem exergy (Susani et al. [36] has a nice discussion of this for physical systems) compared with the one accounted for by the evolution. However, such a term could be important when two actual ecosystems or the same ecosystem and two different stages of its development are compared. In fact, Ulanowicz [26, 27] developed an intriguing measure of *ecosystem information* by using an entire pattern of organization called *ascendency*. The ascendency approach, based on the network organization and not on the organism's internal genetic organization, has an advantage in that it is a holistic measure. Analysis shows a strong correlation between increasing EEx and ascendancy [53, 54], but the role each species plays in the network may have more relevance for dealing with the ecosystem change on the temporal scale of restoration than the genetic character of the species.

18.5.3 Eco-Exergy for Open Ecosystems

The preceding calculation provides a quantity that is analogous to the Gibbs free energy of a closed chemical system. The analogy could be extended to an open ecosystem at steady state, because also in this case the total mass is constant. However, it would also be interesting to consider the case of an open ecosystem that is in contact with a much larger one: The latter could represent a reservoir of biotic species, which can be exchanged between the two ecosystems. For open systems, assuming that the temperature and pressure of the system and the reference are equal, the exergy becomes

$$ EEx = \sum_{i=1}^{m} n_i \left(\mu_i - \mu_{iR} \right), $$

in which μ_{iR} represents the chemical potential of the component i in the reservoir, not in the inorganic soup. Expanding again the chemical potentials, we obtain

$$ EEx = \sum_{i=1}^{m} n_i \left(\mu_i^* - \mu_i^* \right) + \sum_{i=1}^{m} n_i RT_R \ln \left(\frac{a_i}{a_{iR}} \right), $$

$$ EEx = \sum_{i=1}^{m} n_i RT_R \ln \left(\frac{a_i}{a_{iR}} \right), $$

and using the approximation $a_i \approx c_i$, we have

$$ \frac{EEx}{V} = \sum_{i=1}^{m} n_i RT_R \ln \left(\frac{c_i}{c_{iR}} \right). \tag{18.34} $$

Equation (18.34) has the same structure as Eq. (18.13), but here c_{iR} does not represent a hypothetical equilibrium concentration, but the actual concentration of the component i in a species reservoir in contact with the ecosystem.

Therefore, it seems reasonable to apply Eq. (18.34) when discussing the short-term evolution of any ecosystem, because the change in its composition is mainly due to the flow of energy and matter, rather than to the true production of a new species by means of natural selection. This approach would be exactly analogous to the ones adopted in the exergy analysis of chemical systems. To this regard, Eq. (18.34) appears particularly suitable for studying the recovery of a perturbed ecosystem.

18.5.4 Restoration versus Evolutionary Time Scales

When one is concerned about ecosystem restoration and recovery after a disturbance, whether human induced or natural, the time scale of recolonization is a critical factor that in turn influences reestablishment of the ecological interactions (network). The time scale of recovery can be extended, perhaps indefinitely, if the conditions for establishing the old community no longer exist. An ecological state is a path-dependent trajectory, and it is possible that the system will not return to its original state. Successful recovery occurs not through regeneration of life processes, but though an influx of surrounding life. Therefore, the probability of a species recolonizing is not a function of their genetic complexity but rather the proximity of a suitable refuge and their dispersal capabilities.

In the previous case from Patricio et al. [40], for example, no one would assume that the restored ecosystem colonizing the empty substrate was the result of random genetic configurations that happened to re-create the life-originating processes. Clearly, those species recolonized from surrounding areas. The standard successional theory, whether it is a primary or secondary succession, assumes that ecosystem establishment is the result of immigration from other life-sustaining refuges. Although we cannot know how many times life did originate, we know that it was at least once, and in general we do not observe this process occurring today.

The coefficients β_i in EEx formulation Eq. (18.20) are taken as a measure of the work done by evolution for constructing a given organism starting from scratch, i.e., from the inorganic soup. The time scale of this process is therefore 10^9 years. However, EEx is then used for comparing the changes that occur in an ecosystem on the time scales of months or even days (i.e., structural dynamic models in which EEx is used as a goal function). This mismatch of the time scale makes it unreasonable to speak about evolution, because the changes in the ecosystem exergy are given by the import and export of species from the surroundings and by the reproduction of existing individuals. Reproduction uses the work done by evolution, but there is no production of a new genetic code. Therefore, the exergy involved in the construction of a new organism is, in fact, the chemical exergy of the food that is necessary to produce a "standard" individual from an embryo.

18.6 Conclusions

The EEx formulation [Eq. (18.20)],

$$\text{EEx}/RT = c_1 + \sum_{i=2}^{N} \beta_i c_i,$$

was obtained based on the assumption that the ecosystem is a closed system or, at least, a system at steady state, in which the total mass of each chemical is constant. When dealing with long-term evolutionary dynamics, the original expression by Jørgensen may be suitable. Ecosystems are open systems that actively exchange chemicals and biological species with the surroundings. Therefore, one should prefer Expression (18.34),

$$\frac{\text{EEx}}{V} = \sum_{i=1}^{m} n_i RT_R \ln\left(\frac{c_i}{c_{iR}}\right),$$

after having properly defined the reference system when investigating the short-term dynamical recovery of an ecosystem. The reference system can change in accordance with the specific situation. Three situations of some interest come to mind:

1. The study of the recovery of a part of a given ecosystem that had been previously disturbed. In this case, the surrounding, nonperturbed, ecosystem seems the more "natural" reference system. Biological species can migrate from the ecosystem to the perturbed subsystem: The analogy with a gas mixture to a solution in contact with a reservoir of chemical species situation is somewhat similar to the analogy with the exchange of chemicals between thermodynamic systems. Indicators that summarize the difference between perturbed and nonperturbed (or "slightly" perturbed) areas within the same ecosystem are badly needed in applied ecology, as the European Water Framework Directory explicitly refers to such a comparison.
2. The comparison between ecosystems of differing spatial scales, which are in contact, for example, a lagoon or a bay with the open sea. Sea species can migrate into the lagoon and vice versa, but the lagoon ecosystem remains different from the pelagic one because of the differences in salinity, habitats, nutrient concentrations, pollutant concentrations, etc.
3. The comparisons of "similar", ecosystems, i.e., ecosystems that are classified as "similar" because they have some features in common (i.e., temperate lakes or Mediterranean coastal lagoons). A pool of species that are typical of these ecosystems can be found, and the reference system could be an ecosystem in which the concentrations of the typical species are fixed to certain values. In this case, there is no actual equilibrium between any of the considered ecosystems and the reference one. However, this choice of the reference system would be in agreement with the guidelines for the selection of the reference state for chemicals given in [50].

18.7 Acknowledgments

The authors are grateful to thoughtful review comments provided by Sven Jørgensen and Simone Bastianoni, which helped clarify several important aspects. And, although Jørgensen does not agree with everything in the manuscript, he welcomes an open discussion on these issues. Both authors were also supported in part by the Dynamic Systems Program of the International Institute for Applied Systems Analysis in Laxenburg, Austria.

REFERENCES

[1] E. P. Odum, "The strategy of ecosystem development," *Science* **164**, 262–270 (1969).
[2] S. A. Levin, "The problem of pattern and scale in ecology," *Ecology* **73**, 1943–1967 (1992).
[3] F. B. Golley, *A History of the Ecosystem Concept in Ecology* (Yale University Press, New Haven, CT, 1993).
[4] K. N. Laland, F. J. Odling-Smee, and M. W. Feldman, "Evolutionary consequences of niche construction and their implications for ecology," *Proc. Natl. Acad. Sci. USA* **96**, 10242–10247 (1999).

[5]   A. Lotka, "Contribution to the energetics of evolution," *Proc. Natl. Acad. Sci. USA* **8**(6), 147–151 (1992).

[6]   L. Boltzmann, *Populäre Schriften* (Barth, Leipzig, Germany, 1905).

[7]   E. Schrödinger, *What is Life?* (Cambridge University Press, Cambridge, 1944; reprinted 1992).

[8]   H. T. Odum and R. C. Pinkerton, "Time's speed regulator: The optimum efficiency for maximum power output in physical and biological systems," *Am. Sci.* **43**, 331–343 (1955).

[9]   H. T. Odum, "Energy systems and the unification of science," in *Maximum Power: The Ideas and Applications of H. T. Odum*, edited by C. S. Hall (Colorado University Press, Boulder, CO, 1995), pp. 365–372.

[10] S. E. Jørgensen, *Integration of Ecosystem Theories: A Pattern* 2nd ed. (Kluwer Academic, Dordrecht, The Netherlands, 1997).

[11] S. E. Jørgensen and H. Mejer, "A holistic approach to ecological modeling," *Ecol. Model.* **7**, 169–189 (1979).

[12] F. Müller and M. Leupelt, editors, *Eco Targets, Goal Functions, and Orientors* (Springer-Verlag, Berlin, 1998).

[13] B. D. Fath, B. C. Patten, and J. S. Choi, "Complementarity of ecological goal functions," *J. Theor. Biol.* **208**, 493–506 (2001).

[14] G. Bendoricchio and L. Palmeri, "Quo vadis ecosystem?" *Ecol. Model.* **184**, 5–17 (2005).

[15] F. Müller and B. D. Fath, "Introduction: The physical basis of ecological goal functions – Fundamentals, problems, and questions," in *Eco Targets, Goal Functions, and Orientors*, edited by F. Müller and M. Leupelt (Springer-Verlag, Berlin, 1998), pp. 15–18.

[16] J. J. Kay, "Self-organization in living systems," Ph.D. thesis (University of Waterloo, Waterloo, Ontario, Canada, 1984).

[17] E. D. Schneider and J. J. Kay, "Life as a manifestation of the second law of thermodynamics," *Math. Comput. Model.* **19**, 25–48 (1994).

[18] E. D. Schneider and J. J. Kay, "Complexity and thermodynamics. Towards a new ecology," *Futures* **26**, 626–647 (1994).

[19] E. D. Schneider and D. Sagan, *Into the Cool: Energy Flow, Thermodynamics, and Life* (University of Chicago Press, Chicago, 2005).

[20] S. E. Jørgensen, "Review and comparison of goal functions in systems ecology," *Vie Milieu* **44**, 11–20 (1994).

[21] I. Prigogine and J. M. Wiame, "Biologie et thermodynamique des phenomenes irreversible," *Experientia*, **II**, 451–453 (1946).

[22] I. Prigogine, *From Being to Becoming: Time and Complexity in the Physical Sciences* (Freeman, San Francisco, 1980).

[23] J. J. Vallino, "Modeling microbial consortiums as distributed metabolic networks," *Biol. Bull.* **204**, 174–179 (2003).

[24] I. Aoki, "Entropy law in aquatic communities and the general entropy principle for the development of living systems," *Ecol. Model.* **215**, 89–92 (2008).

[25] H. J. Morowitz, *Energy Flow in Biology: Biological Organization as a Problem in Thermal Physics* (Academic, New York, 1968).

[26] R. E. Ulanowicz, *Growth and Development, Ecosystems Phemomenology* (Springer-Verlag, New York, 1986).

[27] R. E. Ulanowicz, *Ecology, The Ascendent Perspective* (Columbia University Press, New York, 1997).

[28] E. F. Cheslak and V. A. Lamarra, "The residence time of energy as a measure of ecological organization," in *Energy and Ecological Modelling*, edited by W. J. Mitsch, R. W. Bossermann, and J. M. Klopatek (Elsevier, Amsterdam, 1981), pp. 591–600.

[29] J .C. Marques, M. A. Pardal, S. N. Nielsen and S. E. Jørgensen, *"Analysis of the Properties of Exergy and Biodiversity along an Estuarini Gradient of Eutrophication,"* Ecol. Model. **102**, 155–167 (1997).

[30] G. N. Hastopoulus and J. H. Keenan, "Principles of general thermodynamics" John Wiley & Sons, New York (1965).

[31] A. Munster, *Classical Thermodynamics* (Wiley, New york, 1970).

[32] Z. Rant, "Exergie, ein neues wort für technische arbeitsfähigkeit," *Furschling Ing.Wesens* **22**(1), 36–37 (1956).

[33] R. B. Evans, "The fromulation of essergy," *Thayer News,* Fall (1968) (Thayer School of Engineering, Dartmouth College, Hanover, NH).

[34] R. B. Evans, "A proof that essergy is the only consistent measure of potential work (for chemical systems)," Ph. D. dissertation (Dartmouth College, Hanover, NH, 1969), University Microfilms, 70–188.

[35] S. E. Jørgensen and S. N. Nielsen, "Thermodynamic orientors: Exergy as a goal function in ecological modelling and as an ecological indicator for the description of ecosystem development," In *Ecotargets, Goal Functions and Orientors,* edited by F. Müller and M. Leupelt (Springer-Verlag, Berlin, 1998), pp. 63–86.

[36] L. Susani, F. M. Pulselli, S. E. Jørgensen, and S. Bastianoni, "Comparison between technological and ecological exergy," *Ecol. Model.* **193**, 447–456 (2006).

[37] S. E. Jørgensen, N. Ladegaard, M. Debeljak, and J. C. Marques, "Calculation of exergy for organism," *Ecol. Model.* **185**, 165–175 (2005).

[38] B. D. Fath, S. E. Jørgensen, B. C. Patten, and M. Straškraba, "Ecosystem growth and development," *Biosystems* **77**, 213–228 (2004).

[39] S. E. Jørgensen and B. D. Fath, "Examination of ecological networks," *Ecol. Model.* **196**, 283–288 (2006).

[40] J. Patrício, F. Salas, M. A. Pardal, S. E. Jørgensen, and J. C. Marques, "Ecological indicators performance during a re-colonisation experiment and its compliance with ecosystem theories," *Ecol. Ind.* **6**, 43–57 (2006).

[41] S. N. Nielsen, "Strategies for structural-dynamical modelling," *Ecol. Model.* **63**, 91–102 (1992).

[42] S. E. Jørgensen, "Parameter estimation and calibration by use of exergy," *Ecol. Model.* **146**, 299–302 (2001).

[43] P. R. Grant, "The feeding of Darwin's finches on *Tribulus cistoides*," *Am. Sci.* **69**, 653–663 (1981).

[44] P. R. Grant, *Ecology and Evolution of Darwin's Finches,* 2nd ed. (Princeton Science Library, Princeton, NJ, 1999).

[45] P. R. Grant and B. R. Grant, "Annual variation in finch number, foraging and food supply on Sala Daphne Major, Galapagos," *Oecologia* **46**, 55–62 (1980).

[46] P. T. Boag and P. R. Grant, "Intense natural selection in a population of Darwin's finches in the Galapagos," *Science* **214**, 82–85 (1981).

[47] P. T. Boag and P. R. Grant, "Darwin's finches on Isle Daphne Major, Galapagos: Breeding and feeding ecology in climatically variable environment," *Ecol. Monogr.* **54**, 463–489 (1984).

[48] S. E. Jørgensen and B. D. Fath, "Modelling the selective adaptation of Darwin's finches," *Ecol. Model.* **176**, 409–418 (2004).

[49] J. Szargut, D. R. Morris, and F. R. Steward, *Exergy Analysis of Thermal, Chemical, and Metallurgical Processes* (Hemisphere, Washington, D.C., 1988).

[50] H. Meier and S. E. Jørgensen, "Exergy and ecological buffer capacity. State of the art in the ecological modelling. Environmental sciences and applications," in *Proceedings of the Seventh Conference on Ecological Modelling* (International Society for Ecological Modelling, Copenhagen, 1979), pp. 831–846.

[51] G. Bendoricchio and S. E. Jørgensen, "Exergy as a goal function of ecosystems dynamic," *Ecol. Model.* **102**, 5–15 (1997).

[52] R. B Evans, G. L Crellin, M. Tribus, "Thermoeconomic considerations of sea water Demineralisation," edited by K. S. Spieglgr, *Principles of Desalination, Academic Press. New York and London* (1966), pp 21–75.

[53] S. E. Jørgensen, *Integration of Ecosystem Theories: A Pattern*, 1st ed. (Kluwer Academic, Dordrecht, The Netherlands, 1992).

[54] R. E. Ulanowicz, S. E. Jørgensen, and B. D. Fath, "Exergy, information and aggradation: An ecosystems reconciliation," *Ecol. Model.* **198**, 520–524 (2006).

Thoughts on the Application of Thermodynamics to the Development of Sustainability Science

*Timothy G. Gutowski, Dušan P. Sekulić,
and Bhavik R. Bakshi*

19.1 Introduction

The term sustainability is used frequently now in many different contexts.[1] For example, in the area of engineering, there have been claims of sustainable products, sustainable manufacturing, sustainable designs, and so forth. Although these uses may be well intended, they actually marginalize the term by implying that just getting better in some way is sustainable. Instead, sustainability needs to be connected to a worldview that encompasses how human society can maintain a good quality of life over a long time. Without this worldview framework, these claims of sustainable this and sustainable that ring hollow. In this context, we are inspired by the work of the authors (mostly economists with biological scientists) of the paper, Arrow et al. [1]. By starting with the well-known statement of sustainability from the UN Brundtland Report [2], they developed a measurable and workable (though controversial) criterion for sustainability.

The Brundtland UN Commission statement on sustainability says, "sustainable development is the development that meets the needs of the present generation without compromising the ability of future generations to meet their own needs." This statement brings up many value-laden issues and at first blush seems unworkable. For example, what is a need for one person could be considered excessive consumption for another. Furthermore, who is to speak for future generations and to articulate their needs? In addition, what *development* means is of crucial importance, in particular, does development require growth, and if so, what kind.

Kenneth Arrow and a group of economists and ecologists took on the task of interpreting this statement into an essentially economics framework. Ultimately, they reinterpret this statement in terms of the maintenance of a certain kind of resource base called *genuine wealth* (this to ensure that the time rate of change of the aggregate utility based on the total consumption must be increasing or at least being equal to zero). Genuine wealth is the stock of all of society's capital assets including (1) *manufactured capital*, including the machines and factories necessary for production, (2) *human capital*, including education and health care, and

---

[1] This chapter borrows heavily from our paper, "Preliminary thoughts on the application of thermodynamics to the development of sustainability criteria," presented at the IEEE International Symposium on Sustainable Systems and Technology, Phoenix, AZ, May 18–20, 2009.

(3) *natural capital*, including minerals, fuels, and the services of ecosystems including plants, animals, insects, etc. As they state,

> This requirement that the productive base (genuine wealth) be maintained does not necessarily entail maintaining any particular set of resources at any given time. Even if some resources such as stocks of minerals are drawn down along some consumptive path, the sustainability criterion could nevertheless be satisfied if other capital assets were accumulated sufficiently to offset the resource decline.

The strength of this criterion is that it is operational. Economists have devised techniques for calculating these types of capital (although with some difficulties and controversy). In fact, the World Bank issues *The Little Green Data Book* each year, which keeps national accounts for many of the countries around the world in terms of genuine investment or a similar metric. According to this book, the world is still behaving in a sustainable way. In spite of depletions in various natural capital resources such as energy and mineral resources and CO_2 damage, these are offset by the additions to manufactured and human capital, with a resulting "Adjusted net savings" in 2005 of 7.4% of the gross national income (GNI) [3]. This feature, that manufactured and human capital can be substituted for natural capital, is, however, extremely controversial. This is the so-called "weak form of sustainability" and it implies, as Herman Daly has put it, "that boats are assumed to be substitutes for fish" [4].

19.2 Resource Maintenance

We agree with the shift in emphasis in the Arrow paper from meeting people's needs, to maintaining a certain resource base. This approach seems tractable, but avoids the fundamental issue of population growth. (More on this in Section 19.4.) Of course, identifying and valuing what is a resource does not come value free either. We can make the best use of thermodynamics, however, by focusing on biophysical resources and using aggregate measures of the second law consequences of the conversion of these resources, such as exergy change and entropy production. By using mass, energy, and entropy balances, we can make statements about the changes in exergy (or available energy) of components of the resource base as well as the entropy production associated with these transformations. Several others have already proposed the thermodynamic treatment of the sustainability problem along these lines. See, for example, the studies of Szargut [5], Sciubba [6, 7], Wall and Gong [8, 9], Odum [10, 11], and Ayres [12, 13].

To start this analysis, the first order of business would be to clearly define the system to be analyzed. This includes selecting the boundaries of the system, and defining the components, parameters, and the internal forces of the system.

A classification of systems with respect to whether their boundary is crossed by mass flow includes (1) a materially closed system and (2) an open system that can exchange mass. Conceptually, the world can be modeled as the closed system, in the first approximation, as a domain of the Earth where human beings live and most biological activity occurs. Although this system is materially closed (if any eventual material exchange with the rest of the universe and remaining constituents of Earth are neglected), it is open to energy transfers, e.g., thermal radiation from the sun and thermal interaction between the core and the Earth's crust, as well as work effects from the gravitational attraction of the moon and other celestial bodies. An equally common scheme is one that partitions the world into the so-called technosphere, including people and their industrial activities, and the so-called ecosphere, containing the "natural system" including ores, ecosystem services, and natural products that are used to enhance and sustain the life of human beings. The ecosphere may further be structured as consisting of biosphere, lithosphere, hydrosphere, and atmosphere – all open subsystems experiencing interactions among themselves and with the technosphere. These systems are materially open and allow for material transfers between these domains. We could easily envision even more complicated arrangements in which specific types of technodomains and ecodomains are identified, and these in turn can exchange materials and energy with each other. This leads us to this question: *Is there a topology of sustainability?* And how does the topology affect the solution? Clearly we want to subdivide the problem to solve it. Alternatively, we may need to consider interactions between different subdomains to address different states of these domains. But could this subdivision preordain the nature of the solution? In an entertaining and extremely useful book, David MacKay reviews the prospects for the UK to supply their own energy needs (taken at their present values) with renewable resources [14]. His disturbing conclusion is that they cannot. He then goes on to explore trading schemes with other parts of the world to ensure their access to renewable fuels. Clearly how one draws the boundaries for this problem matters a great deal. A related question then is this: *How do we partition the problem, and could a solution be made up of the sum of the parts?* These are very fundamental and important questions. An equally relevant question would be this: Do we need to test every proposition compared with a global optimum? Many of the participants in this dilemma can act on only a small part of the problem. A mechanism that aligns local optimum behavior with our global optimum goals would be very helpful.

19.3 A Closed System

Consider the closed-system interpretation of our system. We start by drawing boundaries around the resource base that is to be maintained. To be concrete about this, we might start out with the system boundary around the world, which contains all life on the surface of the Earth extending up into the reaches of the atmosphere that can be affected by human activity (perhaps of the order of 50–100 km) and to the depth of the Earth and oceans that are accessible by humans (perhaps of the order of 10–20 km). For our purposes this is a materially closed system. Here we are obviously ignoring volcanism and interplanetary dust and travel. These are issues that are not our immediate concern. There are a few advantages to this perspective and also disadvantages, which will be discussed shortly. For example, Ayres uses the consequences of an analysis exercised on a closed system to refute Georgescu-Roegen's so-called "fourth law" (on material resources unavailability on Earth) [15]. That is, in general, the materials used by humans on Earth are not lost. They can become diluted or react with other components of the closed system to lower their exergy, but this problem could be solved by the appropriate use of energy resources.

But there will be a lot going on in this closed system. It is not homogeneous and it is not in equilibrium. It does have certain resources that are to be maintained. Note that if the components of this system can spontaneously react with each other, these reactions must involve energy interactions – but these must be across the respective subsystem's boundary. So, to analyze these interactions, some structure must be proclaimed – that is, the closed system must be a composite of subsystems.

The full definition of these subsystems will then allow for the calculation of changes in material integrity, in particular the chemical exergy of the materials components of the biosphere. In the aggregate then, these spontaneous or human-inflicted reactions will result in a net production of entropy and a net change in the cumulative exergy of the components of this system. If this were an isolated system, the question would be, at what rate is this resource base degrading? For the closed system, however, the long-term, so-called renewable energy inputs are available to restore the quality of this resource base and even improve on it. This is, of course, exactly what nature does by increasing the chemical and physical exergies of various material systems to support the biogeochemical cycles such as the hydrological cycle and the carbon cycle. For example, the exergy lost by atmospheric condensation of water vapor, rain, and flow to the oceans is restored by evaporation and transport powered by the sun. Likewise the chemical exergy lost when carbon is oxidized, be it by autotrophic or heterotrophic respiration or even biomass burning, can be restored when carbon dioxide is upgraded to biologically useful carbohydrates by the process of photosynthesis. From this perspective, it is the use of the renewable resources to restore, and even upgrade, the chemical exergy of certain chemical components of the environment that translates into sustainability.

In particular, we believe that one could look at the rates of destruction versus the rates of restoration as a principal indicator of sustainability. To be sure, this is a vast proposal with many details yet to be explored, but one can get a general sense of where things are going by considering the basic events that affect the chemical exergy of our chemical stores (and any eventual physical exergy involved): (1) The reactive environment leads to a destruction of exergy, and this can be accelerated by activity (both human and not) that stirs up materials and brings reactive substances to the surface (e.g., mining, plowing, storms, etc.); (2) humans further decrease the exergy of mineral resources by depleting the sources of highest concentrations; but (3) at the same time, humans increase the exergy of minerals and metals for example by refining and smelting; (4) however, these human activities are powered almost exclusively by use of fossil fuels that can result in a net reduction in chemical exergy for the entire system if the carbon cycle becomes unbalanced as it currently appears to be; and finally, (5) currently, the primary mechanism for chemical exergy increase for the closed system is net primary productivity (NPP). We believe that the net effect of these events is an ongoing net destruction of chemical exergy dominated by the huge losses in the carbon cycle. For a preliminary accounting of global exergy flows see [16].

To see where we are going, and the biophysical limits on this path, consider a scenario that brings 9 billion people up to the per-capita energy used in the United States (350 GJ/yr). This total, about 3.2 ZJ/yr, or 100 TW, is almost exactly equal to estimates of the total NPP [17, 18]. But we already appropriate about 40% of terrestrial NPP, mostly for food [18], and many of us are not at all keen on using the rest of the NPP as an energy source. Of course, there are other energy options, but this increase is enormous, and most proposals for how to meet such a demand come with their own sets of dilemmas.

Indeed there are claims that it is technically feasible to beat the efficiency of the NPP, but these claims are generally based on drawing quite narrow boundaries for the accounting scheme. The area of sustainability science would welcome serious proposals that include an extended-boundaries full life cycle accounting for the resources used. This is discussed further in Section 19.6 of this chapter.

This leads us to perhaps the number one core issue: *Can technology solve the sustainability problem?* In particular, can technology be used to save, maintain, or substitute for ecosystem services? *An equally important question is this: To what extent can technology substitute for behavioral change on the part of humans?*

19.4 The Role of Humans

The complexity of the role of humans on sustainability can be illustrated by asking this simple question: Does the sustainability of humans increase or decrease as their numbers increase? If humans are a resource, we would want more of them. If they are consumers, we would want less. In the weak-sustainability criterion proposed by Arrow et al., investment in human education, training, and health care are clear candidates to be counted in the resource base. In particular, in both the paper by Arrow et al. and the World Bank's calculation, investments in education are counted in the resource base. So, indeed, many economists would count them as a resource. For an entertaining and rather myopic presentation of this point of view, see the last chapter of Tim Harford's book *The Logic of Life* [19]. Here he refers (only half tongue in cheek) to a character (Ted Baxter) in the old television show, *The Mary Tyler Moore Show*, who plans to have six children in the hope that one of them would solve the overpopulation problem.

An alternative approach could count all other biophysical resources and seek to maintain these on a per-capita basis. This line of reasoning was, of course, famously pursued in the late 1700s by the economist Thomas Malthus, with the well-know result. Ever since then, most economists have been loath to acknowledge any biophysical constraints on our future prosperity. Nevertheless, as the human population nears 7 billion, leaving less than 2 hectares of habitable land per person, and with our CO_2 emissions at a sufficient magnitude to alter the chemical composition of the atmosphere and the ocean, it seems fitting to revisit this proposition.

To evaluate the role of humans from a thermodynamic perspective, it seems that a couple of alternatives are possible. At the most elementary level, human beings, whether well educated or not, would probably be considered, not elements of, but consumers of the resource base. This would be shown, for example, if we were to model the resource base as an open thermodynamic system with the humans in the so-called technosphere and the resources in the ecosphere. The human contribution in this case would be to extract resources from the ecosphere and to improve the efficiency of the conversion processes for the resources into useful goods and services in the technosphere. Additionally, and perhaps more important, humans would figure in both the generation of wastes deposits to the ecosphere and the development of renewable energy sources to power these processes. At the same time, the number and appetites of the humans would determine the magnitude of this enterprise.

There is an additional perspective that comes from the ecological thermodynamics community that seeks to develop a thermodynamic measure to account for the additional work potential in a biological form that has life [20]. This is a question that goes back at least to the booklet produced by Erwin Schrödinger in 1944 titled "What is Life?" Currently there are researchers who propose a rather controversial ecological exergy measure that attempts to account for this additional potential in terms of the genetic complexity of the life form. If this perspective were included in the thermodynamic accounting scheme, several issues would need to be addressed immediately, including human population growth and species extinction. It would also raise serious questions about substitutability between species.

19.5 Open Systems

Energy interactions on the global scale of a biosphere model would lead to great difficulties in specifying causality of events within the complex structure of the closed system, even if strong correlations between the set of properties and interactions may be identified. Therefore a meaningful approach to defining sustainability criteria must be in (1) defining a set of essential interacting open subsystems that constitute the ecosphere, (2) identifying the states of these subsystems in conjunction with established interactions and their properties, and (3) formulating resources flows as material and energy interactions among the subsystems. In principle, each of these subsystems may have well-defined end states (while at the same time we may not need to be aware of all intermediate states). In general, each subsystem's state of a global model of sustainability would not be in a steady state. Also, depending on the state and interaction conditions, it is feasible to assume that each subsystem would not be in equilibrium.

Following the previously formulated minimum set of considerations, an example of a reasonably simple global model of an ecosphere interacting with a technosphere that includes humans involving their production base and institutions would consist of four, energy- and material-interacting, thermodynamically well-defined subsystems, as illustrated in Fig. 19.1.

This implies that open, nonisolated subsystems in respective given states interact through energy and material interactions, thus leading to simultaneous changes of mass, energy, and entropy of subsystems (the other properties may or may not change as well). The rate of change of mass, energy, and entropy can be represented as changes of exergy of each subsystem. This change of exergy may be interpreted as an equivalent change of available energy for subsystem's interactions expressed in terms of thermodynamically allowable work interactions (but, in the given context, it may not necessarily be a work interaction). The magnitude of this time rate change of *nonconserved* system exergy may be used as *one of the metrics* for sustainability of a particular system. This would be justified under the assumption that the available energy expresses an inherent potential of a given subsystem to offer a driving force in interactions needed for the overall system's change of state. If one assumes that certain aspects of sustainability of each subsystem can be measured by a positive or

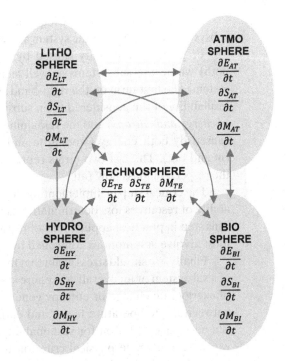

Figure 19.1. Technosphere and four subsystems of ecosphere interacting among themselves through energy and material interactions (arrows indicate interactions). Note that the interactions between the subsystems and the rest of the universe are not presented.

in a limit equal to zero-exergy rate change, it should be clear that such a change must be a consequence of both (1) interactions and (2) related subsystems's change of state (i.e., ultimately would depend on the efficiency of internal transformations). It seems clear that the sustainability of a single subsystem can be neither necessary nor sufficient for establishing the sustainability of a combined set of subsystems or the overall system. However, although exergy is not conserved, it is additive, and certain sustainability features of the interacting subsystems may be measured in terms of an aggregate exergy change. Note also that the overall system, in general, interacts through energy (dominant) and material (weak) interactions with the rest of the universe.

This framework allows the determination of the exergy rate of supply and the exergy rate of use of both material and energy flows. Hence, one may, at least in principle, determine entropy generation (or exergy loss) of each of the subsystems. In the case of the technosphere, one may argue that the loss of energy availability can be correlated through exergy-conversion efficiencies to the state of the current state of development through, say, the level of technology development.

For a transient, open system, under the specified assumptions one can show that the following relationship holds [21]:

$$T_0 \dot{S}_{gen} = -\left(\frac{dE}{dt} - T_0 \frac{dS}{dt}\right) - \sum_j \dot{W}_j + \sum_k \dot{Q}_k \left(1 - \frac{T_0}{T_k}\right)$$
$$+ \sum_{in} \dot{m}(h - T_0 s) - \sum_{out} \dot{m}(h - T_0 s)$$
$$+ \sum_{in} \left[\frac{1}{M}(\mu^* - \mu_0)\dot{m}\right]_i - \sum_{out} \left[\frac{1}{M}(\mu^* - \mu_0)\dot{m}\right]_j . \quad (19.1)$$

The left-hand side of the equation represents the lost exergy in any of the subsystems of Fig. 19.1 or the system as a whole. This loss represents the compounded loss or gain of availability caused by resources removal or waste deposition (or both) within the subsystems as well as all irreversible transformations within the system. The first term on the right-hand side indicates the time rate of change of the availability of the considered open subsystem as a consequence of the *rate changes of energy and entropy,* and the remaining four terms account for exergy interactions caused by both energy (work rate and heat rate) and material flow interactions (in and out). The last two terms represent the rates of chemical exergy carried by the respective mass flow rates.

The critical step in implementing this approach to resources balancing and availability of resources loss determination is the definition of a state of each subsystem. This step in practical applications of the approach is essential, it is not trivial, and it must involve assumptions controlled by the level of rigor imposed.

Finally, we should stress again that the proposed framework for the sustainability of a subsystem or an overall ecosphere system in thermodynamic terms, i.e., in terms of exergy rate change or entropy generation in terms of energy units (the product between the temperature factor and entropy generation), *is not* a sufficient, but it is a necessary, criterion for sustainability. In other words, we are not arguing that we can use a single physical criterion to evaluate sustainability of the ecosphere (that would involve the technosphere with complex nonphysical interactions). We are arguing that, just as we cannot expect the weak sustainability criterion based on the time rate of change of genuine investment in the economics framework to be a single criterion, the physical criterion based on thermodynamics (such as entropy generation or exergy destruction, or both) cannot be considered as a sole criterion. It appears that multiple criteria must be used.

19.6 Technology and Ecosystem Services

It is commonly believed that technology is an essential part of the solution for achieving sustainability. In this section, we consider this claim from a thermodynamic point of view based on the open system described in Section 19.5. The purpose of almost all technological and manufacturing activities is to decrease the entropy generation (increase the exergy) within the technosphere. As per the second law and exergy-balance equation (19.1), this must result in an even larger increase in entropy generation (reduction in exergy) of the surroundings that is due to energy-conversion inefficiencies. Thus all technological activities, regardless of whether they rely on renewable or nonrenewable resources, cause the entropy generation within the surroundings to increase. In most technological systems, this entropy-generation increase commonly manifests itself as environmental impact. The current approach of technological development is such that this environmental impact is usually large and negative. Thus it may be argued that any technological development will result in negative environmental impact, and therefore achieving sustainability through technological development may be impossible [22]. Fortunately, as argued in the rest of this section, thermodynamic insight into ecological systems indicates that shifting the paradigm to consider networks of technological–ecological systems may provide a path to sustainability.

The laws of thermodynamics also apply to ecological systems. Because these systems have sustained themselves for millennia, it is worth considering how they deal with the implications of the second law. Like technological systems, ecological systems also aim to decrease their entropy generation [23]. Thus they also cause the entropy generation within their surroundings to increase and cause environmental impact. However, in self-sustaining systems, this impact is not harmful because intermediate systems have evolved to utilize the exergy loss from other systems to result in an exergy cascade. Thus, although individual activities in an ecosystem are not necessarily efficient – consider the waste created in the fall by the leaves of deciduous trees or the highly "wasteful" eating habits of squirrels or monkeys – the overall system tends to be surprisingly efficient. Individual ecological activities can get away with being wasteful because that waste is put to good use by some other species or activity. In addition, this exergy cascade is such that various activities reinforce each other to result in an autocatalytic system. And all of this runs on renewable energy. Of course, there are situations in which this quasi-equilibrium of nature is disturbed by natural or anthropogenic events, in which case the system has some ability to recover its structure and function or move to another state.

This brief summary of the thermodynamic aspects of ecosystems indicates some approaches that could be used to guide technology development. The ecosystem analogy has already been a motivation for ideas such as industrial ecology and biomimicry. The former field encourages the development of industrial symbiotic networks in which "waste = food," so that the efficiency of the overall industrial ecosystem is enhanced and its environmental impact is reduced. For example, the ash from a coal-burning power plant may be used as an additive to concrete, and waste tires may be used to fuel a cement kiln. Biomimicry aims to emulate ecological systems such as Velcro mimicking barbs on a seed and materials mimicking the structure of a horseshoe crab shell. However, industrial ecology does not go far enough in the development of self-sustaining networks because it focuses mainly on networks of technological systems, and neither does biomimicry because it may still result in products that use highly toxic materials and nonrenewable resources.

Ecological systems play a crucial role in supporting all planetary activities by providing essential goods and services. However, the role of ecosystem services is often ignored, even in approaches for enhancing sustainability including industrial ecology and life-cycle assessment. Thus, although an industrial ecosystem is likely to lead to an exergy cascade, the exergy loss from such a system is still likely to be large and the system is still not likely to be self-sustaining. For example, even after using the waste materials from a coal-burning power plant in other processes, as shown in Fig. 19.2, the symbiotic network still cannot be claimed to be sustainable as it still has emissions of carbon dioxide, relies on fossil resources, large quantities of water, and relies on or deteriorates many degraded ecosystem services. Overcoming these important shortcomings requires connecting industrial ecology with ecosystem ecology [24] by considering networks, not just of industrial systems, but of industrial and ecological systems. Such networks could be developed by greater interaction between industrial ecology and ecological engineering [25], because the purpose of the latter is to engineer ecosystems for providing the goods and services needed for satisfying human and industrial needs. Such a view could result in the consideration of the ecological processes shown in Fig. 19.2 for sequestering the carbon dioxide

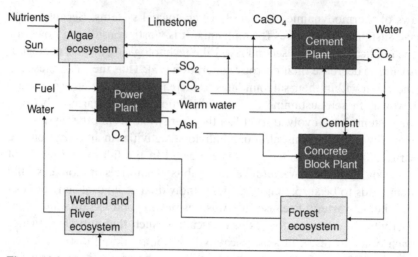

Figure 19.2. Network of technological and ecological systems as a step toward sustainability. The darker boxes are industrial systems, and the lighter boxes are ecological systems [25].

and providing oxygen for combustion and using wetland and riverine ecosystems for treating and reusing industrial water. Algae from an ecosystem that uses warm waste water and air emissions, such as carbon and sulfur dioxides, could also reduce the reliance on fossil fuels. In addition, accounting for ecosystem services in life-cycle assessment is also essential [26].

Thermodynamics indicates that the common notion that technology can lead to sustainability may be true, but only if the definition of technology is broadened to consider the supporting technological and ecological systems because, without these supporting services, the entropy, generation increase within the surroundings is very likely to have a harmful impact. The current approach of focusing on a single or handful of technologies, such as solar, nano or biomass, is unlikely to lead to sustainability. Thermodynamics can play an essential role in understanding the function of ecosystems in the life cycle of economic products. It can also assist in the development of technological–ecological networks by identifying processes with the highest exergy losses and guiding the design of efficient and self-sustaining technological–ecological networks that operate within ecological constraints.

19.7 Conclusion

In this chapter, we attempt to outline some of the issues involved in making statements about sustainable development based on thermodynamics. Although it is too early to definitively declare any metasustainability criteria based on thermodynamics, we believe that thermodynamics provides the right tools for an aggregate assessment of our sustainability from a biophysical perspective. That is, the thermodynamics framework does offer a system-based, rigorous approach to defining loss of availability of resources.

Additionally, thermodynamics offers theoretical limits to assessments of technological advancements and their contribution to solutions of sustainability problems. In fact, often with quite elementary analyses, one can get to the essence of many proposals for technology solutions to sustainability issues. This capability, parts of which are routinely included in the education of chemical and mechanical engineers, should also become standard practice for all new engineering disciplines meant to address sustainability.

Although we have not dwelled on the available data, and have offered only a few examples here, we believe that the biophysical evidence is quite contrary to the neoclassical economics conclusion; we quite clearly are not behaving in a sustainable way at present. At the same time, we fully admit that thermodynamic metrics and derivative criteria must be only a complementary set of metrics or criteria for an assessment of sustainability. No single metric (regardless of how well aggregated) or derivative criterion is able to offer a completely satisfactory solution for all situations.

Finally, although we have taken only the first steps in considering the partitioning of the sustainability problem, it is already quite apparent that claims about "sustainable products" and other small pieces of the problem are vacuous without a credible connection to a convincing worldview of sustainability.

REFERENCES

[1] K. Arrow, P. Dasgupta, L. Goulder, G. Daily, P. Ehrlich, G. Heal, S. Levin, K.-G. Mäler, S. Schneider, D. Starrett, and, B. Walker "Are we consuming too much?," *J. Econ. Perspect.* **18**(3), 147–172 (2004).

[2] *Our Common Future*, Report of the World Commission on Environment and Development (World Commission on Environment and Development, 1987).

[3] *The Little Green Data Book* (World Bank, Washington, D.C., 2007).

[4] H. Daly, "Economics in a full economy," *Sci. Am.* **293**(3), 100–105 (Sept. 2005).

[5] J. Szargut, A. Ziebik, and W. Stanek, "Depletion of the non-renewable natural exergy resources as a measure of the ecological cost," *Energy Convers. Manage.* **43**, 1149–1163 (2002).

[6] E. Sciubba, "A novel exergetic costing method for determining the optimal allocation of scarce resources," in *Proceedings of the Conference on Contemporary Problems of Thermal Engineering* (Technical University of Silesia, Institute of Thermal Technology, Gliwice, Poland, 1998), pp. 311–324.

[7] E. Sciubba, "Cost analysis of energy conversion systems via a novel resource based quantifier," *Energy* **28**, 457–477 (2003).

[8] G. Wall and M. Gong, "On exergy and sustainable development – Part 1: Conditions and concepts," *Exergy Int. J.* **1**(3), 128–145 (2001).

[9] M. Gong and G. Wall, "On exergy and sustainable development – Part 2: Indicators and methods," *Exergy Int. J.* **1**(4), 217–233 (2001).

[10] H. T. Odum, "Self-organization, transformity and information," *Science* **242**, 1132–1139 (1988).

[11] H. T. Odum, *Environmental Accounting: Emergy and Environmental Decision Making* (Wiley, New York, 1996).

[12] R. U. Ayres, "Eco-thermodynamics: Economics and the second law," *Ecol. Econ.* **26**, 189–209 (1998).

[13] R. U. Ayres, "On the life-cycle metaphor: Where ecology and economics diverge," *Ecol. Econ.* **48**, 425–438 (2004).

[14] D. MacKay, *Sustainable Energy – Without the Hot Air* (UIT Cambridge Press, 2008).

[15] R. U. Ayres, "The second law, the fourth law, recycling and limits to growth," *Ecol. Econ.* **29**, 473–483 (1999).

[16] W. A. Hermann, "Quantifying global exergy resources," *Energy* **31** 1685–1702 (2006).

[17] V. Smil, *Energy in Nature and Society* (MIT Press, Cambridge, MA, 2008), p. 73.

[18] S. L. Pimm, *A Scientist Audits the Earth* (Rutgers University Press, New Brunswick, NJ, 2001).

[19] T. Harford, *The Logic of Life* (Random House, New York, 2008).

[20] R. Pastres and B. D. Fath, "Exergy use in ecosystem analysis: Background and challenges," in *Thermodynamics and the Destruction of Resources*, edited by B. Bakshi, T. Gutowski, and D. Sekulić (Cambridge University Press, New York, 2009).

[21] A. Bejan, *Advanced Engineering Thermodynamics* (Wiley, Hoboken, NJ, 2006).

[22] M. H. Huesemann, "The limits of technological solutions to sustainable development," *Clean Technol. Environ. Policy* **5**, 21–34 (2003).

[23] B. D. Fath, "Complementarity of ecological goal functions," *J. Theor. Biol.* **208**, 493 (2001).

[24] D. R. Tilley, "Industrial ecology and ecological engineering: Opportunities for symbiosis," *J. Ind. Ecol.* **7**(2), 13–32 (2003).

[25] R. A. Urban, A. Baral, G. F. Grubb, B. R. Bakshi, W. J. Mitch, "Towards the sustainability of engineered processes: Designing self-reliant networks of technological-ecological systems," *Comput. Chemi. Eng.* **34**, 9, 1413–1420 (2010).

[26] Y. Zhang, S. Singh, and B. R. Bakshi, "Accounting for ecosystem services in life cycle assessment, Part I: A critical review," *Environ. Sci. Technol.* **44**, 2232–2242 (2010).

APPENDIX

Standard Chemical Exergy

Substance	State	Molecular mass M (kg/kmol)	Enthalpy of devaluation $D°$ (kJ/mol)	Standard chemical exergy $e°_{x,ch}$ (kJ/mol)
Al	s	26.9815	930.9	795.7
Al_4C_3	s	143.959	4694.51	4216.2
$AlCl_3$	s	133.3405	467.18	352.2
Al_2O_3	s. α corundum	101.9612	185.69	15.0
$Al_2O_3 \bullet H_2O$	s. boermite	119.9765	128.35	9.4
$Al_2O_3 \bullet 3H_2O$	s. gibbsite	156.0072	24.13	24.1
Al_2S_3	s	150.155	3313.81	2705.3
$Al_2 (SO_4)_3$	s	342.148	596.80	344.3
Al_2SiO_5	s. andalusite	162.046	28.03	9.2
Al_2SiO_5	s. kyanite	162.046	25.94	12.9
Al_2SiO_5	s. sillimanite	162.046	0	15.3
$Al_2SiO_5 \bullet (OH)_4$	s. kaolinite	258.1615	68.25	12.0
$3Al_2O_3 \bullet 2SiO_2$	s. mullite	426.0536	630.11	63.2
Ba	s. II	137.34	747.77	775.1
$BaCO_3$	s. II	197.35	−75.18	53.3
$BaCl_2$	s	208.25	48.69	88.7
BaO	s	153.34	194.15	252.0
BaO_2	s	169.34	113.38	196.7
$Ba(OH)_2$	s	171.36	45.93	160.3
BaS	s	169.40	1012.88	929.0
$BaSO_4$	s. barite	233.40	0	30.7
C	s. graphite	12.01115	393.509	409.87
C	s. diamond	12.01115	395.406	412.77
CCl_4	l	153.823	578.95	472.7
C_2N_2	g. cyanogen	52.0357	1096.14	1118.1
CH_4	g. methane	16.04303	802.3	831.2
C_2H_6	g. ethane	30.0701	1427.8	1495.0
C_3H_8	g. propane	44.172	2045.4	2152.8
C_4H_{10}	g. n-butane	58.1243	2658.4	2804.2
C_5H_{12}	g. n-pentane	72.1514	3274.4	3461.3
C_2H_4	g. ethylene	28.0542	1323.1	1360.3
C_3H_6	g. propylene	42.0813	1927.7	2002.7

(*continued*)

(continued)

Substance	State	Molecular mass M (kg/kmol)	Enthalpy of devaluation $D°$ (kJ/mol)	Standard chemical exergy $e°_{x,ch}$ (kJ/mol)
C_2H_2	g. acetylene	26.0382	1255.6	1265.0
C_6H_6	g. benzene	78.1147	3171.6	3301.3
C_6H_6	l. benzene	78.1147	3137.7	3296.2
C_7H_8	l. methylbenzene	92.1418	3736.4	3928.3
C_8H_{10}	l. ethylbenzene	106.1689	4347.7	4584.8
$C_{10}H_8$	s. naphtalene	128.1753	4984.2	5251.1
$C_{14}H_{10}$	s. anthracene	178.2358	6850.9	7212.6
CH_2O_2	l. formic acid	46.0259	213.0	291.3
C_2H_6O	l. ethylalcohol	46.0695	1235.9	1356.9
$C_2H_4O_2$	l. acetic acid	60.0529	786.6	907.2
C_3H_6O	l. acethone	58.0807	1659.6	1797.3
C_6H_6O	s. phenol	94.1141	2925.9	3126.2
$C_2H_2O_4$	s. oxalic acid	90.0358	202.7	367.9
CH_4ON_2	s. urea	60.0558	544.7	688.6
CO	g	28.0105	282.984	274.71
CO_2	g	44.0095	0	19.48
CS_2	l	76.139	2934.09	1694.3
Ca	s. II	40.08	813.57	729.5
CaC_2	s	64.10	1541.18	1484.6
$CaCO_3$	s. aragonite	100.09	0	16.3
$CaCO_3•MgCO_3$	s. dolomite	184.411	0	32.2
$CaCl_2$	s	110.99	178.21	105.0
$CaFe_2O_4$	s	215.77	161.07	121.1
$Ca_2Fe_2O_4$	s	271.85	321.00	212.2
$Ca_2Mg_5Si_8O_{22}(OH)_2$	s. tremolite	812.41	425.49	79.7
$Ca(NO_3)_2$	s	164.0898	−124.90	−1.0
CaO	s	56.08	178.44	127.3
$CaO•Al_2O_3$	s	158.04	351.66	123.1
$CaO•2Al_2O_3$	s	260.00	541.71	138.8
$3CaO•Al_2O_3$	s	270.20	716.72	382.6
$12CaO•7Al_2O_3$	s	1386.68	3415.71	1546.7
$CaO•Al_2O_3•SiO_2$	s. anortite	218.125	273.92	66.0
$Ca(OH)_2$	s	74.09	69.04	70.8
$Ca_3(PO_4)_2$	s. α	310.18	0	37.3
CaS	s	72.14	1056.57	861.7
$CaSO_4$	s. anhydrite	136.14	104.88	25.3
$CaSO_4•1/2H_2O$	s. α	145.15	83.16	29.2
$CaSO_4•2H_2O$	s. gypsum	172.17	0	25.7
$CaSiO_3$	s. volastonite	116.16	90.24	40.7
Ca_2SiO_4	s. β	172.4	232.28	129.9
Ca_3SiO_5	s	282.2	424.94	271.1
Cd	s. α	112.40	357.10	293.8
Cd	s. ν	112.40	356.51	293.2
$CdCO_3$	s	172.41	0	40.2
$CdCl_2$	s	183.31	126.04	73.4
CdO	s	128.40	98.95	67.3
$Cd(OH)_2$	s	146.41	38.26	59.5
CdS	s	144.46	920.60	746.9
$CdSO_4$	s	208.46	149.24	88.6
$CdSO_4•H_2O$	s	226.48	84.79	80.6

Substance	State	Molecular mass M (kg/kmol)	Enthalpy of devaluation $D°$ (kJ/mol)	Standard chemical exergy $e°_{x,ch}$ (kJ/mol)
Cl_2	g	70.906	160.44	123.6
Cl	g	35.453	201.90	87.1
Cr	s	51.996	569.86	584.7
Cr_3C_2	s	180.010	2415.85	2492.2
Cr_7C_3	s	400.005	5007.63	5155.5
$CrCl_2$	s	122.902	361.91	352.2
$CrCl_3$	s	158.355	281.05	301.9
Cr_2O_3	s	151.990	0	117.2
Cu	s	63.54	201.59	134.2
$CuCO_3$	s	123.55	0	31.1
$CuCl$	s	98.99	144.57	76.2
$CuCl_2$	s	134.45	151.95	82.1
$CuFe_2O_4$	s	239.23	60.62	36.1
CuO	s	79.54	44.27	6.5
Cu_2O	s	143.08	234.56	124.4
$Cu(OH)_2$	s	97.55	−6.37	15.3
CuS	s	95.00	873.87	690.3
Cu_2S	s	159.14	1049.10	791.8
$CuSO_4$	s	159.60	155.65	89.8
Cu_2SO_4	s	223.14	377.15	253.6
D_2	g	4.02946	249.199	263.8
D_2O	g	20.02886	0	31.2
D_2O	l	20.02886	−45.401	22.3
Fe	s. α	55.847	412.12	374.3
Fe_3C	s. α cementite	179.552	1654.97	1553.5
$FeCO_3$	s. siderite	115.856	65.06	123.4
$FeCl_2$	s	126.753	230.77	195.5
$FeCl_3$	s	162.206	253.29	228.1
$FeCr_2O_4$	s	223.837	107.10	207.8
$Fe_{0.947}O$	s. wustite	68.8865	124.01	111.3
FeO	s	71.846	140.16	124.9
Fe_2O_3	s. hematite	159.692	0	12.4
Fe_3O_4	s. magnetite	231.539	117.98	116.3
$Fe(OH)_3$	s	106.869	−48.14	37.5
FeS	s	87.911	1037.54	883.5
FeS_2	s. pyrite	119.075	1684.72	1426.6
$FeSO_4$	s	151.909	209.11	170.9
$FeSi$	s	83.933	1249.42	1155.5
$FeSiO_3$	s	131.931	118.07	159.9
$FeSiO_4$	s. fyalite	203.778	255.30	232.3
$FcTiO_3$	s	151.75	118.90	129.6
H_2	g	2.01594	241.818	236.09
H	g	1.00797	338.874	331.3
HCl	g	36.461	108.82	84.5
HDO	g	19.0213	0.21	18.8
HDO	l	19.0213	−44.38	10.0
HNO_3	l	63.0129	−53.19	43.5
H_2O	g	18.01534	0	9.5
H_2O	l	18.01534	−44.012	0.9
H_3PO_4	s	98.0013	−76.26	89.6

(continued)

(continued)

Substance	State	Molecular mass M (kg/kmol)	Enthalpy of devaluation $D°$ (kJ/mol)	Standard chemical exergy $e°_{x,ch}$ (kJ/mol)
H_2S	g	34.080	946.61	812.0
H_2SO_4	l	98.077	153.25	163.4
K	s	39.102	356.63	366.6
$KAlSi_3O_8$	s. adulare	278.337	66.26	7.4
K_2CO_3	s	138.213	−43.58	84.7
KCl	s	75.555	0	19.6
$KClO_4$	s	138.553	6.67	136.0
$K_2Cr_2O_7$	s	294.184	-190.4	34.3
KNO_3	s	101.1069	−135.90	−19.4
K_2O	s	94.203	350.04	413.1
KOH	s	56.109	52.72	107.6
K_2S	s	110.268	1024.40	943.0
K_2SO_3	s	158.266	300.47	302.6
K_2SO_4	s	174.266	4.62	35.0
K_2SiO_3	s	154.288	75.9	138.2
Mg	s	24.312	725.71	626.1
$MgAl_2O_4$	s. spinel	142.273	274.17	45.3
$MgCO_3$	s	84.321	23.43	29.8
$MgCl_2$	s	95.218	244.65	158.2
$MgFeO_4$	s	200.004	121.53	68.1
MgO	s	40.311	124.38	59.1
$Mg(OH)_2$	s	58.327	42.73	33.2
$Mg(NO_3)_2$	s	148.3218	−64.34	49.7
$Mg_3(PO_4)_2$	s	262.879	76.59	78.1
MgS	s	56.376	1105.11	893.9
$MgSO_4$	s	120.374	166.22	73.0
$MgSiO_3$	s	100.396	87.73	14.8
Mg_2SiO_4	s	140.708	188.35	59.8
$Mg_3Si_2O_5(OH)_4$	s. chrysilite	277.134	117.06	38.8
$Mg_3Si_2O_{10}(OH)_2$	s. talc	379.298	140.26	14.8
Mg_2TiO_4	s	160.52	231.48	119.3
Mn	s. α	54.9381	520.03	487.7
Mn_3C	s	176.82545	1958.20	1878.1
$MnCO_3$	s	114.9475	19.42	86.8
$MnCl_2$	s	124.844	199.18	170.8
$MnFe_2O_4$	s	230.630	118.36	122.6
MnO	s	70.9375	134.81	124.8
MnO_2	s	86.0369	0	26.5
Mn_2O_3	s	157.8744	81.09	100.2
Mn_3O_4	s	228.8119	172.26	187.8
$Mn(OH)_2$	s. amorphous	88.9528	66.47	112.7
MnS	s. green	87.002	1031.23	878.9
$MnSO_4$	s	151.000	180.20	147.8
$MnSiO_3$	s	131.022	110.08	108.0
N_2	g	28.0134	0	0.72
N_2 atmospheric	g	28.1541	0	0.69
NH_3	g	17.0305	316.62	337.9
NH_4Cl	s	53.491	249.43	331.3
NH_4NO_3	s	80.04348	118.08	294.8
$(NH_4)_2SO_4$	s	132.138	511.84	660.6

Substance	State	Molecular mass M (kg/kmol)	Enthalpy of devaluation $D°$ (kJ/mol)	Standard chemical exergy $e°_{x,ch}$ (kJ/mol)
NO	g	30.0061	90.25	88.9
NO_2	g	46.0055	33.18	55.6
N_2O	g	44.0128	82.05	106.9
N_2O_4	g	92.0110	9.163	106.5
N_2O_5	g	108.0104	11.30	125.7
Na	s	22.9898	330.90	336.6
$NaAlO_2$	s	81.9701	128.40	67.15
$NaAlSi_2O_6 \bullet H_2O$	s. analcime	220.055	35.41	20.3
$NaAlSi_3O_8$	s. low albite	262.2245	72.75	21.9
Na_2CO_3	s	105.9891	−75.62	41.1
NaCl	s	58.443	0	14.3
$NaHCO_3$	s	84.0071	-101.94	21.6
$NaNO_3$	s	84.9947	−135.62	−22.7
Na_2O	s	61.9790	243.82	296.2
NaOH	s	39.9972	23.79	74.9
Na_2S	s	78.044	1014.84	921.4
Na_2SO_3	s	126.042	297.63	287.5
Na_2SO_4	s	142.041	0	21.4
Na_2SiO_3	s	122.064	11.31	66.4
Na_2Si2O_5	s	182.149	13.28	68.2
Na_4SiO_4	s	184.043	151.45	256.9
Ni	s	58.71	239.74	232.7
Ni_3C	s	188.14	1180.09	1142.5
$NiCO_3$	s	118.72	−49.93	36.0
$NiCl_2$	s	129.62	94.85	97.2
NiO	s	74.71	0	23.0
$Ni(OH)_2$	s	92.72	−48.13	25.5
NiS	s	909.77	883.15	762.8
Ni_3S_2	s	240.26	1967.14	1720.2
$NiSO_4$	s	154.77	92.25	90.4
$NiSO_4 \bullet 6H_2O$	s. α, tetragonal. green	262.86	-266.75	53.6
O_2	g	31.9988	0	3.97
O	g	15.9994	249.17	233.7
O_3	g	47.9982	142.67	168.1
P	s. α, white	30.9738	840.06	861.4
P	s. red, triclinic	30.9738	822.49	849.2
P_4O_{10}	s. hexagonal	283.8892	376.21	767.7
Pb	s	207.2	305.64	232.8
$PbCO_3$	s	257.20	0	23.1
$PbCl_2$	s	278.10	106.67	42.3
PbO	s. yellow	223.19	88.32	46.9
PbO	s. red	223.19	86.65	45.9
PbO_2	s	239.19	28.24	19.4
Pb_3O_4	s	685.57	198.53	72.2
$Pb(OH)_2$	s	241.20	32.48	20.6
PbS	s	239.25	930.64	743.7
$PbSO_4$	s	303.25	111.12	37.2
$PbSiO_3$	s	283.27	70.88	31.5
Pb_2SiO_4	s	506.46	159.07	75.8

(*continued*)

(continued)

Substance	State	Molecular mass M (kg/kmol)	Enthalpy of devaluation $D°$ (kJ/mol)	Standard chemical exergy $e°_{x,ch}$ (kJ/mol)
S	s. rhombic	32.064	725.42	609.6
SO_2	g	64.0628	428.59	313.4
SO_3	g	80.0622	329.70	249.1
Si	s	28.086	910.94	854.9
SiC	s. α, hexagonal	40.097	1241.69	1204.5
$SiCl_4$	l	169.898	544.81	482.2
SiO_2	s. α. quartz	60.085	0	2.2
SiO_2	s. α, cristobalite	60.085	1.46	3.1
SiO_2	s. amorphous	60.085	7.45	8.2
SiS_2	s	92.214	2149.23	1866.6
Sn	s. I. white	118.69	580.74	558.7
Sn	s. II. gray	118.69	578.65	558.8
$SnCl_2$	s	189.60	416.08	400.3
SnO	s	134.69	294.97	303.8
SnO_2	s	150.69	0	43.0
SnS	s	150.75	1205.74	1070.0
SnS_2	s	182.82	1863.8	1618.5
Ti	s	47.90	944.75	907.2
TiC	s	59.91	1154.16	1136.6
TiO	s	63.90	425.14	419.2
TiO_2	s. rutile	79.90	0	21.7
$TiO3$	s	143.80	368.66	386.1
Ti_3O_5	s	223.70	375.10	414.1
TiS_2	s	112.03	2060.45	1876.2
U	s	238.03	1230.10	1196.6
UCl_3	s	344.39	577.35	556.0
UCl_4	s	379.84	499.39	491.1
UCl_5	s	415.30	536.93	519.5
UO_2	s	270.03	145.19	168.9
UO_3	s	286.03	0	49.8
U_3O_8	s	842.085	115.49	236.2
W	s	183.85	842.87	827.5
WC	s	195.86	1195.84	1199.1
WO_2	s	215.85	253.18	297.5
WO_3	s	231.85	0	69.3
WS_2	s	249.98	2084.51	1796.6
Zn	s	65.37	419.27	339.2
$ZnCO_3$	s	125.38	0	23.5
$ZnCl_2$	s	136.28	583.93	93.4
$ZnFe_2O_4$	s	241.06	74.08	32.2
ZnO	s	81.37	70.99	22.9
$Zn(OH)_2$	s.β	99.38	1918	25.7
ZnS	s. sphalerite	97.43	938.71	747.6
$ZnSO_4$	s	161.43	161.87	82.3
Zn_2SiO_4	s	222.82	112.74	18.1

Notes: Data in this table are updated vs. the latest English translation: J. Szargut, *Exergy Method. Technical and Ecological Applications* (WIT Press, Southampton, UK, 2005) (more than 70 entries have changed).
$T_0 = 298.15$ K, $p_0 = 101.325$ kPa. s, solid; l, liquid; g, gas. *Source:* J. Szargut, *Egzergia. Poradnik Obliczania i Stosowania* (Widawnictwo Politechniki Shlaskej, Gliwice, Poland, 2007).

Index

Abiotic depletion potential, 89
Activity, 457
Advanced exergetic analysis, 388
Advanced exergoeconomic analysis, 392
Aggregate metrics, 311
Aggregation, 100
Allocation, 95, 307
Allocation matrix, 340
Allocation methods, 94
Ammonium nitrate process, 244
Analogy, 60, 436
Ancillary input, 341
Ascendency, 471
Atmosphere, 483
Atmospheric condensation, 480
Australia, 221
Automobile production, 176
Auxiliary equations, 385
Availability function, 70
Availability, 33, 67, 274
Available energy, 4, 21, 49, 80, 478
Available work, 4
Avoidable endogenous cost, 393
Avoidable endogenous exergy destruction,
    391
Avoidable exergy destruction, 391
Avoidable exogenous cost, 393
Avoidable exogenous exergy destruction,
    391

Binary solution, 430
Bioenergy, 222
Biomass, 222, 223
Biomimicry, 485
Biophyscial methods, 292
Biosphere, 482, 483
Body forces, 64
Breeders, 222
By-products, 164

Caloric equation of state, 440
Carbon cycle, 480

Carbon nanotubes, 171, 186
Carbon single-walled nanotubes (SWNTs), 171
Carbon, 171
Carnot cycle, 439
Characteristic function, 456
Chemical energy, 63, 64
Chemical exergy, 166, 180, 181, 378, 480
Chemical flow exergy, 71
Chemical potential, 70, 460
Chemical reaction, 64
Chemical-etching process, 182
Chemical-vapor deposition (CVD), 178
China, 221
Chip, 173
Clean room, 196, 206, 208
Closed system, 478, 479
Coal, 101, 324
Colony, 433, 444
Community (of a population), 59
Competitive strategy, 430
Composition of waste materials, 344
Conservation, 227
Conserved property, 47
Constituent, 54, 59
Constitutive relation, 59
Consumer choice, 371
Consumption, 265
Control volume, 56
Convexification, 447
Copper Ore, 272
Corn ethanol, 103
Cost formation process, 386
Cost of Ownership (CoO), 202
Cost rate, 387
Cost sources, 386
Cumulative degree of perfection (CDP), 91, 99,
    102
Cumulative exergy consumption, 90, 91, 98, 251,
    293

Dalton's law, 437
Data reconciliation, 237

Data rectification, 235
Dead state, 6, 458
Degree of fabrication, 343
Degree of perfection, 180, 183, 184, 185
Dematerialization, 190
Desalination, 148, 151
Design trends, 126
Destruction (available energy), 68
Destruction (energy), 68
Destruction (resources), 480
Destructive processes, 181
Diffusion, 64
Direct metal deposition (DMD), 178
Direct requirement matrix, 297
Disability Adjusted Life Years (DALY), 311
Drink containers, 367
Dry-etching process, 185

ECEC-money ratio, 318
Eco-exergy, 454, 460
Eco-Indicator, 99, 395
Ecological cost, 271
Ecological cumulative exergy consumption, 98, 103, 294
Ecological exergy, 482
Ecological footprint (EF), 139, 214
Ecological goal functions, 455
Ecological system, 299, 365, 374, 484, 486
Ecological Thermodynamics, 453
Ecologically based life cycle assessment (Eco-LCA), 307
Ecology, 453
Economic equilibrium, 452
Economic input-output (EIO), 293
Economic Input-Output Life Cycle Assessment (EIOLCA), 90, 201, 298
Economic valuation, 294
Ecosystem ecology, 453
Ecosystem information, 470
Ecosystem Services, 1, 87, 301, 484, 485
Ecosystem, 88, 403, 405, 407, 413, 415, 416, 418, 419, 420, 421
Efficiency (process), 80
Efficiency rebound, 228
Efficiency, 40, 179, 439
Elastic deformation, 175
Elastic-plastic deformation, 175
Elastic-plastic material, 173
Electric car, 214
Electric induction furnace, 170
Electric induction melting, 184
Electric power requirement, 177
Electrical discharge machining (EDM), 178, 182
Electricity, 101
Electricity from space, 225
Electricity prices, 371
Electricity requirement, 178

Electricity-generation systems, 324
Electrodeionization, 152
Electrodyalysis, 152
Embodied energy, 219
Emerging markets, 321
Emergy, 92, 103
Emergy analysis, 93, 100, 239
Emergy yield ratio, 93
Employment, 229
End-of-Life materials, 271
Endogenous exergy destruction, 389
Energy, 4, 46, 50, 58, 61, 90, 103, 365
Energy (definition), 62
Energy (flow of), 62
Energy balance (overall), 66
Energy balance analysis, 75
Energy balance, 18, 77, 165
Energy carrier, 46, 381
Energy change, 66
Energy conservation, 367
Energy consumption, 144
Energy conversion system, 164
Energy conversion, 77
Energy costing, 214
Energy efficiency, 91
Energy equation, 63
Energy flow, 5
Energy intensity, 369
Energy interaction, 45, 58, 64, 66, 482
Energy matrix error, 368
Energy mode, 63, 64
Energy R&D budget, 225
Energy resource, 46, 51, 80, 212
Energy resources consumption, 163
Energy return on investment (EROI), 90, 104
Energy tax, 372
Energy use (physical limit), 150
Energy-labor tradeoff, 367, 371
Entropy, 441
Entropy balance analysis, 75
Entropy balance, 23, 78, 165
Entropy change, 149
Entropy flow, 68
Entropy generation minimization, 271
Entropy generation, 134, 147, 149, 156, 157, 159, 484
Entropy of mixing, 197, 443
Entropy production, 267, 279, 283, 285
Environmental impact rate, 396
Environmental impact, 395, 484
Environmental loading ratio, 93, 316
Environmental performance index, 139
Equation of state, 436, 438
ERG (Energy Research Group), 368
Errors, 235
Ethanol, 223
Europe, 221
European Union (EU), 227
Evolution, 430

Evolutionary equilibrium, 452
Exchange (of energy), 51
Exchange matrix, 366
Exergetic efficiency, 183, 251, 380
Exergetic life cycle assessment, 255, 293
Exergoeconomic evaluation, 383
Exergoeconomic factor, 387
Exergoeconomic variables, 387
Exergoeconomics, 377
Exergoenvironmental analysis, 378, 394
Exergoenvironmental evaluation, 395
Exergoenvironmental factor, 396
Exergy (balance), 167, 380
Exergy (history), 67
Exergy (physical), 70
Exergy (rate), 69
Exergy (symbolism), 67, 69
Exergy (termoinology), 69
Exergy accounting, 181
Exergy analysis, 38, 98, 101, 249, 378
Exergy balance analysis, 75
Exergy breeding factor, 91
Exergy costing, 384
Exergy destruction, 380, 384, 387, 396, 484
Exergy efficiency, 179
Exergy flow, 5, 155
Exergy loss, 4, 270, 384
Exergy modes, 74
Exergy rate balance, 72
Exergy requirement, 174
Exergy, 4, 33, 80, 101, 116, 165, 218, 378, 455, 458, 478
Exergy-costing principle, 377
Exergy-destruction ratio, 380
Exogenous exergy destruction, 389
Extended exergy accounting, 293
Extended exergy analysis, 94
Extensive variable, 67
Extraction, 115, 119, 125
Extrusion, 172

F principle, 386
Federal choices, 370
Figure of merit, 180
First law of sociothermodynamics, 438, 440
First Law, 50
Flow (of a property), 62
Food, 222, 429
Forging, 172
Forming, 172
Fourth Law, 454
Fossil fuel, 214
Free energy, 436
Fuel cell, 214
Fuel cost, 383
Fugacity, 457
Full fuel cycle analysis, 90
Fusion, 224

Gain, 432
Game theory, 430
Gasoline, 103
GDP/GNP, 365
Genuine progress indicator, 139
Genuine wealth, 477
Geothermal, 324
Gibbs equation, 443
Gibbs free energy of mixing, 11
Gibbs free energy, 70, 145, 165, 441, 456
Goal functions, 455
Gouy-Stodola equation, 274
Gouy-Stodola theorem, 68
Grassmann diagram, 5, 80, 81
Greenhouse-gas, 223
Gross domestic product (GDP), 138, 146
Gross errors, 235

Habitat, 436
Hawk-Dowe population, 430
Heat capacity rate ratio, 156
Heat exchange, 51
Heat exchanger effectiveness, 156
Heat exchanger, 156
Heat exergy, 71
Heat interaction, 28, 66
Heat transfer, 58
Heat, 58
Heating, 440
High pressure carbon monoxide process (HiPCO), 171
Human capital, 477
Human development index (HDI), 139
Human valuation, 292
Humans, 481
Hybrid analysis, 323
Hybrid car, 214
Hybrid LCA, 201
Hydroelectric, 324
Hydrogen, 214, 224
Hydropower, 221
Hydrosphere, 483

Ideal gas, 169
Idle power, 176
Import/export, 371
Incompressible deformation, 174
Incompressible substance, 169
India, 221
Indicator, 136, 137
Induction furnace (Heel), 170
Industrial CEC, 294
Industrial cumulative exergy consumption (ICEC), 98, 294
Industrial ecology, 485, 252
Industrial system, 486
Inefficiency, 185
Information Theory, 121
Input-output analysis, 295

Input-Output, 365
Integration, 429, 448
Intensive variable, 67
Interaction (mass flow), 60
Interaction (non-mass), 60
Interaction, 45, 54, 59, 67
Interconversion (irreversible), 64
Interconversion (reversible), 64
Internal energy, 440
Investment cost, 384, 385
Ion exchange, 152, 202
Ion-beam machining, 182
Iron ore, 373
Iron, 170
Irreversibility, 24, 68, 440
Irreversible conversion, 74
Irreversible exergy use, 175

Japan, 227

Kinetic energy, 63, 64
Kinetic exergy, 74, 378
Knowledge, 414, 415, 417, 420, 421

Laser machining, 182
Leontief inverse matrix, 343
Life cycle assessment (LCA), 185, 200, 220, 235,
    255, 278, 394
Life cycle assessment uncertainty, 220
Life cycle inventory (LCI), 200, 235
Life-cycle impact assessment, 89
Liquid, 440
Lithosphere, 308, 483
Living Planet Index, 214
Loading ratio, 105
Lost work, 68

Maintenance expenses, 383
Management, 406, 408, 420, 421, 422
Manufactured capital, 477
Manufacturing process, 163, 167, 177, 178, 185
Manufacturing system, 164
Manufacturing, 76, 163
Marginal intensity, 371
Mass balance, 56, 77, 165
Mass filter, 342
Mass flow interaction (bulk), 66
Mass flow interaction, 54, 57
Material flow analysis, 88, 275, 292
Material intensity, 144
Material interaction, 482
Materials extraction, 76, 77
Materials mixing, 122
Materials processing system, 164
Materials processing, 76
Materials transformations, 167
Materials, 405, 407, 408, 409, 411–419
Materials-Composition Matrix, 343
Maximum empower principle, 94, 100, 294
Maximum power principle, 454

Melting, 170
Metal cutting, 172
Metric, 133, 134, 136, 137, 482
Mexogenous exergy destruction, 392
Micropower systems, 224
Millennium Ecosystem Assessment, 87
Minimal Gibbs free energy, 443
Minimum work, 167, 168
Miscibility gap, 434
Mixing and contamination of materials, 335
Mixing entropy, 115, 129
Mixture, 450
Molal chemical exergy, 75
Molding, 169
Monetary valuation, 89
Multistage flash, 153

Natural capital, 87, 478
Natural gas, 324
Net emergy, 93
Net energy, 90
Net energy analysis, 90, 101, 292
Net primary productivity (NPP), 480, 481
Net transport of exergy, 74
Net transport, 64
Network algebra, 96
New technologies, 321
Non-conserved property, 47
Non-energy generating system, 49
Nonphysical interactions, 484
Nuclear fuel, 221
Number of transfer units (NTU), 156

Oil, 324
Oil shale, 219
OPEC impact, 370
Open system, 268, 478, 482
Operating expenses, 383
Optimization problem, 239
Orientors, 455
Orthogonal machining, 173

Peak oil, 219
Performance, 381
Phase change, 169
Phase diagram, 434
Phase, 59, 430
Photosynthesis, 480
Physical exergy, 166, 180, 181, 378
Physical input-output table (PIOT), 335
Physical limits, 148
Plasma-enhanced CVD (PECVD), 178, 183,
    184
Plasma-etching process, 182
Plastic deformation, 175
Plastic work, 172
Policy, 405–407
Population, 429, 448
Population (integration), 429
Population (segregation), 429, 450

Population growth, 216, 217, 478
Population pressure, 436
Potential exergy, 74, 378
Potential utility, 265, 268
Preference-based methods, 292
Pressure effects on exergy, 168
Pressure, 27
Primary input, 341
Principle of energy conservation, 19
Principle, 386
Printed circuit board (PCB), 178
Process (real), 68
Process engineering, 235
Process metrics, 134
Process rate, 177
Process-model-based life-cycle analysis,
    90
Product cost, 382
Product, 163
Production rate, 176
Property (independent), 136
Property (intensive), 63
Property, 45, 136, 137

Quality (conversion), 68
Quality (of energy), 50, 52, 101
Quality, 60
Quantity (of energy), 50, 51, 52
Quantity, 60

Rake angle, 173
Random errors, 235
Recycling, 76
Reference environment, 378
Reference state, 458
Refrigeration machine, 381
Relative cost difference, 387
Relative error covariance metric, 240
Removal process (destructive), 183
Renewability indicator, 91, 259
Renewable, 258
Renewable energy, 212, 480
Renewable resources, 102
Research directions, 227
Reserve-to-production ratio (R/P), 212
Reservoir, 21, 28, 31
Resource accounting, 87, 180, 181
Resource aggregation, 87
Resource base, 481
Resource consumption, 267, 275
Resource maintenance, 478
Resource quality, 88, 101
Resources use, 148
Resources, 3, 429, 436
Restoration (resources), 480
Restricted dead state, 71, 73
Restricted state, 70
Reverse osmosis, 153, 202
Reversibility, 218
Reversible conversion, 74

Reversible exergy use, 175
Rolling, 172

Sankey diagram, 5, 80, 81
Scarcity mask, 373
Science of Sustainability, 133, 477
Second law of sociothermodynamics, 440
Second law, 442, 456
Segregation, 429
Semi conductor processes, 178
Semi conductors, 190
Separation, 113, 120, 125
Shear stress, 64
Sherwood Plot, 119
Shortfall, 438
SI, 105
Silicon wafers, 193, 207
Single walled nano tube (SWNT), 178
Social behavior, 429
Social strategy, 59
Society, 59, 403, 405, 414, 419, 420, 421
Sociobiological model, 430
Sociothermodynamics, 59, 436
Solar power, 222
Solar transformity, 92
Solution, 434
Species, 59, 436, 448
Specific cutting energy, 174
Spontaneous Process, 17
Stamping, 172
State of sustainability, 134
State principle, 136
State variable, 59
State, 6, 61, 436
Steel industry, 373
Strain energy, 63, 64
Strain exergy, 74
Strategy diagram, 433, 444, 447
Strategy, 431
Stress-strain behavior, 173
Stress-strain diagram, 175
Substitutability, 88, 100
Subsystem, 480
Superconductors, 229
Surplus energy, 89
Surprise, 406, 422
Surroundings (thermodynamic), 68
Surroundings, 52, 53, 61
Sustainability (analysis), 217
Sustainability (indicators), 480
Sustainability (metrics), 133, 158, 249, 323, 482
Sustainability (state), 134, 158
Sustainability (topology), 479
Sustainability (weak), 478
Sustainability change, 135
Sustainability index, 93, 316
Sustainability indicators systems, 140
Sustainability properties, 135
Sustainability science, 477
Sustainability, 49, 133, 214, 477

Sustainable development, 133, 158, 486
Sustainable development (metrics), 134
System (closed), 55
System (open), 55
System (physical), 60
System (social), 60
System boundary, 52, 54, 62
System definition, 52
System property, 45, 58, 61, 67, 134
System well defined, 47
System, 52, 54, 436, 480
Systems of indicators, 140

Tar sands, 219
Technological development, 484
Technology (evolution), 150
Technology (outdated), 137
Technology evolution time, 150
Technology selection, 153
Technology, 484
Technosphere, 483
Temperature effects on exergy, 168
Temperature, 26, 440
Thermal energy, 63, 64
Thermal equation of state, 440
Thermal exergy, 74
Thermocompression, 152
Thermodynamic cycle, 439
Thermodynamic environment, 378
Thermodynamic inefficiencies, 381
Thermodynamic input-output analysis, 294
Thermodynamic metric, 134, 158
Thermodynamic system, 60
Thermodynamics and resources, 3
Thermodynamics, 3, 218
Tool (machining), 173
Topology of sustainability, 479
Total exergy use along the life cycle, 146
Total molal flow exergy, 71
Total primary energy consumption, 146
Total-requirement matrix, 297
Toxics Release Inventory, 306
Trading, 440
Traffic management, 228

Tramp elements, 334
Transformational technology, 133, 134
Transformities, 102, 300

Ultimate dead state, 71, 73
Ultrapure chemicals, 195
Ultrapure gases, 195, 205
Ultrapure water, 194
UN sustainability metrics (UNCSD), 141
Unavoidable exergy destruction, 389
Unavoidable investment cost, 389
Uncertainty, 235, 406, 414, 420, 422
UNCSD indicators, 139
Unemployment impact, 370
United States DOE, 215

Vapor, 440

Waste generation coefficients, 340
Waste proliferation risk, 221
Waste storage, 221
Waste, 6, 181, 485
Water, 219
Weak form of sustainability, 322, 478
Weight Process, 17
Well being indicator, 139
Wind, 101, 324
WIO-MFA, 336
Work exergy, 71
Work Interaction, 28, 66
Work rate (total), 176
Work rate, 166
Work, 58, 61
Workable design, 387
Working, 438, 440
Workpiece, 173

Yield ratio, 316, 341

Zero avoidable waste, 6
Zero-ELCA, 255
Zero-exergy-emission, 258
Zero emission, 6

Printed in the United States
By Bookmasters